Advances in Neurobiology

Volume 39

Advances in Neurobiology covers basic research in neurobiology and neurochemistry. It provides in-depth, book-length treatment of some of the most important topics in neuroscience including molecular and pharmacological aspects. The main audiences of the series are basic science researchers and graduate students as well as clinicians including neuroscientists (neurobiologists and neurochemists) and neurologists. *Advances in Neurobiology* is indexed in PubMed, Google Scholar, and the Thompson Reuters Book Citation Index.

Elena Blanco-Suarez • Isabella Farhy-Tselnicker
Editors

Astrocyte-Neuron Interactions in Health and Disease

Springer

Editors
Elena Blanco-Suarez
Department of Neuroscience and
Department of Neurological Surgery
Thomas Jefferson University
Philadelphia, PA, USA

Isabella Farhy-Tselnicker
Department of Biology
Texas A&M University
College Station, TX, USA

ISSN 2190-5215 ISSN 2190-5223 (electronic)
Advances in Neurobiology
ISBN 978-3-031-64838-0 ISBN 978-3-031-64839-7 (eBook)
https://doi.org/10.1007/978-3-031-64839-7

This Springer imprint is published by the registered company Springer Nature Switzerland AG
The registered company address is: Gewerbestrasse 11, 6330 Cham, Switzerland

Preface

It is now well established that astrocytes, a type of glia, are crucial for proper function of the central nervous system (CNS). In recent decades, research has expanded our understanding of the interactions between astrocytes and neurons and their impact on numerous CNS disorders when those interactions go awry. Edited by Drs. Elena Blanco-Suarez and Isabella Farhy-Tselnicker, this book contains a collection of review chapters written by top researchers in the field that highlight recent work on multiple aspects of astrocyte-neuron communications. We begin with an evolutionary perspective on how interactions between neurons and glia were established. We then provide an overview of the astrocyte-neurons interactions across multiple species, including invertebrates such as *C. elegans*, as well as mammals like rodents and humans. The first few chapters provide an overview of different types of astrocyte-neuron interactions as they occur under healthy conditions as during brain development and at the synapse. The following chapters focus on the roles of astrocytes and their impact on neurons in a variety of CNS disorders and pathological conditions, including stress, drug addiction, pain, spinal cord injury, and stroke, as well as neurodevelopmental disorders such as autism, and neurodegenerative disorders including Parkinson's disease, Alzheimer's disease, and amyotrophic lateral sclerosis.

We hope this book provides the reader with a comprehensive overview of the types of neuron-astrocyte interactions across multiple aspects of CNS function and across multiple species. Although we have made significant progress in understanding astrocyte biology since they were first described, many open questions about their role in the CNS remain. A deeper understanding of how astrocytes function and their relationship with neurons will enhance our comprehension of the central nervous system in both health and disease, potentially leading to new treatments.

The brain is a complex multicellular organ. Within this complex environment, cells interact with each other to ensure correct brain function, which is critical for the survival of the organism.

Neurons connect with each other to pass information including decoding sensory stimuli and calculating the behavioral response.

In addition to neurons, there is a large class of glia cells with specific roles that are critical for normal brain function.

Astrocytes are a major type of glia and interact with neurons to modulate multiple aspects of neuronal function including synapse activity, metabolism, inflammation, and more.

Though discovered in 1800, astrocytes have been understudied as active participant in brain function, regarded as passive supporters of brain homeostasis.

Seminal work several decades ago showed that astrocytes have instructive role in modulating multiple aspects of neuronal function, affecting survival and activity, and thus, impacting animal behavior. Furthermore, studies revealed that astrocytes play active roles during brain pathology, in some cases being the culprit and in others, the protector. However, much work is still needed to thoroughly characterize their disease-specific roles. And due to their central roles, they have become attractive candidates for therapeutics.

Philadelphia, PA, USA — Elena Blanco-Suarez
College Station,TX, USA — Isabella Farhy-Tselnicker

Contents

About the Editors

Elena Blanco-Suarez, PhD, is an Assistant Professor in the Departments of Neuroscience and Neurological Surgery at the Vickie and Jack Farber Institute for Neuroscience in Thomas Jefferson University, Philadelphia.

Isabella Farhy-Tselnicker, PhD, is an Assistant Professor in the Department of Biology in Texas A&M University, College Station, Texas.

Drs. Blanco-Suarez and Farhy-Tselnicker trained together as postdoctoral fellows, becoming experts in astrocyte-neuron interactions during development and synaptic maturation in mouse models. In 2020, both started their independent careers as Assistant Professors. Dr. Blanco-Suarez's research focuses on the study of astrocyte-mediated neuroplasticity in the context of ischemic stroke, while Dr. Farhy-Tselnicker's lab investigates astrocyte-neuron interactions necessary to shape synapse development and activity. With this book, they brought in experts from across the globe to give current insights into how the crosstalk between astrocytes and neurons is essential for shaping the central nervous system and constitutes a very interesting target for future investigations in disease and injury.

Evolution of Astrocyte-Neuron Interactions Across Species

Caterina Ciani, Maria Ayub, and Carmen Falcone

Abstract Proper functioning of the central nervous system depends on various tightly regulated phenomena, among which astrocyte-neuron interactions are of critical importance. Various studies across the species have highlighted the diverse yet crucial roles of astrocytes in regulating the nervous system development and functions. In simpler organisms like worms or insects, astrocyte-like cells govern basic functions such as structural support to neurons or regulation of extracellular ions. As the species complexity increases, so does the functional and morphological complexity of astrocytes. For example, in fish and amphibians, these cells are involved in synaptic development and ion homeostasis, while in reptiles and birds, astrocytes regulate synaptic transmission and plasticity and are reported to be involved in complex behaviors. Other species like those belonging to mammals and, in particular, primates have a heterogeneous population of astrocytes, exhibiting region-specific functional properties. In primates, these cells are responsible for proper synaptic transmission, neurotransmitter release and metabolism, and higher cognitive functions like learning, memory, or information processing. This chapter highlights the well-established and somewhat conserved roles of astrocytes and astrocyte-neuron interactions across the evolution of both invertebrates and vertebrates.

Keywords Astrocytes · Evolution · Vertebrates · Invertebrates

Abbreviations

ATP	Adenosine triphosphate
CNS	Central nervous system
GCM	Glia-conditioned media

C. Ciani · M. Ayub · C. Falcone (✉)
Neuroscience Department, Scuola Internazionale di Studi Avanzati (SISSA), Trieste, Italy
e-mail: cfalcone@sissa.it

E. Blanco-Suarez, I. Farhy-Tselnicker (eds.), *Astrocyte-Neuron Interactions in Health and Disease*, Advances in Neurobiology 39,
https://doi.org/10.1007/978-3-031-64839-7_1

GFAP	Glial fibrillary acidic protein
Glast	Glutamate/aspartate transporter
Gs	Glutamate synthetase
hPSCs	Human pluripotent stem cells
iPSCs	Induced pluripotent stem cells
MANF	Mesencephalic astrocyte-derived neurotrophic factor
PSCs	Pluripotent stem cells
RG	Radial glia
TSP	Thrombospondin

1 Introduction: The Ancient Origins of Astrocytes

The challenge of defining astrocytes across species arises from the lack of a comprehensive description and the tremendous heterogeneity exhibited by these cells. In fact, astrocytes have been primarily investigated in mammals and have only later been comparatively studied in diverse species, considering factors such as morphology, gene expression, and functions in both normal and stress contexts. While some old foundational studies rely heavily on morphology to identify astrocytes, contemporary research utilizes advanced technologies to enhance our understanding of these cells [1]. The term "astrocytes" was coined by Lenhossek in 1895 to replace the term "glia," which was considered inadequate to capture the multiple functions of these cells [2]. Del Río Hortega's classification in 1932 further categorized mammalian astrocytes into protoplasmic and fibrous types based on morphology and central nervous system (CNS) location, a classification that persisted for decades until Duncan in 1965 highlighted its oversimplification in capturing astrocyte heterogeneity [3, 4].

Astrocytes, a highly heterogeneous cell population indeed, display diverse morphologies, molecular profiles, and distributions across different anatomical regions, with varying physiology and functions across species [5]. The absence of a universal marker makes it challenging to establish criteria for identifying astrocytes across evolution. However, attempts have been made to describe astroglia across nervous systems and identify primitive astrocyte forms in both ancient vertebrate and invertebrate species.

In an evolutionary context, the appearance of a CNS necessitated supportive cells, known as glial cells, with ependymal, radial glial, and astrocyte-like cells being the most ancient ones (Fig. 1) [5, 6]. The phylogenetic relationships of glial cells between invertebrates and vertebrates remain a subject of debate, with studies revealing different genes and conserved pathways governing glial differentiation. The emergence of glial cells is evident in Acoelomorpha, evolving into more complex forms in species like *Caenorhabditis elegans* and species from the Annelida taxon, exhibiting features reminiscent of both mammalian astrocytes and oligodendrocytes [5]. In deuterostomes, the introduction of radial glial (RG) cells marked a significant milestone. In early Chordata, RG cells became the predominant glial cell type, exhibiting radial morphology, elongated cell bodies, and typical intermediate

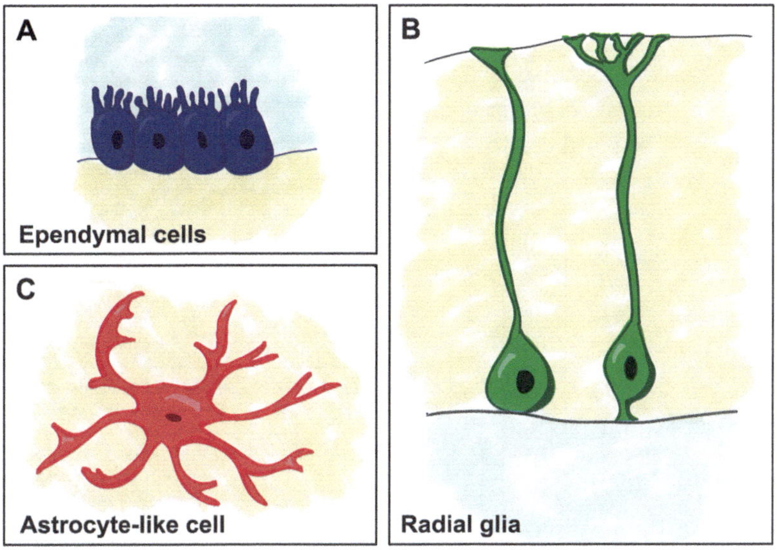

Fig. 1 Schematic diagram of cell types considered as predecessors of mammalian astrocytes. (**a**) Ependymal cells. (**b**) Radial glial cells. (**c**) Astrocyte-like cells (e.g., as seen in reptiles). (Adapted from Falcone et al. [5])

filament expression [6]. These cells, similar to mammalian astrocytes in terms of functions, persist throughout the life of many early vertebrates, playing roles such as glutamate and water homeostasis [6]. The observed increase in CNS complexity across evolution necessitated the specialization and diversification of neuroglia, particularly astrocytes. This complexity and heterogeneity have become notably pronounced in mammals and within mammals in primates, highlighting the crucial role astrocytes play in shaping the complexity of the nervous system across the evolutionary timeline [6]. Here, we provide a recent table summarizing what we know so far about the presence of astrocytes and astrocyte-like cells across invertebrates and vertebrates. However, we should consider that such presence is often still debated in non-mammalian species (Fig. 2).

In this chapter, we will focus on summarizing the astrocyte-neuron interactions across the evolution of both invertebrates and vertebrates, focusing on astrocytic functions in neuronal support, synaptic transmission and plasticity, neuroprotection, and energy metabolism.

2 The Role of Astrocytes in Neuronal Support Mechanisms in Invertebrates

Invertebrate's nervous system is comprised primarily of two types of cells: neurons and glia, arranged in a diffused network of nerve plexus and dorsal or ventral nerve cord [7, 8]. These structures are responsible for receiving and processing internal or

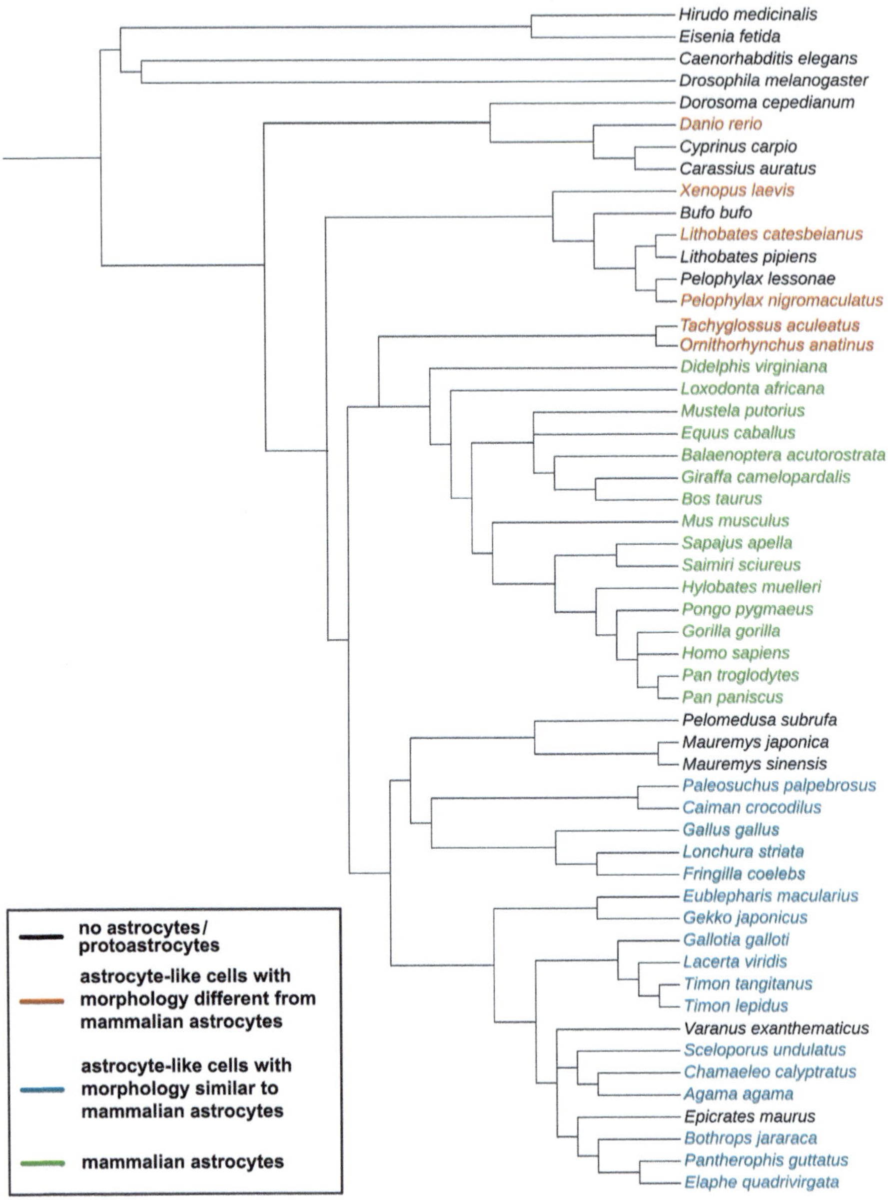

Fig. 2 Evolutionary tree of the species discussed in this study. The cladogram was created on https://phylot.biobyte.de/, based on the NCBI Taxonomy database. Color code: black = no astrocytes and/or proto-astrocytes; orange = species showing astrocyte-like cells with morphology different from mammalian astrocytes; blue = species showing astrocyte-like cells with morphology similar to mammalian astrocytes; green = species showing mammalian astrocyte morphology. (Adapted from Falcone et al. [5])

external stimuli and enabling the organism to respond to the cues accordingly [9]. The common pathways for these functions are somewhat conserved across evolution, and like vertebrates, invertebrates regulate basic or advanced biological demands via changes in the neurotransmitter expression levels or ionic gradients. Moreover, like vertebrates, neuronal cells of invertebrates are mainly responsible for signal transduction or synaptic transmission between themselves and among various parts of the organism [10]. Astrocyte nomenclature is debated in invertebrates; however, in this group, glial cells exert many of the same functions in which vertebrate astrocytes are specialized. For example, although invertebrate glial cells have been classically considered mainly as supporting cells of the nervous system (i.e., they ensheathe the neurons and provide physical protection), these cells also provide metabolic fuel for proper neuronal functioning and are involved in the maintenance of closely monitored neurotransmission and in the overall maintenance of homeostatic environment in the nervous system [11, 12]. Glial cells also participate in the formation of blood-brain barrier-like structures, where present, and help in the neuronal cell migration to specific target sites by providing guidance cues [11, 13].

With technological advances, the emerging evidence suggests new crucial roles of glial cells in diverse nervous system functions. For example, both invertebrate and vertebrate glial cells can secrete neurotrophic factors, are responsible for the phagocytosis of cellular debris, play a significant role in neurite outgrowth, and facilitate synapse formation, maturation, and plasticity [14–16]. Much of the primary work to explore the glial cell functionality as supporting mechanisms is done in marine animals, particularly echinoderms, due to their close proximity with chordates and hemichordates. Mashanov et al. investigated the role of radial glial cells in adult echinoderms and reported multifaceted roles of these cells [17]. These radial glial cells participate in neuroprotective mechanisms, like neurogenesis after an injury, are responsible for nervous system secretory activities, facilitate the formation of the scaffold of the neuroepithelium, and lastly promote the phagocytosis of the cellular debris, thus maintaining homeostatic environment for neuronal growth [17]. The arthropods from the species *L. polyphemus* are reported to employ their glial cells as osmoregulators, and these cells form junctional complexes that modulate the transport of ions and molecules [18]. Glial cells in the lobster *Homarus vulgaris* provide nutritional support to the synaptic regions, thus facilitating neuronal activity [19]. In multiple marine species such as *Glyptonotus antarcticus*, *Panulirus argus*, and crab *Carcinus*, glial cells serve the function of wrapping the sensory neurons and their cell bodies both for neuroprotective and signaling purposes [19–22].

Phylum Annelida constituents represent a slightly more advanced and centralized nervous system comprising glia-like mesenchymal cells with long processes connecting to the nervous cells [5]. *Eisenia fetida* (earthworm) consists of more complex and specialized neuroglia, which include supporting-nutrifying and myelin sheath-forming cells, neurilemmal cells, and sub-neurilemmal cells [11, 23]. Among these cells, the supporting-nutrifying glial cells are GFAP-immunoreactive and have similar nutritive and supporting functions as mammalian astrocytes [11, 23].

Morphology of *Eisenia fetida* glial cells highlights the presence of granular and vesicular systems in their cell bodies and filopodia. Close contacts between glial filopodia and neuronal cell membranes facilitate the emptying of glial granular and vesicular components. These molecular entities may in turn facilitate neuronal survival, provide nutritional support, and support the neuronal terminals [6, 23]. Another well-known model organism that has been investigated for the presence of glia is *Caenorhabditis elegans* (*C. elegans*), a transparent worm of around 1 mm, whose genome is completely sequenced, which shares 60–80% of homology with mammals [24]. The nervous system of *C. elegans* has been mapped in detail and consists of 302 neurons and 56 supporting cells (neuroglia) of both ectodermal and mesodermal origins. A specialized class of astrocytes called proto-astrocytes, which are similar to mammalian astrocytes in numerous functions, have been identified in *C. elegans.* These neuroglial cells are responsible for the formation of sensory organ sensilla, which is responsible for animal locomotion and survival [25, 26], modulate perisynaptic ionic homeostasis, and are involved in neuronal development and morphogenesis [27]. Other functions encompass the establishment of the location of neuronal structures, modulation of neuronal activity, Ca^{2+} transport through voltage-gated channels, K^+ buffering, and neurotransmitter metabolism [28].

Drosophila melanogaster (fruit fly), a model organism from phylum of arthropods, has been extensively explored in both neurodevelopmental and neurodegeneration studies. The proto-astrocytes reported in *C. elegans* have also been observed in *D. melanogaster,* and these cells display a higher degree of specialization in terms of both morphology and physiology [7, 29–31]. The surface glia in Drosophila is responsible for the formation and maintenance of the brain-hemolymph barrier, and cortex glia comprises processes contacting multiple neurons at a time and monitoring Ca^{2+} oscillations [7, 29–31]. Like mammals, Drosophila also contain ensheathing astrocytes like glia called neuropil glia, which serve the function of wrapping of axons and synapses, thus facilitating synaptogenesis and synaptic transmission [7, 29–31]. Similarly to mammalian astrocytes, these glial cells are engaged in Ca^{2+} signaling, they are coupled by gap junctions, and they are involved in the formation of non-overlapping territories otherwise known as tiling, which is crucial for uniform coverage of the brain [30, 31]. Comparative genomic studies between flies, mice, and humans have highlighted the conservation of genetic components pointing to the common evolutionary origin of neuroglia among these species. Among the genes are transcription factors, genes for ion channels, and transporters important in mammalian astrocyte physiology [32]. Among such factors, there is an evolutionary conserved mesencephalic astrocyte-derived neurotrophic factor (MANF) or DmMANF in Drosophila. This neurotrophic factor has well-conserved protein sequences from *C. elegans* to humans. DmMANF is critical for the maturation of nervous system, growth of dopaminergic neurites, and survival of dopaminergic neuronal cell population in Drosophila [33]. The findings discussed above highlight the importance of glial and astrocyte-like cells in neuronal development and maturation, survival, regeneration after an injury, synaptic transmission, release of neurotrophic factors, and overall homeostasis of the nervous system in the invertebrates.

3 Evolution of the Astrocytes' Role in Neuroprotection: From Invertebrates to Non-mammalian Vertebrates

Among all of the functions for which astrocytes are known, one of the most important ones is the neuroprotective function, which is a basic requirement for the evolution of a more complex nervous system. However, astrocyte-driven neuroprotection is also present in non-mammalian organisms, suggesting it is part of the fundamental mechanisms underlying the nervous system function. Indeed, astrocytes can respond to injuries and stress by activating a series of processes known as *reactive astrogliosis*. Astrogliosis is a fascinating and controversial reaction since it can lead to either a neuroprotective role or the release of neurotoxic molecules [5, 34, 35].

A deep knowledge of the astrocytic neuroprotective roles in both physiological and pathological contexts in non-mammals is also important for refining and reducing animal experimentation in future research, by using animal models different from the classic ones, with a great improvement in costs and ethics. An excellent example is *C. elegans*, which is one of the most ancient organisms presenting proto-astrocytes with roles analogous to those of mammals. These proto-astrocytes not only regulate the neuronal development but also offer neuroprotection to neuronal cells in physiology by providing nutrient support and by controlling ion homeostasis [25]. However, *C. elegans* does not present the glia limitans, thus lacking the mechanical astrocytic protective role that has arisen later in evolution [5]. *Drosophila melanogaster* (and the rest of the order of the *Insects*) are still characterized by proto-astrocytes, which could be divided into several subpopulations. In particular, the *Drosophila limitans glia* forms a mechanical support and protection that would develop in vertebrates into the classical blood-brain barrier [7, 29–31]. In fishes, the situation is quite varied: Carps and *Dorosoma* show the predominance of ependymal cells, while rays and sharks already have astroglia-like shaped cells [5, 36]. In zebrafish (*Danio rerio*), a recent study [37] shows the presence of astrocyte-like cells behaving similarly to those of mice. These cells develop from radial glial cells 2 days after fertilization and get in touch with neurons and vessels, expressing markers such as glutamine synthetase, respecting the tiling features typical of astrocytes, and producing Ca^{2+} waves that recap in total those of the mammalian astrocytes [37]. Interestingly, when there is a cerebral wound, there is not an active proliferation of astrocyte-like cells around the trauma. At the same time, there is an increased proliferation of neurons that do last in time, suggesting astrogliosis could help in the formation of a scar, limiting the replacement of the damaged neurons [36]. This observation is confirmed in reptiles, in particular in the gecko (*Gekko japonicus*). Geckos are studied for their abilities in regeneration and wound healing, which seem to be linked to the presence of immature astrocytes and attenuated glial activation in the presence of trauma [38]. RNA-seq analysis showed the expression of genes comparable to those of mammals but in an embryonic phase [5, 38]. Among amphibians, *Xenopus laevis* is the most studied since it is often used as an animal model. Eldridge et al. [39] performed an interesting study on the consequences of brain traumatic injuries in Xenopus, showing how astrocytes can activate a limited astrogliosis analogous to the mammalian one. Indeed, frog astrocytes develop an ameboid morphology and migrate toward the site of the trauma, where they phagocytize the damaged material leading to a first neuroprotective response and a subsequent expression of a series of pro-inflammatory modulators, helping the creation of an edema [39]. Although there are very few studies performed on birds, it is known that their astrocytes show characteristics resembling those of mammals: They present a round soma, with GFAP-expressing processes able to contact the blood vessels and wrap the synapses. Therefore, it is hypothesized that even the roles should be similar to those in mammals [5]. However, an interesting study performed by Pedersen et al. shows an important neuroprotective role of reactive astrocytes in the zebra finch by upregulation of estradiol synthesis in the presence of a brain injury [40]. Such neuroprotective effect could be responsible for the absence of a second wave of degeneration after a brain trauma, which is responsible for neural apoptosis in mammals [40].

4 Evolution of Astrocytes' Role in Synaptic Transmission and Plasticity Across Vertebrates

Healthy brain network organization and connectivity are critical for daily activities of an organism, which in turn are dependent on closely monitored synaptic transmission and plasticity. Synaptic transmission simply refers to the phenomenon of inter-neuronal communication, where presynaptic neuronal terminals release some neurotransmitters that are recognized by postsynaptic endings resulting in the exchange of signals. On the other hand, synaptic plasticity implies an activity-dependent modification of efficacy and strength of synaptic transmission. These two phenomena are crucial for neurophysiological and cognitive functions and are critical for the reorganization of brain networks after damage [15]. For decades, synaptic plasticity has been attributed to be only a neuronal cell-based phenomenon, with glial cells completely out of the picture. Advances in glia research have uncovered the key roles of these diverse cells in modulating synapse numbers, transmission, and plasticity across the species [15]. During early postnatal phases of both vertebrate and invertebrates' nervous system development, synapses undergo the processes of pruning, engulfment, and strengthening. Many studies have reported the role of glial cell-mediated pruning pathways that are employed by both the vertebrates and the invertebrates to undergo these phenomena and are suggested to be conserved over the species, such as CED-1/Draper/Megf10 pathway. Draper is an ortholog of *C. elegans* CED-1 (transmembrane receptor responsible for the engulfment of the cells undergoing programmed death), an essential gene expressed by glial cells for the clearance of cellular debris and apoptotic neurons in the Drosophila embryonic central nervous system [41]. Draper, apart from its role in the CNS, is also reported to be expressed by neuromuscular junction glia, where it is involved in the phagocytosis of excess synaptic processes on motor neurons during developmental stages [42]. MEGF10, an ortholog of Drosophila Draper and *C. elegans* CED-1, is found in higher order species like rodents and primates, where it is reported to play the similar role of phagocytosis of cells and cellular debris. MEGF10 has been observed to modulate phagocytosis both in vitro and in vivo, and its activity requires other proteins such as MerTK. Chung et al. reported that astrocytes utilize MEGF10 and MerTK pathways in an activity-dependent manner to engulf synapses and this process continues till the adult/old stage of the mice [43].

Radial glial cells/astrocyte-like cells of zebrafish have long processes spanning the parenchyma and neurons and have been debated to play a significant role in zebrafish behavior [37]. Chen et al. explored the dynamics of Ca^{2+} signaling in zebrafish astrocytes and reported that in vivo Ca^{2+} imaging of these cells displays spontaneous microdomain Ca^{2+} transients in their long fine processes upon norepinephrine application and might be regulating local neural circuitry. They argued that these Ca^{2+} dynamics are remarkably similar to those reported in behaving mice on norepinephrine administration [44]. They also explored the astrocyte tiling behavior in zebrafish and found out that—like mammalian astrocytes—these cells in zebrafish also tend to be closely associated with the synapses, tile with one another to occupy unique territories, and express markers like glutamate synthetase (Gs) and glutamate/aspartate transporter (Glast) [37].

In the early 1990s, the use of in vitro models gained quite a spotlight, which led to the uncovering of neuroglial cell roles in regulating the synapse formation and transmission. By employing a two-step panning protocol, nearly 100% of the pure population of retinal ganglion cells can be isolated. Pfrieger et al. reported, in these highly purified neuronal cultures, a reduced number of synapses [45]. Treatment of these pure cultures with the glia-conditioned media (GCM) resulted not only in the increased number of synapses but also enhanced synapse maturity as compared to the untreated/pure neuronal cultures [46]. Araque lab investigated neuron-astrocyte interactions and their crosstalk in synaptic transmission. They reported that neuronal activity can elicit Ca^{2+} spikes and waves in the adjacent astrocytes, which in turn can result in the release of gliotransmitters. These chemical transmitters can lead to feedback inhibition of neuronal activity and synaptic transmission [47]. This study subsequently led to the term "tripartite synapses," i.e., the functional integration of pre- and postsynaptic neuronal membranes and their association with the surrounding glia (particularly astrocytes) and combined contribution of these three components in the synaptic transmission. In vitro studies by Durkee et al. in Araque group and other groups reported the effect of neuronal glutamate on the release of astrocyte-derived chemical transmitters like D-serine, glutamate, and adenosine triphosphate. Neurotransmitters can bind to G protein-coupled receptors on the astrocytes triggering calcium oscillations. This rise in astrocyte calcium levels can in turn lead to further synthesis and release of gliotransmitters, which can subsequently have an impact on synaptic transmission and neuronal activity [48, 49].

Further in vitro and in vivo studies have provided key factors and molecular pathways governed by astrocytes in regulating synapse formation, synapse engulfing during early developmental phases, and transmission [15, 50]. One such key factor is cholesterol; many studies have reported that cholesterol is an essential component of the lipid bilayer and synaptic vesicles and synaptosomes are highly enriched with cholesterol [51]. Seminal studies have shown that neurons can synthesize cholesterol to grow, form a limited number of synapses, and survive, but must uptake astrocytic cholesterol for the formation of the required number of functional synapses [46, 52]. Mauch et al. reported that lowering the cholesterol concentration in GCM results in diminished synaptic activity, which, upon adding cholesterol, is fully restored. They also found that treating glia-free neuronal cultures with cholesterol results in enhanced spontaneous excitatory postsynaptic currents [53]. Both GCM and cholesterol are reported to increase the glutamate receptor subunit 2/3 (GluA2/3) expression levels resulting in increased postsynaptic currents, while cholesterol-free conditions diminish GluA2/3 expression along with dendrites [51, 54]. Among the extracellular matrix proteins, thrombospondins (TSPs) are one of the major synaptogenic factors and consist of five members (TSP1-TSP5) [55]. TSPs are secreted by astrocytes and have been reported to increase the number of synapses, and their removal from the conditioned media results in the loss of synaptic activity [55]. To further verify the importance of TSPs in in vivo systems, TSP1/2 double-knockout mice were generated, and these knockout mice displayed fewer and defective cortical excitatory synapses [55]. Hevin, another astrocytic synaptogenic factor, is shown to localize to synaptic clefts and is involved in the formation and maturation of synaptic connections. Hevin null mice displayed impaired synaptic connections in the superior colliculus, further highlighting the importance of astrocytic secreted factors in synapse formation and activity [56].

Much of our understanding of astrocytes' function in humans comes from postmortem, human pluripotent stem cells (hPSCs), or humanized chimeric mouse studies. The most abundant form of astrocytes in humans is the protoplasmic astrocytes that reside in the deeper cortical layers (layers 2–6). Cytoarchitecturally speaking, these astrocytes are much larger and far more elaborate than rodent astrocytes, and their processes can cover blood vessels, neuronal cell bodies, and numerous synapses [57]. The synaptic density of human cortex is approximately 1100 million synapses/mm^3 [58], and it is estimated that a single protoplasmic astrocyte roughly supports and modulates the function of 2 million synapses [57, 58]. Human astrocytes derived from donor cells have been used as tools of implantation in the mouse brain to study their functionality in vivo [59, 60]. Han et al. reported that when human astrocytes are implanted in the mouse brain, they result in the generation of protoplasmic astrocytes of large diameter and varicose astrocytes with long unbranched processes. These humanized chimeric mice display higher long-term potentiation and enhanced learning abilities than the control mice, pointing toward the potential role of astrocytes in higher cognitive functions [59]. Altogether, these pieces of evidence demonstrated that astrocytes are responsible for the secretion of synaptogenic factors/proteins, are involved in the formation, maturation, and survival of synapses and synaptic transmission, and might be an essential contributing factor for advanced cognitive activities.

5 The Role of Astrocytes in Synaptic Homeostasis and Energy Metabolism in Mammals

Astrocytes evolved their functions through evolution, as a consequence of their fundamental roles in the brain. In mammals, they have the ability to be responsible for the most heterogeneous functions: from molecular to metabolic homeostasis and energy metabolism. They became more and more essential for the whole-brain activities and, as recently hypothesized, for the cognitive functions together with neurons [6, 14, 61]. Synaptic activity is assured thanks to the astrocytic transport of ions and protons, the removal and catabolism of neurotransmitters, the release of precursors used by neurons to produce neurotransmitters and the scavengers of reactive oxygen species, and the regulation of the metabolic homeostasis through the synthesis of glycogen and energy substrates [14, 61, 62]. Astrocytes are also responsible for the uptake, metabolism, and storage of glucose, the main fuel of the brain [14]. To understand the importance of this role, we could just consider that all the synaptic events of the brain require an impressive amount of ATP (e.g., for the restoration of ionic gradients, synaptic cargo transport, synapse assembly and maintenance, and Ca^{2+} homeostasis), which derive from glycolysis (limited to astrocytes) and mitochondrial oxidative phosphorylation (prevalently in neurons) [62, 63]. The exchange of energy is performed under the form of an astrocyte-neuron lactate shuttle, in which glucose is released by neurons in the synapses, uptaken by astrocytes, and transformed in lactate, which goes back to neurons where it is converted into fuel by the oxidative phosphorylation [62, 63]. The brain consumes up to 20% of the total body energy [63], and a problem in the energy supply would impact the synaptic activity, with huge consequences in terms of neuronal function. Astrocytes can actively cooperate with special neurons sensitive to glucose levels, helping to generate the proper metabolic answer [61]. For instance, in the case of hypoglycemia, astrocytes are the only cells able to use fatty acids as substrates for energy, by transforming them into ketone bodies [61]. Their involvement in energy homeostasis is witnessed also by their active synthesis of cholesterol and by the expression of receptors for hormones involved in energy homeostasis such as leptin, insulin-like growth factor-1, thyroid hormone, and insulin, which—all together—regulate the astrocytic production of glucose and the interaction with neurons [61]. While the importance of astrocytes in metabolism is becoming more and more recognized in mammals, unfortunately, we know very little about the role of astrocytes in energy metabolism in non-mammalian species. Various studies in the fly show roles for glia in metabolism similar to the ones exerted in mammals. Different subpopulations of glia seem to cooperate to provide support to neuronal metabolism in *D. melanogaster*. Here, glia can take up the main circulating sugar, trehalose, degrade it to produce alanine and lactate, and finally release lactate to be used by neurons to fuel their activity [64]. Therefore, it has been suggested that a metabolic compartmentalization of the fly brain be similar to the astrocyte-neuron lactate shuttle model (ANLS), which has been proposed for mammals [64–66].

6 Human Astrocytes: Specialization and Uniqueness

Human astrocytes exhibit remarkable diversity and distinctive anatomical characteristics, setting them apart in the landscape of glial cells. Notably, these astrocytes display enlarged soma, intricate and elongated processes, and a unique transcriptomic profile, emphasizing their specialized nature [57, 63, 67].

In comparison with their rodents' counterparts, human astrocytes not only harbor numerous conserved genes but also manifest thousands of differentially expressed genes, underscoring the evolutionary divergence [63]. Divergent gene expression profiles between human and mouse astrocytes reveal distinctive functional emphases. Mouse astrocytes predominantly upregulate genes associated with metabolism and mitochondrial respiration, while human astrocytes seem to prioritize defense response genes and those linked to the extracellular space and secreted factors [63]. This intricate molecular divergence extends to pluripotent stem cell-derived astrocytes, as evidenced by significant interspecies differences in gene expression related to cellular respiration, glucose and lactate transport, and pyruvate utilization—highlighting the heightened metabolic capacity in human astrocytes compared to their chimpanzee counterparts [68]. Notably, studies in postmortem brains not only showed that a specific class of primate astrocytes—the interlaminar astrocytes—share similar expression of certain genes with their murine counterparts, such as Sox2, Pax6, Nestin, Aqp4, and Glast, but also showed differential expression in genes such as S100b, Cryab, and Hopx, emphasizing the unique molecular signature of some subpopulations of human astrocytes [69]. Beyond their traditional roles in homeostasis, plasticity, and injury response observed in rodents, human astrocytes may play specific roles in primate behavior and cognition [59]. Engraftment of human glial progenitor cells into the forebrain of mice enhances learning, while human astrocytes contribute to improved learning, memory, and social skills in mice, mirroring cognitive traits more akin to primates than rodents [59].

This intricate interplay between astrocytes and neurons across mammalian evolution, especially in primates, holds key insights into the emergence of distinct cognitive abilities within this phylogenetic group. Unraveling the unique characteristics of astrocytes in humans not only contributes to our understanding of the coevolution with neurons but also sheds light on astrocyte-specific human pathologies.

7 Astrocytes and the Coevolution with Neurons

Intellectual abilities result from a deep and intricate interaction between neurons and astrocytes. Indeed, evolution passed from a diffuse nervous system formed by a few neurons interconnected one with another, to a complex and well-organized net in which neurons are organized in ganglia [70]. This type of organization pushed toward the development of cells responsible for the homeostasis and compartmentalization of these neuronal conglomerates. Therefore, astrocyte-like cells and glia in general appeared early in evolution, as a development of ancestral epithelial cells (the ependymoglial cells), leading to more complex cognitive abilities and behaviors. This hypothesis is confirmed by the increasing astrocyte-to-neuron ratio, which increases from leech (1:20) to nematodes (1:5), becoming more abundant in *Drosophila* (in which glia is 25% of all brain cells) and in mammals (with 60% of glial cells in the brain) [71, 72]. Humans have an important presence of astrocytes with a number of processes ten times higher than all the other mammals [71, 72].

Invertebrates are a good example of how a higher complexity of the nervous system leads to more refined cognitive performances [70]. Coelenterates are characterized by a simple diffuse nervous system. Neurons use conduction instead of synapses and peptides as neurotransmitters, with no need for specialized cells with supportive roles, for an extremely slow adaptive behavior [5, 6, 70, 73]. Platyhelminthes show a central nervous system with neurons compartmentalized, supported, and electrically isolated by a primitive glia, leading to a rapid and coordinated, even if stereotyped, response to the environmental stimuli [5, 6, 70, 73]. In Annelida, glia is subdivided into distinctive subpopulations with supportive and trophic functions, leading to advanced behaviors such as food capture and manipulation [5, 6, 70, 73]. In insects, there is a developed nervous system (which requires a corresponding evolved glial system) that controls a sophisticated social behavior [5, 70]. Cephalopods are the only invertebrate that shows the presence of glial cells, which are similar to the mammalian astrocytes (e.g., complex behaviors and social interactions), with cognitive abilities that are also analogous to those of complex vertebrates [70, 73].

In vertebrates, astrocytes became the most diffuse glia, evolving their functions over time. However, studies show that astrocytes in the brains of some vertebrates (as in turtles) cannot be induced in astrocytosis or form glial scars in the presence of a wound [6, 70, 74]. This suggests the first role of these cells in homeostasis and autoimmune protection (forming the blood-brain barrier) that evolved during the evolution for more flexible and adaptive behaviors. Studies in Squamata [5, 75] show a great variety of astroglial presence between species and between different areas of the brain, suggesting a parallel evolution between species. In mammals and birds, there is further specialization in the architectural organization of astrocytes and their heterogeneity, with surprising analogies between the two classes [5]. Indeed, birds have a glia-to-neuron ratio of 0.4 and different astrocytic morphotypes, able to connect synapses with the vascular network. The increment of heterogeneity of astrocytes reflects new brain abilities such as fantastic navigational skills, social interactions, and ability to perform complex cognitive tasks. In mammals, there is a large variety of morphology according to the species. However, astrocytes of primates show main differences compared to those of the lower vertebrates, from the morphological differentiation (protoplasmic astrocytes, interlaminar astrocytes, varicose projection astrocytes, and fibrous astrocytes) to high specialization in roles and an intimate relationship with neurons [5, 70]. Indeed, astrocytes become the predominant cell in the brain, compartmentalizing, processing, and integrating information [5, 73, 74]. However, the appearance of astrocytes in a few reptile species (evolutionary earlier than the appearance of mammals and birds) leads to the hypothesis that astrocytes could have promoted (more than to be a consequence) the increase of brain complexity.

The human brain needs a special mention as the most evident example of the astrocytic increase in functional competencies: Astrocytes are characterized by a bigger volume and an enhanced number of branches compared to all the other primates and mammals, with subsequent higher connectivity with the surrounding cells [5]. In particular, it was shown that there was an increased density of astrocytes belonging to layer I and the white matter, which correlates with the brain size, compared to other non-human primates [76]. Interestingly, this increase is not relevant in all the other areas of the brain, suggesting a direct involvement of these areas in higher metabolic and cognitive activities [77]. Moreover, the injection of astrocytic precursors in the first layer of the cortex of chimeric mice leads to the formation of well-connected human-mouse astrocytic nets, with faster learning and better memory consequences in the behavioral tests, compared to the wild-type counterparts [72].

8 Conclusion

The study of astrocyte evolution across species sheds light on the intricate relationship between these glial cells and the development of the nervous system. From their ancient origins as primitive supportive cells to their complex and diverse forms in mammals, astrocytes have undergone significant evolutionary adaptations and underscore the dynamic nature of the nervous system evolution. By understanding the evolution of astrocytes, we gain a deeper appreciation for their crucial contributions to neuronal support, synaptic transmission, neuroprotection, and energy metabolism across the evolutionary timeline.

Despite the challenges in defining astrocytes across species, recent advancements in technology and comparative studies will provide valuable insights into their roles and functions. In fact, moving forward, continued research in this field by using more recent techniques such as spatial transcriptomics, comparative genomics, transcriptomics, and proteomics, and the use of iPSC-derived models (both in vitro and with induced pluripotent stem cells (iPSC)-derived cells transplanted in chimeric mice), promises to unravel further mysteries surrounding astrocyte-neuron interactions across evolution.

Acknowledgments We thank the grant Human Technopole Early Career Fellowship awarded to C.F. from the Human Technopole (Milan, Italy).

References

1. King JS. A comparative investigation of neuroglia in representative vertebrates: a silver carbonate study. J Morphol. 1966;119:435–65.
2. Lenhossék MV. Der Feinere Bau Des Nervensystems Im Lichte Neuester Forschung. Berlin: Fischer's Medicinische Buchhandlung H. Kornfield; 1895.
3. Del Rio-Hortega P. In: Penfield W, editor. Cytology and cellular pathology of the nervous system, vol. 2. New York: Hoeber; 1932. p. 482–534.
4. Duncan D. Light and electron microscopic study of neuroglia in the normal spinal cord of the rat. Anat Rec. 1965;151:345.
5. Falcone C. Evolution of astrocytes: from invertebrates to vertebrates. Front Cell Dev Biol. 2022;10:931311.
6. Verkhratsky A, Ho MS, Parpura V. Evolution of neuroglia. Adv Exp Med Biol. 2019;1175:15–44.
7. Edwards TN, Meinertzhagen IA. The functional organisation of glia in the adult brain of Drosophila and other insects. Prog Neurobiol. 2010;90:471–97.
8. Ortega A, Olivares-Bañuelos TN. Neurons and glia cells in marine invertebrates: an update. Front Neurosci. 2020;14:121.
9. Hickman C, Larson A, Keen S, I'Anson H, Eisenhour D. Integrated principles of zoology. McGrawHill (USA). 2023.
10. Watson AHD. Presynaptic modulation of sensory afferents in the invertebrate and vertebrate nervous system. Comp Biochem Physiol A Physiol. 1992;103:227–39.
11. Coles JA. In: Squire LR, editor. Glial cells: invertebrate. Amsterdam: Elsevier; 2009. p. 749–59.
12. Brand AH, Livesey FJ. Neural stem cell biology in vertebrates and invertebrates: more alike than different? Neuron. 2011;70:719–29.
13. Leise EM. Modular construction of nervous systems: a basic principle of design for invertebrates and vertebrates. Brain Res Brain Res Rev. 1990;15:1–23.
14. Verkhratsky A, Nedergaard M. Physiology of astroglia. Physiol Rev. 2018;98:239–389.
15. Südhof TC. Towards an understanding of synapse formation. Neuron. 2018;100:276–93.
16. Gradisnik L, Velnar T. Astrocytes in the central nervous system and their functions in health and disease: a review. World J Clin Cases. 2023;11:3385–94.
17. Mashanov VS, Zueva OR, García-Arrarás JE. Radial glial cells play a key role in echinoderm neural regeneration. BMC Biol. 2013;11:49.
18. Harrison JB, Lane NJ. Lack of restriction at the blood-brain interface in Limulus despite atypical junctional arrangements. J Neurocytol. 1981;10:233–50.
19. Hámori J, Horridge GA. The lobster optic lamina. IV Glial cells. J Cell Sci. 1966;1:275–80.
20. Laverack MS, Ardill DJ. The innervation of the aesthetasc hairs of Panulirus argus. J Cell Sci. 1965;s3-106:45–60.
21. Spencer M, Linberg KA. Ultrastructure of aesthetasc innervation and external morphology of the lateral antennule setae of the spiny lobster Panulirus interruptus (Randall). Cell Tissue Res. 1986;245:69–80.
22. Chaigneau J, et al. Organ of Bellonci of an Antarctic crustacean, the marine isopod Glyptonotus antarcticus. J Morphol. 1991;207:119–28.
23. Csoknya M, Dénes V, Wilhelm M. Glial cells in the central nervous system of earthworm, Eisenia fetida. Acta Biol Hung. 2012;63(Suppl 1):114–28.
24. Aschner M, et al. Gene-environment interactions: neurodegeneration in non-mammals and mammals. Neurotoxicology. 2010;31:582–8.
25. Oikonomou G, Shaham S. The glia of Caenorhabditis elegans. Glia. 2011;59:1253–63.
26. Oikonomou G, Shaham S. On the morphogenesis of glial compartments in the sensory organs of Caenorhabditis elegans. Worm. 2012;1:51–5.
27. Bacaj T, Tevlin M, Lu Y, Shaham S. Glia are essential for sensory organ function in C. elegans. Science. 2008;322:744–7.
28. Stout RF, Verkhratsky A, Parpura V. Caenorhabditis elegans glia modulate neuronal activity and behavior. Front Cell Neurosci. 2014;8:67.

29. Freeman MR, Doherty J. Glial cell biology in Drosophila and vertebrates. Trends Neurosci. 2006;29:82–90.
30. Parker RJ, Auld VJ. Roles of glia in the Drosophila nervous system. Semin Cell Dev Biol. 2006;17:66–77.
31. Hartenstein V. Morphological diversity and development of glia in Drosophila. Glia. 2011;59:1237–52.
32. Yang Y, Jackson R. Astrocyte identity: evolutionary perspectives on astrocyte functions and heterogeneity. Curr Opin Neurobiol. 2019;56:40–6.
33. Palgi M, et al. Evidence that DmMANF is an invertebrate neurotrophic factor supporting dopaminergic neurons. Proc Natl Acad Sci. 2009;106:2429–34.
34. Liddelow SA, Barres BA. Reactive astrocytes: production, function, and therapeutic potential. Immunity. 2017;46:957–67.
35. Sofroniew MV. Astrocyte reactivity: subtypes, states, and functions in CNS innate immunity. Trends Immunol. 2020;41:758–70.
36. Baumgart EV, Barbosa JS, Bally-Cuif L, Götz M, Ninkovic J. Stab wound injury of the zebrafish telencephalon: a model for comparative analysis of reactive gliosis. Glia. 2012;60:343–57.
37. Chen J, Poskanzer KE, Freeman MR, Monk KR. Live-imaging of astrocyte morphogenesis and function in zebrafish neural circuits. Nat Neurosci. 2020;23:1297–306.
38. Gu Y, et al. Different astrocytic activation between adult Gekko japonicus and rats during wound healing in vitro. PLoS One. 2015;10:e0127663.
39. Spruiell Eldridge SL, et al. A focal impact model of traumatic brain injury in xenopus tadpoles reveals behavioral alterations, neuroinflammation, and an astroglial response. Int J Mol Sci. 2022;23:7578.
40. Pedersen AL, Brownrout JL, Saldanha CJ. Neuroinflammation and neurosteroidogenesis: reciprocal modulation during injury to the adult zebra finch brain. Physiol Behav. 2018;187:51–6.
41. Freeman MR, Delrow J, Kim J, Johnson E, Doe CQ. Unwrapping glial biology: Gcm target genes regulating glial development, diversification, and function. Neuron. 2003;38:567–80.
42. Fuentes-Medel Y, et al. Glia and muscle sculpt neuromuscular arbors by engulfing destabilized synaptic boutons and shed presynaptic debris. PLoS Biol. 2009;7:e1000184.
43. Chung W-S, et al. Astrocytes mediate synapse elimination through MEGF10 and MERTK pathways. Nature. 2013;504:394–400.
44. Ding F, et al. α1-Adrenergic receptors mediate coordinated Ca2+ signaling of cortical astrocytes in awake, behaving mice. Cell Calcium. 2013;54:387–94.
45. Pfrieger FW. Role of glial cells in the formation and maintenance of synapses. Brain Res Rev. 2010;63:39–46.
46. Ullian EM, Sapperstein SK, Christopherson KS, Barres BA. Control of synapse number by glia. Science. 2001;291:657–61.
47. Araque A, Navarrete M. Glial cells in neuronal network function. Philos Trans R Soc Lond Ser B Biol Sci. 2010;365:2375–81.
48. Durkee CA, et al. Gi/o protein-coupled receptors inhibit neurons but activate astrocytes and stimulate gliotransmission. Glia. 2019;67:1076–93.
49. Kofuji P, Araque A. Astrocytes and behavior. Annu Rev Neurosci. 2021;44:49–67.
50. Lyon KA, Allen NJ. From synapses to circuits, astrocytes regulate behavior. Front Neural Circuits. 2022;15:786293.
51. Goritz C, Mauch DH, Pfrieger FW. Multiple mechanisms mediate cholesterol-induced synaptogenesis in a CNS neuron. Mol Cell Neurosci. 2005;29:190–201.
52. Pfrieger FW, Barres BA. Synaptic efficacy enhanced by glial cells in vitro. Science. 1997;277:1684–7.
53. Mauch DH, et al. CNS synaptogenesis promoted by glia-derived cholesterol. Science. 2001;294:1354–7.
54. Craig AM, Blackstone CD, Huganir RL, Banker G. The distribution of glutamate receptors in cultured rat hippocampal neurons: postsynaptic clustering of AMPA-selective subunits. Neuron. 1993;10:1055–68.

55. Christopherson KS, et al. Thrombospondins are astrocyte-secreted proteins that promote CNS synaptogenesis. Cell. 2005;120:421–33.
56. Kucukdereli H, et al. Control of excitatory CNS synaptogenesis by astrocyte-secreted proteins Hevin and SPARC. Proc Natl Acad Sci USA. 2011;108:E440–9.
57. Oberheim NA, et al. Uniquely hominid features of adult human astrocytes. J Neurosci. 2009;29:3276–87.
58. DeFelipe J, Alonso-Nanclares L, Arellano JI. Microstructure of the neocortex: comparative aspects. J Neurocytol. 2002;31:299–316.
59. Han X, et al. Forebrain engraftment by human glial progenitor cells enhances synaptic plasticity and learning in adult mice. Cell Stem Cell. 2013;12:342–53.
60. Padmashri R, et al. Modeling human-specific interlaminar astrocytes in the mouse cerebral cortex. J Comp Neurol. 2021;529(4):802–10. https://doi.org/10.1002/cne.24979.
61. Camandola S. Astrocytes, emerging stars of energy homeostasis. Cell Stress. 2018;2:246–52.
62. Chen Z, et al. Brain energy metabolism: astrocytes in neurodegenerative diseases. CNS Neurosci Ther. 2023;29:24–36.
63. Li J, et al. Conservation and divergence of vulnerability and responses to stressors between human and mouse astrocytes. Nat Commun. 2021;12:3958.
64. Volkenhoff A, et al. Glial glycolysis is essential for neuronal survival in *Drosophila*. Cell Metab. 2015;22:437–47.
65. Pellerin L, Magistretti PJ. Glutamate uptake into astrocytes stimulates aerobic glycolysis: a mechanism coupling neuronal activity to glucose utilization. Proc Natl Acad Sci. 1994;91:10625–9.
66. De Backer J-F, Grunwald Kadow IC. A role for glia in cellular and systemic metabolism: insights from the fly. Curr Opin Insect Sci. 2022;53:100947.
67. Falcone C, Martínez-Cerdeño V. Astrocyte evolution and human specificity. Neural Regen Res. 2023;18:131–2.
68. Zintel TM, Pizzollo J, Claypool CG, Babbitt CC. Astrocytes drive divergent metabolic gene expression in humans and chimpanzees. Genome Biol Evol. 2024;16:evad239.
69. Falcone C, et al. Cortical interlaminar astrocytes are generated prenatally, mature postnatally, and express unique markers in human and nonhuman primates. Cereb Cortex. 2021;31(1):379–95. https://doi.org/10.1093/cercor/bhaa231.
70. Verkhratsky A, Arranz AM, Ciuba K, Pękowska A. Evolution of neuroglia. Ann N Y Acad Sci. 2022;1518:120–30.
71. Verkhratsky A. Physiology of neuronal-glial networking. Neurochem Int. 2010;57:332–43.
72. Robertson JM. Astrocytes and the evolution of the human brain. Med Hypotheses. 2014;82:236–9.
73. Laming PR, et al. Neuronal–glial interactions and behaviour. Neurosci Biobehav Rev. 2000;24:295–340.
74. Kálmán M, et al. Absence of post-lesion reactive gliosis in elasmobranchs and turtles and its bearing on the evolution of astroglia. J Exp Zool B Mol Dev Evol. 2013;320:351–67.
75. Lõrincz D, Kálmán M. Distribution of GFAP in squamata: extended immunonegative areas, astrocytes, high diversity, and their bearing on evolution. Front Neuroanat. 2020;14:49.
76. Falcone C, et al. Cortical interlaminar astrocytes across the therian mammal radiation. J Comp Neurol. 2019;527:1654–74.
77. Munger EL, et al. Comparative analysis of astrocytes in the prefrontal cortex of primates: insights into the evolution of human brain energetics. J Comp Neurol. 2022;530:3106–25.

Glia in Invertebrate Models: Insights from *Caenorhabditis elegans*

Maria D. Purice, Liza J. Severs, and Aakanksha Singhvi

Abstract Glial cells modulate brain development, function, and health across all bilaterian animals, and studies in the past two decades have made rapid strides to uncover the underlying molecular mechanisms of glial functions. The nervous system of the invertebrate genetic model *Caenorhabditis elegans* (*C. elegans*) has small cell numbers with invariant lineages, mapped connectome, easy genetic manipulation, and a short lifespan, and the animal is also optically transparent. These characteristics are revealing *C. elegans* to be a powerful experimental platform for studying glial biology. This chapter discusses studies in *C. elegans* that add to our understanding of how glia modulate adult neural functions, and thereby animal behaviors, as well as emerging evidence of their roles as autonomous sensory cells. The rapid molecular and cellular advancements in understanding *C. elegans* glia in recent years underscore the utility of this model in studies of glial biology. We conclude with a perspective on future research avenues for *C. elegans* glia that may readily contribute molecular mechanistic insights into glial functions in the nervous system.

Keywords Invertebrate · Glia · *C. elegans* · Sensory perception · Aging and stress

Maria D. Purice and Liza J. Severs contributed equally.

M. D. Purice · L. J. Severs
Division of Basic Sciences, Fred Hutchinson Cancer Center, Seattle, WA, USA

A. Singhvi (✉)
Division of Basic Sciences, Fred Hutchinson Cancer Center, Seattle, WA, USA

Department of Biological Structure, University of Washington School of Medicine, Seattle, WA, USA
e-mail: asinghvi@fredhutch.org

E. Blanco-Suarez, I. Farhy-Tselnicker (eds.), *Astrocyte-Neuron Interactions in Health and Disease*, Advances in Neurobiology 39,
https://doi.org/10.1007/978-3-031-64839-7_2

Abbreviations

aECM	Apical extracellular matrix
AMsh	Amphid sheath glia
AMso	Amphid socket glia
C. elegans	*Caenorhabditis elegans*
Ca^{2+}	Calcium
CEPsh	Cephalic sheath glia
CEPso	Cephalic socket glia
Cl^-	Chloride
D. melanogaster	*Drosophila melanogaster*
D. rerio	*Danio rerio*
DEG/ENaC	Degenerin/epithelial sodium channel
DELM-1 and DELM-2	Degenerin-like channel mechanosensory linked 1 and 2
DR	Dietary restriction
DTS	Developmentally timed sleep
GLR	Glia-like cells of the nerve ring
GLT-1/EAAT2	Glutamate transporter 1/excitatory amino acid transporter 2
GPCR	G protein-coupled receptor
HOsh	Hook sheath glia
HOso	Hook socket glia
HSF-1/HSF1	Heat shock factor 1
IAA	Isoamyl alcohol
ILsh	Inner-labial sheath glia
ILso	Inner-labial socket glia
IRK-2/KCNJ	Inward rectifier potassium channel-2/potassium inwardly rectifying channel subfamily J member 12
miRNA	MicroRNA
NHR	Nuclear hormone receptor
NMJ	Neuromuscular junction
NPR-28	Neuropeptide receptor 28
NRE	Neuron receptive ending
OLQso	OLQ socket glia
PCh	Post-cloacal hypodermal cell
PCsh	Post-cloacal sheath glia
PCso	Post-cloacal socket glia
PHsh	Phasmid sheath glia
PHso1	Phasmid socket 1 glia
PHso2	Phasmid socket 2 glia
PSR-1/PSR	Phosphatidylserine receptor
$R_{(n)}st$	Ray structural cell, where n dictates the ray number (1–9)
RGBA-1	Regulatory-gene-for-behavioral-aging-1
RNAi	RNA interference

scRNA-seq	Single-cell RNA-sequencing
sh	Sheath
SIS	Stress-induced sleep
so	Socket
SPsh	Spicule sheath glia
SPso	Spicule socket glia
UPR	Unfolded protein response
UPR^{ER}	Endoplasmic reticulum unfolded protein response
UPR^{mt}	Mitochondrial unfolded protein response
VGCC	Voltage-gated calcium channel
XBP-1	X-box-binding protein 1
XME	Xenobiotic-metabolizing enzyme

1 Introduction

Glial cells modulate neuronal development, function, and health across both invertebrates and vertebrates. Most nervous systems of tractable model organisms contain both neurons and glia. However, while neurons are studied in some detail, much less is understood about glial functions and properties at the molecular detail [1–4]. The critical need to address this knowledge gap in neuroscience is underscored by the growing implications for glia in neurological impairments [5–8].

The field of glial cell and molecular biology has made exponential strides in recent years, aided by studies across genetic model systems, including the nematode, *C. elegans*. The nervous system of an adult *C. elegans* animal is composed of 302 neurons and 56 glia in the hermaphrodite and 387 neurons and 90 glia in the male [9–12]. Each neuron and glial cell originate from an invariant developmental lineage and make stereotypic contacts with each other, which means that the cellular anatomy of glia:neuron contacts in the nervous system is identical across individual animals [9, 11]. The development and function of the *C. elegans* nervous system have striking molecular conservation with that of other species, revealing it as a facile platform to uncover molecular mechanisms of neural functions [13–15]. Studies in this model have contributed discovery of mechano-transduction pathways [16], Notch signaling [17], Par complex signaling [18], axon guidance genes (Netrin/DCC) [19], synaptic function genes [20], and discovery of metazoan RNAi [21]. Further, while this animal's genome does not encode for ion channels that drive sodium-dependent action potentials like in mammals, some of its neurons utilize voltage-gated CaV1 **ca**lcium (Ca^{2+}) channels to fire calcium-mediated action potentials [22, 23]. On the other hand, the *C. elegans* nervous system contrasts with those of other model organisms in three cellular features. First, it has a smaller number of neurons and glia compared with, for example, the approximate 100,000 neurons in adult fly *D. melanogaster,* or larval fish *D. rerio,* or ~86 billion neurons and glia each in humans [24–27]. Second, unlike the invariant developmental cell lineages of *C. elegans,* the development and cell:cell contacts of neurons and glia in

	C. elegans	*Drosophila*	Mammals
Glial properties			
Interact with sensory cells/neurons			
Interact with epi/endothelia			
Associate with synapses (tripartite synapse)			
Enwrap axons (non-myelinating glia)			
Glia:glia tiling			
Glial intracellular calcium dynamics			
Barrier function (create privileged/BBB spaces)	?		
Glial impact on neuronal properties			
Act as neuronal progenitor cell			
Neuropil positioning / axon guidance			
Axon-dendrite specification			
Synapse positioning			
Synaptic function regulation			
Engulfment of neuron fragments			
Stress response signaling to neurons			
Regulate neural aging			
Regulate neurotransmitter diffusion			
Regulate neurons with single-cell specificity		?	?
Trophic support for neurons	x		

Table Key	Yes	(?) Maybe/ Unknown	(x) No

Fig. 1 A list of well-defined glial molecular or morphological phenotypes and functions conserved across mammals, flies, and worms (indicated for each animal by green boxes). The question mark indicates that this has yet to be studied in *C. elegans*. X indicates that this function is not conserved in *C. elegans*. See text for details. (Updated and modified from Freeman and Rowitch [28])

other species are stochastic at single-cell resolution. Third, *C. elegans* neurons survive without associated glia, indicating that their survival does not depend on glial trophic or metabolic support (Fig. 1) [29, 30]. Importantly, however, loss of glial functions in *C. elegans* does lead to structural and functional deficits in associated neurons (Table 1) (see below) [31]. Taken together, these features suggest that this model holds unique advantages for examining glial regulatory functions independent of the confound of glial trophic or metabolic support functions.

In addition, multiple features and tools in this model also lend power to studying the impacts of glia on nervous system function and organismal attributes like aging and inter-tissue dynamics, with single glia-neuron resolution [14, 30, 32]. This includes the fact that *C. elegans* boasts easy and powerful genetic toolkits, a short

Table 1 *C elegans* glia, associated neurons, and behaviors affected by respective glia-neuron interactions

Organ	Glia	Associated neuron(s)	Number of sensilla	Documented functions of glia-neuron interactions	Animal behavior regulated by glia-neuron interactions
Sensilla glia					
Amphid	AMso, AMsh	Contact AMsh only: AFD, AWA, AWB, AWC AMsh-AMso channel neurons: ASE, ASG, ASH, ASI, ASK, ASJ, ADF, ADL	2 (L/R)	AMsh: NRE shape and plasticity, glial compartment, dendrite outgrowth, ion and pH regulation, engulfment, epithelia-glia mechanobiology. AMso: adult neurogenesis	Chemosensation, gustation, mechanosensation, thermosensation, pathogen recognition
Phasmid	PHsh, **PHso1**, PHso2	PHA, PHB, PHC Contact PHso only: PQR	2 (L/R)	PHso: adult neurogenesis	n.d.
Anterior deirid	ADEsh, ADEso	ADE	2 (L/R)	n.d.	n.d.
Posterior deirid	PDEsh, PDEso	PDE	2 (L/R	n.d.	n.d.
Inner labial	ILsh, ILso	IL1, IL2 Contact ILso only: BAG, URX, FLP	6 (L/R, D/V/L)	IL/OL nerve outgrowth, neuropil placement	ILso: Mechanosensation
Outer labial	OL(Q/L)sh, OL(Q/L)so	OLQ or OLL	6 (L/R, D/V/L)	n.d.	OLQso: Mechanosensation
Cephalic	CEPsh, CEPso	CEP (+ CEM in males)	4 (D/V, L/R)	CEPsh: epithelia-glia mechanobiology; CEPso: aECM pore formation	n.d.
Hook	HOsh, HOso	HOA, HOB	1	n.d.	Male mating
Ray	R(N)st	R(N)A, R(N)B	18 (L/R), (N) = 1–9	n.d.	Male-mating
Post-cloacal	PCsh, PCso	PCA PCsh only: PCB, PCC	2 (L/R)	n.d.	Male-mating
Spicule	SPsh syncytium (2 cells), SPso syncytium (4 cells)	SPD, SPV	2 (L/R)	NRE development. SPso: extra-synaptic source of dopamine	Male-mating
Neuropil glia					
Neuropil	CEPsh	Many neuropil axons	4 (D/V, L/R)	Neuropil assembly, AIY synapse positioning, neurotransmitter clearance	Locomotion during sleep, repetitive behavior, swimming induced paralysis
Mesodermal-lineage glia					
Mesodermal	GLR	RME, others (?)	6 (D/V/L, L/R)	RME axon specification	Locomotion

Modified from Singhvi and Shaham [30]
Sensilla with sex-specific glia are noted in bold. PHso1 glia (magenta) are present only in hermaphrodites; and all glia of hook, spicule, ray, and post-cloacal sensilla (orange) are only present in males

organismal lifespan, and is optically transparent. Furthermore, the adult neural connectome is mapped in both *C. elegans* sexes [11, 12, 33–36]. In addition, many of the neural circuits within have been mapped for specific and measurable animal behaviors, with many of the component neurons already having undergone extensive analysis through genetic and transcriptomic methods [12, 13, 37–40]. Finally, a single nuclear RNA-seq molecular atlas of all glia of the *C. elegans* nervous system has recently been compiled and validated in vivo for both sexes (Purice and Singhvi unpublished).

Leveraging the unique properties of this model, studies in recent years have made rapid strides in the study of *C. elegans* glial biology, including their roles in sensory processing and animal behaviors, organismal aging, and sexual dimorphism. Along with relevant glial anatomy, this chapter discusses the current understanding of glial biology and functions derived from studies in *C. elegans,* as well as knowledge gaps and future perspectives. This overview focuses exclusively on the post-development animal, and readers are directed elsewhere for discussions of glial development or glial roles in neurodevelopment [29, 30, 41, 42].

2 Glial Anatomy

C. elegans glia were initially described by light and electron microscopy studies. Specifically, they were described as cells that contact neurons within sense organ "sensilla" and had shared ultrastructural morphological characteristics with neurons but lacked presynaptic structures, thus making them distinct from neurons [11, 43, 44]. Of the 56 glial cells in the *C. elegans* hermaphrodite (somatically female), 50 are of neuroectodermal origin (i.e., neuron/glia/epithelial progenitors) and six are of mesodermal origin [9, 10, 41, 45]. All neuroectodermal-lineage glia in *C. elegans* reside within the animal's 11 sense organs and are named accordingly (Fig. 2 and Table 1). Males have an additional 34 neuroectodermal-lineage glia (Fig. 2) [68]. All *C. elegans* glia are listed in Table 1 (reproduced from Singhvi and Shaham [30]), and their anatomy is summarized below.

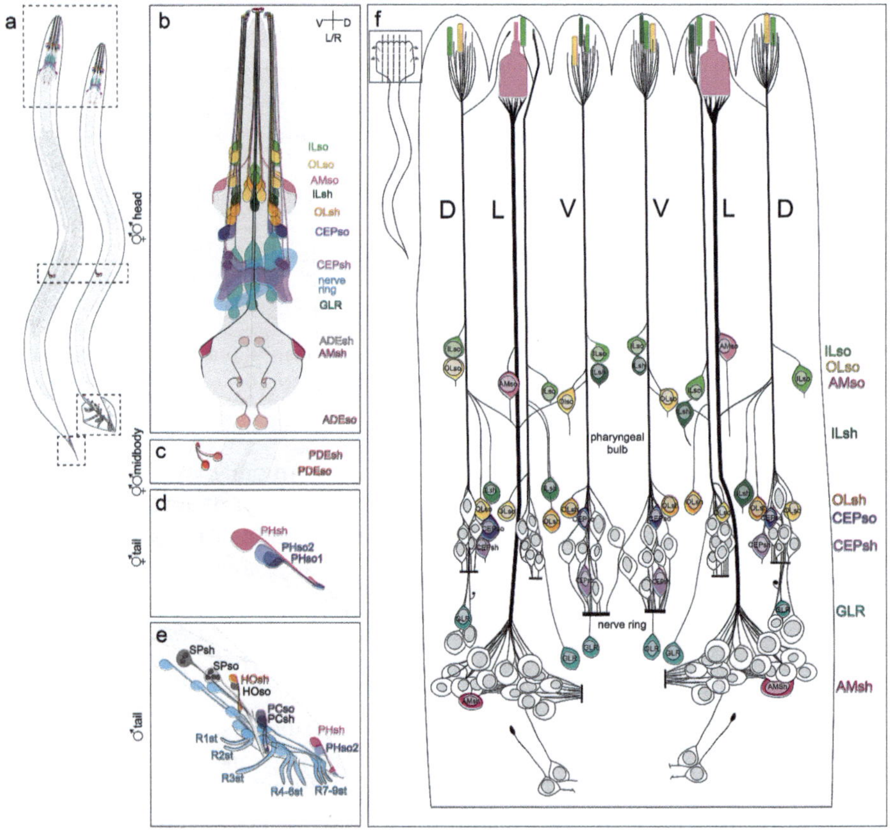

Fig. 2 Anatomical map of *C. elegans* glia across sexes. The position for each panel is denoted by the dashed box in the *C. elegans* schematic in (**a**). Shared glia of the hermaphrodite and male in the (**b**) head and (**c**) midbody. Glia in the tail of (**d**) hermaphrodite and (**e**) male tail. (**f**) Fillet preparation of the adult male and hermaphrodite head showing the position of *C. elegans* glia. The glia are color- and position-coded to match panel (**b**). Position of neurons is colored in grayscale. (**e** is adapted from Ward et al. [11])

2.1 Sex-Shared Glia

Neuroectoderm glia are anatomically of two types, **sh**eath (**sh**) and **so**cket (**so**). The animal has 7 sex-shared sense organs, each consisting of 1–12 neurons, one sheath glia, and 1–2 socket glia. Thus, hermaphrodites total 24 sheath and 26 socket glia, and 60 sensory neurons. All sense organs except two (anterior/posterior deirid) are located at either the head region or the tail region of the animal (Fig. 2a–e). Within each sense organ, each sensory neuron dendrite tip terminates in a specialized structure called the **n**euron **r**eceptive **e**nding (NRE), which houses modality-specific molecular machinery for sensory transduction (Figs. 2f and 3a–e). Both sheath and socket glia send cellular processes to the periphery, to contact the NRE and neuronal dendrite, and each other (Figs. 2 and 3). Here, the terminal process endings of sheath and socket glia together either make a channel through which NREs traverse, or encase or contact the sensory NRE, or embed them in the animal cuticle (Figs. 2 and 3b–e) [11, 46, 47]. This is the only place of glia:neuron and glia:glia physical contact in the sense organ.

Sheath glia form electron-dense apical junctions with both contacting neuron dendrites (at the NRE base) and partner sense organ socket glia (Fig. 3c, d) [48]. Socket glia also form cell:cell junctions with the overlying epithelia (also called hypodermis) [33, 46]. The anatomy and **a**pical **e**xtra**c**ellular **m**atrix (aECM) of socket glia are thought to dictate whether NREs project outward through a cellular pore or embed into the animal cuticle (Fig. 3c) [11]. For example, **ceph**alic **so**cket glia (CEPso) with closed aECM sheet embed mechanosensory CEP NREs into the cuticle, while **am**phid **so**cket glia (AMso) form a channel pore through which chemosensory NRE sample cues from the environment. Like neurons and epithelia, *C. elegans* sheath glia exhibit hallmarks of cell polarity. Besides the junctions noted above [11, 33, 46], this includes segregation of apical-basal markers. Thus, glial apical membranes contact NREs, while basolateral membranes contact epithelia [48–50]. Furthermore, at least one glia, the **am**phid **sh**eath (AMsh), is shown to compartmentalize its apical membrane into separate molecular microdomains around distinct NRE contact sites [46, 50–52] and secrete dense matrix-laden vesicles into the glial channel lumen.

The four **ceph**alic **sh**eath (CEPsh) glia also extend posterior sheath-like processes to the nerve ring or "brain" neuropil of the animal, which contains most of the animal's neuronal synapses (Figs. 2b and 3f). Analogous to vertebrate or *Drosophila* astrocytes, CEPsh posterior sheaths infiltrate the neuropil, enwrap at least one synapse in a classic tripartite configuration [12], serve as diffusion barrier for some neurotransmitters [53, 54], and enwrap axons and synapses en masse [12, 53]. Interestingly, while the *C. elegans* genome does not encode for myelin protein like most invertebrates, CEPsh glia express *hlh-17/Olig2*, which activates the expression of myelin-associated genes in myelinating oligodendrocyte glia of mammals [55, 56].

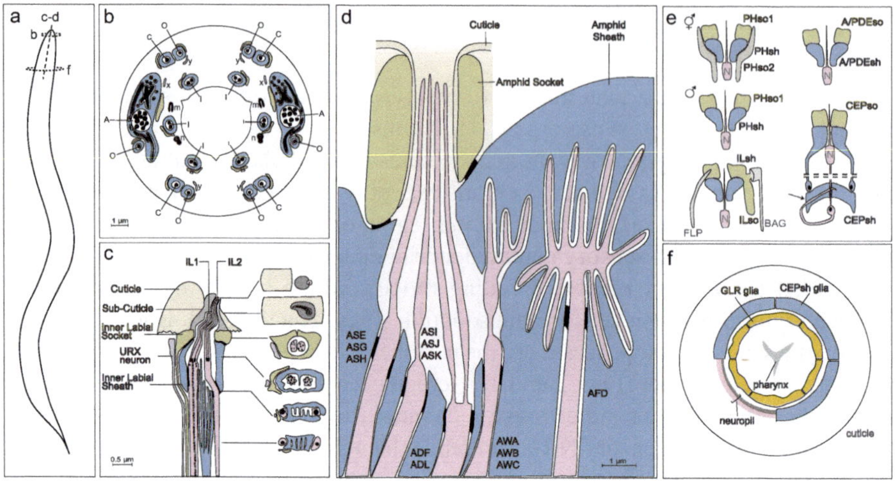

Fig. 3 Detailed view of glia-neuron interactions in *C. elegans*. The locations depicted for panels B, C, D, and F are denoted by the dashed line in the *C. elegans* schematic in (**a**). Color code is maintained throughout the figure as follows: sense organ neurons (pink), socket glia (green), sheath glia (blue), and cuticle (light brown). (**b**) Cross section near the nose tip showing a summary of the arrangement of neurons within the sensilla in the head. Letters represent each sensilla, where **i**nner labial (I), **o**uter labial (O), **c**ephalic (C), **a**mphid (A), not in sensilla URX/URY (x/y), and inner and outer label neurons (m/n). All sensilla are shown as neurons surrounded by sheath cells with socket cells adjacent. Ciliated neurons are black, and unciliated neurons are light purple (e.g., URX/URY (x/y) neurons). (**c**) The inner-labial sensillum in longitudinal (left) and transverse sections (right) depicting neuron-glia interactions with embedded (IL1) and channel (IL2) neurons. The outlines of the cells were simplified from the EM tracings, and the number of membrane invaginations in the sheath cell was reduced for clarity. The black regions between cells represent tight junctions. Lines in the IL1 neuron represent microtubules. Vesicles are shown near the basal body of the IL1 neuron. The socket cell and the anterior part of the sheath cell are surrounded by a labial epidermal cell at which the socket cells make additional electron-dense cell:cell junctions (not shown). (**d**) The amphid sensillum in a longitudinal section depicting embedded neurons in the amphid glia (AFD, AWA, AWB, AWC) and channel neurons in the socket glia (ASE, ASG, ASH, ASI, ASJ, ASK, ADF, ADL). Drawing is simplified from EM tracings for clarity. The black regions between cells represent tight junctions. The socket cell and the anterior part of the sheath cell are surrounded by an epithelial cell to which the socket glia make additional cell:cell junctions (not shown). (**e**) Schematic highlighting sheath and socket glia:neuron interactions within individual sense organs. The CEP sense organ also depicts sheath glia:neuron interactions between both the sheath and axon (arrow). (**f**) A cross section through the nerve ring showing the relative location of neuropil GLR glia (yellow) and CEPsh glia (blue) within this region. (**b**, **c** are adapted from Ward et al. [11]. **d** is adapted from Perkins et al. [46]. **f** is modified from Singhvi and Shaham [30])

2.2 *Mesoderm Lineage GLR Glia*

The six mesoderm-derived **gli**a-**l**ike cells of the nerve **r**ing (GLR) are anatomically organized into three pairs (dorsal, ventral, or lateral L/R), with cell soma posterior to the nerve ring (Fig. 2b, f). They express the *C. elegans* homolog of myoD (**myo**blast **d**etermination protein 1)/MYF5/MYF6, HLH-1, implicated in muscle growth in *C. elegans* and humans [57–59]. GLR glia interact with **neuromuscular junctions** (NMJs) and form gap junction with the RME motor neurons and muscle [12]. GLR processes wrap the nerve ring proximally (from the inside) [12], while CEPsh do so distally (from the outside, Fig. 3f). Thus, it is possible that these glia together create a specialized pseudocoelomic environment for the neuropil, analogous to astrocytic blood-brain barrier functions [29, 30, 52].

2.3 Sexually Dimorphic Glia

The *C. elegans* hermaphrodite is somatically female [60]. Four hermaphrodite glia are anatomically sexually dimorphic: **am**phid **so**cket (AMso), **ceph**alic **sh**eath and **so**cket (CEPsh and CEPso, respectively), and **ph**asmid **so**cket **1** (PHso1). The two AMso glia in males divide to generate MCM interneurons [61]. This post-embryonic asymmetric stem cell-like division is reminiscent of neurogenic vertebrate radial glial cell divisions [61, 62]. CEPsh and CEPso **cep**halic sense organ glia interact with NREs of CEP neurons in both sexes (embeds in the cuticle) [11, 46, 63]. In males, they also interact with chemosensory CEM neurons, whose NRE traverses a glial channel [10] to sense pheromones for male mating behaviors [64, 65]. To anatomically enable putative chemosensory functions of CEM neurons, CEPso glia undergo a transient sex-specific switch as males reach sexual maturity, which transforms their closed-pore nubbin for the sex-shared CEP to also form an **a**pical **e**xtracellular **m**atrix (aECM) pore through which the chemosensory CEM projects to sample the external environment [66]. Lastly, in males, the PHso2 functions like the hermaphrodite PHso1, for socket glial function in the **ph**asmid sense organs [67, 68]. Male PHso1 transdifferentiate, or in some cases divides, into PHD neurons [69].

2.4 Male-Specific Glia

The 36 male-specific glia reside in four copulatory sense organs in the animal tail—ray, spicule, hook, and post-cloacal sensilla (Fig. 2e), required for male mating behaviors [70]. These glia have variations on the classic sheath:socket glia organization of sex-shared sensilla [52, 71].

There are nine bilateral ray sense organs, each comprising two sensory neurons. NREs of these neurons sample the external world through a channel formed by a single glial cell (not two) known as the **r**ay **st**ructural cell [$R(_n)$st, where **n** indicates the ray number (1–9)]. The R_nst is presumed to be a hybrid sheath-socket glial cell. Like sheath glia, R_nst contact ray NREs and has numerous vesicles believed to secrete material into the ray lumen [52, 71]. Like socket glia, R_nst connect to overlying epithelia (hyp7 syncytium) via adherens junctions [52, 71]. In the bilateral spicule sense organ, both the **sp**icule **sh**eath cell (SPsh) and **sp**icule **so**cket cell (SPso) are syncytial, i.e., derived from the fusion of cells during development [52, 63, 71,

72]. Thus, the SPso glia contains four nuclei, and the SPsh contains two [52, 63, 71]. The SPso secretes a cuticle that covers the outside of the prong-like spicule structure used during copulation and also forms a channel with SPsh glia through which spicule NRE senses environmental cues [63, 70–72]. Finally, the hook and post-cloacal sense organs uniquely include epithelial cells. The shield-shaped **ho**ok contains the processes of two sensory neurons, a **sh**eath cell (HOsh), a **so**cket cell (HOso), and four epithelial (hypodermal) cells [52, 71] and is the only unilateral sense organ in the animal. Each **p**ost-**c**loacal sensillum contains NREs of three sensorimotor neurons, **sh**eath (PCsh) and **so**cket glia (PCso), and a binucleate syncytial epithelial cell (PCh) [52, 71].

3 Molecular Mechanisms of Glial Communication

C. elegans glia regulate single-neuron properties and organismal aging through multiple molecular signaling mechanisms, as summarized below.

3.1 Neuropeptide Signaling

Two studies suggest that *C. elegans* glia communicate with neurons and other tissues via neuropeptides to regulate organismal aging and male mating behavior (also see Section 5.2) [73, 74]. Corroborating this, **s**ingle-**c**ell **RNA-seq**uencing (scRNA-seq) atlases have uncovered multiple neuropeptides [75, 76]. However, little is known about glial neuropeptide storage, processing, or release mechanisms in either *C. elegans* or other species [77].

3.2 Gliotransmission

It was recently reported that AMsh glia respond to external stimuli independent of neurons and release GABA to modulate neuronal sensory adaption [78–80] (see Section 4 below). In addition, the male SPso glia are dopaminergic and express CAT-2/tyrosine hydroxylase, a rate-limiting dopamine biosynthetic enzyme (see Section 4.7 below). Whether other sense organ glia release GABA, dopamine, or other neurotransmitters, or how this is regulated, remains to be determined.

3.3 Gap Junction Signaling

Gap junctions mediate the exchange of ions and molecules/second messengers <1 kDa [81].

Invertebrate gap junctions, including those of *C. elegans*, are comprised of pannexin-related innexins [82–86]. Of the 25 encoded innexin genes in the *C. elegans* genome, the expression of at least eight is detected in varying subsets of glia as mapped in vivo by transcriptional reporters [85]. Some express in individual glia (e.g., *inx-11* in CEPsh and *inx-18* in ILso glia), while others have a broad expression in many glia [85]. Curiously, no glia:neuron gap junction has yet been reported for neuroectodermal glia by ultrastructural studies [85, 87] and thus far innexins are molecularly identified in only one of the cells of sense organ glia:glia pair [85]. Functional roles of innexins in sense organ glia await inquiry.

GLR glia make gap junctions to head muscle cells and to RME motor neurons that innervate these muscles and express at least 11 innexins in interesting patterns [85]. Five of the 11 innexins express in the two lateral GLRs, but not dorsal or ventral pairs, including *inx-22* [85]. Barring a role for GLR-RME gap junctions in RME axon specification in development [81], roles for GLR innexins in adult neural functions remain to be explored.

3.4 Transporters and Ion Channels

Glia express varying subsets of transporters and ion channels, with restricted subcellular localization of these to discrete neuronal contact sites [50]. A role for some of these in regulating neural functions and animal behavior is reported (see Section 4) [30, 50].

4 Glial Roles in Neural Functions and Animal Behavior

Studies below reveal roles for different sense organ sheath glia in regulating NRE properties and activity of contacting sensory neurons, with an impact on animal sensory behaviors.

4.1 Chemosensation

Chemosensory perception is one of the best characterized sensory behaviors in the worm, with 32 of the 302 hermaphrodite neurons implicated thus far in mediating this sensory modality [88]. The animal uses chemosensory cues to dictate many behaviors like food seeking, mating, discernment of repulsive and attractive cues, and switching to the alternate developmental state of dauer [78, 88–92]. Sensory neurons driving this function reside in the amphid, phasmid, inner-labial sensory organs [88], and male-specific sense organs [93]. Studies of glial roles in chemosensation have focused primarily on AMsh glia, and roles for AMso, **ph**asmid (PHsh, PHso), or **i**nner-**l**abial (ILsh, Ilso) sense organ glia in chemosensation have yet to be systematically examined. Furthermore, chemosensory [78] and mechanosensory [79, 94] systems are the only examples described in which glial cells, specifically the AMsh, are able to autonomously respond to sensory cues.

The animal's major chemosensory organ, the **am**phid, is located at its nose tip, where the **sh**eath glia (AMsh) interact with 12 neurons and the **so**cket glia (AMso) interact with eight neurons (Fig. 3d). Of the four neurons that contact the AMsh glia, three sense volatile olfactory cues (AWA, AWB, and AWC). All eight neurons that interact with both the AMsh and AMso glia sense water-soluble chemicals (gustatory) [31, 52, 95]. AMsh glia-ablated animals have deficits in AWA- and AWC-mediated chemotaxis and exhibit changes in cilia morphology in AWA, AWC, and AFD neurons [31]. The severity of these deficits is reminiscent of mutant animals lacking NRE cilia [31, 78], suggesting that AMsh glia impact chemosensation, at least in part, by regulating component NRE morphologies.

AMsh glia also deploy multiple ion regulators to regulate animal chemotaxis behaviors and neuronal response properties independent of modulating NRE shape. This includes glial K^+ channels (KQT-2/KCNQ5, EGL-36/KCNC), **ch**l**oride** (Cl^-) channels (BEST-9/BEST1, CLH-3/CLCN2), K^+-Cl^- transporters (KCC-3/SLC12A6, KCC-1/SLC12A4), and nucleoside transporters (ENT-4/hENT3) [50, 95]. The combination of ionic regulators varies by individual sensory cue or neuron. Independently, the L-type **v**oltage-**g**ated **c**alcium **c**hannel (VGCC) EGL-19 is also shown to impact shape and function of the polymodal ASH sensory neuron and its adaption to repetitive chemosensory stimuli [92].

In addition to affecting animal behavior and sensory neuron activity, these ion regulators also impact Ca^{2+} responses of AMsh glia to sensory cues. For example, AMsh glia exhibit Ca^{2+} responses to the odorant octanol, which requires the regulators noted above, with varying kinetics: While *best-9* mutants had normal Ca^{2+} responses to secondary exposure of octanol, mutants in *kqt-2*, *egl-36*, or *kcc-1* had impaired adaptive responses to secondary exposure, and *kqt-2* mutants displayed altered Ca^{2+} response kinetics [80, 95]. Similarly, AMsh Ca^{2+} responses to the odorant **i**so**a**myl **a**lcohol (IAA) were severely attenuated in *egl-19* mutants [95].

AMsh glia also regulate Na^+ and Cl^- transport to modulate pH sensing of the worm's environment. The **deg**enerin/**e**pithelial sodium (**Na**$^+$) **c**hannel (DEG/ENaC) ACD-1/ACCN5 was initially thought to act in neurons but subsequently shown to act in AMsh glia for animal chemotaxis to tastants and for pH avoidance with TAX-2/CNGB1-3 and DEG-1/SCNN1G channels [96, 97]. The inward rectifying ClC chloride channel CLH-1/CLCN1 in AMsh glia also regulates glial intracellular pH/acidification, which in turn impacts ACD-1 activity [98].

AMsh glia are also reported to respond directly to high concentrations of volatile chemosensory cues with intracellular Ca^{2+} dynamics, at thresholds fold higher than those activating neurons. Briefly, high doses of the odorant IAA (attractive at low concentrations with response mediated by AWC neuron, aversive at high concentrations with response mediated by nociceptive/polymodal ASH neurons), elicit an increase in intracellular Ca^{2+} in AMsh glia [78]. This response persists when ASH is ablated or in dissociated AMsh glia in culture, indicating that it is likely independent of ASH response properties [78]. Further, AMsh glia require molecular pathways distinct from neurons to respond to odorants. Where the ASH neuron utilizes STR-61 to detect IAA, the AMsh glia utilizes the GPCR SRH-79, and the knockdown of *srh-79* leads to defects in the glial Ca^{2+} response [78]. Downstream of *srh-79,* GPA-3/GPCR, OSM-9/TRPV channel, and ITR-1/IP3R also regulate glial responses to IAA [78]. Finally, glial chemosensory responses are cue-specific, as not all cues elicit this (e.g., no response to heavy metals, high salt, osmolarity, detergents, or alkaloids) [78].

Together, these studies uncover the molecular complexity by which one glial cell regulates associated neurons by impacting their shape, activity, ionic milieu, or a combination of these. Additionally, multiple regulators impact one stimulus response or one ion, and each regulator has varying impacts on each neuron/stimulus. Further, not all regulators also impact AMsh intracellular ionic flux. In corollary, some components are shared across sensory modalities, like the AMsh glial EGl-19/VGCC impacting both mechanosensation and chemosensation simultaneously [78–80, 94]. Lastly, not all contacting chemosensory neurons are regulated equivalently by AMsh glia. Thus, in summary, AMsh glia regulate individual NREs, cues, and modalities with exquisite single-neuron specificity. How might they do so? Emerging evidence reveals that one mechanism guiding specificity is glial localization of regulatory cues to molecular microdomains around individual neuron contact sites [50]. This study showed that AMsh glia make at least three molecular microdomains. Intriguingly, for at least one cue, KCC-3/SLC12A6, chemosensory neurons like AWC repel it through yet-unidentified ciliary signals, which renders it localized only around thermosensory AFD NRE contacts. Tracking this, the mislocalization of the KCC-3 microdomain localization impairs functions of AWC and associated animal chemosensory behaviors, in addition to thermotaxis behaviors with kcc-3 loss [50, 112].

Thus, one model is that AMsh glia can dictate specificity in regulating different chemosensory neurons/cues by localizing a combinatorial bouquet of regulatory cues to each NRE contact site. This will be interesting to interrogate in future studies. Regardless, we posit that such specificity in signaling may be an emergent property of glia-neuron interactions also beyond sense organ glia. Uncovering the underlying molecular network will be key, followed by mechanistically assessing the molecular and functional logic for such glia-neuron specificity and its impact on neural health and disease. The invariant glia-neuron interactions, neuron shapes and manipulability, quantifiable animal behavior correlates, and clear distinctions of individual stimuli-driven circuits, along with the molecular parts list already defined above, suggest that AMsh may be a powerful prototype glial cell in which to untangle how they cooperatively impact global sensory responses.

4.2 Mechanosensation

Besides male mating sense organs [99], six of the 11 sex-shared sense organs mediate mechanosensation [16, 52]. These are the amphid, cephalic, inner and outer labial, and anterior and posterior deirid sense organs [16, 52]. Glia associate with 25 of the 45 touch-sensing neurons [79, 87, 100] and express multiple mechanosensitive channels [79, 96, 101, 102]. While mechanosensation by neurons is explored in some depth [103, 104], glial roles in this sensory perception and behavior are only recently being revealed. Further, like chemosensation, recent work in *C. elegans* also implicates glia as autonomous mechanosensors [79, 92, 101, 102].

DEG/ENaC Na^+ channels were discovered in *C. elegans* [96, 105]. Two DEG/ENaC subunits DELM-1 and DELM-2 (**de**generin-**l**ike channel **m**echanosensory linked-**1** and **de**generin-**l**ike channel **me**chanosensory linked-**2**) are required in glia for normal nose touch sensitivity. Briefly, DELM-1 and DELM-2 loss in **OLQ** and **IL so**cket (OLQso and ILso) glia lead to defective nose touch responses and affects the neurons' ability to adapt to stimuli [101], a phenotype that mimics loss of TRP channel TRPA-1 in OLQ and IL1 neurons that drive *C. elegans* nose touch behaviors [106]. Knockout of DELM-1 can be rescued by overexpressing the **in**ward **r**ectifying potassium (**K**$^+$) channel IRK-2/KCNJ in OLQ and IL2 glia, suggesting that these channels may aid in setting basal excitability levels of associated sensory neurons [101]. Further implicating glial maintenance of Na^+/K^+ homeostasis in mechanosensation, the Na^+/K^+ ATPases EAT-6/ATP1A3 and CATP-1/ATP1A2 are also required in OLQ and IL socket glia to support the function of nose touch neurons [102].

Like in mammals, the CIC Cl^- channels CLH-1/CLCN1 maintain Cl^- and bicarbonate HCO_3^- homeostasis [79, 107]. This is required for regulating mechanosensory responses of the polymodal ASH neuron to touch stimuli, with *clh-1* glial-specific knockout mutant animals exhibiting an inability to respond to nose touch stimulation [79]. CLH-1 localizes in the AMsh to membranes around ASH NRE, where it activates $GABA_A$ channels to lower neuronal membrane potential and provide inhibitory regulation [79]. CLH-1-dependent negative feedback prevents excess glutamate signaling in ASH, to enable an effective adaptive response to touch. *C. elegans* CLH-1 appears to have a conserved function to its mammalian ortholog CLCN-1/2 (also called CLC-1/2), raising the notion that roles for glial ClC channels in mechanosensation may be conserved.

As for chemosensation, the AMsh glia also respond to external mechanical forces through the L-type VGCC EGL-19/CACNA1 with intracellular Ca^{2+} dynamics upon gentle mechanical touch [92]. This glial function could be tied to morphologic changes observed in both neurons and glia in *egl-19* mutants, with both cells exhibiting aberrantly elongated cell processes and cell bodies [92]. Further studies will be needed to determine how this change in glial Ca^{2+} activity or morphology impacts the mechanosensory behaviors of the worm.

How these different molecular modules integrate mechanical response properties in a single glia remains an open question. It also remains to be examined if glia of all mechanosensory sense organs (ILsh/ILso, OLsh/OLso, AMso, male sense organs) impact animal mechanosensory behaviors. Given the molecular conservation of all genes identified thus far, the study of *C. elegans* glial roles in mechanosensory perception may inform not only how glia impact touch and proprioception but may also inform if glia dynamically track ambient mechanical forces as from hypertension (astrocyte glia) or enteric distension (enteric glia). In line with this notion, recent work (see Section 5 below) shows that some (e.g., AMsh, CEPsh) but not all (e.g., PHsh) glia are mechano-biologically coupled with contacting epithelia in post-embryonic animals [49].

To date, the only glial cell in *C. elegans* reported to respond to touch stimuli as a mechanosensory cell independent of associated sensory neurons is the AMsh glia [79, 94]. This was first documented by patch clamp recordings of the AMsh glia upon tactile stimulation of the animal nose tip and observing inward currents in AMsh glia with robust increases in intracellular Ca^{2+} [94]. While touch-evoked currents in the ASH neuron were inhibited by Na^{+} channel blocker amiloride, AMsh currents were insensitive, indicating distinct mechanisms [94]. Furthermore, AMsh glia appeared more touch-sensitive than its partner polymodal/mechanosensory neuron, ASH, and currents observed in the AMsh were proportionate to the intensity of the stimulus applied [94].

That AMsh glia respond to both chemosensory and mechanosensory cues raises the question of functional relevance: Why do glia respond to sensory cues, and why through molecular pathways distinct from neurons? Given that these responses are seen only under high sensory stimulation, we posit that this may be a threshold mechanism for the system to delimit hyperactivation and prevent neuronal damage caused by intense or sustained stimuli. It may thus be one mode to gate the operating range for stimuli that are physiologically relevant but potentially harmful. While this hypothesis awaits formal test, these findings also raise a second tantalizing question of whether all sense organ glia in *C. elegans* exhibit similar functions.

4.3 Thermosensation

C. elegans exhibit remarkable sensitivity to temperature changes, with an ability to detect 0.01 °C variations within the physiological range of 14–26 °C [111]. The thermotaxis memory of *C. elegans*, or ability to "learn" new temperatures, is an example of a long-lasting sensory memory [100, 108–111]. Amphid organ AMsh glia ensheath the microvilli-rich NREs of the animal's primary thermosensory neuron, the AFD [11] (Fig. 3d), while CEPsh glia contact AFD axons and synapses with downstream interneurons of the thermotaxis circuit like AIY in the brain neuropil [12]. Ablation of AMsh and acute block of AMsh glial secretion both result in deficits in the ability of animals to detect temperature appropriately and loss of AFD NRE shape [31, 112].

AMsh glia impact AFD NRE shape and function by at least three independent mechanisms, each of which impacts animal thermosensory memory and perception [14, 31, 32, 50, 112–115].

One, AMsh glia localize the glia-specific K^+/Cl^- co-transporter KCC-3/SLC12A6 to apical membranes around the AFD NRE, through a process independent of the AFD neuron itself [50, 112]. KCC-3 regulates the concentration of Cl^- ions in the intercellular space between AMsh glia and AFD NRE. Cl^- ions directly inhibit the neuronal receptor guanylyl cyclase GCY-8, which, in turn, regulates GMP levels in the AFD neuron. This process regulates the neuronal actin polymerizing factor WSP-1/nWASp, to control the shape of glia contacting the AFD NRE [112]. Thus, glial *kcc-3* mutants have altered AFD NRE shape and thereby neuronal activity and animal thermosensory behavior [112, 114].

In a second and independent mechanism, it was shown that AMsh glia dynamically modulate AFD NRE shape, and thereby animal behavior, by engulfing/pruning AFD NRE fragments throughout life [14, 113]. This study provided the first evidence that a *C. elegans* glia executes the conserved and disease-relevant function of glial engulfment/pruning, establishing *C. elegans* and this glia-neuron pair as a setting to genetically probe mechanisms of glial engulfment. It also demonstrated that a peripheral sense organ glia engulfs NREs to impact animal sensory behaviors. Briefly, the study uncovered that glial engulfment is regulated by neuronal TAT-1/ATP8A and AFD neuronal activity. Further, AMsh glia repurpose the apoptotic corpse phagocytosis machinery, also discovered in *C. elegans,* to engulf fragments of a living neuron. This includes AMsh glial small GTPase CED-10/Rac1, ternary GEF complex (CED-2/CrkII, CED-5/DOCK180, CED-12/ELMO), glial PAT-2/*α*-integrin, and putative **p**hosphatidyl**s**erine **r**eceptor PSR-1/PSR [14, 113]. Strikingly, this study also found that glial CED-10 levels alone could dictate the amount of engulfment, revealing that the AMsh glia regulate this actively and independently of neurons. Finally, this study provided single-cell resolution evidence that glial engulfment of a single NRE impacts animal thermotaxis behaviors. The engulfment machinery described for AMsh is reminiscent of glial engulfment/pruning by vertebrate RPE glial-like cells and *Drosophila* astrocytes [116–121].

Finally, in a third mechanism, AMsh glial secretion regulated by post-embryonic expression of PROS-1/Prox/Prospero also impacts AFD NRE shape and thermosensory behaviors [115]. That *pros-1* knockdown has a stronger effect than either *kcc-3* or *ced-10* manipulations, and that it does not regulate KCC-3 expression (AS, personal communication), implies that additional glial factors regulating AFD NRE await discovery.

Like chemo- and mechano-sensation, AMsh glia also directly track temperature to regulate temperature-dependent expression of the glial VER-1/FLT1-4/VEGFR receptor tyrosine kinase (on at 25 °C, off at 15 °C). For this, the glial TTX-1/Otx/OTD transcription factor binds to the *ver-1* promoter and regulates its transcription independent of AFD thermosensory neuron signaling [122]. How AMsh glia sense temperature and regulate TTX-1 is currently unknown.

4.4 Locomotion

The name *"elegans"* is derived from the animal's *elegan*t sinusoidal movement [123], driven by five motor neuron types [12, 124, 125]. *C. elegans* locomotion is influenced by the animal's sensory preferences [126, 127], egg-laying [128, 129], mating [130], foraging [131], defecation [132], pharyngeal pumping [133, 134], and sleep [123, 135]. *C. elegans* motor neuron:NMJ contacts are not associated with neuroectoderm lineage glia, but instead with the GLR glia.

Two glia, CEPsh and GLR, impact animal locomotion (Figs. 2b and 3f) [53, 135, 136]. GLR glia connect to muscles in the worm head through gap junctions [12]. Post-embryonic loss of GLR glia by genetic ablations induces severe reduction in locomotion rates and increased turning frequency, reversal rates, and pausing [136]. Further, LGC-55/amine-gated Cl^- channels are expressed in muscles, a subset of neurons, and GLR glia and are required in GLR or neurons to control head movements [137]. Similarly, post-embryonic ablation of CEPsh glia increases the number of reversals animals make while exploring their environment [53]. This is regulated by glutamate level control by CEPsh glial **gl**utamate **t**ransporter GLT-1/EAAT2. Mutations with presynaptic block of glutamate release (*eat-4*/VGLUT) [138] decrease reversals, while mutations that cause overactivation of glutamate receptors [131] increase reversals [53]. Interestingly, while ablating CEPsh glia causes multiple locomotor defects, *glt-1* mutants only exhibit excess reversals [53]. This suggests that CEPsh glia exert complex regulation on animal locomotion, with one mechanism being glutamate-dependent reversals. How GLR and CEPsh glia cooperatively regulate coherent locomotion motor output has not yet been examined. Independently, steroid metabolism in GLR glia via NTA-1/hydroxysteroid dehydrogenase regulates animal disperal (phoretic) behavior [139].

4.5 Neurotransmitter Diffusion

Like mammalian glia [140–148], *C. elegans* glia also buffer extracellular glutamate through the glutamate transporter GLT-1/EAAT2 to regulate neuronal excitability [53, 54]. Loss of CEPsh glia or glial GLT-1 causes increased glutamate levels. This was shown to impact the activity of the AVA command interneuron that drives backward locomotion, leading to aberrant and repetitive reversal behaviors [53]. In line with this, a mutation in MGL-2/mGluR5 receptors ameliorates *glt-1* mutant defects. Further, the study showed that MGL-2 acts extrasynaptically in neurons presynaptic to AVA (RIM, AIB, RIA, ADA, and AIM neurons) and requires the G proteins EGL-30/Gαq and GOA-1/Gαo to regulate glutamate-driven reversals [53].

CEPsh glia also regulate dopamine signaling through SWIP-10/metallo-β-lactamase. Briefly, loss of CEPsh glial *swip-10* impairs the ability of the animal to sustain thrashing in liquids, a phenotype associated with hyper-dopamine activation [54, 149]. *swip-10* mutant animals also move aberrantly on bacterial lawns and exhibit increased dopamine vesicle fusion and neuronal dopamine activation when presented with food [54]. Interestingly, glial SWIP-10 regulates levels of extracellular glutamate arising from changes in AAT-1/SLC7A5-mediated glutamate/cysteine exchange or VGLU-3/SLC17A6-mediated vesicular glutamate release. The locomotor effects in *swip-10* mutant animals are reduced in mutants with defects in vesicular glutamate release or receptor expression, and phenocopied by mutations in *glt-1*/EAAT2 glutamate transporters [54]. Based on these findings, the study inferred that SWIP-10 limits dopaminergic signaling and behaviors by modulating reciprocal glutamate buffering [54].

4.6 Sleep State

C. elegans have sleep-like states characterized by canonical sleep features like reduced neural activity, changes in body posture, decreased responsiveness to external stimuli, behavioral reversibility, and sleep drive [150–152]. Furthermore, sleep and circadian clock genes are conserved in *C. elegans* [153–159] and sleep drive increases with deprivation and in response to stress [160]. *C. elegans* sleep occurs in two forms. Developing larval animals undergo periodic **d**evelopmentally **t**imed **s**leep (DTS) prior to entering molts [151], with ~75% of all neurons globally reducing their activity brain-wide [152]. Adult *C. elegans* exposed to stressors like UV or heat exhibit **s**tress-**i**nduced **s**leep (SIS) [160–165]. SIS is also required for the consolidation of olfactory memory and synapse remodeling [166]. DTS and SIS sleep states are mediated by distinct neuronal and molecular pathways [151, 161].

CEPsh glia are implicated in both DTS and SIS [135, 167]. Ablating CEPsh glia post-embryonically results in precocious, intermittent, and prolonged DTS bouts, delayed development, and locomotion defects (see above). CEPsh impact on DTS is at its tripartite interaction at the synapse between ALA neuron and the command interneuron AVE [52]. CEPsh glia inhibit ALA activity, which inhibits AVE. Thus, in animals where CEPsh glia are absent, AVE experiences sustained inhibition, which decouples input-output signals in the AVE. Ablation of ALA neuron or loss of the general sleep-regulator EGL-4/PKG mitigates sleep deficits of CEPsh-ablated animals [135]. The role of CEPsh glia in SIS is less defined, as animals without CEPsh glia had SIS deficits only in one of three triggers tested (UV, but not toxin or heat) [167]. Nonetheless, this shows that like molecular and neuronal pathways, CEPsh glial roles in the two *C. elegans* sleep states are also distinct.

4.7 Male Mating

Adult males prioritize mate-seeking over food seeking and exhibit sexual conditioning, where the presence of a mate overrides the effects of starvation during chemosensory learning [168]. This behavior requires the male-specific MCM neurons.

These are born of AMso glia during the L3 larval stage through a cell-intrinsic developmental program and become integrated into preexisting circuits [61]. Sexual conditioning behaviors also require the male-specific CEM sensory neurons, which are born embryonically, synapse with MCM interneurons, and associate with CEPsh and CEPso glia [11, 61, 168, 169] (Fig. 2b). PHD neurons are male-specific proprioceptive neurons in the phasmid sense organ that drive a coordinated backward movement along the hermaphrodite body during mating based on proprioceptive and other sensory cues [69]. These derive from trans-differentiation (79% incidence) or asymmetric division (21% incidence) of male PHso1 glia in the chemosensory phasmid organ [69] at the late-larval to adult sexual maturation and integrate into copulatory circuits in the animal's "brain" neuropil (Fig. 3e).

Despite a strong mating drive, males also exhibit a period of sexual disinterest following ejaculation, similar to the refractory period in response to sperm release in other organisms. The neural circuit involved in this process has been identified which includes a critical role for **sp**icule **so**cket (SPso) glia (Fig. 2e) [170]. The SPso glia are dopaminergic and express CAT-2/tyrosine hydroxylase, a rate-limiting dopamine biosynthetic enzyme. Dopamine released from the SPso glia promotes ejaculation and increases the length of the refractory period [170]. Interestingly, SPso dopamine may act hormonally, as these cells do not express the dopamine reuptake transporter DAT-1 [170]. How glial neurotransmitters may act hormonally to regulate male-specific behaviors remains to be explored.

5 Glial Roles in Stress, Aging, and Neurodegeneration

Mammalian glia are increasingly implicated in stress-dependent inflammation, neural aging, and neurodegeneration, but molecular underpinnings of these remain unclear at molecular detail [171–174]. Recent studies have made exciting molecular revelations that *C.* elegans glia regulate organismal stress responses, animal lifespan, and sensorineural aging.

5.1 Stress Response

Under unfavorable conditions, juvenile *C. elegans* trigger an alternate developmental stage called dauer. Dauers have extensively remodeled neuroanatomy, which presumably enables their ability to withstand extended periods of scarcity [90, 175, 176]. For example, the anterior endings of the bilateral AMsh glia expand and fuse to result in a binucleate glial cell [176]. This glial remodeling occurs independent of neurons and requires REMO-1/GPCR in glia, which regulates downstream expression of the stress- or temperature-induced gene *ver-1/FLT1-4/VEGFR* [90, 122].

C. elegans must defend against pathogens such as molds, bacteria, and fungi. To accomplish this, *C. elegans* glial cells employ detoxification proteins, such as

xenobiotic-**m**etabolizing **e**nzymes (XMEs), which are also involved in detoxification processes in the mammalian liver [177, 178]. Briefly, in response to the fungus *Penicillium brevicompactum,* the **n**uclear **h**ormone **r**eceptors NHR-45/HNF4A and NHR-156/NR1D1 activate mycotoxin-degrading XMEs in the intestine and glia by transcriptional changes [178]. Defects in this stress-response pathway led to mitochondrial stress and developmental delay. Considering that AMsh glial cells modulate chemosensory behavior (see Section 4), it is possible that the induction of XMEs in AMsh glia could be linked to the observed animal's behavioral response changes upon pathogen exposure [179]. Mammalian olfactory sustentacular glial cells also express XMEs abundantly, suggesting that aspects of this biology may be conserved [180].

Recent work also shows that some glia (AMsh, CEPsh) are mechano-biologically coupled with epithelia, which impacts the ability of glia and associated neuron to maintain cell shape against external mechanical stress in adult animals [49]. Curiously, this coupling occurs through a previously unknown mechanical transduction pathway involving the HSP70 co-chaperone BAG2/3, FGFR, and Spectrins. Briefly, loss of epithelial UNC-23/BAG-2 induces glial loss of apical SMA-1/βH-Spectrin through glial EGL-15/FGFR signaling, which deforms glial cell shape under ambient mechanical stress from locomotion. This consequently causes age-dependent progressive loss of shape of glia-associated neuron endings and altered neuronal temperature response profiles in adult animals. The study further showed that this epithelia-glia mechano-biological coupling is both spatially and temporally heterogeneous. Loss of *unc-23* specifically during a developmentally critical period between L4-adult transition molt impacts glia-neuron shape in adult animals progressively. Further, *unc-23* mutant animals have NRE defects in some but not all AMsh-associated neurons (all embedded neurons but no channel neurons; Fig. 3d). Moreover, while multiple glia in the animal head region are impacted by *unc-23* loss, glia in the animal body or tail are unaffected even though UNC-23 is broadly expressed throughout the animal body. This revealed that different glial cells have distinct and dynamically regulated responses to external mechanical stress [49].

Finally, the **u**nfolded **p**rotein **r**esponse (UPR) is known to regulate protein folding fidelity under environmental stressors like heat shock [181, 182]. UPR-deficient cells die upon protein folding stress, while overexpression of UPR activators leads to increased health span and lifespan [183, 184]. In *C. elegans*, global overexpression of the transcription factor **h**eat **s**hock **f**actor (HSF-1/HSF1), a component of the UPR, not only enhances the animal's temperature tolerance but also extends its lifespan [182]. Interestingly, animals that overexpress HSF-1 in CEPsh glia have increased lifespan, enhanced resilience to heat stress, and decreased protein aggregation [182]. Regulation of this glial heat shock response by CEPsh glia appears to occur independent of neuropeptide signaling, the canonical thermosensory circuitry, and known neurotransmitters. Further, glial mitochondrial unfolded protein response (UPRmt) is shown to signal via UNC-13/Munc13 dependent small clear vesicle signaling to regulate neuronal protein homeostasis/aggregation [185].

5.2 Organismal Aging

Twenty-five years ago, Apfeld and Kenyon (1999) showed that ablating AMsh glia extends animal lifespan by ~50%, similar to decreasing sensory neuron functions [186]. This already hinted at the relationship between glia, sensory perception, and the aging process, paving the way for a deeper exploration of glia's role in organismal longevity. Since then, several lines of evidence have linked multiple glia and glial molecular pathways to organismal lifespan.

One line of evidence pointing to an independent mechanistic link between *C. elegans* glia, stress resistance, and lifespan is the phosphagen energy buffering system [187]. Briefly, a decrease in S6 kinase RSKS-1/S6K or overexpression of glial-specific ARGK-1/creatine kinase extends animal lifespan, acting partly through the energy sensor AAK-2/AMPK [187]. However, ARGK-1 functions only in the context of RSKS-1 deficiency and regulates animal body size and stress resistance [187].

Additionally, glia are associated with aging-related functions regulated by **d**ietary **r**estriction (DR) and **micro****RNAs** (miRNAs) [188, 189]. In *C. elegans*, the transcription factors PHA-4/FoxA and SKN-1/Nrf2 are necessary for DR-induced lifespan extension, and regulation of pan-glial miR-228, whose expression is promoted by DR [188, 190]. PHA-4 promotes *miR-228* expression, while SKN-1 antagonizes *miR-228* expression [188]. *miR-228* also then functions in feedback loops to target both *pha-4* and *skn-1* [188, 191, 192]. Animals mutant for *mir-228* are resistant to heat stress and have longer lifespans [188]. While compelling, given also that miR-228 belongs to the conserved glia-enriched miR-96/182/183 family [193], it must be noted that miR-228 also expresses in non-glial excretory canal, vulva, rectum, and tail hypodermis. Thus, specific roles of PHA-4, SKN-1, and miR-228 in glia remain to be determined.

C. elegans glia are also shown to impact organismal physiology and aging via neuropeptide signaling. Briefly, glial RGBA-1 neuropeptide (**r**egulatory-**g**ene-for-**b**ehavioral-**a**ging-1) acts on neuronal receptor NPR-28/GPCR, which drives SIR-2.1/Sirtuin-dependent activation of **mit**ochondrial **u**nfolded **p**rotein **r**esponse (UPR^{mt}) pathway to modulate age-related decline in mating, feeding, and locomotion [73]. RGBA-1 was identified by HPLC/tandem mass spectrometry; its function in glia was ascertained by tissue-specific deletion, cell-specific rescue, and expression pattern analyses [73]. Briefly, it is expressed by CEPsh and other glia. NPR-28-dependent regulation of UPR^{mt} was mapped to serotonergic and dopaminergic neurons. This study also proposed that polymorphisms in either *rgba-1* or *npr-28* are causal to the observed natural variations in age-related decline in mating behavior [73]. Interestingly, while *rgba-1* and *npr-28* regulate the age-dependent decline of specific animal behaviors, they do not affect lifespan, implying that glial molecular pathways regulating age-dependent behavioral decline (i.e., health span) are distinct from those regulating longevity [73].

Independently, modulation of **e**ndoplasmic **r**eticulum **u**nfolded **p**rotein **r**esponse (UPR^{ER}) in CEPsh glia also regulates longevity by neuropeptide signaling to distal intestinal cells, although the identity of specific neuropeptides remains unknown in this case [74]. Specifically, a key component of UPR^{ER} is the conserved transcription factor **X**-box-**b**inding **p**rotein 1 (XBP-1) [194]. Overexpression of XBP-1 in CEPsh glia is sufficient to control UPR^{ER} activation in distal intestinal cells, promoting animals' resistance to protein aggregation and chronic ER stress, and increasing lifespan. Further, disrupting neuropeptide processing via *egl-3*/PCSK2 or dense core vesicle release via *unc-31*/CAPS suppresses UPR^{ER} activation in intestinal cells [74].

Additionally, defects in epithelia-glia mechanobiology coupling within a critical developmental period (L4 larva-adult transition) cause progressive age-dependent deterioration in the shape of glial-associated NREs, reminiscent of premature neural aging. Loss of this coupling also leads to a dramatically reduced adult lifespan [49, 195]. These defects are not mediated by the canonical DAF-2/InsR-DAF-16/FOXO signaling, suggesting additional glial mechanisms regulating organismal and neural aging are at play [49]. Finally, loss of AMsh regulatory cues like KCC-3 and defects in glial engulfment all have age-dependent effects on the AFD NRE shape and associated thermosensory functions through yet-unknown aging pathways [112, 113].

Furthermore, studies of glial SWIP-10/MBLAC1 reveal a role for CEPsh glia in not only dopamine neuron hyperexcitability but also their progressive neurodegeneration. Briefly, an imbalance of extracellular dopamine/glutamate levels is thought to induce a pathological increase in neuronal Ca^{2+}, leading to neuronal death with age [54].

Taken together, it is apparent that the link between glia, age, and stress-response pathways is likely complex molecular networks, and our current understanding of these is only beginning to scratch the surface. That different glial cells and stress/aging-response pathways are activated under various environmental stress conditions, raises several questions on how they are all regulated, whether they are each specific to certain glia or stressors, and how their activation tracks animal age or sex. Further, it remains to be determined how manipulation of glial UPR^{ER} impacts intestinal cells, systemic homeostasis, and animal lifespan. Future mechanistic studies, readily executed in this powerful genetic and short-lived animal model, will be exciting to define how glia across the nervous system regulate aging and health span, systemic protein homeostasis, and stress-response pathways.

6 Concluding Remarks and Future Outlook

It is now well established that glia are critical contributors to neural functions and aging. Dissecting the underlying molecular glia-neuron interactions is key, then, to understanding nervous system functions in health and developing interventions in diseases where these are perturbed. The speed with which deep molecular insights into glia are being uncovered in *C. elegans* offers a compelling argument that this model is a powerful and genetically tractable experimental platform for molecular and genetic interrogations at single-cell resolution. Given the extensive conservation and disease relevance of different molecular modules reported thus far, it is likely that insights derived here will be broadly relevant cross-species.

One emerging contribution of these studies is that even a single glia deploys multiple mechanisms to interact with individual neurons and other tissues for function and aging, and this can track context or cue. How any one glia and different glia regulate variable expression of various cues (including ion channels, transporters, gap junctions, and neuropeptides) and how they together modulate neural functions are ripe for mechanistic inquiry. The genetic tractability of this animal sets up a powerful platform to explore the underlying molecular logic of glial impact on neural functions like sensory behavior, male mating, and aging.

Across modalities, *C. elegans* glia regulate sensory processing both by regulating sensory neuron properties and by acting as sensory cells. It remains unknown whether all glia execute these functions and, if so, by which molecular mechanisms. Further, it also raises the tantalizing question of whether glia in other species also act as sensory cells. Alluding to this being likely a conserved glial property at least for one modality, mammalian non-neuronal Merkel glia express mechanosensory channels and impact touch sensations [196–199]; and Schwann cells are mechanosensitive and respond to both positive and negative changes in force [200, 201].

Transcriptomic profiling, both bulk cell-specific and global scRNA-seq, has uncovered markers that make many of the previously understudied *C. elegans* glia now molecularly accessible. While not all glia have single-cell-specific markers from these studies, these molecular insights are foundational to developing genetic tools that target individual glia reproducibly [202]. This will enable tighter control and sophistication to screen glial contributions to different circuit functions and neuronal properties. We envision that the development of such tools would also be fast-tracked by advanced bioinformatic tools to compare cross-study datasets for rapid insight and power analyses.

In summary, the fully mapped connectome, tools to target individual glial subtypes, and short lifespan make *C. elegans* a unique platform to rapidly interrogate the intricate relationships between glia and neurons at single-cell, circuit, and brain-wide levels, by age or sex, at and with single-cell and molecular mechanistic detail. The last two decades have rapidly expanded our knowledge of *C. elegans* glia and glia-neuron interactions in neural function and aging, with deep evolutionary conservation and disease relevance (Fig. 1). Excitingly, this is the tip of the iceberg, and the field is poised to exponentially offer insights into how these mystery cells of our brains molecularly modulate neural health and disease.

Acknowledgments This work was funded by Esther A. & Joseph Klingenstein Fund and the Simons Foundation Award in Neuroscience (227823), Brain Research Foundation Seed Grant (BRFSG-2023-10), and NIH/NINDS funding (NS114222) to AS. AS acknowledges prior support from the Glenn Foundation for Medical Research and AFAR Junior Faculty Grant and from the Simons Foundation for Autism Research (SFARI) Bridge to Independence Award. AS sincerely thanks philanthropic support to her laboratory. This work was also supported by the Washington Research Fellowship, the T32 Interdisciplinary Training in Cancer Research (5T32CA80416-25) to MP, and Biological Mechanisms of Healthy Aging T32 training grant (5T32AG066574-04) to LS. We apologize to those whose work was not cited due to our oversight or space considerations.

References

1. Barres BA. The mystery and magic of glia: a perspective on their roles in health and disease. Neuron. 2008;60(3):430–40.
2. Allen NJ, Barres BA. Neuroscience: glia – more than just brain glue. Nature. 2009;457(7230):675–7.
3. Khakh BS, Sofroniew MV. Diversity of astrocyte functions and phenotypes in neural circuits. Nat Neurosci. 2015;18(7):942–52.
4. Allen NJ, Eroglu C. Cell biology of astrocyte-synapse interactions. Neuron. 2017;96(3):697–708.
5. Kaminsky N, et al. Connecting malfunctioning glial cells and brain degenerative disorders. Genomics Proteomics Bioinformatics. 2016;14(3):155–65.
6. Chung WS, et al. Do glia drive synaptic and cognitive impairment in disease? Nat Neurosci. 2015;18(11):1539–45.
7. Zuchero JB, Barres BA. Glia in mammalian development and disease. Development. 2015;142(22):3805–9.
8. Gleichman AJ, Carmichael ST. Glia in neurodegeneration: drivers of disease or along for the ride? Neurobiol Dis. 2020;142:104957.
9. Sulston JE, Horvitz HR. Post-embryonic cell lineages of the nematode, *Caenorhabditis elegans*. Dev Biol. 1977;56(1):110–56.
10. Sulston JE. Neuronal cell lineages in the nematode Caenorhabditis elegans. Cold Spring Harb Symp Quant Biol. 1983;48(Pt 2):443–52.
11. Ward S, et al. Electron microscopical reconstruction of the anterior sensory anatomy of the nematode Caenorhabditis elegans.?2UU. J Comp Neurol. 1975;160(3):313–37.
12. White JG, et al. The structure of the nervous system of the nematode Caenorhabditis elegans. Philos Trans R Soc Lond Ser B Biol Sci. 1986;314(1165):1–340.
13. Bargmann CI. Neurobiology of the Caenorhabditis elegans genome. Science. 1998;282(5396):2028–33.
14. Raiders S, et al. Engulfed by glia: glial pruning in development, function, and injury across species. J Neurosci. 2021;41(5):823–33.
15. Hobert O. The neuronal genome of Caenorhabditis elegans. WormBook. 2013:1–106.
16. Goodman MB. Mechanosensation. WormBook. 2006:1–14.
17. Greenwald I, Kovall R. Notch signaling: genetics and structure. WormBook. 2013:1–28.
18. Gonczy P, Rose LS. Asymmetric cell division and axis formation in the embryo. WormBook. 2005:1–20.
19. Lundquist EA. Small GTPases. WormBook. 2006:1–18.
20. Jin Y. Synaptogenesis. WormBook. 2005:1–11.
21. Lee MH, Schedl T. RNA-binding proteins. WormBook. 2006:1–13.
22. Jiang J, et al. C. elegans enteric motor neurons fire synchronized action potentials underlying the defecation motor program. Nat Commun. 2022;13(1):2783.
23. Liu Q, et al. C. elegans AWA olfactory neurons fire calcium-mediated all-or-none action potentials. Cell. 2018;175(1):57–70 e17.
24. Scheffer LK, et al. A connectome and analysis of the adult Drosophila central brain. Elife. 2020;9:e57443.

25. Lyons DA, Talbot WS. Glial cell development and function in zebrafish. Cold Spring Harb Perspect Biol. 2014;7(2):a020586.
26. Hinsch K, Zupanc GK. Generation and long-term persistence of new neurons in the adult zebrafish brain: a quantitative analysis. Neuroscience. 2007;146(2):679–96.
27. von Bartheld CS, Bahney J, Herculano-Houzel S. The search for true numbers of neurons and glial cells in the human brain: a review of 150 years of cell counting. J Comp Neurol. 2016;524(18):3865–95.
28. Freeman MR, Rowitch DH. Evolving concepts of gliogenesis: a look way back and ahead to the next 25 years. Neuron. 2013;80(3):613–23.
29. Oikonomou G, Shaham S. The glia of Caenorhabditis elegans. Glia. 2011;59(9):1253–63.
30. Singhvi A, Shaham S. Glia-neuron interactions in Caenorhabditis elegans. Annu Rev Neurosci. 2019;42:149–68.
31. Bacaj T, et al. Glia are essential for sensory organ function in C. elegans. Science. 2008;322(5902):744–7.
32. Ray S, Singhvi A. Charging up the periphery: glial ionic regulation in sensory perception. Front Cell Dev Biol. 2021;9:687732.
33. Doroquez DB, et al. A high-resolution morphological and ultrastructural map of anterior sensory cilia and glia in Caenorhabditis elegans. Elife. 2014;3:e01948.
34. Jarrell TA, et al. The connectome of a decision-making neural network. Science. 2012;337(6093):437–44.
35. Varshney LR, et al. Structural properties of the Caenorhabditis elegans neuronal network. PLoS Comput Biol. 2011;7(2):e1001066.
36. Cook SJ, et al. Whole-animal connectomes of both Caenorhabditis elegans sexes. Nature. 2019;571(7763):63–71.
37. Kaletsky R, et al. Transcriptome analysis of adult Caenorhabditis elegans cells reveals tissue-specific gene and isoform expression. PLoS Genet. 2018;14(8):e1007559.
38. Packer JS, et al. A lineage-resolved molecular atlas of C. elegans embryogenesis at single-cell resolution. Science. 2019;365(6459):eaax1971.
39. Yemini E, et al. NeuroPAL: a multicolor atlas for whole-brain neuronal identification in C. elegans. Cell. 2021;184(1):272–288 e11.
40. Taylor SR, et al. Molecular topography of an entire nervous system. Cell. 2021;184(16):4329–4347 e23.
41. Shaham S. Glia-neuron interactions in the nervous system of Caenorhabditis elegans. Curr Opin Neurobiol. 2006;16(5):522–8.
42. Singhvi A, Shaham S, Rapti G. Glia development and function in the nervous system of Caenorhabditis elegans, vol. 7. Cold Spring Harb Perspect Biol; 2023. p. a020578.
43. Thomas JH. The mind of a worm. Science. 1994;264(5166):1698–9.
44. Stout RF Jr, Verkhratsky A, Parpura V. Caenorhabditis elegans glia modulate neuronal activity and behavior. Front Cell Neurosci. 2014;8:67.
45. Horvitz HR, Sulston JE. Isolation and genetic characterization of cell-lineage mutants of the nematode Caenorhabditis elegans. Genetics. 1980;96(2):435–54.
46. Perkins LA, et al. Mutant sensory cilia in the nematode Caenorhabditis elegans. Dev Biol. 1986;117(2):456–87.
47. Albert PS, Brown SJ, Riddle DL. Sensory control of dauer larva formation in Caenorhabditis elegans. J Comp Neurol. 1981;198(3):435–51.
48. Low IIC, et al. Morphogenesis of neurons and glia within an epithelium. Development. 2019;146(4):dev171124.
49. Martin C, Bent JS, Hill T, Topalidou I, Singhvi A. Epithelial UNC-23 limits mechanical stress to maintain glia-neuron architecture in C. elegans. Dev Cell. 2024;59:1–21.
50. Ray S, et al. Neuron cilia constrain glial regulators to microdomains around distal neurons. Cell Rep. 2024;43(3):113844.
51. Perens EA, Shaham S. C. elegans daf-6 encodes a patched-related protein required for lumen formation. Dev Cell. 2005;8(6):893–906.

52. Altun, Z.F., Herndon, L.A., Wolkow, C.A., Crocker, C., Lints, R., Hall, D.H., WormAtlas. 2002–2023.
53. Katz M, et al. Glutamate spillover in C. elegans triggers repetitive behavior through presynaptic activation of MGL-2/mGluR5. Nat Commun. 2019;10(1):1882.
54. Gibson CL, et al. Glial loss of the metallo beta-lactamase domain containing protein, SWIP-10, induces age- and glutamate-signaling dependent, dopamine neuron degeneration. PLoS Genet. 2018;14(3):e1007269.
55. McMiller TL, Johnson CM. Molecular characterization of HLH-17, a C. elegans bHLH protein required for normal larval development. Gene. 2005;356:1–10.
56. Yoshimura S, et al. mls-2 and vab-3 control glia development, hlh-17/Olig expression and glia-dependent neurite extension in C. elegans. Development. 2008;135(13):2263–75.
57. Krause M, et al. Elements regulating cell- and stage-specific expression of the C. elegans MyoD family homolog hlh-1. Dev Biol. 1994;166(1):133–48.
58. Legerlotz K, Smith HK. Role of MyoD in denervated, disused, and exercised muscle. Muscle Nerve. 2008;38(3):1087–100.
59. Chen L, et al. The Caenorhabditis elegans MYOD homologue HLH-1 is essential for proper muscle function and complete morphogenesis. Development. 1994;120(6):1631–41.
60. Zarkower D. Somatic sex determination. WormBook. 2006:1–12.
61. Sammut M, et al. Glia-derived neurons are required for sex-specific learning in C. elegans. Nature. 2015;526(7573):385–90.
62. Noctor SC, et al. Cortical neurons arise in symmetric and asymmetric division zones and migrate through specific phases. Nat Neurosci. 2004;7(2):136–44.
63. Sulston JE, White JG. Regulation and cell autonomy during postembryonic development of Caenorhabditis elegans. Dev Biol. 1980;78(2):577–97.
64. Chasnov JR, et al. The species, sex, and stage specificity of a Caenorhabditis sex pheromone. Proc Natl Acad Sci USA. 2007;104(16):6730–5.
65. White JQ, et al. The sensory circuitry for sexual attraction in C. elegans males. Curr Biol. 2007;17(21):1847–57.
66. Fung W, et al. A sex-specific switch in a single glial cell patterns the apical extracellular matrix. Curr Biol. 2023;33(19):4174–4186 e7.
67. Klass MR. Aging in the nematode Caenorhabditis elegans: major biological and environmental factors influencing life span. Mech Ageing Dev. 1977;6(6):413–29.
68. Sulston JE, Albertson DG, Thomson JN. The Caenorhabditis elegans male: postembryonic development of nongonadal structures. Dev Biol. 1980;78(2):542–76.
69. Molina-Garcia L, et al. Direct glia-to-neuron transdifferentiation gives rise to a pair of male-specific neurons that ensure nimble male mating. Elife. 2020;9:e48361.
70. Lints R, Hall DH. Handbook of C. elegans male anatomy. WormAtlas; 2005. Available from: https://www.wormatlas.org/male/malehomepage.htm.
71. Lints R, Hall DH. Reproductive system, overview. In: Herndon LA, editor. WormAtlas; 2009.
72. Jiang LI, Sternberg PW. An HMG1-like protein facilitates Wnt signaling in Caenorhabditis elegans. Genes Dev. 1999;13(7):877–89.
73. Yin JA, et al. Genetic variation in glia-neuron signalling modulates ageing rate. Nature. 2017;551(7679):198–203.
74. Frakes AE, et al. Four glial cells regulate ER stress resistance and longevity via neuropeptide signaling in C. elegans. Science. 2020;367(6476):436–40.
75. Roux AE, et al. Individual cell types in C. elegans age differently and activate distinct cell-protective responses. Cell Rep. 2023;42(8):112902.
76. Ghaddar A, et al. Whole-body gene expression atlas of an adult metazoan. Sci Adv. 2023;9(25):eadg0506.
77. Ubink R, Calza L, Hokfelt T. 'Neuro'-peptides in glia: focus on NPY and galanin. Trends Neurosci. 2003;26(11):604–9.
78. Duan D, et al. Sensory glia detect repulsive odorants and drive olfactory adaptation. Neuron. 2020;108(4):707–721 e8.

79. Fernandez-Abascal J, et al. A glial ClC Cl(−) channel mediates nose touch responses in C. elegans. Neuron. 2022;110(3):470–485 e7.
80. Graziano B, et al. Glial KCNQ K+ channels control neuronal output by regulating GABA release from glia in C. elegans. Neuron. 2024;112(11):1832–1847.e7. https://doi.org/10.1016/j.neuron.2024.02.013
81. Meng L, et al. Regulation of neuronal axon specification by glia-neuron gap junctions in C. elegans. Elife. 2016;5:e19510.
82. Goodenough DA, Paul DL. Gap junctions. Cold Spring Harb Perspect Biol. 2009;1(1):a002576.
83. Baranova A, et al. The mammalian pannexin family is homologous to the invertebrate innexin gap junction proteins. Genomics. 2004;83(4):706–16.
84. Pannasch U, et al. Astroglial networks scale synaptic activity and plasticity. Proc Natl Acad Sci USA. 2011;108(20):8467–72.
85. Altun ZF, et al. High resolution map of Caenorhabditis elegans gap junction proteins. Dev Dyn. 2009;238(8):1936–50.
86. Giaume C, et al. Connexin and pannexin hemichannels in brain glial cells: properties, pharmacology, and roles. Front Pharmacol. 2013;4:88.
87. Hall DH, Altun Z. C. elegans atlas. New York: Cold Spring Harbor Laboratory Press; 2008.
88. Bargmann CI. Chemosensation in C. elegans. WormBook. 2006:1–29.
89. Wexler LR, Miller RM, Portman DS. C. elegans males integrate food signals and biological sex to modulate state-dependent chemosensation and behavioral prioritization. Curr Biol. 2020;30(14):2695–2706 e4.
90. Lee IH, et al. Stress-induced neural plasticity mediated by glial GPCR REMO-1 promotes C. elegans adaptive behavior. Cell Rep. 2021;34(2):108607.
91. Bargmann CI, Hartwieg E, Horvitz HR. Odorant-selective genes and neurons mediate olfaction in C. elegans. Cell. 1993;74(3):515–27.
92. Chen D, et al. The voltage-gated calcium channel EGL-19 acts on glia to drive olfactory adaptation. Front Mol Neurosci. 2022;15:907064.
93. Barr MM, Garcia LR. Male mating behavior. WormBook. 2006:1–11.
94. Ding G, et al. In vivo tactile stimulation-evoked responses in Caenorhabditis elegans amphid sheath glia. PLoS One. 2015;10(2):e0117114.
95. Wang L, et al. Glial regulators of ions and solutes required for specific chemosensory functions in Caenorhabditis elegans. iScience. 2022;25(12):105684.
96. Wang Y, et al. A glial DEG/ENaC channel functions with neuronal channel DEG-1 to mediate specific sensory functions in C. elegans. EMBO J. 2008;27(18):2388–99.
97. Wang Y, D'Urso G, Bianchi L. Knockout of glial channel ACD-1 exacerbates sensory deficits in a C. elegans mutant by regulating calcium levels of sensory neurons. J Neurophysiol. 2012;107(1):148–58.
98. Grant J, Matthewman C, Bianchi L. A novel mechanism of pH buffering in C. elegans glia: bicarbonate transport via the voltage-gated ClC Cl- Channel CLH-1. J Neurosci. 2015;35(50):16377–97.
99. Portman DS. Sexual modulation of sex-shared neurons and circuits in Caenorhabditis elegans. J Neurosci Res. 2017;95(1–2):527–38.
100. Goodman MB, Sengupta P. The extraordinary AFD thermosensor of C. elegans. Pflugers Arch. 2018;470(5):839–49.
101. Han L, et al. Two novel DEG/ENaC channel subunits expressed in glia are needed for nose-touch sensitivity in Caenorhabditis elegans. J Neurosci. 2013;33(3):936–49.
102. Johnson CK, et al. The Na(+)-K(+)-ATPase is needed in glia of touch receptors for responses to touch in C. elegans. J Neurophysiol. 2020;123(5):2064–74.
103. Handler A, Ginty DD. The mechanosensory neurons of touch and their mechanisms of activation. Nat Rev Neurosci. 2021;22(9):521–37.
104. Schafer WR. Mechanosensory molecules and circuits in C. elegans. Pflugers Arch. 2015;467(1):39–48.

105. Kaulich E, et al. Physiological insight into the conserved properties of Caenorhabditis elegans acid-sensing degenerin/epithelial sodium channels. J Physiol. 2023;601(9):1625–53.
106. Kindt KS, et al. Caenorhabditis elegans TRPA-1 functions in mechanosensation. Nat Neurosci. 2007;10(5):568–77.
107. Jentsch TJ, Pusch M. CLC chloride channels and transporters: structure, function, physiology, and disease. Physiol Rev. 2018;98(3):1493–590.
108. Garrity PA, et al. Running hot and cold: behavioral strategies, neural circuits, and the molecular machinery for thermotaxis in C. elegans and Drosophila. Genes Dev. 2010;24(21):2365–82.
109. Ito H, Inada H, Mori I. Quantitative analysis of thermotaxis in the nematode Caenorhabditis elegans. J Neurosci Methods. 2006;154(1–2):45–52.
110. Mori I. Genetics of chemotaxis and thermotaxis in the nematode Caenorhabditis elegans. Annu Rev Genet. 1999;33:399–422.
111. Ramot D, et al. Thermotaxis is a robust mechanism for thermoregulation in Caenorhabditis elegans nematodes. J Neurosci. 2008;28(47):12546–57.
112. Singhvi A, et al. A glial K/Cl transporter controls neuronal receptive ending shape by chloride inhibition of an rGC. Cell. 2016;165(4):936–48.
113. Raiders S, et al. Glia actively sculpt sensory neurons by controlled phagocytosis to tune animal behavior. Elife. 2021;10:e63532.
114. Yoshida A, et al. A glial K(+)/Cl(−) cotransporter modifies temperature-evoked dynamics in Caenorhabditis elegans sensory neurons. Genes Brain Behav. 2016;15(4):429–40.
115. Wallace SW, et al. PROS-1/Prospero is a major regulator of the glia-specific secretome controlling sensory-neuron shape and function in C. elegans. Cell Rep. 2016;15(3):550–62.
116. Young RW. The renewal of photoreceptor cell outer segments. J Cell Biol. 1967;33(1):61–72.
117. Young RW, Bok D. Participation of the retinal pigment epithelium in the rod outer segment renewal process. J Cell Biol. 1969;42(2):392–403.
118. Lakkaraju A, et al. The cell biology of the retinal pigment epithelium. Prog Retin Eye Res. 2020;78:100846. https://doi.org/10.1016/j.preteyeres.2020.100846.
119. Kevany BM, Palczewski K. Phagocytosis of retinal rod and cone photoreceptors. Physiology (Bethesda). 2010;25(1):8–15.
120. Chung WS, Allen NJ, Eroglu C. Astrocytes control synapse formation, function, and elimination. Cold Spring Harb Perspect Biol. 2015;7(9):a020370.
121. Scott-Hewitt N, et al. Local externalization of phosphatidylserine mediates developmental synaptic pruning by microglia. EMBO J. 2020;39(16):e105380.
122. Procko C, Lu Y, Shaham S. Glia delimit shape changes of sensory neuron receptive endings in C. Elegans. Development. 2011;138(7):1371–81.
123. Riddle D, et al. Introduction: the neural circuit for locomotion. In: C. elegans II. New York: Cold Spring Harbor Laboratory Press; 1997.
124. Richmond JE, Jorgensen EM. One GABA and two acetylcholine receptors function at the C. elegans neuromuscular junction. Nat Neurosci. 1999;2(9):791–7.
125. Riddle DL, Blumenthal T, Meyer BJ, et al. Section I, Introduction: the neural circuit for locomotion. In: Harbor CS, editor. C. elegans II. New York: Cold Spring Harbor Laboratory Press: Cold Spring Harbor; 1997.
126. Choi S, et al. Sensory neurons arouse C. elegans locomotion via both glutamate and neuropeptide release. PLoS Genet. 2015;11(7):e1005359.
127. Zhan X, et al. Locomotion modulates olfactory learning through proprioception in C. elegans. Nat Commun. 2023;14(1):4534.
128. Hardaker LA, et al. Serotonin modulates locomotory behavior and coordinates egg-laying and movement in Caenorhabditis elegans. J Neurobiol. 2001;49(4):303–13.
129. Sherlekar AL, et al. The C. elegans male exercises directional control during mating through cholinergic regulation of sex-shared command interneurons. PLoS One. 2013;8(4):e60597.
130. Koo PK, et al. The robustness of Caenorhabditis elegans male mating behavior depends on the distributed properties of ray sensory neurons and their output through core and male-specific targets. J Neurosci. 2011;31(20):7497–510.

131. Zheng Y, et al. Neuronal control of locomotion in C. elegans is modified by a dominant mutation in the GLR-1 ionotropic glutamate receptor. Neuron. 1999;24(2):347–61.
132. Nagy S, et al. Caenorhabditis elegans exhibit a coupling between the defecation motor program and directed locomotion. Sci Rep. 2015;5:17174.
133. Trojanowski NF, Raizen DM, Fang-Yen C. Pharyngeal pumping in Caenorhabditis elegans depends on tonic and phasic signaling from the nervous system. Sci Rep. 2016;6:22940.
134. Spanier B, et al. Caenorhabditis elegans neprilysin NEP-1: an effector of locomotion and pharyngeal pumping. J Mol Biol. 2005;352(2):429–37.
135. Katz M, et al. Glia modulate a neuronal circuit for locomotion suppression during sleep in C. elegans. Cell Rep. 2018;22(10):2575–83.
136. Stefanakis N, et al. LET-381/FoxF and UNC-30/Pitx2 control the development of C. elegans mesodermal glia that regulate motor behavior. EMBO J. 2024;43:956–92.
137. Ringstad N, Abe N, Horvitz HR. Ligand-gated chloride channels are receptors for biogenic amines in C. elegans. Science. 2009;325(5936):96–100.
138. Bellocchio EE, et al. Uptake of glutamate into synaptic vesicles by an inorganic phosphate transporter. Science. 2000;289(5481):957–60.
139. Yang H, et al. Glial expression of a steroidogenic enzyme underlies natural variation in hitchhiking behavior. Proc Natl Acad Sci U S A. 2024;121(28):e2320796121. https://doi.org/10.1073/pnas.2320796121
140. Fields RD, Stevens-Graham B. New insights into neuron-glia communication. Science. 2002;298(5593):556–62.
141. Ransom BR. Glial modulation of neural excitability mediated by extracellular pH: a hypothesis revisited. Prog Brain Res. 2000;125:217–28.
142. Alvarez-Maubecin V, et al. Functional coupling between neurons and glia. J Neurosci. 2000;20(11):4091–8.
143. Ciappelloni S, et al. Dynamics of surface neurotransmitter receptors and transporters in glial cells: single molecule insights. Cell Calcium. 2017;67:46–52.
144. Araque A, Navarrete M. Glial cells in neuronal network function. Philos Trans R Soc Lond Ser B Biol Sci. 2010;365(1551):2375–81.
145. Anderson CM, Swanson RA. Astrocyte glutamate transport: review of properties, regulation, and physiological functions. Glia. 2000;32(1):1–14.
146. Mahmoud S, et al. Astrocytes maintain glutamate homeostasis in the CNS by controlling the balance between glutamate uptake and release. Cells. 2019;8(2):184.
147. Parpura V, et al. Glutamate-mediated astrocyte-neuron signaling. Nat Neurosci. 1994;369:744–7.
148. Henn FA, Goldstein MN, Hamberger A. Uptake of the neurotransmitter candidate glutamate by glia. Nature. 1974;249(458):663–4.
149. McDonald PW, et al. Vigorous motor activity in Caenorhabditis elegans requires efficient clearance of dopamine mediated by synaptic localization of the dopamine transporter DAT-1. J Neurosci. 2007;27(51):14216–27.
150. Trojanowski NF, Raizen DM. Call it worm sleep. Trends Neurosci. 2016;39(2):54–62.
151. Raizen DM, et al. Lethargus is a Caenorhabditis elegans sleep-like state. Nature. 2008;451(7178):569–72.
152. Nichols ALA, et al. A global brain state underlies C. elegans sleep behavior. Science. 2017;356(6344):eaam6851.
153. Lakowski B, Hekimi S. Determination of life-span in Caenorhabditis elegans by four clock genes. Science. 1996;272(5264):1010–3.
154. Banerjee D, et al. Developmental timing in C. elegans is regulated by kin-20 and tim-1, homologs of core circadian clock genes. Dev Cell. 2005;8(2):287–95.
155. Hristova M, et al. The Caenorhabditis elegans heterochronic regulator LIN-14 is a novel transcription factor that controls the developmental timing of transcription from the insulin/insulin-like growth factor gene ins-33 by direct DNA binding. Mol Cell Biol. 2005;25(24):11059–72.

156. Monsalve GC, Van Buskirk C, Frand AR. LIN-42/PERIOD controls cyclical and developmental progression of C. elegans molts. Curr Biol. 2011;21(24):2033–45.
157. Van Buskirk C, Sternberg PW. Epidermal growth factor signaling induces behavioral quiescence in Caenorhabditis elegans. Nat Neurosci. 2007;10(10):1300–7.
158. Singh K, et al. Deep conservation of genes required for both Drosphila melanogaster and Caenorhabditis elegans sleep includes a role for dopaminergic signaling. Sleep. 2014;37(9):1439–51.
159. Huang H, et al. Genome-wide screen for genes involved in Caenorhabditis elegans developmentally timed sleep. G3 (Bethesda). 2017;7(9):2907–17.
160. Hill AJ, et al. Cellular stress induces a protective sleep-like state in C. elegans. Curr Biol. 2014;24(20):2399–405.
161. Nath RD, et al. C. elegans stress-induced sleep emerges from the collective action of multiple neuropeptides. Curr Biol. 2016;26(18):2446–55.
162. da Costa Souza A, Ribeiro S. Sleep deprivation and gene expression, vol. 25. Curr Top Behav Neurosci; 2015. p. 65–90.
163. McEwen BS, Karatsoreos IN. Sleep deprivation and circadian disruption stress, allostasis, and allostatic load. Sleep Med Clin. 2022;17(2):253–62.
164. Gonzales DL, et al. A microfluidic-induced C. elegans sleep state. Nat Commun. 2019;10(1):5035.
165. DeBardeleben HK, et al. Stress-induced sleep after exposure to ultraviolet light is promoted by p53 in Caenorhabditis elegans. Genetics. 2017;207(2):571–82.
166. Chandra R, et al. Sleep is required to consolidate odor memory and remodel olfactory synapses. Cell. 2023;186(13):2911–2928.e20.
167. Soto R, Van Buskirk C. The C. elegans CEPsh glia are largely dispensable for stress-induced sleep, vol. 2020. microPublication Biology; 2020.
168. Sakai N, et al. A sexually conditioned switch of chemosensory behavior in C. elegans. PLoS One. 2013;8(7):e68676.
169. Srinivasan J, et al. A blend of small molecules regulates both mating and development in Caenorhabditis elegans. Nature. 2008;454(7208):1115–8.
170. LeBoeuf B, et al. Caenorhabditis elegans male sensory-motor neurons and dopaminergic support cells couple ejaculation and post-ejaculatory behaviors. Elife. 2014;3:e02938.
171. Droge W, Schipper HM. Oxidative stress and aberrant signaling in aging and cognitive decline. Aging Cell. 2007;6(3):361–70.
172. Sofroniew MV. Molecular dissection of reactive astrogliosis and glial scar formation. Trends Neurosci. 2009;32(12):638–47.
173. Napoli I, Neumann H. Microglial clearance function in health and disease. Neuroscience. 2009;158(3):1030–8.
174. Hanslik KL, Marino KM, Ulland TK. Modulation of glial function in health, aging, and neurodegenerative disease. Front Cell Neurosci. 2021;15:718324.
175. Cassada RC, Russell RL. The dauerlarva, a post-embryonic developmental variant of the nematode Caenorhabditis elegans. Dev Biol. 1975;46(2):326–42.
176. Albert PS, Riddle DL. Developmental alterations in sensory neuroanatomy of the Caenorhabditis elegans dauer larva. J Comp Neurol. 1983;219(4):461–81.
177. Hartman JH, et al. Xenobiotic metabolism and transport in Caenorhabditis elegans. J Toxicol Environ Health B Crit Rev. 2021;24(2):51–94.
178. Wallace SW, et al. Nuclear hormone receptors promote gut and glia detoxifying enzyme induction and protect C. elegans from the mold P. brevicompactum. Cell Rep. 2021;37(13):110166.
179. Kim DH, Flavell SW. Host-microbe interactions and the behavior of Caenorhabditis elegans. J Neurogenet. 2020;34(3–4):500–9.
180. Heydel JM, et al. Odorant-binding proteins and xenobiotic metabolizing enzymes: implications in olfactory perireceptor events. Anat Rec (Hoboken). 2013;296(9):1333–45.
181. Read A, Schroder M. The unfolded protein response: an overview. Biology (Basel). 2021;10(5):384.
182. Gildea HK, et al. Glia of C. elegans coordinate a protective organismal heat shock response independent of the neuronal thermosensory circuit. Sci Adv. 2022;8(49):eabq3970.

183. Taylor RC, Dillin A. XBP-1 is a cell-nonautonomous regulator of stress resistance and longevity. Cell. 2013;153(7):1435–47.
184. Douglas PM, et al. Heterotypic signals from neural HSF-1 separate thermotolerance from longevity. Cell Rep. 2015;12(7):1196–204.
185. Bar-Ziv R, et al. Glial-derived mitochondrial signals affect neuronal proteostasis and aging. Sci Adv. 2023;9(41). https://doi.org/10.1126/sciadv.adi1411
186. Apfeld J, Kenyon C. Regulation of lifespan by sensory perception in Caenorhabditis elegans. Nature. 1999;402(6763):804–9.
187. McQuary PR, et al. C. elegans S6K mutants require a creatine-kinase-like effector for lifespan extension. Cell Rep. 2016;14(9):2059–67.
188. Smith-Vikos T, et al. MicroRNAs mediate dietary-restriction-induced longevity through PHA-4/FOXA and SKN-1/Nrf transcription factors. Curr Biol. 2014;24(19):2238–46.
189. Uno M, Nishida E. Lifespan-regulating genes in C. elegans. NPJ Aging Mech Dis. 2016;2:16010.
190. Pandit A, et al. PHA-4/FOXA-regulated microRNA feed forward loops during Caenorhabditis elegans dietary restriction. Aging (Albany NY). 2014;6(10):835–55.
191. Panowski SH, et al. PHA-4/Foxa mediates diet-restriction-induced longevity of C. elegans. Nature. 2007;447(7144):550–5.
192. Bishop NA, Guarente L. Two neurons mediate diet-restriction-induced longevity in C. elegans. Nature. 2007;447(7144):545–9.
193. Vora M, et al. Deletion of microRNA-80 activates dietary restriction to extend C. elegans healthspan and lifespan. PLoS Genet. 2013;9(8):e1003737.
194. Taylor, R.C. and A. Dillin, Aging as an event of proteostasis collapse. Cold Spring Harb Perspect Biol, 2011. 3(5) a004440.
195. Rahmani P, Rogalski T, Moerman DG. The C. elegans UNC-23 protein, a member of the BCL-2-associated athanogene (BAG) family of chaperone regulators, interacts with HSP-1 to regulate cell attachment and maintain hypodermal integrity. Worm. 2015;4(2):e1023496.
196. Ikeda K, et al. The respiratory control mechanisms in the brainstem and spinal cord: integrative views of the neuroanatomy and neurophysiology. J Physiol Sci. 2017;67(1):45–62.
197. Woo SH, Lumpkin EA, Patapoutian A. Merkel cells and neurons keep in touch. Trends Cell Biol. 2015;25(2):74–81.
198. Maricich SM, et al. Rodents rely on Merkel cells for texture discrimination tasks. J Neurosci. 2012;32(10):3296–300.
199. Maksimovic S, et al. Epidermal Merkel cells are mechanosensory cells that tune mammalian touch receptors. Nature. 2014;509(7502):617–21.
200. O'Hagan R, Chalfie M, Goodman MB. The MEC-4 DEG/ENaC channel of Caenorhabditis elegans touch receptor neurons transduces mechanical signals. Nat Neurosci. 2005;8(1):43–50.
201. Abdo H, et al. Specialized cutaneous Schwann cells initiate pain sensation. Science. 2019;365(6454):695–9.
202. Fung W, Wexler L, Heiman MG. Cell-type-specific promoters for C. elegans glia. J Neurogenet. 2020;34(3–4):335–46.

Astrocyte Development in the Rodent

Yajun Xie, Corey C. Harwell, and A. Denise R. Garcia

Abstract Astrocytes have gained increasing recognition as key elements of a broad array of nervous system functions. These include essential roles in synapse formation and elimination, synaptic modulation, maintenance of the blood-brain barrier, energetic support, and neural repair after injury or disease of the nervous system. Nevertheless, our understanding of mechanisms underlying astrocyte development and maturation remains far behind that of neurons and oligodendrocytes. Early efforts to understand astrocyte development focused primarily on their specification from embryonic progenitors and the molecular mechanisms driving the switch from neuron to glial production. Considerably, less is known about postnatal stages of astrocyte development, the period during which they are predominantly generated and mature. Notably, this period is coincident with synapse formation and the emergence of nascent neural circuits. Thus, a greater understanding of astrocyte development is likely to shed new light on the formation and maturation of synapses and circuits. Here, we highlight key foundational principles of embryonic and postnatal astrocyte development, focusing largely on what is known from rodent studies.

Keywords Astrocyte · Glia · Gliogenesis · Cell fate · Morphogenesis

Y. Xie · C. C. Harwell
Department of Neurology, University of California, San Francisco, San Francisco, CA, USA

Eli and Edythe Broad Center of Regeneration Medicine and Stem Cell Research, San Francisco, CA, USA

A. D. R. Garcia (✉)
Department of Neurobiology and Anatomy, Drexel University College of Medicine, Philadelphia, PA, USA
e-mail: adg82@drexel.edu

E. Blanco-Suarez, I. Farhy-Tselnicker (eds.), *Astrocyte-Neuron Interactions in Health and Disease*, Advances in Neurobiology 39,
https://doi.org/10.1007/978-3-031-64839-7_3

Abbreviations

ALG	Astrocyte-like glia
AMPA	α-Amino-3-hydroxy-5-methyl-4-isoxazolepropionic acid
BDNF	Brain-derived neurotrophic factor
bHLH	Basic helix-loop-helix
BLBP	Brain lipid-binding protein
BMP	Bone morphogenetic protein
CAM	Cell adhesion molecule
CNS	Central nervous system
CNTF	Ciliary neurotrophic factor
CT-1	Cardiotrophin-1
FGF	Fibroblast growth factor
GABA	Gamma-aminobutyric acid
GFAP	Glial fibrillary acidic protein
JAK-STAT	Janus kinase/signal transducer and activator of transcription
LIF	Leukemia inhibitory factor
mGluR5	Metabotropic receptor glutamate 5
NF1	Nuclear factor-1
NMDA	N-methyl-D-aspartate
Shh	Sonic hedgehog
SVZ	Subventricular zone

1 Introduction

Astrocytes are derived from the same pool of multipotent radial glial cells that produce neurons. The switch toward the production of astrocytes and the repression of alternative fates is regulated by a complex network of transcription factors and chromatin remodeling factors whose precise mechanism of function is only beginning to be uncovered. Defining the roles of these spatial and temporal molecular factors will be important for understanding how the appropriate complement of heterogeneous astrocytes arises during the course of development.

2 Astrocyte Development in the Embryo

2.1 Astrogenic Progenitor Competence

Radial glia serve as the primary progenitor cells of the developing central nervous system. Radial glia proceed through temporal competence states, first producing neurons and then switching to produce glial cells. In mice, neurogenesis initiates around embryonic days (E) 9–10 within the caudal spinal cord and then progresses rostrally and along ventral-dorsal and lateral-medial gradients in the brain [1]. The transition to gliogenic competence states occurs around E12.5 in the developing spinal cord and between E16 and E18 in the cerebral cortex [2]. Culturing isolated stem cells from embryonic mouse cerebral cortex in clonal conditions produces first neurons followed by glial cells after 10 days, suggesting a significant role for intrinsically determined temporal competence and glial fate switching [3]. Many studies demonstrate that the temporal switch toward gliogenesis incorporates intrinsic mechanisms governed by genetic and epigenetic regulators [2, 4] and secreted extrinsic signals derived from local environment (Fig. 1). Intrinsically determined fate switching relies on dynamic chromatin remodeling and DNA methylation to control the expression of genes essential for astrocyte production [2, 4, 5]. In the cortex, Polycomb group complex genes including *Ezh2* regulate dynamic changes in H3K27me3, a chromatin modification associated with transcriptional silencing. Loss of *Ezh2* in neurogenic radial glia leads to prolonged neurogenesis and a delay in the onset of astrogenesis [6]. DNA methylation is a covalent epigenetic modification that functions by influencing DNA-protein interactions, often leading to transcriptional repression. Cytosine nucleotide methylation often in the form of 5-methylcytosine (5mC) is typically associated with transcriptional repression and

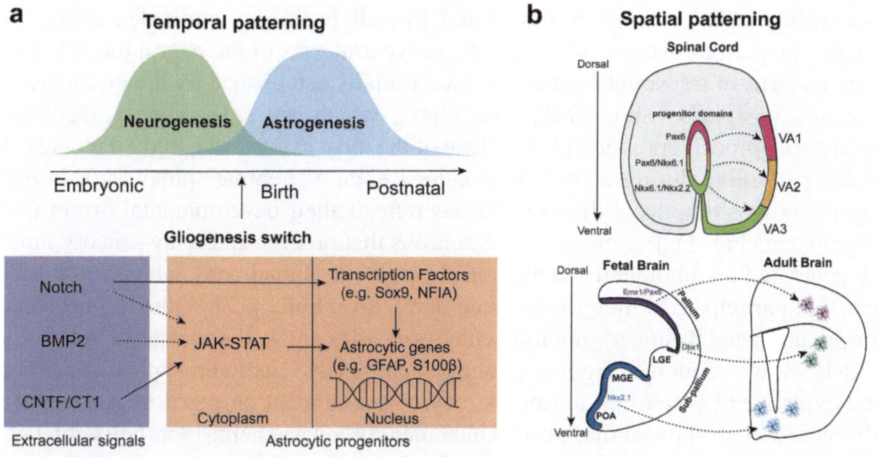

Fig. 1 Astrogenesis undergoes spatial and temporal patterning starting at the embryonic stage. (**a**) Molecular mechanisms underlying astrogenesis switch. Several extracellular signals converge on JAK-STAT signaling, which regulates the expression of astrocyte genes. Some signals, such as Notch, directly modulate epigenetic signals that regulate the expression of astrocyte genes. (**b**) Distinct progenitor domains contribute to spatially distinctive astrocyte populations in both the spinal cord and the brain. Pax6+, Pax6+Nkx6.1+ and Nkx6.1+Nkx2.2+ progenitors give rise to three subtypes of astrocytes in ventral spinal cord along dorsal-ventral axis. Emx1+/Pax6+, Dbx1+, and Nkx2.1+ progenitors generate distinctive regions of astrocytes in the brain. VA1, VA2, and VA3 indicate ventral astrocyte types 1, 2, and 3. LGE lateral ganglionic eminence, MGE medial ganglionic eminence, POA preoptic area

is known to be important for gliogenic competence [7]. Radial glial cells are known to undergo two successive waves of demethylation during neural development: first to demethylate neurogenic genes at the onset of neurogenesis and then to demethylate gliogenic genes marking the onset of astrogliogenesis. During specification, glial cells undergo mass de novo methylation at the sites of genes involved in neuronal function to solidify their identity. In addition to these intrinsic mechanisms, neuron-derived cardiotrophin-1 (CT-1) is essential for the initiation of astrogenesis [8]. Like ciliary neurotrophic factor (CNTF) and leukemia inhibitory factor (LIF), CT-1 activates the JAK-STAT pathway in precursor cells, leading to transcriptional upregulation of two astrocytic genes, *gfap* and *s100b* [9, 10]. Moreover, developmental cues such as BMPs and Notch collaboratively facilitate astrogenesis, partially mediated through the JAK-STAT pathway [11–13]. Both nuclear factor-I (NFI) and Sox9 transcriptional factors are important for gliogenic switch within the developing spinal cord [14–16].

2.2 *Spatial Patterns of Astrogenesis*

The neural tube is segmented into distinct neuroepithelial progenitor domains along the neuraxis that produce distinct neuron types. The expression of transcription factors within these domains is modulated by cell position coordinates along the anterior-posterior, rostral-caudal, and dorsal-ventral axes of the neural tube [1, 17]. Such patterns of transcriptional activity are initially established by the gradients of morphogen signals (Shh, BMP, and Wnts), which are subsequently regulated through reciprocal repression [1, 17]. One of the most extensively studied models of spatial patterning during neural development is the vertebrate spinal cord. In this region, the positioning of diverse neurons reflects their developmental origin to a large extent (Fig. 1). For instance, the neurons that process and relay sensory input are generated predominantly in the dorsal part of the spinal cord, whereas the neurons that participate in motor output are derived ventrally [17]. Although neuronal specification and its underlying molecular principles are well understood, relatively little is known about the origin of glia in the entire CNS, including spinal cord. The first evidence of patterning during astrocyte development emerged in 2005, when Muroyama et al. showed that the establishment of a basic helix-loop-helix (bHLH) code is required for the specification of V2 astrocytes within the p2 progenitor domain [18], which differs from transcription factors with pan-glial roles in the embryonic spinal cord, such as Sox9 [16]. Combinatorial expression of Pax6, Nkx6.1, and Nkx2.2 transcription factors specified three subtypes of ventral white matter astrocytes [19]. These astrocyte subtypes are systematically arrayed within the ventral white matter domain along the dorsal-ventral axis, reflecting the spatial organization of their progenitor domains [20, 21]. In the developing brain, regionally defined progenitor populations also contribute to the formation of spatially distinct astrocytes (Fig. 1). For instance, the Olig2 lineage is responsible for generating astrocytes predominantly in the dorsal pallium, but not in the basal forebrain [22].

Lineages marked by Emx1/Pax6, Dbx1, and Nkx2.1 give rise to astrocytes in the dorsal, intermediate, and ventral regions of the forebrain, respectively [23–25]. Such findings suggest that developmental origins confer astrocyte regional identities. Yet, it remains to be determined whether these intrinsic mechanisms retain their influence on the diversity of astrocytes once progenitor cells complete their relocation.

Although neurons and astrocytes originate from the same progenitor pools, these progenitors do not invariably differentiate into both cell types in a region-specific manner. For example, cortical protoplasmic astrocytes are generated in a spatially restricted fashion from precursors that also give rise to developmental columns of pyramidal neurons [26]. Conversely, neurons derived from Nkx2.1+ progenitors exhibit a distribution across more varied regions compared to these astrocytes from the same progenitor pools [27]. This suggests a differential regulatory mechanism guiding the spatial patterning and distribution of these cell types during brain development.

3 Postnatal Astrocyte Development

Although specification of astrocyte progenitor cells occurs in the developing embryo, the production and maturation of astrocytes occur predominantly during early postnatal development [28]. Notably, this coincides with the development of synapses and refinement of neural circuit [29]. Indeed, the recognition that astrocytes are "more than just nerve glue" has been driven in part by the identification of these cells as key regulators of synapse formation and maturation [30, 31]. Understanding the mechanisms regulating astrocyte production and maturation will facilitate a better understanding of how these cells cooperate with neurons to achieve brain function.

3.1 Postnatal Astrocyte Progenitors

The generation of astrocytes is most robust during the first postnatal week. The prevailing model for astrocyte production comes from pioneering work in primate brains demonstrating that in the early postnatal cortex, there is a progressive decline in cells with radial glial morphologies with a concomitant appearance of cells with transitional morphologies that eventually gives rise to cells with more complex, mature astrocyte characteristics (Fig. 2) [32–34]. Radial glial cells are the first cells to differentiate from neuroepithelia in the early neural tube and are characterized by their distinct morphology in which cell bodies are localized to the ventricular zone while extending a long process radially toward the pial surface and a shorter apical process toward the ventricle [35]. In addition to their characteristic morphology, radial glia express several markers associated with astrocytes, including GFAP,

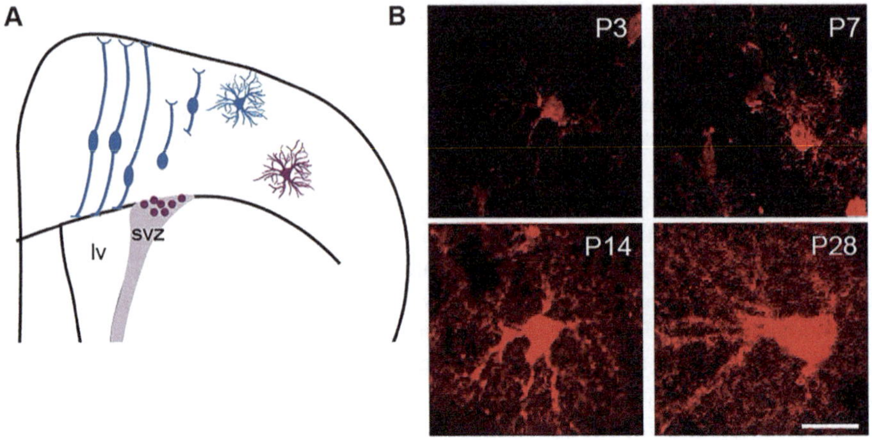

Fig. 2 Astrocyte growth throughout development. (**a**) In the perinatal brain, radial glia transition from radial morphologies in protoplasmic astrocytes (blue). A second population of glial progenitors resides in the dorsolateral corner of the SVZ (purple), which migrates into the cortex, proliferates locally, and generates additional protoplasmic astrocytes. Mature astrocytes have also been shown to undergo proliferation and contribute to the expansion of the cortical astrocyte population. lv, lateral ventricle; svz, subventricular zone. (**b**) Cortical astrocytes labeled at P0 and photographed at various ages throughout the first month of postnatal development. Astrocyte maturation takes place over the first 4 weeks of postnatal development, with dramatic increases in size and morphological complexity between P7 and P14, and further elaboration of small branchlets until P28 and beyond. Scale bar: 10 μm. Photomicrographs courtesy of Kathryn Markey

brain lipid-binding protein (BLBP), and the astrocyte-specific glutamate transporter, GLAST [36–38]. In the embryo, radial glia proliferate and generate neurons that use their parent radial glial fiber to migrate to their laminar position in the cortex [39]. However, after the completion of neurogenesis, radial glia generate astrocytes. Lineage-tracing studies have demonstrated mixed clonal clusters comprised of both projection neurons and parenchymal astrocytes that are organized in a columnar manner [26, 40].

Later studies using retroviral labeling of postnatal day 2 (P2) rodent brains demonstrated that a population of glial progenitor cells also resides in the perinatal subventricular zone (SVZ) [41]. Analysis of labeled cells at short-term time points showed that cells were restricted largely to the dorsolateral corner of the SVZ and had small, round morphologies with thin or no processes (Fig. 2) [41, 42]. Long-term chase experiments showed that labeled cells were found in the cortex and overlying white matter and exhibited morphologies consistent with protoplasmic and fibrous astrocytes, respectively. The dorsolateral SVZ progenitors comprise molecularly diverse populations and are distinct from radial glial progenitors [43]. A subpopulation of these SVZ cells expresses aldolase C/zebrin II [44], and more recently, genetic labeling experiments showed that some dorsolateral SVZ cells express the transcription factor *Gli1* [45, 46], indicating high levels of Sonic hedgehog (Shh) signaling. Fate mapping from *Gli1*-expressing progenitors demonstrated that they produce cortical astrocytes [45] and oligodendrocytes in the overlying white matter [46], suggesting these cells correspond to retrovirally labeled cells at P2. The precise relationship between the aldolase C/zebrin II cells and those with Shh activity is not known. Because astrocytes within the *Gli1* lineage comprise half of the cortical astrocyte population, it is tempting to speculate whether the *Gli1* and aldolase C/zebrin II progenitors comprise two complementary populations, each responsible for generating distinct cortical astrocyte lineages.

How the expansion of the astrocyte population is accomplished is not well understood. Labeling with proliferation markers shows that astrocyte proliferation is most robust during the first postnatal week [45, 47]. Recent viral labeling studies demonstrated that the expansion of the astrocyte population is accomplished by the local proliferation of cells within the cortex (Fig. 2) [48]. Interestingly, these cells exhibit the morphological and electrophysiological properties of mature astrocytes, including complex, bushy processes, large K^+ currents, and gap junction coupling [48]. Thus, early-born astrocytes derived from radial glia and SVZ progenitors seed the cortex, differentiate, and then proceed to proliferate facilitating local expansion of the astrocyte population. However, clonal analysis demonstrated that astrocytes within a single clone can be widely dispersed [47, 49] (discussed further below), suggesting that neighboring astrocytes within a defined region may arise from both local proliferation and long-distance migration of cells generated from distal regions.

3.2 *Developmental Trajectory of Astrocytes*

In contrast to the other primary cells of the nervous system, neurons, and oligodendrocytes, the developmental trajectory of astrocytes remains poorly understood. Whereas specific stages of oligodendrocyte development, from precursor to premyelinating to fully differentiated cells, can be identified by the expression of defined genes [50], discrete stages of astrocyte development and corresponding markers to identify them are not well characterized. A major challenge to accomplishing this is the fact that several genes expressed in astrocyte progenitors persist into mature stages. At birth, radial glia express GFAP and several populations of astrocytes exhibit high levels of GFAP expression in the adult brain, including in the hippocampus, hypothalamus, and Bergmann glia. This is further complicated by the fact that reactive astrocytes after injury upregulate GFAP expression, thus making GFAP a poor marker to distinguish between astrocyte progenitors and mature astrocytes. Similarly, the transcription factor, Sox9, is expressed in astrocyte progenitor cells and adult astrocytes [16, 51, 52]. Early microarray studies identified four markers that exhibit temporally regulated expression in the spinal cord between E14.5 and P7 [53]. Though astrocyte production is largely complete by P7, astrocytes continue to undergo considerable developmental maturation between P7 and P42, when they elaborate their processes and begin interacting with nascent synapses. The identification of markers that distinguish between proliferating astrocyte progenitor cells, post-mitotic, but immature astrocytes, and mature, fully differentiated astrocytes would facilitate studies to deepen our understanding of astrocyte development and their interactions with synapses and nascent circuits.

3.3 Astrocyte Migration

Astrocytes are distributed throughout the parenchyma of the CNS. How astrocytes migrate out of their germinal zones and occupy their positions in the parenchyma is not well understood. The prevailing view is that astrocytes, like neurons, utilize radial glia to guide their migration. This is supported by several observations. Retroviral labeling of SVZ cells at P2 and P14 shows that cells migrate dorsally into cortex or overlying white matter, with limited migration along the anteroposterior axis [42]. Retroviral labeling and genetic labeling produce cells with a columnar distribution in the cortex [26, 42]. These observations are supported by recent studies in the spinal cord and cortex demonstrating that astrocyte migration is restricted to the territory directly overlying their germinal zone [19, 23]. Whereas both neurons and oligodendrocytes migrate long distances, in some cases crossing dorsal-ventral or anterior-posterior boundaries, these observations indicate that astrocytes exhibit more limited migration, observing strict boundaries that are yet to be defined. Indeed, even injury or ablation of cells from a defined germinal zone fails to recruit astrocytes from a neighboring germinal zone [23], suggesting that cell lineage is a strong driver of positional identity.

However, emerging evidence demonstrates a broader dispersion of astrocytes along the anteroposterior axis than previously appreciated. Using multiclonal lineage tracing with combinatorial markers together with large volume 3D imaging, clonal analysis reveals that while most clones were comprised of cells with radial dispersion, some clones included cells distributed along the mediolateral and anteroposterior axes [47]. Some clones showed a dispersion as large as 1.86×10^6 μm^3. Key to these novel observations is the analysis of much larger volumes of tissue than earlier studies, together with the application of genetic labeling strategies enabling more definitive identification of clones.

The highly organized layering of the cortex prompts the question of whether astrocytes exhibit layer specificity. During neurogenesis, the production of neurons is highly stereotyped. Early-born neurons occupy deep layers, and late-born neurons must migrate past these older neurons to occupy more superficial layers [54]. Whether cortical astrocytes observe similar principles in their production and migration is not well understood. Clonal analysis of embryonic progenitors identified a similar temporal pattern for cortical astrocytes. Clones labeled at E16.5 were found predominantly in deep layers, while those labeled at E17.5 were found in upper layers at P30 [49]. However, labeling perinatal progenitors in the SVZ generated clones dispersed throughout all cortical layers with no apparent preferential distribution [47]. One possibility is that embryonic and perinatal progenitors comprise two different progenitor pools, each with distinct behaviors that contribute distinct astrocyte lineages to the cortex. Alternatively, the behaviors of embryonic and perinatal progenitors may be governed by intrinsic timing that shifts the production of cells destined for specific cortical locations. Advancing our understanding of these apparently conflicting behaviors of astrocyte progenitor cells will require a greater understanding of molecular identifiers of each population. Interestingly, using spatial transcriptomics, cortical astrocytes were mapped onto layers based on patterns of gene expression [55, 56]. These layers can be defined by the expression patterns of several genes, including *Chrdl1*, *Scel*, *Il3*, *Id3*, and several members of the glypican family, *gpc4*, *gpc5*, and *gpc6*. Expression of these genes is distributed across the upper, mid-, and deep layers, and these patterns were observed as early as P14 and persisted into adulthood. Astrocyte layers diverge from conventional neuronal layers, but nevertheless observe neuronal cues. Disturbing normal neuronal layer patterning also disturbed the patterning of astrocyte layers. How these laminar patterns of gene expression emerge remains to be discovered.

3.4 Sculpting the Astrocyte Population

How the CNS establishes the appropriate number of astrocytes is unknown. In addition to cell proliferation, developmental elimination is a fundamental mechanism by which tissues are sculpted. It is well established that both neurons and oligodendrocytes are initially overproduced followed by a period of elimination mediated by apoptosis [57, 58]. Whether astrocytes experience such elimination is not well understood, though emerging reports indicate that astrocyte cell death occurs during early postnatal periods. In the retina, astrocyte number increases progressively from birth until P5, but declines rapidly thereafter until reaching adult numbers at P14 [59]. Interestingly, retinal astrocytes engulfed by microglia are not undergoing apoptosis, suggesting that cells targeted for elimination are otherwise viable [59]. Perturbation of microglial-mediated astrocyte cell death led to extravasation of blood cells into the extracellular space, suggesting that dysregulation of astrocyte cell number interferes with vascular integrity. Microglial engulfment of astrocytes has also been observed in the postnatal amygdala [60], suggesting that this may be a general mechanism throughout the CNS by which astrocyte cell number is regulated. How cells are targeted for engulfment or survival is not known.

3.5 Morphogenesis

One of the defining features of astrocytes is their complex morphology, characterized by densely ramified processes that interact intimately with synapses and the vasculature. The elaboration of these processes occurs primarily over the course of the first 3–4 postnatal weeks [61]. Indeed, astrocytes show a dramatic 12-fold increase in the size of their anatomical domain between P7 and P21 [62]. There is only minimal overlap between astrocyte processes, creating a tiling organization of cells with distinct anatomical territories [63]. These processes do not express GFAP and instead are visualized only by using dye fills or labeling membranes [63, 64], which has long limited their careful study. Another major limitation to studying these processes is their size, which is beyond diffraction limits of conventional light microscopy approaches. It has been estimated that a single astrocyte can contact up to 100,000 synapses in the rodent brain [65]. These interactions make them well positioned to exert highly localized modulation of synapses in multiple circuits within their territory. Yet, the precise mechanisms governing their morphological maturation and characteristics are only beginning to be understood.

A growing body of evidence now shows that neuronal activity regulates astrocyte morphogenesis. Early work in the cat visual cortex demonstrated that cats dark reared during the period coinciding with critical period development show a reduction in astrocyte maturation [66]. However, a limitation of this study is the reliance on GFAP immunostaining. Nevertheless, more recent work using single-cell labeling of astrocytes in the mouse visual cortex shows that mice raised in the dark have smaller astrocytes with less complex morphologies compared to mice raised in typical light conditions [67]. The influence of neuronal activity on astrocyte morphology persists into adulthood. Ultrastructural analysis reveals that astrocyte interactions with synapses increase following whisker stimulation [68]. Glutamatergic signaling through mGluR5 receptors promotes astrocyte growth [62]. Expression of mGluR5 declines after P8 [69], suggesting that glutamate acts on other receptors. Astrocytes express both AMPA and NMDA receptors, and their relative expression increases over the first 2–4 postnatal weeks [69, 70], coinciding well with astrocyte morphogenesis. Recently, a role for inhibitory activity on morphogenesis has also been reported [71]. Astrocyte-specific deletion of $GABA_B$ receptors at P1 decreases morphological complexity at P28 in various brain regions, including the forebrain, cerebellum, and brainstem, suggesting that inhibitory signaling directly in astrocytes is a general mechanism influencing astrocyte morphogenesis. Precisely how glutamatergic and GABAergic activities mediate morphological maturation is not well understood. Expression of the endothelial receptor B (*Ednrb*) is reduced in astrocyte-specific mutants of $GABA_B$ receptors, and targeted deletion of *Ednrb* in astrocytes shows similar deficits in astrocyte morphology [71]. The Endrb ligand, endothelin, is found in neurons, glia, and endothelial cells [72, 73], suggesting astrocyte morphogenesis is a function of cooperative signaling between multiple cell types.

Several molecular mechanisms involved in astrocyte morphogenesis have recently been discovered, though our understanding of these players and the precise ways in which they direct this process is still in its early stages. Among the first such mechanisms identified was growth factor signaling, which was identified in *Drosophila*. Also known as neuropil glia or astrocyte-like glia (ALG), *Drosophila* possess a class of glial cells that share key morphological, biochemical, and functional characteristics with vertebrate astrocytes [74] (and for further review, see also [75, 76]). ALG exhibit complex, highly ramified morphologies, associate intimately with synapses, and can be identified by high expression of both glutamate and

GABA transporters, Eaat1 and Gat, respectively [74, 77]. ALG in mutants that lack the *Drosophila* FGF receptor Heartless (*Htl*), or Htl ligands, Pyramus (*Pyr*), and Thisbe (*Ths*), fail to fully elaborate processes into the synaptic neuropil. The requirement for FGF signaling is conserved in vertebrate astrocytes [78]. Brain-derived neurotrophic factor (BDNF) also plays a key role in astrocyte morphogenesis. This is accomplished through a truncated TrkB receptor, TrkB.T1, which lacks the tyrosine kinase domain [79]. TrkB.T1 is highly enriched in astrocytes, compared to full-length TrkB. Addition of BDNF to astrocyte cultures significantly increased astrocyte complexity, and both in vitro and in vivo experiments demonstrate a requirement for BDNF signaling specifically through the truncated TrkB.T1 receptor for astrocyte morphogenesis.

Membrane proteins are increasingly being implicated as important regulators of astrocyte morphogenesis. Neuroligins comprise a family of cell adhesion molecules that interact with neurexins. Such interactions are well studied in neurons and have well-characterized roles in synapse function [80]. However, astrocytes express high levels of both neurexins and neuroligins [69, 81, 82], among several other cell adhesion molecules, such as ephrins, neuronal cell adhesion molecule (NrCAM), and hepatocyte cell adhesion molecule (hepaCAM, also known as GlialCAM) [83–86]. Knockdown of astrocyte neuroligins or neuronal neurexins dramatically reduces morphological complexity of astrocytes in vitro and in vivo [67]. Similarly, knockdown of hepaCAM at birth produces smaller astrocytes with less complex morphologies at P7 that are more pronounced at P21 [87]. These observations suggest that the initial elaboration of primary and other major astrocyte processes is mechanistically uncoupled from those governing the full elaboration of finer, higher order processes that characterize mature, fully differentiated astrocytes. HepaCAM also plays a role in astrocyte tiling [87]. Recently, a requirement for a lipid-binding G protein-coupled receptor, sphingosine-1-phosphate receptor 1 (S1p1r), in astrocyte development was discovered in *zebrafish* [88]. As with the neuroligin, neurexin, and hepaCAM mutants, astrocytes carrying mutations in *S1p1r* show dramatically reduced volumes and simpler morphologies, with no differences in astrocyte number or position [88], further supporting the idea that astrocyte morphogenesis may be regulated by distinct mechanisms underlying discrete steps or phases in the development of astrocyte morphology. Recent work demonstrates that ablation of Sonic hedgehog signaling in cortical neurons reduces astrocyte complexity significantly, resulting in diminished ensheathment of synapse clefts. This finding substantiates the influence of extrinsic signaling pathways in dictating astrocyte morphogenesis [89].

4 Perspectives

Despite recent progress on elucidating molecular mechanisms underlying astrocyte morphogenesis, large gaps remain in understanding several events in astrocyte development. Of particular interest is the question of how astrocyte diversity emerges. The morphological diversity of astrocytes was recognized by Cajal and further described in recent studies using cutting-edge labeling and imaging approaches [47, 90], whereas transcriptional profiling studies have demonstrated diverse molecular signatures [56, 91–93]. Yet, how such diversity arises remains largely unknown. Growing evidence points to a key role for local environmental cues, derived largely from neurons [55, 89, 94, 95]. However, genetic fate-mapping experiments also demonstrate that astrocytes are derived from regionally and molecularly distinct progenitor lineages [23, 45]. The precise relationships between morphology, transcriptional signature, environmental cues, and lineage remain a rich area for further interrogation. Mechanisms regulating astrocyte proliferation, expansion, and migration also pose several interesting questions. While local proliferation of mature astrocytes has been shown to expand the cortical astrocyte population, it remains unknown whether astrocytes in other regions are similarly expanded. And if proliferation of mature astrocytes is a general mechanism for expanding astrocytes, what triggers them to enter a post-mitotic state? Finally, a major limitation to advancing our understanding of astrocyte development is the lack of molecular tools. While many studies have relied on markers, such as GFAP, Sox9, and Aldh1l1, each has distinct limitations and none specifically identify discrete stages of astrocytes along their developmental trajectory. Thus, the discovery of novel markers that can accurately define stage-specific astrocytes is crucial for elucidating the embryonic and postnatal mechanisms underlying astrocyte development. Developing deeper insights into the fundamental principles of astrocyte development will be an important focus for the future.

References

1. Rowitch DH, Kriegstein AR. Developmental genetics of vertebrate glial-cell specification. Nature. 2010;468:214–22. https://doi.org/10.1038/nature09611.
2. Freeman MR. Specification and morphogenesis of astrocytes. Science. 2010;330:774–8. https://doi.org/10.1126/science.1190928.
3. Qian X, et al. Timing of CNS cell generation: a programmed sequence of neuron and glial cell production from isolated murine cortical stem cells. Neuron. 2000;28:69–80. https://doi.org/10.1016/s0896-6273(00)00086-6.
4. Zarei-Kheirabadi M, Vaccaro AR, Rahimi-Movaghar V, Kiani S, Baharvand H. An overview of extrinsic and intrinsic mechanisms involved in astrocyte development in the central nervous system. Stem Cells Dev. 2020;29:266–80. https://doi.org/10.1089/scd.2019.0189.
5. Adam MA, Harwell CC. Epigenetic regulation of cortical neurogenesis; orchestrating fate switches at the right time and place. Curr Opin Neurobiol. 2020;63:146–53. https://doi.org/10.1016/j.conb.2020.03.012.
6. Hirabayashi Y, et al. Polycomb limits the neurogenic competence of neural precursor cells to promote astrogenic fate transition. Neuron. 2009;63:600–13. https://doi.org/10.1016/j.neuron.2009.08.021.
7. Noack F, et al. Assessment and site-specific manipulation of DNA (hydroxy-)methylation during mouse corticogenesis. Life Sci Alliance. 2019;2:10.26508/lsa.201900331.
8. Barnabe-Heider F, et al. Evidence that embryonic neurons regulate the onset of cortical gliogenesis via cardiotrophin-1. Neuron. 2005;48:253–65. https://doi.org/10.1016/j.neuron.2005.08.037.

9. Bonni A, et al. Regulation of gliogenesis in the central nervous system by the JAK-STAT signaling pathway. Science. 1997;278:477–83. https://doi.org/10.1126/science.278.5337.477.
10. Namihira M, Nakashima K, Taga T. Developmental stage dependent regulation of DNA methylation and chromatin modification in a immature astrocyte specific gene promoter. FEBS Lett. 2004;572:184–8. https://doi.org/10.1016/j.febslet.2004.07.029.
11. Gomes WA, Mehler MF, Kessler JA. Transgenic overexpression of BMP4 increases astroglial and decreases oligodendroglial lineage commitment. Dev Biol. 2003;255:164–77. https://doi.org/10.1016/s0012-1606(02)00037-4.
12. Kageyama R, Ohtsuka T, Hatakeyama J, Ohsawa R. Roles of bHLH genes in neural stem cell differentiation. Exp Cell Res. 2005;306:343–8. https://doi.org/10.1016/j.yexcr.2005.03.015.
13. Wu Y, Liu Y, Levine EM, Rao MS. Hes1 but not Hes5 regulates an astrocyte versus oligodendrocyte fate choice in glial restricted precursors. Dev Dyn. 2003;226:675–89. https://doi.org/10.1002/dvdy.10278.
14. Deneen B, et al. The transcription factor NFIA controls the onset of gliogenesis in the developing spinal cord. Neuron. 2006;52:953–68. https://doi.org/10.1016/j.neuron.2006.11.019.
15. Kang P, et al. Sox9 and NFIA coordinate a transcriptional regulatory cascade during the initiation of gliogenesis. Neuron. 2012;74:79–94. https://doi.org/10.1016/j.neuron.2012.01.024.
16. Stolt CC, et al. The Sox9 transcription factor determines glial fate choice in the developing spinal cord. Genes Dev. 2003;17:1677–89. https://doi.org/10.1101/gad.259003.
17. Briscoe J, Novitch BG. Regulatory pathways linking progenitor patterning, cell fates and neurogenesis in the ventral neural tube. Philos Trans R Soc Lond Ser B Biol Sci. 2008;363:57–70. https://doi.org/10.1098/rstb.2006.2012.
18. Muroyama Y, Fujiwara Y, Orkin SH, Rowitch DH. Specification of astrocytes by bHLH protein SCL in a restricted region of the neural tube. Nature. 2005;438:360–3. https://doi.org/10.1038/nature04139.
19. Hochstim C, Deneen B, Lukaszewicz A, Zhou Q, Anderson DJ. Identification of positionally distinct astrocyte subtypes whose identities are specified by a homeodomain code. Cell. 2008;133:510–22. https://doi.org/10.1016/j.cell.2008.02.046.
20. Molofsky AV, et al. Astrocytes and disease: a neurodevelopmental perspective. Genes Dev. 2012;26:891–907. https://doi.org/10.1101/gad.188326.112.
21. Bayraktar OA, Fuentealba LC, Alvarez-Buylla A, Rowitch DH. Astrocyte development and heterogeneity. Cold Spring Harb Perspect Biol. 2014;7:a020362. https://doi.org/10.1101/cshperspect.a020362.
22. Ono K, et al. Regional- and temporal-dependent changes in the differentiation of Olig2 progenitors in the forebrain, and the impact on astrocyte development in the dorsal pallium. Dev Biol. 2008;320:456–68. https://doi.org/10.1016/j.ydbio.2008.06.001.
23. Tsai HH, et al. Regional astrocyte allocation regulates CNS synaptogenesis and repair. Science. 2012;337:358–62. https://doi.org/10.1126/science.1222381.
24. Maekawa M, et al. Pax6 is required for production and maintenance of progenitor cells in postnatal hippocampal neurogenesis. Genes Cells. 2005;10:1001–14. https://doi.org/10.1111/j.1365-2443.2005.00893.x.
25. Stoykova A, Treichel D, Hallonet M, Gruss P. Pax6 modulates the dorsoventral patterning of the mammalian telencephalon. J Neurosci. 2000;20:8042–50. https://doi.org/10.1523/JNEUROSCI.20-21-08042.2000.
26. Magavi S, Friedmann D, Banks G, Stolfi A, Lois C. Coincident generation of pyramidal neurons and protoplasmic astrocytes in neocortical columns. J Neurosci. 2012;32:4762–72. https://doi.org/10.1523/JNEUROSCI.3560-11.2012.
27. Torigoe M, Yamauchi K, Zhu Y, Kobayashi H, Murakami F. Association of astrocytes with neurons and astrocytes derived from distinct progenitor domains in the subpallium. Sci Rep. 2015;5:12258. https://doi.org/10.1038/srep12258.
28. Sauvageot CM, Stiles CD. Molecular mechanisms controlling cortical gliogenesis. Curr Opin Neurobiol. 2002;12:244–9. https://doi.org/10.1016/s0959-4388(02)00322-7.

29. Farhy-Tselnicker I, Allen NJ. Astrocytes, neurons, synapses: a tripartite view on cortical circuit development. Neural Dev. 2018;13:7. https://doi.org/10.1186/s13064-018-0104-y.
30. Stogsdill JA, Harwell CC, Goldman SA. Astrocytes as master modulators of neural networks: synaptic functions and disease-associated dysfunction of astrocytes. Ann N Y Acad Sci. 2023;1525:41–60. https://doi.org/10.1111/nyas.15004.
31. Ullian EM, Sapperstein SK, Christopherson KS, Barres BA. Control of synapse number by glia. Science. 2001;291:657–61. https://doi.org/10.1126/science.291.5504.657.
32. Choi BH, Lapham LW. Radial glia in the human fetal cerebrum: a combined Golgi, immunofluorescent and electron microscopic study. Brain Res. 1978;148:295–311. https://doi.org/10.1016/0006-8993(78)90721-7.
33. Misson JP, Edwards MA, Yamamoto M, Caviness VS Jr. Identification of radial glial cells within the developing murine central nervous system: studies based upon a new immunohistochemical marker. Brain Res Dev Brain Res. 1988;44:95–108. https://doi.org/10.1016/0165-3806(88)90121-6.
34. Schmechel DE, Rakic P. A Golgi study of radial glial cells in developing monkey telencephalon: morphogenesis and transformation into astrocytes. Anat Embryol (Berl). 1979;156:115–52. https://doi.org/10.1007/BF00300010.
35. Malatesta P, Appolloni I, Calzolari F. Radial glia and neural stem cells. Cell Tissue Res. 2008;331:165–78. https://doi.org/10.1007/s00441-007-0481-8.
36. Feng L, Hatten ME, Heintz N. Brain lipid-binding protein (BLBP): a novel signaling system in the developing mammalian CNS. Neuron. 1994;12:895–908. https://doi.org/10.1016/0896-6273(94)90341-7.
37. Levitt P, Rakic P. Immunoperoxidase localization of glial fibrillary acidic protein in radial glial cells and astrocytes of the developing rhesus monkey brain. J Comp Neurol. 1980;193:815–40. https://doi.org/10.1002/cne.901930316.
38. Shibata T, et al. Glutamate transporter GLAST is expressed in the radial glia-astrocyte lineage of developing mouse spinal cord. J Neurosci. 1997;17:9212–9. https://doi.org/10.1523/JNEUROSCI.17-23-09212.1997.
39. Noctor SC, Martinez-Cerdeno V, Ivic L, Kriegstein AR. Cortical neurons arise in symmetric and asymmetric division zones and migrate through specific phases. Nat Neurosci. 2004;7:136–44. https://doi.org/10.1038/nn1172.
40. Shen Z, et al. Distinct progenitor behavior underlying neocortical gliogenesis related to tumorigenesis. Cell Rep. 2021;34:108853. https://doi.org/10.1016/j.celrep.2021.108853.
41. Levison SW, Goldman JE. Both oligodendrocytes and astrocytes develop from progenitors in the subventricular zone of postnatal rat forebrain. Neuron. 1993;10:201–12. https://doi.org/10.1016/0896-6273(93)90311-e.
42. Levison SW, Chuang C, Abramson BJ, Goldman JE. The migrational patterns and developmental fates of glial precursors in the rat subventricular zone are temporally regulated. Development. 1993;119:611–22.
43. Marshall CAG, Suzuki SO, Goldman JE. Gliogenic and neurogenic progenitors of the subventricular zone: who are they, where did they come from, and where are they going? Glia. 2003;43:52–61. https://doi.org/10.1002/glia.10213.
44. Staugaitis SM, Zerlin M, Hawkes R, Levine JM, Goldman JE. Aldolase C/zebrin II expression in the neonatal rat forebrain reveals cellular heterogeneity within the subventricular zone and early astrocyte differentiation. J Neurosci. 2001;21:6195–205. https://doi.org/10.1523/JNEUROSCI.21-16-06195.2001.
45. Gingrich EC, Case K, Garcia ADR. A subpopulation of astrocyte progenitors defined by Sonic hedgehog signaling. Neural Dev. 2022;17:2. https://doi.org/10.1186/s13064-021-00158-w.
46. Tong CK, et al. A dorsal SHH-dependent domain in the V-SVZ produces large numbers of oligodendroglial lineage cells in the postnatal brain. Stem Cell Rep. 2015;5:461–70. https://doi.org/10.1016/j.stemcr.2015.08.013.
47. Clavreul S, et al. Cortical astrocytes develop in a plastic manner at both clonal and cellular levels. Nat Commun. 2019;10:4884. https://doi.org/10.1038/s41467-019-12791-5.

48. Ge WP, Miyawaki A, Gage FH, Jan YN, Jan LY. Local generation of glia is a major astrocyte source in postnatal cortex. Nature. 2012;484:376–80. https://doi.org/10.1038/nature10959.
49. Zhang X, et al. Clonal analysis of gliogenesis in the cerebral cortex reveals stochastic expansion of glia and cell autonomous responses to Egfr dosage. Cells. 2020;9:2662.
50. Akay LA, Effenberger AH, Tsai LH. Cell of all trades: oligodendrocyte precursor cells in synaptic, vascular, and immune function. Genes Dev. 2021;35:180–98. https://doi.org/10.1101/gad.344218.120.
51. Molofsky AV, et al. Expression profiling of Aldh1l1-precursors in the developing spinal cord reveals glial lineage-specific genes and direct Sox9-Nfe2l1 interactions. Glia. 2013;61:1518–32. https://doi.org/10.1002/glia.22538.
52. Sun W, et al. SOX9 is an astrocyte-specific nuclear marker in the adult brain outside the neurogenic regions. J Neurosci. 2017;37:4493–507. https://doi.org/10.1523/JNEUROSCI.3199-16.2017.
53. Chaboub LS, et al. Temporal profiling of astrocyte precursors reveals parallel roles for Asef during development and after injury. J Neurosci. 2016;36:11904–17. https://doi.org/10.1523/JNEUROSCI.1658-16.2016.
54. Greig LC, Woodworth MB, Galazo MJ, Padmanabhan H, Macklis JD. Molecular logic of neocortical projection neuron specification, development and diversity. Nat Rev Neurosci. 2013;14:755–69. https://doi.org/10.1038/nrn3586.
55. Bayraktar OA, et al. Astrocyte layers in the mammalian cerebral cortex revealed by a single-cell in situ transcriptomic map. Nat Neurosci. 2020;23:500–9. https://doi.org/10.1038/s41593-020-0602-1.
56. Farhy-Tselnicker I, et al. Activity-dependent modulation of synapse-regulating genes in astrocytes. Elife. 2021;10:e70514. https://doi.org/10.7554/eLife.70514.
57. Oppenheim RW. Cell death during development of the nervous system. Annu Rev Neurosci. 1991;14:453–501. https://doi.org/10.1146/annurev.ne.14.030191.002321.
58. Wong FK, Marín O. Developmental cell death in the cerebral cortex. Annu Rev Cell Dev Biol. 2019;35:523–42. https://doi.org/10.1146/annurev-cellbio-100818-125204.
59. Puñal VM, et al. Large-scale death of retinal astrocytes during normal development is non-apoptotic and implemented by microglia. PLoS Biol. 2019;17:e3000492. https://doi.org/10.1371/journal.pbio.3000492.
60. VanRyzin JW, et al. Microglial phagocytosis of newborn cells is induced by endocannabinoids and sculpts sex differences in juvenile rat social play. Neuron. 2019;102:435–449 e436. https://doi.org/10.1016/j.neuron.2019.02.006.
61. Bushong EA, Martone ME, Ellisman MH. Maturation of astrocyte morphology and the establishment of astrocyte domains during postnatal hippocampal development. Int J Dev Neurosci. 2004;22:73–86. https://doi.org/10.1016/j.ijdevneu.2003.12.008.
62. Morel L, Higashimori H, Tolman M, Yang Y. VGluT1+ neuronal glutamatergic signaling regulates postnatal developmental maturation of cortical protoplasmic astroglia. J Neurosci. 2014;34:10950–62. https://doi.org/10.1523/JNEUROSCI.1167-14.2014.
63. Bushong EA, Martone ME, Jones YZ, Ellisman MH. Protoplasmic astrocytes in CA1 stratum radiatum occupy separate anatomical domains. J Neurosci. 2002;22:183–92. https://doi.org/10.1523/JNEUROSCI.22-01-00183.2002.
64. Yu X, Nagai J, Khakh BS. Improved tools to study astrocytes. Nat Rev Neurosci. 2020;21:121–38. https://doi.org/10.1038/s41583-020-0264-8.
65. Halassa MM, Fellin T, Takano H, Dong JH, Haydon PG. Synaptic islands defined by the territory of a single astrocyte. J Neurosci. 2007;27:6473–7. https://doi.org/10.1523/JNEUROSCI.1419-07.2007.
66. Muller CM. Dark-rearing retards the maturation of astrocytes in restricted layers of cat visual cortex. Glia. 1990;3:487–94. https://doi.org/10.1002/glia.440030607.
67. Stogsdill JA, et al. Astrocytic neuroligins control astrocyte morphogenesis and synaptogenesis. Nature. 2017;551:192–7. https://doi.org/10.1038/nature24638.

68. Genoud C, et al. Plasticity of astrocytic coverage and glutamate transporter expression in adult mouse cortex. PLoS Biol. 2006;4:e343. https://doi.org/10.1371/journal.pbio.0040343.
69. Boisvert MM, Erikson GA, Shokhirev MN, Allen NJ. The aging astrocyte transcriptome from multiple regions of the mouse brain. Cell Rep. 2018;22:269–85. https://doi.org/10.1016/j.celrep.2017.12.039.
70. Cheng S, et al. Vision-dependent specification of cell types and function in the developing cortex. Cell. 2022;185:311–327 e324. https://doi.org/10.1016/j.cell.2021.12.022.
71. Cheng Y-T, et al. Inhibitory input directs astrocyte morphogenesis through glial GABABR. Nature. 2023;617:369–76. https://doi.org/10.1038/s41586-023-06010-x.
72. Adams KL, et al. Endothelin-1 signaling maintains glial progenitor proliferation in the postnatal subventricular zone. Nat Commun. 2020;11:2138. https://doi.org/10.1038/s41467-020-16028-8.
73. Kuwaki T, et al. Physiological role of brain endothelin in the central autonomic control: from neuron to knockout mouse. Prog Neurobiol. 1997;51:545–79. https://doi.org/10.1016/S0301-0082(96)00063-9.
74. Stork T, Sheehan A, Tasdemir-Yilmaz OE, Freeman MR. Neuron-glia interactions through the Heartless FGF receptor signaling pathway mediate morphogenesis of Drosophila astrocytes. Neuron. 2014;83:388–403. https://doi.org/10.1016/j.neuron.2014.06.026.
75. Freeman MR. Drosophila central nervous system glia. Cold Spring Harb Perspect Biol. 2015;7:a020552. https://doi.org/10.1101/cshperspect.a020552.
76. Markey KM, Saunders JC, Smuts J, von Reyn CR, Garcia ADR. Astrocyte development-more questions than answers. Front Cell Dev Biol. 2023;11:1063843. https://doi.org/10.3389/fcell.2023.1063843.
77. Huang Y, Ng FS, Jackson FR. Comparison of larval and adult Drosophila astrocytes reveals stage-specific gene expression profiles. G3 Genes|Genomes|Genetics. 2015;5:551–8. https://doi.org/10.1534/g3.114.016162.
78. Chen J, Poskanzer KE, Freeman MR, Monk KR. Live-imaging of astrocyte morphogenesis and function in zebrafish neural circuits. Nat Neurosci. 2020;23:1297–306. https://doi.org/10.1038/s41593-020-0703-x.
79. Holt LM, et al. Astrocyte morphogenesis is dependent on BDNF signaling via astrocytic TrkB.T1. Elife. 2019;8:e44667. https://doi.org/10.7554/eLife.44667.
80. Sudhof TC. Synaptic neurexin complexes: a molecular code for the logic of neural circuits. Cell. 2017;171:745–69. https://doi.org/10.1016/j.cell.2017.10.024.
81. Cahoy JD, et al. A transcriptome database for astrocytes, neurons, and oligodendrocytes: a new resource for understanding brain development and function. J Neurosci. 2008;28:264–78. https://doi.org/10.1523/JNEUROSCI.4178-07.2008.
82. Zhang Y, et al. Purification and characterization of progenitor and mature human astrocytes reveals transcriptional and functional differences with mouse. Neuron. 2016;89:37–53. https://doi.org/10.1016/j.neuron.2015.11.013.
83. Favre-Kontula L, et al. GlialCAM, an immunoglobulin-like cell adhesion molecule is expressed in glial cells of the central nervous system. Glia. 2008;56:633–45. https://doi.org/10.1002/glia.20640.
84. Murai KK, Nguyen LN, Irie F, Yamaguchi Y, Pasquale EB. Control of hippocampal dendritic spine morphology through ephrin-A3/EphA4 signaling. Nat Neurosci. 2003;6:153–60. https://doi.org/10.1038/nn994.
85. Saint-Martin M, Goda Y. Astrocyte–synapse interactions and cell adhesion molecules. FEBS J. 2023;290:3512–26. https://doi.org/10.1111/febs.16540.
86. Takano T, et al. Chemico-genetic discovery of astrocytic control of inhibition in vivo. Nature. 2020;588:296–302. https://doi.org/10.1038/s41586-020-2926-0.
87. Baldwin KT, et al. HepaCAM controls astrocyte self-organization and coupling. Neuron. 2021;109:2427–2442.e2410. https://doi.org/10.1016/j.neuron.2021.05.025.

88. Chen J, et al. Astrocyte growth is driven by the Tre1/S1pr1 phospholipid-binding G protein-coupled receptor. Neuron. 2024;112:93–112 e110. https://doi.org/10.1016/j.neuron.2023.11.008.
89. Xie Y, et al. Astrocyte-neuron crosstalk through Hedgehog signaling mediates cortical synapse development. Cell Rep. 2022;38:110416. https://doi.org/10.1016/j.celrep.2022.110416.
90. Lanjakornsiripan D, et al. Layer-specific morphological and molecular differences in neocortical astrocytes and their dependence on neuronal layers. Nat Commun. 2018;9:1623. https://doi.org/10.1038/s41467-018-03940-3.
91. Batiuk MY, et al. Identification of region-specific astrocyte subtypes at single cell resolution. Nat Commun. 2020;11(1):1220. https://doi.org/10.1038/s41467-019-14198-8.
92. Bayraktar OA, et al. Astrocyte layers in the mammalian cerebral cortex revealed by a single-cell in situ transcriptomic map. Nat Neurosci. 2020;23(4):500–9. https://doi.org/10.1038/s41593-020-0602-1.
93. Chai H, et al. Neural circuit-specialized astrocytes: transcriptomic, proteomic, morphological, and functional evidence. Neuron. 2017;95(3):531–549.e9. https://doi.org/10.1016/j.neuron.2017.06.029.
94. Farmer WT, et al. Neurons diversify astrocytes in the adult brain through sonic hedgehog signaling. Science. 2016;351:849–54. https://doi.org/10.1126/science.aab3103.
95. Hasel P, et al. Neurons and neuronal activity control gene expression in astrocytes to regulate their development and metabolism. Nat Commun. 2017;8:15132. https://doi.org/10.1038/ncomms15132.

Neuron-Astrocyte Interactions: A Human Perspective

Taylor Pio, Emily J. Hill, Nardos Kebede, Jimena Andersen, and Steven A. Sloan

Abstract This chapter explores the intricate interactions between neurons and astrocytes within the nervous system with a particular emphasis on studies conducted in human tissue or with human cells. We specifically explore how neuron-astrocyte interactions relate to processes of cellular development, morphology, migration, synapse formation, and metabolism. These findings enrich our understanding of basic neurobiology and how disruptions in these processes are relevant to human diseases.

The study of human neuron-astrocyte interactions is made possible because of transformative in vitro advancements that have facilitated the generation and sustained culture of human neural cells. In addition, the rise of techniques like sequencing at single-cell resolution has enabled the exploration of numerous human cell atlases and their comparisons to other animal model systems. Thus, the innovations outlined in this chapter illuminate the convergence and divergence of neuron-astrocyte interactions across species. As technologies progress, continually more sophisticated in vitro systems will increasingly reflect in vivo environments and deepen our command of neuron-glial interactions in human biology.

Keywords Neuron-astrocyte Interactions · Cellular development · Human tissue modeling · In vitro systems · Comparative neurobiology

Abbreviations

2D	Two-dimensional
3D	Three-dimensional
ApoE	Apolipoprotein E

Taylor Pio and Emily J. Hill contributed equally with all other contributors.

T. Pio · E. J. Hill · N. Kebede · J. Andersen · S. A. Sloan (✉)
Department of Human Genetics, Emory University School of Medicine, Atlanta, GA, USA
e-mail: sasloan@emory.edu

E. Blanco-Suarez, I. Farhy-Tselnicker (eds.), *Astrocyte-Neuron Interactions in Health and Disease*, Advances in Neurobiology 39,
https://doi.org/10.1007/978-3-031-64839-7_4

ASD	Autism spectrum disorder
ATP	Adenosine triphosphate
BBB	Blood-brain barrier
BMP	Bone morphogenetic protein
Ca^{2+}	Calcium
CT-1	Cardiotrophin-1
CNS	Central nervous system
CNTF	Ciliary neurotrophic factor
DKK	Dickkopf WNT signaling pathway inhibitor
FABP	Fatty acid binding protein family
FACS	Fluorescence-activated cell sorting
FGF	Fibroblast growth factor
FMRP	Fragile X messenger ribonucleoprotein
GABA	Gamma-aminobutyric acid
gIPC	Glial intermediate progenitor cell
HI	Hypoxia-ischemia
hiPSCs	Human-induced pluripotent stem cells
IEM	Inborn errors of metabolism
IL	Interleukin
JAK-STAT	Janus kinase/signal transducers and activators of transcription
LIF	Leukemia inhibitory factor
mGluR5	Metabotropic glutamate receptor 5
mTORC1	Mechanistic target of rapamycin complex 1
NDD	Neurodevelopmental disorders
NFIA	Nuclear factor IA
NGN	Neurogenins
nIPCs	Neuronally restricted intermediate progenitors
NL1	Neuroligin
NLGN	Neuroligin
NPC	Neural progenitor cell
OCD	Obsessive-compulsive disorder
OPC	Oligodendrocyte progenitor cell
oRG	Outer radial glia
PAPs	Peripheral astrocyte processes
PGC-1α	Peroxisome proliferator-activated receptor gamma co-activator 1α
RA	Retinoic acid
ROS	Reactive oxygen species
SCZ	Schizophrenia
SREBP2	Sterol regulatory element-binding protein 2
SVZ	Subventricular zone
TCA	Tricarboxylic acid
TGF	Transforming growth factor
TSLP	Thymic stromal lymphopoietin
VGluT1	Vesicular glutamate transporter 1
VZ	Ventricular zone
ZIKV	Zika virus

1 Introduction

Rodent and primate models have long been used to understand how the human brain develops. Indeed, these studies have provided a wealth of data that informed our understanding of key developmental timelines, as well as relevant intrinsic and extrinsic regulatory factors. However, recent advances, driven in part by the advent of single-cell technologies, have highlighted possible human-specific signatures and interactions that since prompted novel human in vitro models. These platforms, and the ability to work directly with human cells, have highlighted how interactions between neurons and astrocytes provide cues that guide cell fate decisions, morphology, synaptic biology, and metabolic coupling.

It was key advances in the isolation, culture, and coculture of neuronal and glial cells that first made many of these findings possible. The purification of different cell types by antibody-based methods like fluorescence-activated cell sorting (FACS), or the gentler immunopanning method, provided a means for the isolation and culture of cells from human primary tissue. Furthermore, advances in two-dimensional (2D) human primary glial cell culture conditions, such as serum-free medium, allowed cells to be maintained in more physiological and non-reactive states [1–3]. In addition, the use of human-induced pluripotent stem cells (hiPSCs) provides options to generate a myriad of cell types that can be cultured either alone or cocultured with others to study specific interactions [4–9]. For example, tripartite cultures of human neurons, astrocytes, and microglia have been recently developed and used to study non-cell autonomous signaling contributions in microglial development and in Alzheimer's disease [10, 11]. These cultures can be relatively simple, where three cell types are mixed and cultured in 2D [10], or more complex, in which cells are cultured separately using a 3D microfluidic platform [11] allowing media exchange between the three cell types.

Finally, 3D cell cultures of the human brain, also known as human brain spheroids or organoids, have more recently been used to study neuron-glial interactions in the developing brain. These 3D cultures are generated from human iPSCs that are allowed to differentiate into the neuroepithelial lineage in an undirected or directed approach, using small molecules and growth factors to pattern the organoids [12–16]. Both protocols recapitulate the cell diversity and neuron-astrocyte interactions of the developing human brain. Region-specific organoid models of the dorsal [14, 17–19] and ventral forebrain [18–20], cerebellum [21–24], midbrain [25–27], spinal cord [28–30], hippocampus [31, 32], hypothalamus [33, 34], thalamus [20, 35, 36], striatum [37, 38], and brain stem [39] have been validated (Table 1). These organoids can also be combined into "assembloids" to model neural circuit development and interregional interactions. Because these organoids follow human developmental trajectories, they provide an important window into human fetal neural development and how it may be perturbed in neurodevelopmental disorders.

Table 1 2D and 3D human stem cell models

Model	Citations	Experimental applications	Contains astrocytes?	Disorder studied
2D (monolayer)				
Astrocyte 2D cultures	Krencik and Zhang [9]	IHC	Yes	
	Tcw et al. [5]	FACS, Bulk RNAseq	Yes	
	Li et al. [8]	IHC, RNAseq,	Yes	
	Canals et al. [7]	IHC, glutamate uptake, calcium imaging, mouse brain engraftment	Yes	
	Tchieu et al. [4]	IHC, FACS, Western, electrophysiology, bulk RNAseq, bulk ATACseq	Yes	
	Barbar et al. [6]	IHC, FACS, scRNAseq, electrophysiology, synaptosome engulfment	Yes	
3D (organoid)				
Dorsal forebrain	Sloan et al. [14]	Bulk RNAseq, scRNAseq	Yes	
	Sloan et al. [17]	IHC, calcium and cell migration live imaging	Yes	
	Velasco et al. [18]	IHC, scRNAseq	Yes	
Ventral forebrain	Birey et al. [19]	IHC, scRNAseq, electrophysiology, live imaging	Yes	Timothy syndrome
	Xiang et al. [20]	IHC, bulk RNAseq, bulk ATACseq, scRNAseq, calcium imaging, electrophysiology	Yes	
	Velasco et al. [18]	IHC, scRNAseq	Yes	
Cerebellum	Muguruma et al. [23]	FACS, IHC, electrophysiology, qPCR	Not investigated	
	Bràs et al. [21]	IHC, calcium imaging, qPCR	Yes	Machado-Joseph disease
	Atamian et al. [24]	IHC, scRNAseq,	Yes	

(continued)

Table 1 (continued)

Model	Citations	Experimental applications	Contains astrocytes?	Disorder studied
Midbrain	Qian et al. [33]	IHC, bulk RNAseq, electrophysiology	Yes	ZIKA virus
	Smits et al. [27]	IHC, bulk RNAseq, electrophysiology	Not investigated	Parkinson's disease
	Fiorenzano et al. [26]	IHC, scRNAseq, electrophysiology, 3D bioengineering	Yes	
	Sabate-Soler et al. [25]	Microglial engraftment, phagocytosis assay, IHC, Western blot, cytokine release assay, mass spectrometry, scRNAseq, electrophysiology	Yes	
Spinal cord	Hor et al. [30]	IHC, qPCR, RNAi, Western blot	Yes	Spinal muscular atrophy
	Andersen et al. [28]	scRNAseq, optogenetics, IHC, electrophysiology, assembloid cultures	Yes	
Hippocampus	Sakaguchi et al. [31]	IHC, electrophysiology, calcium imaging	Yes	
	Pomeshchik et al. [32]	IHC, Western blot, electrophysiology, mass spectrometry	Yes	Alzheimer's disease
Hypothalamus	Qian et al. [33]	IHC, bulk RNAseq, electrophysiology	Yes	ZIKA virus
	Sarrafha et al. [34]	IHC, FACS, Western blot, scRNAseq, MEA	Yes	
Thalamus	Ozone et al. [35]	IHC, qPCR, electron microscopy, hormone secretion assays	Not investigated	
	Xiang et al. [20]	IHC, scRNAseq, electrophysiology, calcium imaging	Yes	
	Kiral et al. [36]	scRNAseq, electrophysiology, calcium imaging	Yes	
Striatum	Miura et al. [37]	SNP array, IHC, scRNAseq, viral labeling, optogenetics, electrophysiology	Yes	
	Chen et al. [38]	IHC, assembloids, scRNAseq, EdU labeling, Sholl analysis	Yes	
Brain stem	Eura et al. [39]	IHC, bulk RNAseq, scRNAseq, mass spectrometry	Possible, further validation needed	

Selection of protocols for both 2D and 3D differentiation of human neural stem cells. Emphasis placed on studies in which both neurons and astrocytes are present and the platform is amenable for investigating neuron-glial interactions

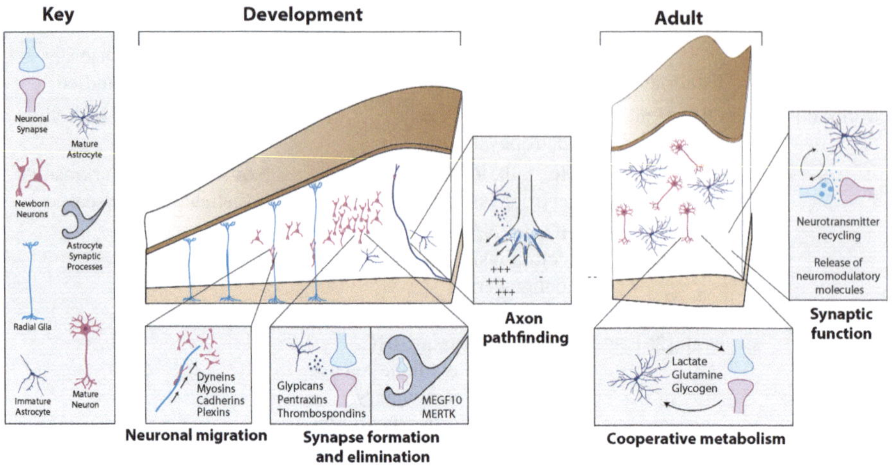

Fig. 1 Astrocyte-neuron interactions in the developing and adult human nervous system. Astrocyte and neuron interactions are critical for multiple developmental processes including neuronal migration, synapse formation and elimination, and axon pathfinding. In the developed nervous system, astrocytes also play active roles in regulating metabolic and synaptic function

In the following sections, we will illustrate discoveries in human neuron-astrocyte interactions, many of which use these modeling techniques to investigate influences on development, morphology, migration, synapse formation, metabolism, and how each of these interactions are disrupted in disease settings (Fig. 1).

2 Formation of Neurons and Astrocytes

The earliest neuron-glia interactions occur during development as neurons and astrocytes begin to emerge from neural stem cells. Both neurons and astrocytes are derived from neuroectodermal cells lining the ventricular zone (VZ) of the neural tube. These neural stem cells form multipotent progenitors called radial glia with astroglial-like features and apical-basal polarity [40]. Early in development, radial glia differentiate exclusively into neurons through symmetric or asymmetric divisions [41]. Radial glia also give rise to neuronally restricted intermediate progenitors (nIPCs) that divide symmetrically in the subventricular zone (SVZ) to produce neurons [42]. As neurons begin to populate the developing cortex and form synaptic connections, radial glia gradually transition to producing astrocytes and oligodendrocytes, a process termed the gliogenic switch. The timing of the gliogenic switch is critical in establishing proper ratios of neuron and astrocyte populations in the developing brain [43]. Furthermore, astrogenesis and the onset of synapse formation are tightly linked as astrocytes regulate mechanisms of synapse formation and function [44].

Radial glia integrate intrinsic and extrinsic cues, which together induce gene expression changes to instruct either neuronal or astrocytic fates. During the neurogenic period, radial glia either maintain their proliferative state through exposure to high levels of NOTCH [45] and fibroblast growth factor (FGF) [46] signaling or become neurogenic through, among other extrinsic drivers, bone morphogenetic protein (BMP) [47, 48], WNT [49], and retinoic acid (RA) [50] signaling. Importantly, these pathways act in time- and context-dependent ways to instruct or inhibit neurogenesis. Intrinsic neurogenic drivers that act downstream of NOTCH signaling have also been defined including the transcription factors neurogenin (NGN) 1, NGN2, and ASCL1 [51]. Astrogenesis is similarly guided by a combination of transcription factors and extrinsic signals. Extrinsic signals were initially identified through mouse studies that observed radial glia differentiating into neurons when cultured on embryonic cortical slices versus astrocytes on postnatal slices [52]. These extrinsic molecules, including ciliary neurotrophic factor (CNTF), leukemia inhibitory factor (LIF), and cardiotrophin-1(CT-1), are part of the interleukin (IL)-6 cytokine family and converge on Janus kinase/signal transducers and activators of transcription (JAK/STAT) signaling [52]. Neurons, themselves, are also critical for instructing the timing of the gliogenic switch. They directly secrete ligands that can induce an astrogenic fate, including molecules like CT-1 [52]. Additional neuronally secreted drivers of astrogenesis have been identified in human studies, including transforming growth factor β2 (TGFβ2), neuroligin 1 (NLGN1), thymic stromal lymphopoietin (TSLP), DKK1, and BMP4, which synergistically act via the mTORC1 pathway to drive astrocyte development [53]. Intrinsic transcription factors regulating astrogenesis have also been characterized including nuclear factor IA (NFIA) [54], SOX9 [55], and NGN3 [56]. While these seminal findings were mainly performed in mouse models, they are likely to hold true in human brain development as well. For example, human iPSC-derived astrocytes can be induced via NFIA and/or SOX9 activation in vitro [4, 8, 57–59].

While many astrogenic signaling cues are conserved between species, there are several intriguing differences in the primate brain that support the idea of human-specific features of astrocyte development. In general, human brain development is prolonged compared to rodents and even other primate species [60]. Additionally, primate-enriched progenitor populations have been identified in recent years and are thought to contribute to the increased cortical size seen in primate brains. For example, outer radial glia (oRG) populate an expanded outer layer found in the primate brain called the outer subventricular zone [61]. These oRGs have been shown to serve as an additional significant source of neurons [61], astrocytes [62], and oligodendrocytes [63]. Another purportedly human-enriched progenitor is a glial intermediate progenitor cell (gIPC) observed at the intersection of astrocyte and oligodendrocyte lineages in human transcriptomic datasets of the developing brain and spinal cord [64–66]. This cell type may also overlap with recently described pre-oligodendrocyte precursor cell (OPC) [63]. In general, these uniquely primate-enriched progenitor populations in the glial lineage [65] suggest that novel extrinsic and intrinsic cues may dictate the timing and efficacy of gliogenesis.

3 Morphology

3.1 Classically Described Morphology

Both astrocytes and neurons display heterogeneous morphologies that can be influenced not only by their region of origin and species, but also by their local environment, indicating that intercellular interactions are key determinants of morphology. Generally, neurons use branched, specialized processes called dendrites to receive or transduce input and extend axonal processes within and across regions of the nervous system to form chemical synapses with other neurons. Astrocytes form a number of characteristic, specialized features including peripheral astrocyte processes (PAPs) that preferentially associate with synapses in an activity-dependent manner [67], branches, branchlets, leaflets [68, 69], and endfeet that contact synapses and blood vessels [68]. Astrocytes and their processes also occupy distinct physical spaces, referred to as tiling [70, 71]. In concert, these specialized morphological traits allow astrocytes to dynamically interact with and alter other cell types of the central nervous system as well as the surrounding vasculature.

Multiple interactions between astrocytes and neurons influence morphology and morphology-dependent function. Cocultures of hiPSC-derived astrocytes and neurons increase the complexity of neuronal dendritic arborization compared to cultures without glia [72]. In turn, the production and secretion of Notch ligands by neurons can impact the maturation of astrocytes, which express cell-surface Notch receptors. This signaling is also sufficient to induce a more mature astrocyte transcriptomic identity [73]. Neuronal cues also shape astrocytic development through contact mediated by cell adhesion molecules like neuroligin 1 (NL1), NL2, and NL3 [74]. It is important to note that astrocyte morphology can also impact neuronal function. Genetically encoded indicators have shown that depolarization of astrocyte PAPs specifically inhibits astrocytic glutamate transport, thereby augmenting neuronal signaling [75]. Subcompartment-specific proteomics in astrocytes has also revealed functional specialization across cell domains. The obsessive-compulsive disorder (OCD)-associated protein SAPAP3, encoded by the gene *DLGAP3*, is concentrated in distinct astrocyte subcompartments where it modulates the organization of the actin cytoskeleton and astrocyte morphology and in turn alters physical interacting domains with neurons [76].

Neuronal activity also plays a critical role in astrocyte development and morphology. Cocultures of rat neurons with human astrocytes in vitro lead to increased astrocytic arborization and expression of mature astrocyte genes, such as glutamate transporters *SLC1A2* and *SLC1A3*, and the canonical marker *ALDH1L1*. In addition, vesicular glutamate transporter 1 (VGluT1) signaling was shown to be important for astrocyte maturation [77]. During the period of astrocytic maturation, astrocytic processes have higher occupancy at VGluT1+ glutamatergic synapses, and loss of VGluT1 signaling leads to a decrease in astrocytic arborization and maturation [77]. Similarly, inhibitory neurotransmission can enhance the complexity of astrocyte processes through the $GABA_A$ receptor [78, 79], and adrenergic signaling can affect both branching and morphology of developing astrocytes [80, 81]. Recent in silico analysis of human neuron and astrocytic transcriptomes also revealed neuronal activity-dependent changes in astrocytic gene expression [73].

3.2 *Specialized Human Morphology*

Descriptors of astrocytes in model systems beg the question, "Do human astrocytes exhibit unique structure?" More than 100 years ago, Santiago Ramón y Cajal characterized the extensive and heterogeneous appearance of astrocytes in human tissue [82]. Foundational studies performed by Oberheim et al. have more recently described human astrocytes as significantly larger and more complex than rodent astrocytes [83–85]. While subsequent studies will need to more extensively characterize primate astrocyte morphology using technologies in living tissues, two of the described subtypes, varicose projection and intralaminar astrocytes, have thus far only been observed in primates [86–89]. These observations have led to speculation that cortical evolution and expansion in humans may be facilitated by a corresponding expansion in astrocytic form and function [84, 87, 90].

Humans may also exhibit specialized cellular distributions of neurons and glia [91]. In the mouse cortex, astrocytes demonstrate layer-specific morphological and molecular differences. There is conflicting evidence as to whether or not these layers are distinct from classic cortical layers as defined by excitatory neurons [92, 93]. However, there is consensus that genetic disruption of cortical neuronal layering is sufficient to alter astrocytic layering, suggesting that neurons, in part, instruct these laminar differences. To support this idea, a recent study has shown that in astrocytes, δ-catenin, originally thought to be a neuron-specific protein, controls neuron-contact-dependent interactions that regulate astrocyte morphology in a layer-specific manner [94].

Regional neuronal and astrocyte morphologies are also heterogeneous outside the cortex. In the spinal cord, cells undergo highly enforced positional patterning. In vivo rodent studies indicate that the spinal cord develops positionally distinct astrocytes that locally proliferate [95, 96]. This is also reflected in single-cell transcriptomics of the developing human spinal cord where astrocyte heterogeneity exists across the dorso-ventral axis and between white and gray matters [66]. The functions accompanying this patterning remain to be defined. Some studies indicate that glial specialization may in part explain the diverse morphology of neurons across the nervous system. For example, mouse dopaminergic neurons cultured with glial cells derived from different central nervous system (CNS) regions exhibit differential morphologies. Neurons plated with mesencephalic glial cells developed highly branched processes, while those plated with striatal glia produced cells with singular, linear neurites [97].

4 Migration

The complex organization of the nervous system arises in part due to cellular migration across neuroanatomical regions during development. Migrating cells interact with their environment dynamically and bi-directionally [98]. Notably, during cortical development, radial glia emerge as pivotal players, functioning as precursors for both neurons and other glial cells. Radial glia extend processes to the surface of the pia, serving as essential scaffolds intimately interacting with neural progenitors and migrating neurons. In rodent models, radial glia-derived B1 integrin signals [99] and basement membrane attachment [100] are essential for accurate neuronal cortical layer formation. Comparative studies suggest that migration of human neural progenitors may be protracted [101]. In addition, astrocytes have been shown to influence neuronal axon pathfinding in a multitude of model systems, guiding circuit formation after neuronal positioning [102]. Adult astrocyte migration is typically studied in the context of disease and injury, and in humans, this migration has been shown to be a result of G protein-coupled receptor activation in astrocytes [103]. Future studies will define the role that cellular migration may plate in region-specific position and morphology.

5 At the Synapse

The neuron-to-neuron model of synaptic communication has, in recent years, been amended to include astrocytes as essential members and is now colloquially coined the 'tri-partite' synapse. There are fantastic reviews concerning the multitude of astrocytic functions at the synapse [44, 104, 105]. The functional contributions of astrocytes, which will be briefly reviewed here, include three major categories: synaptic formation, elimination, and function. Though the bulk of this knowledge is rodent-derived, this section will highlight human studies wherever possible.

5.1 Synaptogenesis

The timeline of astrocyte generation in development overlaps with synaptogenesis in both humans and rodents [106–108]; this is likely no coincidence. Cocultures of rodent neurons and astrocytes provided the first evidence that astrocytes provide synaptogenic cues [109–111]. Since then, glia have been implicated in synapse formation across many studies and model systems [112, 113].

Further studies in rodent models have identified specific factors released by astrocytes and their corresponding neuronal receptors that are necessary to promote synapse formation. Astrocyte-derived signals can act as both positive and negative synaptic regulators [114], exerting precise control over the number and even the type of synapse being formed. Among the first identified cues was the family of extracellular matrix protein thrombospondins, which have been shown to increase glutamatergic synapse formation [115–117]. Additional molecules, including cholesterol [118], glypicans [113, 119], neuroligins [74], and chordins [108, 120], have also been shown to play essential and active roles in synaptic development. These molecules are abundant in human astrocytes as well, suggesting they likely play a similar role [121].

Some of these synaptogenic mechanisms have been experimentally validated in human studies using primary tissue and stem cell-derived models. Human astrocytes have been shown to strongly promote neuron survival and provide synaptogenic cues, as originally identified in rodents [2, 14]. Human-derived astrocytes have also been shown to increase the speed [122], number [111], and maturity [72] of human synapses formed in vitro. In a human iPSC model, astrocytes were also shown to be necessary for spontaneous network activity and synaptic maturation [72]. It remains an important and open question whether there are additional human-specific astrocyte factors that act on synaptic formation and/or specification.

5.2 Synapse Elimination

Astrocytes, microglia, and oligodendrocyte precursors assist in the extensive pruning of excess synapses throughout normal development and into adulthood, and this has been primarily demonstrated in both rodent and drosophila models [123–127]. Astrocytes engulf both excitatory and inhibitory synapses [14, 123], and synaptic elimination has been shown to be activity-dependent [123]. Complement cascade proteins C1q and C3 are important in tagging specific synapses for elimination [128]. Moreover, astrocyte-produced cytokines like IL-33 promote microglial synaptic engulfment [129]. While most functional synaptic elimination studies have been performed in animal models, primary human astrocytes and iPSC-derived human astrocytes are also capable of engulfing isolated synaptosomes [2, 14]. Synaptic elimination is a normal process in healthy cells [14, 123]; however, excessive elimination has also been shown to be a feature of reactive astrocytes in adult disease [130, 131], as well as early-life stress [132].

5.3 Synaptic Function

During and after synaptic maturation, astrocytes play key roles in synaptic maintenance. In drosophila, neuron-astrocyte interactions are required for developmental neural activity [133, 134]. Positional astrocyte identity in the rodent spinal cord is necessary for aspects of normal motor circuit development [135]. In human iPSC-derived cells, astrocytes enhance spontaneous network activity [72]. However, it has been proposed that human astrocytes exhibit distinct neurotransmitter receptor signatures compared to their rodent counterparts and thus respond uniquely to developmental glutamatergic and purinergic receptor agonists [2, 84].

One of the major synaptic functions of astrocytes is neurotransmitter uptake, which confines signaling, primarily excitatory, to active synapses. Astrocytes accomplish this via glutamate transporters, predominantly GLAST and GLT-1, also known as EAAT1 and EAAT2, respectively. Knockdown studies in rodents have indicated that the majority of glutamate transport occurs via these two transporters, as their removal results in increased glutamate levels and excitotoxicity [136]. Interestingly, studies in rodent astrocytes show that astrocytic expression of glutamate transporters is heterogeneous across regions of the CNS. For example, GLT-1 expression in the rodent brain is ten-fold higher than that of the spinal cord [137].

In addition to controlling the spread of neurotransmitter signals at synapses, astrocytes uptake neurotransmitters to undergo a host of conversion and recycling efforts; experiments in hiPSC and primary human cells confirm robust transporter function in human astrocytes [2, 14]. Astrocytes are considered the primary source of glutamine, the precursor for both glutamate and gamma-aminobutyric acid (GABA), in the nervous system [138, 139]. Astrocytes are also essential players in neuromodulation. They have been experimentally shown to release transmitters that modulate synaptic transmission, molecules colloquially termed "gliotransmitters" [140–142]. The identity of these molecules includes D-serine, adenosine triphosphate (ATP), and other neuroactive signaling factors that may modulate synaptic activity. The precise mechanisms by which gliotransmitters are released from an astrocyte require further investigation. Proposed pathways include vesicular release, transporter reversal, or gliotransmitter-specific shuttles. A recent study using single-cell transcriptomics in the rodent brain identified a small subpopulation of astrocytes that are specialized to express glutamate exocytic machinery [143].

6 Metabolic Coupling

Glucose is the primary energetic substrate for the adult human brain. Its metabolism is necessary for both energy production and also critical synthetic steps involved in neurotransmitter production and antioxidant generation in both neurons and astrocytes [144]. However, astrocytes and neurons exhibit quite distinct, and complimentary, metabolic profiles. While astrocytes present a more pronounced glycolytic existence, neurons preferentially rely on oxidative metabolism via mitochondrial oxidative phosphorylation [145]. Even glucose uptake and utilization are a cooperative interaction between neurons and astrocytes. Present on astrocytic endfeet are GLUT1 glucose transporters, by which glucose enters the brain from the bloodstream [146, 147]. Since astrocytic endfeet is a core component of the blood-brain barrier (BBB) and ensheathe over 99% of capillaries [148], astrocytes are the primary gatekeepers for circulating glucose to enter the nervous system. Importantly, the GLUT1 transporter is activated by astrocytic uptake of glutamate, making neuronal activity and subsequent neuron-astrocytic communication essential for glucose and other nutrients to pass through the BBB to neurons [146–148]. One consequence of neuronal reliance on oxidative phosphorylation for energy needs is the dependence on lactate as an energetic substrate. It has been proposed and generally accepted that astrocytes (and oligodendrocytes) shuttle lactate to neurons, which can then be metabolized into pyruvate to run through the tricarboxylic acid (TCA) cycle and subsequent oxidative phosphorylation for ATP generation [149–152].

An important debate that arises often is how the evolutionary expansion of the human brain is tethered to a nonlinear increase in energy demand [153]. This is because the brain has a disproportionate energy demand relative to its size. It contributes only 2% of the body weight but requires 20% of the blood flow, from which it extracts about 50% of available oxygen and 20% of available glucose [154]. One potential consequence of this increased energy demand is that the cells within the human brain (neurons and astrocytes) have been forced to adapt in new ways to handle energetic substrates. Studies using in vitro human astrocyte cultures and human cortical slice cultures suggest some potential differences between mouse and human astrocyte metabolism. For example, lower intracellular concentrations of TCA intermediates suggest that human astrocytes may exhibit diminished TCA flux and oxidative phosphorylation [155, 156]. However, metabolic studies in human cells have been difficult to execute and relatively sparse because in vitro culture conditions rarely recapitulate the precise metabolic dynamics of the in vivo environment. Future approaches, including organoids, human xenograft models, and organ-on-a-chip type culture systems may be required to achieve accurate metabolic profiling of human neurons and astrocytes.

6.1 Astrocyte Metabolism in Synaptogenesis

Astrocytes are the primary source of glycogen in the brain and also actively synthesize, store, and release various lipids including sterols, fatty acids, triglycerides, phospholipids, and sphingolipids [157]. These lipid moieties are particularly critical for modulating synaptogenesis. Exposure of rat retinal ganglion cell cultures to astrocyte-derived cholesterol, transported by apolipoprotein E (APOE), is sufficient to form functional synapses, increasing both the number and activity of synapses [118]. However, cholesterol levels in the brain act in a Goldilocks scenario—either too little or too much cholesterol can lead to dysfunction. A decrease in astrocytic levels of the transcription factor sterol regulatory element-binding protein 2 (SREBP2, encoded by the gene *SREBF2*) results in a decrease in overall secreted brain cholesterol and has also been associated with diabetes onset [158]. Human astrocytes have been shown to produce high levels of both APOE and SREBP2, and this phenomenon is likely recapitulated in human development [2]. Astrocytes also actively uptake lipids secreted by neurons, where they are metabolized to prevent fatty acid toxicity [159].

6.2 Mitochondrial Biogenesis

Mitochondrial biogenesis and metabolism in astrocytes are also important for neuronal-glial metabolic coupling in the developing brain [160]. Impaired mitochondrial biogenesis in developing astrocytes leads to both defects in astrocytic arborization and a decrease in both the number and activity of nearby synapses [77, 161]. Specifically, peroxisome proliferator-activated receptor gamma co-activator 1α (PGC-1α) upregulation, which is regulated by metabotropic glutamate receptor 5 (mGluR5), is required for astrocyte mitochondrial biogenesis [161, 162]. mGluR5 is highly expressed early in astrocytic development and increases PGC-1α expression, which leads to a transient period of higher mitochondrial cytosolic occupancy [161]. Astrocytic mitochondrial function remains critical for brain homeostasis after development as well. Disruption in mitochondrial function, leading to deficiencies in astrocytic metabolism, has been implicated in neurodegenerative disorders like Parkinson's and Alzheimer's diseases. Importantly, astrocytic mitochondria are responsible for the detoxification of reactive oxygen species (ROSs) and fatty acids taken up from the synapse, and loss can lead to neuroinflammation [163]. It has even been recently suggested that astrocytes may play neuroprotective roles in stroke by extracellularly transferring functional mitochondria into neurons via a calcium-dependent mechanism involving CD38 signaling [164].

7 Neurodevelopmental Disorders

The human brain is complex and susceptible to a myriad of developmental aberrations that alter neural network formation. Here, we highlight several mechanisms that can lead to dysfunction in neuron-astrocyte interactions and discuss how neurodevelopmental disorders can be studied using human model systems.

7.1 Dysfunction in Timing

The gliogenic switch requires precise timing to ensure that adequate amounts of both neurons and glia are produced at the right time and place. Early or delayed transition from neuron-to-glial differentiation can lead to reduction or enlargement in the size of the developing brain [165]. Recent studies conducted in cynomolgus monkeys showed that a loss of the autism-linked gene CDH8 that is present in neural progenitors leads to increased numbers of glial cells and, ultimately, macrocephaly in the primate brain [166]. Disrupted neuron to glial ratios have also been observed in human samples derived from patients with Fragile X syndrome [167]. Defects in developmental timing can also result in fetal microcephaly. Zika virus (ZIKV), a mosquito-transmitted arbovirus, can be passed to the developing fetus through maternal transmission across the placenta and has been studied in many non-human primate models [16, 168–171], hiPSC-derived 2D cultures of neuronal lineage cells [172–174], and human brain organoids [174–180]. In human brain organoid models, ZIKV has been shown to preferentially infect neural progenitor cells (NPCs), leading to either apoptosis or premature NPC differentiation [33, 177, 181, 182].

7.2 Dysfunction in Synapse Formation/Morphology

Disruption of synapse formation or maintenance can lead or contribute to the pathogenesis of a wide range of neurodevelopmental disorders (NDD). Some NDDs, such as autism spectrum disorder (ASD) and schizophrenia (SCZ), are hypothesized to reflect an imbalance of excitatory and inhibitory synapses in the brain [183–185]. This imbalance could result from disrupted synaptic formation, elimination, morphology, or function, all of which are influenced by neuron-astrocyte interactions. hiPSCs derived from ASD patients and differentiated into neuronal cultures were shown to have impaired synaptogenesis when cocultured with ASD patient-derived astrocytes, but not with control astrocytes [186]. ASD astrocytes have also been shown to secrete IL-6, which decreases the number of synaptic puncta in neuronal cultures—a phenomenon that can be rescued by abrogating IL-6 secretion and restoring synaptogenesis to control levels [186].

Studies in human cortical organoids derived from ASD individuals have shown that ASD astrocytes exhibit increased Ca^{2+} signaling in response to ATP [187]. Ca^{2+} release in astrocytes is promoted by ATP, glutamate, and GABA uptake and is upstream of the release of neuromodulators such as glutamate, ATP, D-serine, and GABA that can modulate postsynaptic activity [188, 189]. Transplantation of human ASD astrocytes into the mouse brain was also reported to evoke ASD-like behavioral symptoms, such as repetitive behaviors and impaired memory [187].

Microtubule dynamics are critical for vesicular trafficking for both neurotransmitter release and signaling between astrocytes and neurons. In one study by Delépine and colleagues, loss of MECP2 in Rett syndrome human astrocytes resulted in impaired microtubule-dependent vesicular trafficking and subsequent neuronal and behavioral phenotypes [190]. In Fragile X, the FMRP protein both binds mRNAs and traffics them via microtubule binding for local transcription at the synapse [191]. However, FMRP is not a neuronal-specific protein and is found at abundant levels in astrocytes. Astrocyte-specific FMRP loss in mice leads to dysregulated synaptogenic, glutamate transporter and receptor levels and behavioral phenotypes in vivo [192–195]. In human cultures, phenotypes arising from loss of FMRP in neurons can be rescued by coculture with conditioned media derived from healthy human astrocytes [196].

7.3 Dysfunction in Metabolism

Metabolic disruptions during brain development can arise from either intrinsic causes or extrinsic causes. Mutations in protein-coding genes involved in key biochemical pathways, referred to as inborn errors of metabolism (IEM), can result in astrocyte metabolic dysfunction and a consequential loss of homeostatic neuroprotection [197]. Mutations in mitochondrial enzymes and oxidative phosphorylation lead to a variety of symptoms including seizures, movement disorders, and developmental delays [198, 199]. Lipid metabolism dysfunction via the fatty acid binding protein family (*FABP*) of genes has also been linked to SCZ and ASD. Frameshift mutations in *FABP3 and FABP7*, which are expressed by astrocytes and oligodendrocytes, were found in a patient cohort of individuals with schizophrenia and bipolar disorder [200]. Elevated levels of *FABP5* mRNA were found in individuals with ASD, and altered fatty acid binding by mutated FABP7 protein has also been observed in SCZ and ASD [200]. In human iPSCs derived from patients with Fragile X, cholesterol metabolism and homeostasis are severely disrupted, suggesting that astrocytic lipid metabolism is critical for maintaining baseline neuronal activity [201].

Neonatal hypoxia-ischemia (HI), or a loss of oxygen to the brain during development, can also cause defects in neuron-astrocyte metabolism. HI leads to defects in astrocyte glutamate uptake from the synapse, leading to subsequent neuronal excitotoxicity [202, 203]. In the presence of HI, astrocytes also contain lower concentrations of TCA cycle intermediates and glutamine, leading to a decrease in extracellular metabolite concentrations and neuronal uptake. Another common disorder characterized by metabolic dysfunction is hepatic encephalopathy, where liver dysfunction leads to toxic accumulation of cerebral ammonia. While many mechanisms have been proposed as to the underlying cause of neurological dysfunction in the presence of toxic ammonia levels, work validated in human models suggests that high ammonia leads to inhibited anaplerotic TCA reactions in astrocytes and subsequent neuronal dysregulation [204].

8 Concluding Remarks

At all levels—development, normal physiological conditions, and disease—interactions between neurons and astrocytes are fundamental. Recently, our understanding of these interactions in the human brain has undergone a transformative shift, propelled by advancements in in vitro modeling techniques. The ability to generate, isolate, and (co-)culture human neurons and astrocytes, marked by species-specific characteristics, has deepened our understanding of how these two cell types interact in terms of development, morphology, migration, synaptic behavior, and metabolism. As technologies become more sophisticated and human in vitro systems inch ever closer to in vivo environments, we will better appreciate the convergence and divergence of neuron-astrocyte interactions across species [205–207].

References

1. Foo LC, et al. Development of a method for the purification and culture of rodent astrocytes. Neuron. 2011;71:799–811.
2. Zhang Y, et al. Purification and characterization of progenitor and mature human astrocytes reveals transcriptional and functional differences with mouse. Neuron. 2016;89:37–53.
3. Prah J, et al. A novel serum free primary astrocyte culture method that mimic quiescent astrocyte phenotype. J Neurosci Methods. 2019;320:50–63.
4. Tchieu J, et al. NFIA is a gliogenic switch enabling rapid derivation of functional human astrocytes from pluripotent stem cells. Nat Biotechnol. 2019;37:267–75.
5. Tcw J, et al. An efficient platform for astrocyte differentiation from human induced pluripotent stem cells. Stem Cell Rep. 2017;9:600–14.
6. Barbar L, et al. CD49f is a novel marker of functional and reactive human iPSC-derived astrocytes. Neuron. 2020;107:436–453.e12.
7. Canals I, et al. Rapid and efficient induction of functional astrocytes from human pluripotent stem cells. Nat Methods. 2018;15:693–6.
8. Li X, et al. Fast generation of functional subtype astrocytes from human pluripotent stem cells. Stem Cell Rep. 2018;11:998–1008.
9. Krencik R, Zhang S-C. Directed differentiation of functional astroglial subtypes from human pluripotent stem cells. Nat Protoc. 2011;6:1710–7.
10. Guttikonda SR, et al. Fully defined human pluripotent stem cell-derived microglia and triculture system model C3 production in Alzheimer's disease. Nat Neurosci. 2021;24:343–54. https://doi.org/10.1038/s41593-020-00796-z.
11. Park J, et al. A 3D human triculture system modeling neurodegeneration and neuroinflammation in Alzheimer's disease. Nat Neurosci. 2018;21:941–51.
12. Xia X, Zhang S-C. Differentiation of neuroepithelia from human embryonic stem cells. In: Gordon D, Scolding NJ, editors. Neural cell transplantation: methods and protocols. Totowa: Humana Press; 2009. p. 51–8. https://doi.org/10.1007/978-1-60327-931-4_4.
13. Lancaster MA, et al. Cerebral organoids model human brain development and microcephaly. Nature. 2013;501:373–9.
14. Sloan SA, et al. Human astrocyte maturation captured in 3D cerebral cortical spheroids derived from pluripotent stem cells. Neuron. 2017;95:779–790.e6.
15. Renner M, et al. Self-organized developmental patterning and differentiation in cerebral organoids. EMBO J. 2017;36:1316–29.
16. Steinbach RJ, et al. A neonatal nonhuman primate model of gestational Zika virus infection with evidence of microencephaly, seizures and cardiomyopathy. PLoS One. 2020;15:e0227676.
17. Sloan SA, Andersen J, Paşca AM, Birey F, Paşca SP. Generation and assembly of human brain region–specific three-dimensional cultures. Nat Protoc. 2018;13:2062–85.
18. Velasco S, et al. Individual brain organoids reproducibly form cell diversity of the human cerebral cortex. Nature. 2019;570:523–7.
19. Birey F, et al. Assembly of functionally integrated human forebrain spheroids. Nature. 2017;545:54–9.
20. Xiang Y, et al. Fusion of regionally-specified hPSC-derived organoids models human brain development and interneuron migration. Cell Stem Cell. 2017;21:383–398.e7.
21. Brás J, et al. Establishment and characterization of human pluripotent stem cells-derived brain organoids to model cerebellar diseases. Sci Rep. 2022;12:12513.

22. Chen Y, et al. Generation of advanced cerebellar organoids for neurogenesis and neuronal network development. Hum Mol Genet. 2023;32:2832–41.
23. Muguruma K, Nishiyama A, Kawakami H, Hashimoto K, Sasai Y. Self-organization of polarized cerebellar tissue in 3D culture of human pluripotent stem cells. Cell Rep. 2015;10:537–50.
24. Atamian A, et al. Human cerebellar organoids with functional Purkinje cells. Cell Stem Cell. 2024;31:39–51.e6.
25. Sabate-Soler S, et al. Microglia integration into human midbrain organoids leads to increased neuronal maturation and functionality. Glia. 2022;70:1267–88.
26. Fiorenzano A, et al. Single-cell transcriptomics captures features of human midbrain development and dopamine neuron diversity in brain organoids. Nat Commun. 2021;12:7302.
27. Smits LM, et al. Modeling Parkinson's disease in midbrain-like organoids. NPJ Park Dis. 2019;5:1–8.
28. Andersen J, et al. Generation of functional human 3D cortico-motor assembloids. Cell. 2020;183:1913–1929.e26.
29. Lee J-H, et al. Production of human spinal-cord organoids recapitulating neural-tube morphogenesis. Nat Biomed Eng. 2022;6:435–48.
30. Hor JH, et al. Cell cycle inhibitors protect motor neurons in an organoid model of spinal muscular atrophy. Cell Death Dis. 2018;9:1–12.
31. Sakaguchi H, et al. Generation of functional hippocampal neurons from self-organizing human embryonic stem cell-derived dorsomedial telencephalic tissue. Nat Commun. 2015;6:8896.
32. Pomeshchik Y, et al. Human iPSC-derived hippocampal spheroids: an innovative tool for stratifying Alzheimer disease patient-specific cellular phenotypes and developing therapies. Stem Cell Rep. 2020;15:256–73.
33. Qian X, et al. Brain-region-specific organoids using mini-bioreactors for modeling ZIKV exposure. Cell. 2016;165:1238–54.
34. Sarrafha L, et al. Novel human pluripotent stem cell-derived hypothalamus organoids demonstrate cellular diversity. iScience. 2023;26:107525.
35. Ozone C, et al. Functional anterior pituitary generated in self-organizing culture of human embryonic stem cells. Nat Commun. 2016;7:10351.
36. Kiral FR, et al. Generation of ventralized human thalamic organoids with thalamic reticular nucleus. Cell Stem Cell. 2023;30:677–688.e5.
37. Miura Y, et al. Generation of human striatal organoids and cortico-striatal assembloids from human pluripotent stem cells. Nat Biotechnol. 2020;38:1421–30.
38. Chen X, et al. Human striatal organoids derived from pluripotent stem cells recapitulate striatal development and compartments. PLoS Biol. 2022;20:e3001868.
39. Eura N, et al. Brainstem organoids from human pluripotent stem cells. Front Neurosci. 2020;14:538.
40. Bentivoglio M, Mazzarello P. The history of radial glia. Brain Res Bull. 1999;49:305–15.
41. Noctor SC, Martínez-Cerdeño V, Ivic L, Kriegstein AR. Cortical neurons arise in symmetric and asymmetric division zones and migrate through specific phases. Nat Neurosci. 2004;7:136–44.
42. Haubensak W, Attardo A, Denk W, Huttner WB. Neurons arise in the basal neuroepithelium of the early mammalian telencephalon: a major site of neurogenesis. Proc Natl Acad Sci. 2004;101:3196–201.
43. Morrow T, Song MR, Ghosh A. Sequential specification of neurons and glia by developmentally regulated extracellular factors. Development (Cambridge, England). 2001;128:3585–94.
44. Chung W-S, Allen NJ, Eroglu C. Astrocytes control synapse formation, function, and elimination. Cold Spring Harb Perspect Biol. 2015;7:a020370.
45. Del Bene F, Wehman AM, Link BA, Baier H. Regulation of neurogenesis by interkinetic nuclear migration through an apical-basal notch gradient. Cell. 2008;134:1055–65.
46. Yoon K, et al. Fibroblast growth factor receptor signaling promotes radial glial identity and interacts with Notch1 signaling in telencephalic progenitors. J Neurosci. 2004;24:9497–506.

47. Li W, Cogswell CA, LoTurco JJ. Neuronal differentiation of precursors in the neocortical ventricular zone is triggered by BMP. J Neurosci. 1998;18:8853–62.
48. Moon B-S, et al. Bone morphogenetic protein 4 stimulates neuronal differentiation of neuronal stem cells through the ERK pathway. Exp Mol Med. 2009;41:116–25.
49. Hirabayashi Y, et al. The Wnt/beta-catenin pathway directs neuronal differentiation of cortical neural precursor cells. Development (Cambridge, England). 2004;131:2791–801.
50. Haushalter C, Asselin L, Fraulob V, Dollé P, Rhinn M. Retinoic acid controls early neurogenesis in the developing mouse cerebral cortex. Dev Biol. 2017;430:129–41.
51. Bertrand N, Castro DS, Guillemot F. Proneural genes and the specification of neural cell types. Nat Rev Neurosci. 2002;3:517–30.
52. Barnabé-Heider F, et al. Evidence that embryonic neurons regulate the onset of cortical gliogenesis via cardiotrophin-1. Neuron. 2005;48:253–65.
53. Voss AJ, et al. Identification of ligand-receptor pairs that drive human astrocyte development. Nat Neurosci. 2023;26:1339–51.
54. Deneen B, et al. The transcription factor NFIA controls the onset of gliogenesis in the developing spinal cord. Neuron. 2006;52:953–68.
55. Stolt CC, et al. The Sox9 transcription factor determines glial fate choice in the developing spinal cord. Genes Dev. 2003;17:1677–89.
56. Lee J, et al. Neurogenin3 participates in gliogenesis in the developing vertebrate spinal cord. Dev Biol. 2003;253:84–98.
57. Caiazzo M, et al. Direct conversion of fibroblasts into functional astrocytes by defined transcription factors. Stem Cell Rep. 2015;4:25–36.
58. Quist E, et al. Transcription factor-based direct conversion of human fibroblasts to functional astrocytes. Stem Cell Rep. 2022;17:1620–35.
59. Leng K, et al. CRISPRi screens in human iPSC-derived astrocytes elucidate regulators of distinct inflammatory reactive states. Nat Neurosci. 2022;25:1528–42.
60. Wallace JL, Pollen AA. Human neuronal maturation comes of age: cellular mechanisms and species differences. Nat Rev Neurosci. 2024;25:7–29.
61. Hansen DV, Lui JH, Parker PRL, Kriegstein AR. Neurogenic radial glia in the outer subventricular zone of human neocortex. Nature. 2010;464:554–61.
62. Allen DE, et al. Fate mapping of neural stem cell niches reveals distinct origins of human cortical astrocytes. Science. 2022;376:1441–6.
63. Huang W, et al. Origins and proliferative states of human oligodendrocyte precursor cells. Cell. 2020;182:594–608.e11.
64. Trevino AE, et al. Chromatin and gene-regulatory dynamics of the developing human cerebral cortex at single-cell resolution. Cell. 2021;184:5053–5069.e23.
65. Ramos SI, et al. An atlas of late prenatal human neurodevelopment resolved by single-nucleus transcriptomics. Nat Commun. 2022;13:7671.
66. Andersen J, et al. Single-cell transcriptomic landscape of the developing human spinal cord. Nat Neurosci. 2023;26:902–14.
67. Lavialle M, et al. Structural plasticity of perisynaptic astrocyte processes involves ezrin and metabotropic glutamate receptors. Proc Natl Acad Sci. 2011;108:12915–9.
68. Khakh BS, Sofroniew MV. Diversity of astrocyte functions and phenotypes in neural circuits. Nat Neurosci. 2015;18:942–52.
69. Aboufares El Alaoui A, Jackson M, Fabri M, de Vivo L, Bellesi M. Characterization of subcellular organelles in cortical perisynaptic astrocytes. Front Cell Neurosci. 2021;14:573944.
70. Bushong EA, Martone ME, Jones YZ, Ellisman MH. Protoplasmic astrocytes in CA1 stratum radiatum occupy separate anatomical domains. J Neurosci. 2002;22:183–92.
71. Bushong EA, Martone ME, Ellisman MH. Maturation of astrocyte morphology and the establishment of astrocyte domains during postnatal hippocampal development. Int J Dev Neurosci. 2004;22:73–86.
72. Klapper SD, et al. Astrocyte lineage cells are essential for functional neuronal differentiation and synapse maturation in human iPSC-derived neural networks. Glia. 2019;67:1893–909.

73. Hasel P, et al. Neurons and neuronal activity control gene expression in astrocytes to regulate their development and metabolism. Nat Commun. 2017;8:15132.
74. Stogsdill JA, et al. Astrocytic neuroligins control astrocyte morphogenesis and synaptogenesis. Nature. 2017;551:192–7.
75. Armbruster M, et al. Neuronal activity drives pathway-specific depolarization of peripheral astrocyte processes. Nat Neurosci. 2022;25:607–16.
76. Soto JS, et al. Astrocyte–neuron subproteomes and obsessive–compulsive disorder mechanisms. Nature. 2023;616:764–73.
77. Morel L, Higashimori H, Tolman M, Yang Y. VGluT1+ neuronal glutamatergic signaling regulates postnatal developmental maturation of cortical protoplasmic astroglia. J Neurosci. 2014;34:10950–62.
78. Matsutani S, Yamamoto N. Neuronal regulation of astrocyte morphology in vitro is mediated by GABAergic signaling. Glia. 1997;20:1–9.
79. Cheng Y-T, et al. Inhibitory input directs astrocyte morphogenesis through glial GABABR. Nature. 2023;617:369–76.
80. Rosenberg MF, et al. β-adrenergic signaling promotes morphological maturation of astrocytes in female mice. J Neurosci. 2023;43:8621–36.
81. Sardar D, et al. Mapping astrocyte transcriptional signatures in response to neuroactive compounds. Int J Mol Sci. 2021;22:3975.
82. Garcia-Lopez P, Garcia-Marin V, Freire M. The histological slides and drawings of Cajal. Front Neuroanat. 2010;4:9.
83. Oberheim NA, Wang X, Goldman S, Nedergaard M. Astrocytic complexity distinguishes the human brain. Trends Neurosci. 2006;29:547–53.
84. Oberheim NA, et al. Uniquely hominid features of adult human astrocytes. J Neurosci. 2009;29:3276–87.
85. Oberheim NA, Goldman SA, Nedergaard M. Heterogeneity of astrocytic form and function. Methods Mol Biol (Clifton, NJ). 2012;814:23–45.
86. Falcone C, et al. Cortical interlaminar astrocytes across the therian mammal radiation. J Comp Neurol. 2019;527:1654–74.
87. Falcone C, et al. Cortical interlaminar astrocytes are generated prenatally, mature postnatally, and express unique markers in human and nonhuman primates. Cereb Cortex. 2021;31:379–95.
88. Falcone C, et al. Redefining varicose projection astrocytes in primates. Glia. 2022;70:145–54.
89. Degl'Innocenti E, Dell'Anno MT. Human and mouse cortical astrocytes: a comparative view from development to morphological and functional characterization. Front Neuroanat. 2023;17:1130729.
90. Krencik R, van Asperen JV, Ullian EM. Human astrocytes are distinct contributors to the complexity of synaptic function. Brain Res Bull. 2017;129:66–73.
91. Li Y, et al. Spatiotemporal transcriptome atlas reveals the regional specification of the developing human brain. Cell. 2023;186:5892–5909.e22.
92. Lanjakornsiripan D, et al. Layer-specific morphological and molecular differences in neocortical astrocytes and their dependence on neuronal layers. Nat Commun. 2018;9:1623.
93. Bayraktar OA, et al. Astrocyte layers in the mammalian cerebral cortex revealed by a single-cell in situ transcriptomic map. Nat Neurosci. 2020;23:500–9.
94. Tan CX, et al. δ-Catenin controls astrocyte morphogenesis via layer-specific astrocyte-neuron cadherin interactions. J Cell Biol. 2023;222:e202303138.
95. Tien A-C, et al. Regulated temporal-spatial astrocyte precursor cell proliferation involves BRAF signalling in mammalian spinal cord. Development (Cambridge, England). 2012;139:2477–87.
96. Hochstim C, Deneen B, Lukaszewicz A, Zhou Q, Anderson DJ. The spinal cord contains positionally distinct astrocyte subtypes whose identities are specified by a homeodomain transcriptional code. Cell. 2008;133:510–22.

97. Denis-Donini S, Glowinski J, Prochiantz A. Glial heterogeneity may define the three-dimensional shape of mouse mesencephalic dopaminergic neurones. Nature. 1984;307:641–3.
98. Silva CG, Peyre E, Nguyen L. Cell migration promotes dynamic cellular interactions to control cerebral cortex morphogenesis. Nat Rev Neurosci. 2019;20:318–29.
99. Belvindrah R, Graus-Porta D, Goebbels S, Nave K-A, Müller U. β1 Integrins in radial glia but not in migrating neurons are essential for the formation of cell layers in the cerebral cortex. J Neurosci. 2007;27:13854–65.
100. Haubst N, Georges-Labouesse E, De Arcangelis A, Mayer U, Götz M. Basement membrane attachment is dispensable for radial glial cell fate and for proliferation, but affects positioning of neuronal subtypes. Development. 2006;133:3245–54.
101. Marchetto MC, et al. Species-specific maturation profiles of human, chimpanzee and bonobo neural cells. Elife. 2019;8:e37527.
102. Rapti G. Regulation of axon pathfinding by astroglia across genetic model organisms. Front Cell Neurosci. 2023;17:1241957.
103. Rutkowska A, Preuss I, Gessier F, Sailer AW, Dev KK. EBI2 regulates intracellular signaling and migration in human astrocyte. Glia. 2015;63:341–51.
104. Allen NJ, Eroglu C. Cell biology of astrocyte-synapse interactions. Neuron. 2017;96:697–708.
105. Farizatto KLG, Baldwin KT. Astrocyte-synapse interactions during brain development. Curr Opin Neurobiol. 2023;80:102704.
106. Silbereis JC, Pochareddy S, Zhu Y, Li M, Sestan N. The cellular and molecular landscapes of the developing human central nervous system. Neuron. 2016;89:248–68.
107. Wang L, et al. A cross-species proteomic map reveals neoteny of human synapse development. Nature. 2023;622:112–9.
108. Farhy-Tselnicker I, et al. Activity-dependent modulation of synapse-regulating genes in astrocytes. Elife. 2021;10:e70514.
109. Pfrieger FW, Barres BA. Synaptic efficacy enhanced by glial cells in vitro. Science. 1997;277:1684–7.
110. Ullian EM, Sapperstein SK, Christopherson KS, Barres BA. Control of synapse number by glia. Science. 2001;291:657–61.
111. Hartley RS, Margulis M, Fishman PS, Lee VM-Y, Tang C-M. Functional synapses are formed between human NTera2 (NT2N, hNT) neurons grown on astrocytes. J Comp Neurol. 1999;407:1–10.
112. Lee E, Chung W-S. Glial control of synapse number in healthy and diseased brain. Front Cell Neurosci. 2019;13:42.
113. Allen NJ, et al. Astrocyte glypicans 4 and 6 promote formation of excitatory synapses via GluA1 AMPA receptors. Nature. 2012;486:410–4.
114. Kucukdereli H, et al. Control of excitatory CNS synaptogenesis by astrocyte-secreted proteins Hevin and SPARC. Proc Natl Acad Sci. 2011;108:E440–9.
115. Christopherson KS, et al. Thrombospondins are astrocyte-secreted proteins that promote CNS synaptogenesis. Cell. 2005;120:421–33.
116. Xu J, Xiao N, Xia J. Thrombospondin 1 accelerates synaptogenesis in hippocampal neurons through neuroligin 1. Nat Neurosci. 2010;13:22–4.
117. Eroglu Ç, et al. Gabapentin receptor α2δ-1 is a neuronal thrombospondin receptor responsible for excitatory CNS synaptogenesis. Cell. 2009;139:380–92.
118. Mauch DH, et al. CNS synaptogenesis promoted by glia-derived cholesterol. Science. 2001;294:1354–7.
119. Farhy-Tselnicker I, et al. Astrocyte-secreted glypican 4 regulates release of neuronal pentraxin 1 from axons to induce functional synapse formation. Neuron. 2017;96:428–445.e13.
120. Blanco-Suarez E, Liu T-F, Kopelevich A, Allen NJ. Astrocyte-secreted chordin-like 1 drives synapse maturation and limits plasticity by increasing synaptic GluA2 AMPA receptors. Neuron. 2018;100:1116–1132.e13.
121. Hodge RD, et al. Conserved cell types with divergent features in human versus mouse cortex. Nature. 2019;573:61–8.

122. Johnson MA, Weick JP, Pearce RA, Zhang S-C. Functional neural development from human embryonic stem cells: accelerated synaptic activity via astrocyte coculture. J Neurosci. 2007;27:3069–77.
123. Chung W-S, et al. Astrocytes mediate synapse elimination through MEGF10 and MERTK pathways. Nature. 2013;504:394–400.
124. Auguste YSS, et al. Oligodendrocyte precursor cells engulf synapses during circuit remodeling in mice. Nat Neurosci. 2022;25:1273–8.
125. Dejanovic B, et al. Complement C1q-dependent excitatory and inhibitory synapse elimination by astrocytes and microglia in Alzheimer's disease mouse models. Nat Aging. 2022;2:837–50.
126. Schafer DP, et al. Microglia sculpt postnatal neural circuits in an activity and complement-dependent manner. Neuron. 2012;74:691–705.
127. Buchanan J, et al. Oligodendrocyte precursor cells ingest axons in the mouse neocortex. Proc Natl Acad Sci. 2022;119:e2202580119.
128. Stevens B, et al. The classical complement cascade mediates CNS synapse elimination. Cell. 2007;131:1164–78.
129. Vainchtein ID, et al. Astrocyte-derived Interleukin-33 promotes microglial synapse engulfment and neural circuit development. Science. 2018;359:1269–73.
130. Liddelow SA, et al. Neurotoxic reactive astrocytes are induced by activated microglia. Nature. 2017;541:481–7.
131. Selkoe DJ. Alzheimer's disease is a synaptic failure. Science. 2002;298:789–91.
132. Wheeler MA, Quintana FJ. Astrocytes on steroids binge on synapses to cope with stress. Immunity. 2023;56:1983–5.
133. Ackerman SD, Perez-Catalan NA, Freeman MR, Doe CQ. Astrocytes close a motor circuit critical period. Nature. 2021;592:414–20.
134. Bajar BT, Phi NT, Randhawa H, Akin O. Developmental neural activity requires neuron–astrocyte interactions. Dev Neurobiol. 2022;82:235–44.
135. Molofsky AV, et al. Astrocyte-encoded positional cues maintain sensorimotor circuit integrity. Nature. 2014;509:189–94.
136. Rothstein JD, et al. Knockout of glutamate transporters reveals a major role for astroglial transport in excitotoxicity and clearance of glutamate. Neuron. 1996;16:675–86.
137. Regan MR, et al. Variations in promoter activity reveal a differential expression and physiology of glutamate transporters by glia in the developing and mature CNS. J Neurosci. 2007;27:6607–19.
138. Hamberger AC, Chiang GH, Nylén ES, Scheff SW, Cotman CW. Glutamate as a CNS transmitter. I. Evaluation of glucose and glutamine as precursors for the synthesis of preferentially released glutamate. Brain Res. 1979;168:513–30.
139. Norenberg MD, Martinez-Hernandez A. Fine structural localization of glutamine synthetase in astrocytes of rat brain. Brain Res. 1979;161:303–10.
140. Navarrete M, et al. Astrocyte calcium signal and gliotransmission in human brain tissue. Cereb Cortex. 2013;23:1240–6.
141. Araque A, et al. Gliotransmitters travel in time and space. Neuron. 2014;81:728–39.
142. Covelo A, Araque A. Neuronal activity determines distinct gliotransmitter release from a single astrocyte. Elife. 2018;7:e32237.
143. de Ceglia R, et al. Specialized astrocytes mediate glutamatergic gliotransmission in the CNS. Nature. 2023;622:120–9.
144. Mulica P, Grünewald A, Pereira SL. Astrocyte-neuron metabolic crosstalk in neurodegeneration: a mitochondrial perspective. Front Endocrinol. 2021;12:668517.
145. Bélanger M, Allaman I, Magistretti PJ. Brain energy metabolism: focus on astrocyte-neuron metabolic cooperation. Cell Metab. 2011;14:724–38.
146. Kacem K, Lacombe P, Seylaz J, Bonvento G. Structural organization of the perivascular astrocyte endfeet and their relationship with the endothelial glucose transporter: a confocal microscopy study. Glia. 1998;23:1–10.

147. Loaiza A, Porras OH, Barros LF. Glutamate triggers rapid glucose transport stimulation in astrocytes as evidenced by real-time confocal microscopy. J Neurosci. 2003;23:7337–42.
148. Mathiisen TM, Lehre KP, Danbolt NC, Ottersen OP. The perivascular astroglial sheath provides a complete covering of the brain microvessels: an electron microscopic 3D reconstruction. Glia. 2010;58:1094–103.
149. Mächler P, et al. In vivo evidence for a lactate gradient from astrocytes to neurons. Cell Metab. 2016;23:94–102.
150. Mangia S, et al. Dynamics of lactate concentration and blood oxygen level-dependent effect in the human visual cortex during repeated identical stimuli. J Neurosci Res. 2007;85:3340–6.
151. Mangia S, Simpson TI, Vannucci SJ, Carruthers A. The in vivo neuron-to-astrocyte lactate shuttle in human brain: evidence from modeling of measured lactate levels during visual stimulation. J Neurochem. 2009;109:55–62. https://doi.org/10.1111/j.1471-4159.2009.06003.x.
152. Magistretti PJ, Pellerin L. Cellular mechanisms of brain energy metabolism and their relevance to functional brain imaging. Philos Trans R Soc Lond B Biol Sci. 1999;354(1387):1155–63. https://royalsocietypublishing.org/doi/10.1098/rstb.1999.0471?url_ver=Z39.88-2003&rfr_id=ori%3Arid%3Acrossref.org&rfr_dat=cr_pub++0pubmed
153. Wang SS-H, et al. Functional trade-offs in white matter axonal scaling. J Neurosci. 2008;28:4047–56.
154. Dienel GA. Brain glucose metabolism: integration of energetics with function. Physiol Rev. 2019;99:949–1045.
155. Westi EW, et al. Divergent cellular energetics, glutamate metabolism, and mitochondrial function between human and mouse cerebral cortex. Mol Neurobiol. 2022;59:7495–512.
156. Li J, et al. Conservation and divergence of vulnerability and responses to stressors between human and mouse astrocytes. Nat Commun. 2021;12:3958.
157. Lee JA, Hall B, Allsop J, Alqarni R, Allen SP. Lipid metabolism in astrocytic structure and function. Semin Cell Dev Biol. 2021;112:123–36.
158. Ferris HA, et al. Loss of astrocyte cholesterol synthesis disrupts neuronal function and alters whole-body metabolism. Proc Natl Acad Sci USA. 2017;114:1189–94.
159. Ioannou MS, et al. Neuron-astrocyte metabolic coupling protects against activity-induced fatty acid toxicity. Cell. 2019;177:1522–1535.e14.
160. Boisvert MM, Erikson GA, Shokhirev MN, Allen NJ. The aging astrocyte transcriptome from multiple regions of the mouse brain. Cell Rep. 2018;22:269–85.
161. Zehnder T, et al. Mitochondrial biogenesis in developing astrocytes regulates astrocyte maturation and synapse formation. Cell Rep. 2021;35:108952.
162. Wu Z, et al. Mechanisms controlling mitochondrial biogenesis and respiration through the thermogenic coactivator PGC-1. Cell. 1999;98:115–24.
163. Fiebig C, et al. Mitochondrial dysfunction in astrocytes impairs the generation of reactive astrocytes and enhances neuronal cell death in the cortex upon photothrombotic lesion. Front Mol Neurosci. 2019;12:40.
164. Hayakawa K, et al. Transfer of mitochondria from astrocytes to neurons after stroke. Nature. 2016;535:551–5.
165. Cloëtta D, et al. Inactivation of mTORC1 in the developing brain causes microcephaly and affects gliogenesis. J Neurosci. 2013;33:7799–810.
166. Li B, et al. CHD8 mutations increase gliogenesis to enlarge brain size in the nonhuman primate. Cell Discov. 2023;9:1–23.
167. Dias CM, et al. Glial dysregulation in the human brain in fragile X-associated tremor/ataxia syndrome. Proc Natl Acad Sci USA. 2023;120:e2300052120.
168. Dudley DM, et al. A rhesus macaque model of Asian-lineage Zika virus infection. Nat Commun. 2016;7:12204.
169. Martinot AJ, et al. Fetal neuropathology in Zika virus-infected pregnant female rhesus monkeys. Cell. 2018;173:1111–1122.e10.
170. Gurung S, et al. Zika virus infection at mid-gestation results in fetal cerebral cortical injury and fetal death in the olive baboon. PLoS Pathog. 2019;15:e1007507.

171. Seferovic M, et al. Experimental Zika virus infection in the pregnant common marmoset induces spontaneous fetal loss and neurodevelopmental abnormalities. Sci Rep. 2018;8:6851.
172. Lanko K, et al. Replication of the Zika virus in different iPSC-derived neuronal cells and implications to assess efficacy of antivirals. Antivir Res. 2017;145:82–6.
173. Ledur PF, et al. Zika virus infection leads to mitochondrial failure, oxidative stress and DNA damage in human iPSC-derived astrocytes. Sci Rep. 2020;10:1218.
174. Zhou T, et al. High-content screening in hPSC-neural progenitors identifies drug candidates that inhibit Zika virus infection in fetal-like organoids and adult brain. Cell Stem Cell. 2017;21:274–283.e5.
175. Xu M, et al. Identification of small-molecule inhibitors of Zika virus infection and induced neural cell death via a drug repurposing screen. Nat Med. 2016;22:1101–7.
176. Li Y, et al. Induction of expansion and folding in human cerebral organoids. Cell Stem Cell. 2017;20:385–396.e3.
177. Garcez PP, et al. Zika virus impairs growth in human neurospheres and brain organoids. Science. 2016;352:816–8.
178. Gabriel E, et al. Recent Zika virus isolates induce premature differentiation of neural progenitors in human brain organoids. Cell Stem Cell. 2017;20:397–406.e5.
179. Li C, et al. Zika virus disrupts neural progenitor development and leads to microcephaly in mice. Cell Stem Cell. 2016;19:120–6.
180. Cugola FR, et al. The Brazilian Zika virus strain causes birth defects in experimental models. Nature. 2016;534:267–71.
181. Retallack H, et al. Zika virus cell tropism in the developing human brain and inhibition by azithromycin. Proc Natl Acad Sci USA. 2016;113:14408–13.
182. Nowakowski TJ, et al. Expression analysis highlights AXL as a candidate Zika virus entry receptor in neural stem cells. Cell Stem Cell. 2016;18:591–6.
183. Rubenstein JLR, Merzenich MM. Model of autism: increased ratio of excitation/inhibition in key neural systems. Genes Brain Behav. 2003;2:255–67.
184. Liu Y, et al. A selective review of the excitatory-inhibitory imbalance in schizophrenia: underlying biology, genetics, microcircuits, and symptoms. Front Cell Dev Biol. 2021;9:664535.
185. Sohal VS, Rubenstein JLR. Excitation-inhibition balance as a framework for investigating mechanisms in neuropsychiatric disorders. Mol Psychiatry. 2019;24:1248–57.
186. Russo FB, et al. Modeling the interplay between neurons and astrocytes in autism using human induced pluripotent stem cells. Biol Psychiatry. 2018;83:569–78.
187. Allen M, et al. Astrocytes derived from ASD individuals alter behavior and destabilize neuronal activity through aberrant Ca2+ signaling. Mol Psychiatry. 2022;27:2470–84.
188. Glaum SR, Holzwarth JA, Miller RJ. Glutamate receptors activate Ca2+ mobilization and Ca2+ influx into astrocytes. Proc Natl Acad Sci. 1990;87:3454–8.
189. Bazargani N, Attwell D. Astrocyte calcium signaling: the third wave. Nat Neurosci. 2016;19:182–9.
190. Delépine C, et al. Altered microtubule dynamics and vesicular transport in mouse and human MeCP2-deficient astrocytes. Hum Mol Genet. 2016;25:146–57.
191. Wang H, et al. Dynamic association of the fragile X mental retardation protein as a messenger ribonucleoprotein between microtubules and polyribosomes. Mol Biol Cell. 2008;19:105–14.
192. Higashimori H, et al. Selective deletion of astroglial FMRP dysregulates glutamate transporter GLT1 and contributes to fragile X syndrome phenotypes in vivo. J Neurosci. 2016;36:7079–94.
193. Men Y, et al. Astroglial FMRP deficiency cell-autonomously up-regulates miR-128 and disrupts developmental astroglial mGluR5 signaling. Proc Natl Acad Sci USA. 2020;117:25092–103.
194. Cheng C, Lau SKM, Doering LC. Astrocyte-secreted thrombospondin-1 modulates synapse and spine defects in the fragile X mouse model. Mol Brain. 2016;9:74.
195. Jacobs S, Doering LC. Astrocytes prevent abnormal neuronal development in the fragile x mouse. J Neurosci. 2010;30:4508–14.

196. Sharma SD, et al. Astrocytes mediate cell non-autonomous correction of aberrant firing in human FXS neurons. Cell Rep. 2023;42:112344.
197. Olivera-Bravo S, et al. Astrocyte dysfunction in developmental neurometabolic diseases. In: von Bernhardi R, editor. Glial cells in health and disease of the CNS. Cham: Springer International Publishing; 2016. p. 227–43. https://doi.org/10.1007/978-3-319-40764-7_11.
198. Seifert G, Schilling K, Steinhäuser C. Astrocyte dysfunction in neurological disorders: a molecular perspective. Nat Rev Neurosci. 2006;7:194–206.
199. Chan F, et al. The role of astrocytes in seizure generation: insights from a novel in vitro seizure model based on mitochondrial dysfunction. Brain. 2019;142:391–411.
200. Shimamoto C, et al. Functional characterization of FABP3, 5 and 7 gene variants identified in schizophrenia and autism spectrum disorder and mouse behavioral studies. Hum Mol Genet. 2014;23:6495–511.
201. Talvio K, et al. An iPSC-derived astrocyte model of fragile X syndrome exhibits dysregulated cholesterol homeostasis. Commun Biol. 2023;6:789.
202. Brekke E, Morken TS, Sonnewald U. Glucose metabolism and astrocyte–neuron interactions in the neonatal brain. Neurochem Int. 2015;82:33–41.
203. Morken TS, et al. Altered astrocyte–neuronal interactions after hypoxia-ischemia in the neonatal brain in female and male rats. Stroke. 2014;45:2777–85.
204. Drews L, et al. Ammonia inhibits energy metabolism in astrocytes in a rapid and glutamate dehydrogenase 2-dependent manner. Dis Model Mech. 2020;13:dmm047134.
205. Hirbec H, et al. Emerging technologies to study glial cells. Glia. 2020;68:1692–728.
206. Lovett ML, Nieland TJF, Dingle Y-TL, Kaplan DL. Innovations in 3D tissue models of human brain physiology and diseases. Adv Funct Mater. 2020;30:1909146.
207. Maiolo L, et al. Glial interfaces: advanced materials and devices to uncover the role of astroglial cells in brain function and dysfunction. Adv Healthc Mater. 2021;10:2001268.

Gene Expression at the Tripartite Synapse: Bridging the Gap Between Neurons and Astrocytes

Gillian Imrie, Madison B. Gray, Vishnuvasan Raghuraman, and Isabella Farhy-Tselnicker

Abstract Astrocytes, a major class of glial cells, are an important element at the synapse where they engage in bidirectional crosstalk with neurons to regulate numerous aspects of neurotransmission, circuit function, and behavior. Mutations in synapse-related genes expressed in both neurons and astrocytes are central factors in a vast number of neurological disorders, making the proteins that they encode prominent targets for therapeutic intervention. Yet, while the roles of many of these synaptic proteins in neurons are well established, the functions of the same proteins in astrocytes are largely unknown. This gap in knowledge must be addressed to refine therapeutic approaches. In this chapter, we integrate multiomic meta-analysis and a comprehensive overview of current literature to show that astrocytes express an astounding number of genes that overlap with the neuronal and synaptic transcriptomes. Further, we highlight recent reports that characterize the expression patterns and potential novel roles of these genes in astrocytes in both physiological and pathological conditions, underscoring the importance of considering both cell types when investigating the function and regulation of synaptic proteins.

Gillian Imrie and Madison B. Gray contributed equally to this work.

Supplementary Information The online version contains supplementary material available at https://doi.org/10.1007/978-3-031-64839-7_5.

G. Imrie · M. B. Gray
Department of Biology, Texas A&M University, College Station, TX, USA
e-mail: grimrie@tamu.edu; madison_gray@tamu.edu

V. Raghuraman
Department of Biology, Texas A&M University, College Station, TX, USA

Department of Computer Science and Engineering, Texas A&M University, College Station, TX, USA
e-mail: vasan2001@tamu.edu

E. Blanco-Suarez, I. Farhy-Tselnicker (eds.), *Astrocyte-Neuron Interactions in Health and Disease*, Advances in Neurobiology 39,
https://doi.org/10.1007/978-3-031-64839-7_5

Keywords Astrocyte · Neuron · Tripartite synapse · Synaptic proteome · Synaptic gene expression

Abbreviations

$A_{1A}R$	Adenosine A1a receptor
$A_{2A}R$	Adenosine A2a receptor
AD	Alzheimer's disease
ADHD	Attention deficit hyperactive disorder
ALS	Amyotrophic lateral sclerosis
AMPAR	α-Amino-3-hydroxy-5-methyl-4-isoxazolepropionic acid receptor
APOE	Apolipoprotein E
ASD	Autism spectrum disorder
AQP4	Aquaporin-4
BCAR1	Breast cancer anti-estrogen resistance protein 1
BP	Bipolar disorder
BP	Biological process (relevant to GO analysis)
CAMKII	Calcium/calmodulin-stimulated protein kinase II
Ca_V	Voltage-gated calcium (Ca^{2+}) channel
CLTA	Clathrin light-chain A
CNS	Central nervous system
CPLX2	Complexin 2
DLG4	Disks large homolog 4
DREADD	Designer receptor exclusively activated by designer drugs
DS	Down syndrome
EM	Electron microscopy
m/e EPSC	Miniature/evoked excitatory postsynaptic current
m/e EPSP	Miniature/evoked excitatory postsynaptic potential
FPKM	Fragments per kilobase of transcript per million mapped reads
FTD	Frontotemporal dementia
FDR	False discovery rate
GABA	γ-aminobutyric acid
$GABA_AR$	GABA type A receptor
$GABA_BR$	GABA type B receptor

I. Farhy-Tselnicker (✉)
Department of Biology, Texas A&M University, College Station, TX, USA

Texas A&M Institute for Neuroscience (TAMIN), Texas A&M University, College Station, TX, USA

Center for Biological Clocks Research, Texas A&M University, College Station, TX, USA
e-mail: ifarhy@bio.tamu.edu

GAT	GABA transporter
GBM	Glioblastoma multiform
GEO	Gene expression omnibus
GFAP	Glial fibrillary acidic protein
GLAST; EAAT1	Glutamate-aspartate transporter/excitatory amino acid transporter type 1
GLT1; EAAT2	Glutamate transporter 1/excitatory amino acid transport type 2
GLUA1-4	Glutamate ionotropic receptor AMPA type subunit 1-4
GLUN1	Glutamate ionotropic receptor NMDA type subunit 1
GLUN2A-D	Glutamate ionotropic receptor NMDA type subunit 2, subtypes A-D
GLUN3A-B	Glutamate ionotropic receptor NMDA type subunit 3, subtypes A-B
GLUK1-4	Glutamate ionotropic receptor kainate type subunit 1-4
GO	Gene ontology
GPCR	G-protein-coupled receptor
GS	Glutamine synthetase
GWAS	Genome-wide association studies
HD	Huntington's disease
ICC	Immunocytochemistry
IHC	Immunohistochemistry
IP3R2	Inositol-1,4,5-trisphosphate receptor type 2
iPSC	Induced pluripotent stem cell
m/e IPSC	miniature/evoked Inhibitory postsynaptic current
m/e IPSP	miniature/evoked Inhibitory postsynaptic potential
K_2P	Two-pore domain potassium (K^+) channel
KAR	Kainate receptor
KO	Knockout
K_V	Voltage-gated potassium (K^+) channel
LTD	Long-term depression
LTP	Long-term potentiation
MAGUK	Membrane-associated guanylate kinase
MGLUR	Metabotropic glutamate receptor
Na_V	Voltage-gated sodium (Na^+) tetrodotoxin-sensitive sodium channel
NCBI	National Center for Biotechnology Information
NMDAR	N-methyl-D-aspartate receptor
NLGN	Neuroligin
NRXN	Neurexin
P	Postnatal day
PFC	Prefrontal cortex
PI3K	Phosphoinositide 3 kinase
PKA	Protein kinase A
PKC	Protein kinase C
PLC	Phospholipase C

PSD95	Postsynaptic density 95
qPCR	Quantitative polymerase chain reaction
SP1	Sphingosine-1-phosphate
S1PR1	Sphingosine-1-phosphate receptor type 1
SAM	Synapse adhesion molecule
SE	Status epilepticus
SHANK1	SH3 and multiple ankyrin repeat domains 1
shRNA	Short hairpin RNA/small hairpin RNA
siRNA	Small interfering RNA/short interfering RNA/silencing RNA
SNAP25	Synaptosome-associated protein 25
SNARE	Soluble N-ethylmaleimide-sensitive factor attachment protein receptor
SST	Somatostatin
SSTR2	Somatostatin receptor type 2
SUD	Substance use disorder
SYT11	Synaptotagmin-11
TEA	Tetraethylammonium
TLE	Temporal lobe epilepsy
VAMP2	Vesicle-associated membrane protein 2
VGLUT1	Vesicular glutamate transporter 1
VGLUT2	Vesicular glutamate transporter 2

1 Introduction

Central nervous system (CNS) function relies on the precise formation and regulation of synapses, the specialized junctions between neurons where electrical impulses are converted to chemical signals to facilitate brain-wide information processing [1]. The neuronal synapse is composed of the axonal presynaptic terminal of one neuron, from which signaling molecules (such as neurotransmitters) are released into the synaptic cleft, and the postsynaptic density at the dendrite of another neuron which contains specialized receptors that bind these signaling molecules and activate appropriate downstream cellular pathways [2]. Over many decades, thousands of synaptic proteins have been identified with distinct functional roles specific to synapse type and compartment [3]. These include (but are not limited to) cell adhesion molecules that hold the pre- and postsynaptic compartments together; scaffolding proteins that organize proteins within their specific synaptic compartment, such as those required for the release of neurotransmitter-containing vesicles from the presynaptic active zones, and postsynaptic neurotransmitter receptors and transporters; proteins responsible for metabolic functions and energy generation; and a host of ion channels that facilitate the changes in membrane potential of both pre- and postsynaptic neurons required for synaptic transmission (for reviews, see [4–6]). Synaptic dysfunction is linked to a variety of neurological disorders, referred to as synaptopathies [7–10] including autism [11–13], epilepsy

[14–16], and neurodegeneration [17–20], making synaptic proteins salient targets for therapeutics. Yet, therapeutic efforts are impeded by the immense complexity of synapse types and regulatory mechanisms, stressing the need to better understand the function of synapse-related genes and proteins.

In addition to neurons, synapses frequently contain astrocytes, comprising a tripartite synapse [21–24]. Astrocytes, a major type of glia, regulate numerous aspects of synaptic function from the onset of synaptogenesis through senescence [25–28]. Astrocytes express many receptors that can sense and respond to neuronal signals released at the synapse including glutamate, γ-aminobutyric acid (GABA), and adenosine [29]. A single astrocyte can contact or ensheathe over 100,000 synapses [30, 31] regulating them through physical contact via cell adhesion molecules; secretion of synapse-modulating factors such as glypicans, thrombospondins, and cholesterol (reviewed in [28, 32, 33]); and providing homeostatic and metabolic support through vascular coupling [34, 35], ion buffering [36–38], and neurotransmitter turnover [39–41]. Another layer of regulation stems from astrocytic heterogeneity, whereupon their synaptic modulation can depend on spatiotemporal context. [42] For example, during mouse visual cortex development, astrocytes in different cortical layers exhibit unique gene enrichment dependent on developmental stage and neuronal activity [43, 44], leading to the layer and age-specific expression of distinct α-amino-3-hydroxy-5-methyl-4-isoxazolepropionic acid receptor (AMPAR) subunits [43]. Likewise, in the adult hippocampal dentate gyrus, specific layers are populated by astrocytes with distinct molecular characteristics including morphology and glutamate currents [45]. Thus, by simultaneously sensing the synaptic microenvironment and the brain metabolic state, astrocytes integrate this information to fine-tune synaptic metabolism, transmission, and plasticity [46] influencing not only neuronal health but also many facets of neurotransmission, circuit function, and behavior [47–49].

Given the incredible versatility of astrocytic roles at the synapse, "omics" and mechanistic studies have identified many astrocytic genes involved in synapse regulation. Interestingly, studies also revealed that astrocytes express many of the same genes critical for synaptic structure and function as neurons, including vesicular transporters [50], neurotransmitter receptors [51–53], scaffolding proteins [54], and adhesion molecules [55–57]. Yet while extensive work has determined the roles of many of these genes in neurons, their roles in astrocytes remain relatively unexplored. Because synaptic dysregulation in brain disorders is centrally linked to mutations in synapse-related genes [58–60], identifying the physiological characteristics and functions of these genes in astrocytes is critical for understanding how interactions between brain cells shape the synaptic environment and for refining therapeutic strategies.

In this book chapter, we combine multiomic meta-analysis with an overview of current literature to provide an updated perspective on the astrocytic expression of synaptic genes classically studied in neurons, and the current understanding of the functions that they mediate. We leverage this approach to address the following questions: (1) How many and which synaptic genes are expressed by both astrocytes and neurons? (2) How do the expression patterns and functions of these genes and their protein products compare between astrocytes and neurons?

2 Meta-Analysis Identifies Numerous Synaptic Genes Expressed by Both Astrocytes and Neurons

The explosion of multiomic studies over recent years has provided access to a rich store of information about cell type-specific gene and protein expression, allowing the thorough analysis of astrocyte and neuronal transcriptomes at unprecedented resolution [27, 43, 44, 61–69]. Moreover, studies focusing on cellular and molecular mechanisms that regulate synapse formation and activity have identified hundreds of synaptic proteins [28, 70] with distinct functional roles. Here, we synthesize these findings to analyze the astrocytic expression patterns and function of synaptic proteins.

To determine which synaptic genes are expressed in both cell types, we utilized publicly accessible datasets of neuronal, astrocytic, and synaptic genes (Fig. 1a). To identify genes encoding for synaptic proteins, we used the synaptic proteome

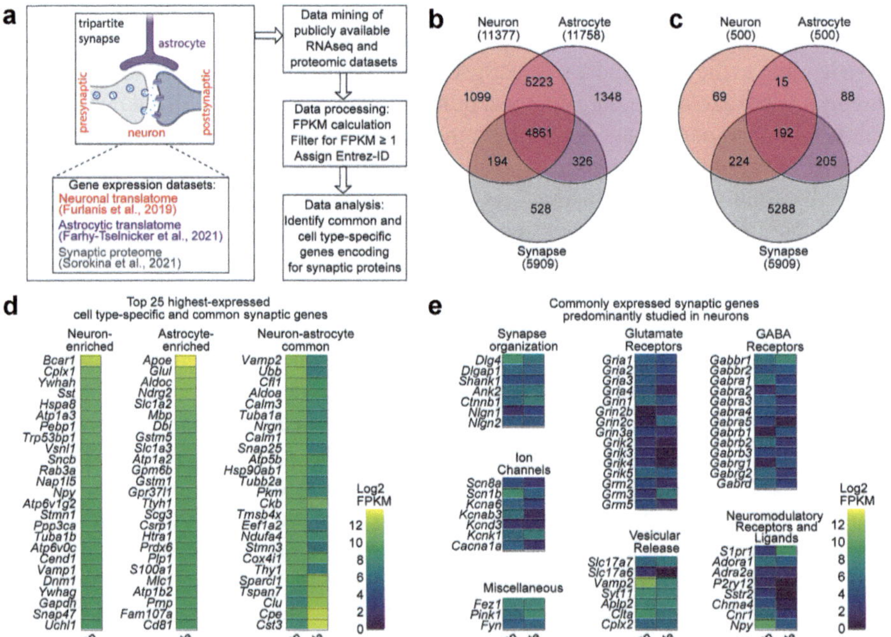

Fig. 1 Astrocytes express many of the same synaptic genes as neurons. (**a**) Diagram of process by which synaptic genes commonly expressed by neurons and astrocytes were determined. Gene lists (FPKM ≥1) were obtained from publicly available datasets and processed to obtain unified gene names and IDs. (**b**) Venn diagram showing the overlap between synaptic proteome, total number of genes expressed by neurons, and astrocytes. (**c**) Same as B but using the top 500 highest expressing genes in each cell type. (**d**) Heatmaps of the top 25 highest expressing genes shown as Log2 FPKM enriched in neurons, astrocytes, or common to both cell types. (**e**) Heatmaps of genes expressed in both neurons and astrocytes, known to play important roles at the synapse that have been predominantly studied in neurons

SQLite database [71], a repository of proteins with validated expression at the synapse within different brain regions and species compiled from proteomic studies published between 2000 and 2020 [72]. A list of neuronal and astrocytic genes was obtained from age and brain region-matched ribosomal profiling (Ribo-tag) studies of neurons [62] and astrocytes [43] from postnatal day (P) 25–28 mouse cortices. Ribo-tag sequencing is particularly advantageous for our analysis as it enables not only profiling of actively translated mRNAs, which suggests the presence of functional proteins within the cell, but also profiling mRNAs from both somata and the fine processes of astrocytes and neurons, thus ensuring that the transcripts localized to the tripartite synapse are represented in the sequencing data [27, 73, 74].

To compare the datasets, we first processed the data to unify gene IDs and represent expression level values in fragments per kilobase of transcript per million mapped reads (FPKM). Astrocyte data in FPKM were obtained from Farhy-Tselnicker et al. [43]. Neuronal data were obtained from Furlanis et al. [62], and FPKM was calculated from raw counts provided in supplemental data using transcript lengths (canonical) from the biomaRt R package [75], and averaged among different cortical neuronal subtypes to generate a pan-neuronal average FPKM. The R package RSQLite [76] was used to connect to the synaptic proteome database to extract a gene list of cortical synaptic protein regions matching the abovementioned neuronal and astrocytic datasets. Entrez IDs (www.ncbi.nlm.nih.gov/gene/) were assigned to all genes using biomaRt R package [75], and *Mus musculus* annotated genome (GRCm39) was accessed through the Ensembl database (www.ensembl.org). Genes missing Entrez ID due to outdated gene names or lack of genome mapping were either manually updated from the National Center for Biotechnology Information (NCBI) database, or excluded from our analysis, while for entries with different gene names (in case of synonyms) but identical Entrez IDs, FPKM values were averaged to generate a unique entry. Overall, 5909 cortical synaptic genes and ~11,000 genes (FPKM $\geq$1) from each cell type were selected for further analysis (Table S1).

To identify the numbers and IDs of synaptic genes overlapping between neurons and astrocytes (Fig. 1b and c), we used "Interactivenn" [77]. Surprisingly, we observed very similar numbers of overlapping genes between each cell type and the synaptic gene list (neurons: 5055 genes; astrocytes: 5187), with 4861 genes (~80% of all the analyzed synaptic genes) common to both cell types. We also identified 326 unique astrocyte-synaptic genes and 194 unique neuron-synaptic genes (Fig. 1b). The presence of astrocytic unique genes within the synaptic proteomic database is not surprising, as astrocytes are known to closely interact with the synaptic compartments, and thus can be inadvertently included in synaptosome preparations used to identify synaptic proteins [78]. Since for many genes the expression levels (in FPKM) vary between the cell types, we also compared the top 500 most highly expressed genes from each cell type with the synaptic gene list and identified 205 astrocyte-enriched, 224 neuronal-enriched, and 192 common genes within this list (Fig. 1c). Overall, there is ~40% overlap between the top 500 neuronal and astrocytic genes (207 genes; Fig. 1c).

To visualize the identified synaptic genes, we plotted the top 25 most highly expressed common and cell type-specific genes as Log2 FPKM (Fig. 1d). Apolipoprotein E (APOE; *Apoe),* a cholesterol transport protein known for its implications in Alzheimer's disease (AD) [79], is the most highly expressed astrocyte-enriched synaptic gene, while in neurons, it is the gene *Bcar1,* which encodes for the scaffolding protein breast cancer anti-estrogen resistance protein 1 (BCAR1) [80]. The most highly expressed synaptic genes common to both neurons and astrocytes include genes, which encode proteins related to vesicular transport (*Vamp2* [81, 82], *Snap25* [82, 83]), metabolism (*Aldoa* [84]), and synaptogenesis (*Sparcl1* [28, 85]). Notably, the expression levels of these genes differ between the cell types. For example, the soluble N-ethylmaleimide-sensitive factor attachment protein receptor (SNARE) complex-associated gene *Snap25* is more highly expressed in neurons (Log2 FPKM in neurons =10.6; in astrocytes = 7.3; Table S1), while carboxypeptidase E (*Cpe*), a secreted protein with multiple roles in neurodevelopment and neuroprotection [86], is more abundant in astrocytes (Log2 FPKM in neurons = 9.6; in astrocytes = 12.8; Table S1). These data are in line with our previously published analysis of astrocytic gene expression and enrichment compared to whole brain input [43].

To further investigate the identities of synaptic genes expressed in both neurons and astrocytes, we reviewed the literature to identify genes that were shown to play critical roles at the synapse such as those involved in synapse structure and organization, e.g., cell adhesion molecules, scaffolding proteins, neurotransmitter and neuromodulator receptors, ion channels, and genes related to vesicular release (Table 1; Fig. 1e). We observed many of these genes expressed in both cell types including disks large homolog 4 (*Dlg4*), which encodes for the protein postsynaptic density 95 (PSD95), glutamate receptor AMPA (*Gria*), and N-methyl-D-aspartate (NMDA; *Grin*) subunits, and vesicular release-related genes including *Vamp2* and *Slc17a7* (Fig. 1e; Table 1). Importantly, for the majority of these genes, we find remarkably fewer publications focusing on understanding their function in astrocytes than in neurons despite the overall comparable expression levels between cell types. We overview these studies in Sect. 3 of this chapter.

To determine which cellular processes are enriched in neuronal, astrocytic, and common gene lists (top 500 genes), we performed gene ontology (GO) analysis (biological processes) using the string database [87] (https://string-db.org/) (Fig. 2). Our analysis shows that neuronal-synaptic genes are enriched in processes related to vesicular release including synaptic vesicle exocytosis, regulation of vesicle fusion, and synaptic vesicle priming (Fig. 2a). On the other hand, astrocytic synaptic genes are strongly enriched in processes related to amino acid transport and neurotransmitter reuptake (Fig. 2b). Further, synaptic genes common to both cell types are enriched in processes associated with cellular metabolic processes, mitochondrial function, and energy generation (Fig. 2c).

Taken together, this analysis demonstrates that at the transcriptomic level, astrocytes express an overwhelming number of genes known to encode proteins present at the synapse, with some being uniquely present (or enriched) compared to neurons. These unique/enriched genes appear to confer cell type-specific functions,

Table 1 Synapse-related genes studied in astrocytes

Category/ gene name	Protein	Synaptic role	Synapse type	Associated diseases	Ref. in astrocytes
Synapse organization					
Dlg4	Postsynaptic density 95	Postsynaptic scaffolding	Excitatory	ASD, epilepsy	[105]
Dlgap1	DLG associated protein 1	Postsynaptic scaffolding	Excitatory	Schizophrenia	[253, 254]
Shank1	SH3 and multiple ankyrin repeat domains protein 1	Postsynaptic scaffolding	Excitatory	ASD, schizophrenia	[56]
Ank2	Ankyrin B	Adaptor	Nonspecific	AD, ASD, schizophrenia	[255, 256]
Ctnnb1	Catenin β1	Adhesion regulation	Nonspecific	GBM	[120, 257]
Nlgn1-3	Neuroligins 1-3	Adhesion	Excitatory/ Inhibitory	AD, anxiety, ASD, schizophrenia	[55, 117]
Voltage-gated ion channels					
Scn8a	Na^+ channel α8	Na^+ conductance	Nonspecific	Epilepsy, scn8a-linked neurodevelopmental disorders	[131]
Scn1b	Na^+ channel β1	Na^+ conductance	Nonspecific	Epilepsy	[132]
Kcna6	K^+ channel shaker-related subfamily member 6	K^+ conductance	Nonspecific	Epilepsy, neurodevelopmental disorders	[135]
Kcnab3	K^+ channel, shaker-related β member 3	K^+ conductance	Nonspecific	Epilepsy, migraine,	[136, 137]
Kcnd3	K^+ channel, Shal-related member 3	K^+ conductance	Nonspecific	Ataxia, epilepsy, neurodevelopmental disorders	[138]
Kcnk1	Two-pore domain K^+ channel member 1	K^+ conductance	Nonspecific	Epilepsy, neurodevelopmental disorders	[142, 258, 259]

(continued)

Table 1 (continued)

Category/ gene name	Protein	Synaptic role	Synapse type	Associated diseases	Ref. in astrocytes
Cacna1a	Ca^{2+} channel α1A	Ca^{2+} conductance	Nonspecific	Ataxia, migraine, neurodevelopmental disorders, neurodegeneration	[148]
Glutamate receptors					
Gria1-4	AMPA receptor 1-4	Fast glutamatergic transmission LTP/ LTD	Excitatory	ASD, learning disorders, AD, epilepsy, schizophrenia, nonspecific syndromic intellectual disability	[173, 260]
Grin1, Grin2b, 2c, 3a	NMDA receptor 1, 2b, 2c, 3a	Glutamatergic transmission, LTP/ LTD	Excitatory	ASD, Grin2b-associated neurodevelopmental disorder, schizophrenia, stroke, BP, epilepsy, schizophrenia	[187, 261, 262]
Grik1-5	Kainate receptor 1-5	Glutamatergic transmission	Excitatory	ADHD, ASD, ataxia, BP, Brugada syndrome epilepsy, megalencephalic leukoencephalopathy, MD, schizophrenia	[193]
Grm2,3,5	Metabotropic glutamate receptor 2, 3, 5	GPCR-mediated signaling	Excitatory	AD, ALS, depression, HD, PD, schizophrenia, SUD	[201, 203, 263]
GABA receptors					
Gabbr1-2	GABA type B receptor 1-2	GPCR-mediated signaling	Inhibitory	ASD, epilepsy, schizophrenia, neuropathic pain, SUD	[214]
Gabra1-5, Gabrb1-3, Gabrg1-2, Gabrd	GABA type A α, β, γ, δ	Chloride conductance, GABAergic transmission	Inhibitory	Anxiety, ASD, epilepsy, sleep disorders, SUD	[264, 265]

Neuromodulatory receptors and ligands					
S1pr1	Sphingosine 1-phosphate receptor 1	GPCR-mediated signaling	Inhibitory	Neuroinflammation	[216, 217]
Adora1	Adenosine A1 receptor	GPCR-mediated signaling	Nonspecific	AD, epilepsy, PD, sleep disorders, stroke	[220, 266]
Adra2a	Adrenergic receptor α2A	GPCR-mediated signaling	Nonspecific (mainly excitatory)	ADHD, mood disorders, SUD	[267, 268]
P2ry12	P2Y receptor	GPCR-mediated signaling	Nonspecific	Neuroinflammation	[269]
Sstr2	Somatostatin receptor 2	GPCR-mediated signaling	Nonspecific	AD, epilepsy, schizophrenia	[231]
Chrna4	Nicotinic receptor α4	Ach-gated Na^+ conduction	Nonspecific	Epilepsy	[270]
Cnr1	Cannabinoid receptor 1	GPCR-mediated signaling	Nonspecific	HD (age at onset), schizophrenia	[271, 272]
Npy	Neuropeptide Y	Signaling neuropeptide	Nonspecific	Anxiety, epilepsy, obesity/metabolic disorders	[273]
Vesicular release					
Slc17a7	Vesicular glutamate transporter 1	Glutamate vesicular packaging, neurotransmission	Excitatory	AD, cognitive disorder, PD	[88]
Slc17a6	Vesicular glutamate transporter 2	Glutamate vesicular packaging, neurotransmission	Excitatory	Epilepsy, movement disorders	[88]
Vamp2	Synaptobrevin-2	Vesicle fusion at the active zone	Nonspecific	AD, epilepsy, PD	[246, 274]
Syt11	Synaptotagmin-11	Calcium-dependent neurotransmitter release	Nonspecific	ASD, AD, PD	[251]
Aplp2	Amyloid beta precursor-like protein 2	Synaptic vesicle recycling	Nonspecific	AD, epilepsy	[275]

(continued)

Table 1 (continued)

Category/ gene name	Protein	Synaptic role	Synapse type	Associated diseases	Ref. in astrocytes
Clta	Clathrin light-chain A	Synaptic vesicle recycling	Nonspecific	AD, ASD, cognitive disorders, HD, PD	[276]
Cplx2	Complexin 2	Synaptic vesicle exocytosis	Nonspecific	Unknown	[253]
Miscellaneous					
Fez1	Fasciculation and elongation protein z1	Axonal transport	Nonspecific	ASD, AD, PD	[277–279]
Pink1	PTEN-induced putative kinase 1	Regulation of mitochondrial dynamics	Nonspecific	PD	[280, 281]
Fyn	FYN proto-oncogene, Src family tyrosine kinase	Plasticity modulation	Nonspecific	AD, PD, schizophrenia	[282–284]

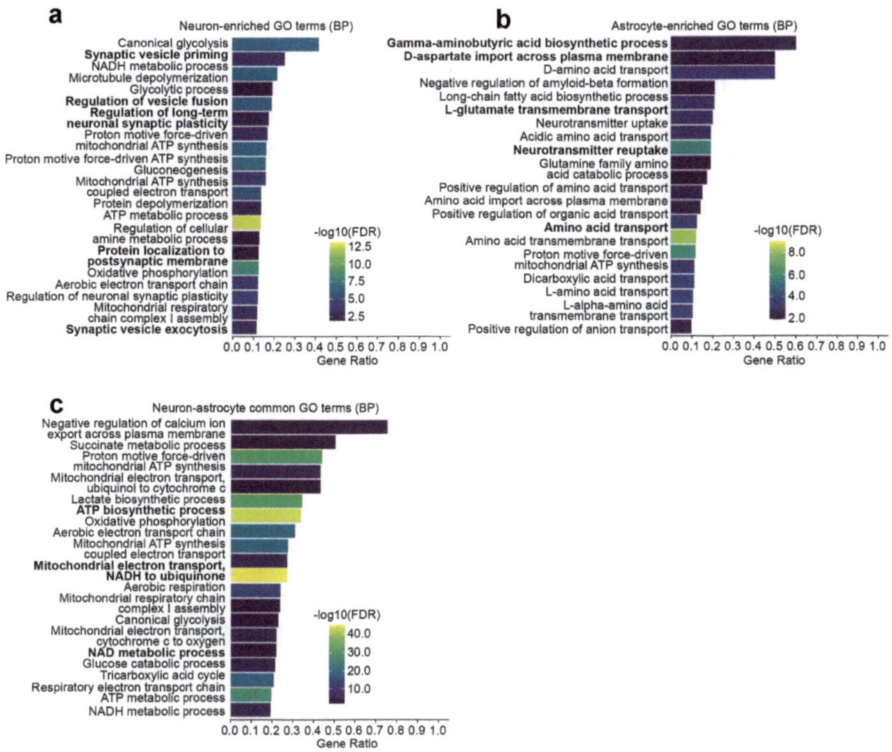

Fig. 2 Astrocytic and neuronal gene enrichment for cell type-specific and common GO biological processes. Gene Ontology (GO) terms analysis of the top 500 most highly expressing synaptic genes in each cell type using String database for biological process (BP) in (**a**) neuron-enriched synaptic genes, (**b**) astrocyte-enriched synaptic genes, and (**c**) common synaptic genes to both cell types. GO terms are plotted by gene ratio value and color-coded by -log10 false discovery rate (FDR). Example GO terms are highlighted in bold

such as neurotransmitter release in neurons, and amino acid transport and reuptake in astrocytes. Furthermore, we show that the expression levels (in FPKM) differ between the two cell types for many of the common genes, which may suggest cell type-specific differences in protein levels or functional roles. Alternatively, this could indicate that a particular gene is expressed in a given cell type at different levels throughout the life span of the animal, or in distinct subpopulations of cells, a conclusion that is not possible to make when analyzing bulk sequencing data. For example, a recent study showed that the gene encoding for vesicular transporter VGLUT1 is expressed by a subpopulation of hippocampal astrocytes [88], while others showed differential spatiotemporal expression of the synapse-regulating genes *Gpc4* and *Chrdl1* in cortical astrocytes [43, 44, 89]. Future studies using multiple additional datasets and comparing transcriptomes of both single cells and different developmental stages will provide critical insight into the expression patterns of synaptic genes in both astrocytes and neurons.

In summary, neurons and astrocytes each have unique and common roles at the synapse, expressing many of the same genes to carry these out. Since targeting synaptic genes is an attractive therapeutic strategy, it is important to keep in mind how the potential treatments will affect not only neurons, but also astrocytes, which may express the same gene but utilize it for different functions. These results also raise the question of whether genes found through genome-wide association studies (GWAS) might play a role in disease states through their function in astrocytes and other glia as well as in neurons. Thus, more work is needed to broaden our understanding of the expression patterns and functional roles of synaptic genes in astrocytes and how these can be harnessed to correct synaptic dysfunction prevalent in disease.

3 The Expression and Function of Synaptic Genes in Astrocytes

The above analysis shows that astrocytes share many thousands of genes with the synaptic transcriptome, including those critical for proper synaptic function. Surprisingly, compared to the large body of work describing their roles in neurons across multiple species, fewer studies to date have investigated their astrocyte-specific expression patterns, functions, and regulatory mechanisms. In this section, we highlight studies in mammals focusing on some of these essential synaptic genes (Fig. 1e, Table 1) to demonstrate the current knowledge regarding their astrocytic protein expression and functions as compared to neurons.

3.1 Astrocytes Express Many of the Same Synapse Organization Genes as Neurons

The pre- and postsynaptic neuronal compartments are held together by a large variety of specialized synapse adhesion molecules (SAMs), which determine synapse organization and specificity [90]. In addition, within each synaptic compartment, scaffolding proteins ensure high levels of organization, which is essential for synaptic transduction such as presynaptic vesicle docking or postsynaptic receptor clustering [91]. It has been shown that astrocytes express some of the same SAMs and scaffolding proteins as neurons, which they utilize to directly contact a neuronal-synaptic compartment or interact with each other [92–94]. Here, we focus on studies that identify the potential astrocytic roles of synapse organizing proteins previously considered predominantly neuronal. To learn about the astrocyte-specific scaffolding and cell adhesion proteins, we refer the reader to these excellent reviews [95, 96].

3.1.1 Organizers of the Excitatory Postsynaptic Density

The excitatory postsynaptic density is a complex matrix of specialized proteins and molecules at the neuronal dendrites, serving as a scaffold for signaling proteins, receptors, and cytoskeletal elements that are integral for proper synaptic function [97, 98]. Our data mining analysis revealed several excitatory postsynaptic density genes expressed in both cell types, and here, we review their protein level expression and function in astrocytes.

Disks large homolog 4 (PSD95; *Dlg4*), better known by its protein product PSD95, plays a key role in the organization and function of excitatory synapses in the CNS [99, 100]. It is part of the membrane-associated guanylate kinase (MAGUK) protein family and in neurons acts as a scaffold facilitating the anchorage and organization of many postsynaptic proteins [99, 101]. Neuronal PSD95 plays an important role in synaptic plasticity and the regulation of synaptic strength [100] due to its differential association with the two primary classes of ionotropic glutamate receptors, NMDA (*Grin*) and AMPA (*Gria*). NMDARs are tightly bound to PSD95, while AMPAR interaction is looser, allowing for flexibility in their membrane expression [102, 103]. PSD95 is dysregulated in Alzheimer's disease (AD), with both increased and decreased expression reported [104, 105]. Interestingly, proteomic analysis and immunohistochemistry (IHC) detected its presence in human astrocytes of the entorhinal cortex, where its functional role remains uncharacterized; however, its reduction is associated with reactive astrogliosis in AD [105]. Furthermore, since AMPARs and NMDARs are also expressed by astrocytes [106, 107] (Fig. 1e), astrocytic PSD95 may interact with these receptors to regulate their membrane insertions in a similar fashion to neurons.

SH3 and multiple ankyrin repeat domains 1 (SHANK1; *Shank1*) are scaffolding proteins with important roles in neural development and cognition, which interact with glutamate receptors and other organizational molecules [108, 109]. In neurons, targeted deletion of SHANK1 leads to deficits in social behavior, reduced spine density, and weaker synaptic transmission (yet enhanced spatial learning) in rodents [110, 111]. In astrocytes, SHANK1 has been identified in cellular fractions as well as by IHC, where it is differentially regulated across development [55]. In vivo, astrocytic SHANK1 is most highly expressed in young mice (around P5), and stabilizes between P15-P60, suggesting a potential developmental role in tripartite synapse organization [22]. While the precise role of SHANK1 in astrocytes is yet to be investigated, its known interactions with NMDAR and metabotropic glutamate receptors (MGLURs; which are also expressed in astrocytes; Fig. 1e, and see subsection 3.3) may suggest a similar function [112].

3.1.2 Synapse Adhesion Molecules (SAMs)

SAMs are a specialized form of cell adhesion molecules that modulate physical interactions between pre- and postsynaptic compartments and the extracellular matrix. Of these, neuroligins (NLGN; *Nlgn*) are widely expressed in neurons across the CNS where they are localized to the postsynaptic compartment. They interact trans-heterophilically with the presynaptic neurexins (NRXN; *Nrxn*) to shape multiple aspects of synapse development and maintenance [113, 114]. Importantly, NLGNs have differential effects on synapse specialization. NLGN1 is predominantly expressed in glutamatergic synapses, NLGN2 is mainly GABAergic, and NLGN3 is expressed in both synapse types, while NLGN4 is expressed in glycinergic synapses [115]. In neurons, it has been shown that *Nlgn1* or *Nlgn2* knockout (KO) results in decreased AMPAR currents and protein expression, as well as disrupted GABAR expression and reduced inhibitory postsynaptic current (IPSC) amplitudes. KO of *Nlgn3* enhances the frequency of IPSCs but not the amplitude and is associated with autism spectrum disorders (ASD) in humans [116]. Astrocytes express NLGNs at similar levels to neurons [55] (Fig. 1e). It has been demonstrated that astrocytic NLGN interactions with neuronal NRXNs are necessary for normal development. In neuron-astrocyte co-cultures, knockdown of NLGN1-3 disrupted astrocyte morphological complexity, and these effects were phenocopied by silencing presynaptic NRXN1 or NRXN2 in vitro and in vivo, implying these molecules as the primary interaction sites for astrocyte NLGNs. Additionally, shRNA knockdown of astrocytic NLGN2 disrupted synapse development and activity [55]. Notably, recent evidence shows that conditional deletion of *Nlgn1-3* from astrocytes during early postnatal days using cre/loxP system has no substantial effects on the synaptic proteome or neuronal activity [117]. These observations may be due to insufficient knockout of NLGN from astrocytes using this approach as was previously reported [55]. These findings underscore the need for employing multiple approaches to characterize the roles of synaptic (and other) proteins in astrocytes, to better understand their cell type-specific function.

β-catenin (CTNNB1; *Ctnnb1*) is a synaptic protein expressed in both neurons and astrocytes [118] (Fig. 1e). In neurons, CTNNB1 plays multifunctional roles in cell-cell adhesion and intracellular signaling, such as NLGN3 clustering, as well as in modulating the localization and size of presynaptic vesicle clusters [119]. In human progenitor-derived astrocyte cultures, CTNNB1 is shown to modulate glutamate uptake and metabolism through the positive regulation of glutamate transporter 1 (GLT1; EAAT2; *Slc1a2*) and glutamine synthetase (GS; *Glul*). While siRNA knockdown of *Ctnnb1* led to a ~70% decrease in astrocytic GLT1 protein, *Ctnnb1* overexpression increased GLT1 protein levels by ~15-fold [120]. These findings were further shown in vivo, where morpholino knockdown of *Ctnnb1* reduced the protein levels of GLT1 and GS as evident from Western blots of prefrontal cortex (PFC) tissue of juvenile mice [120].

Altogether, there is compelling evidence that synapse organization proteins canonically characterized in neurons, which are required for appropriate synaptic morphology and function, are expressed in astrocytes. While some of the overviewed studies offered clues about the potential roles these proteins may play in astrocytes, many unanswered questions remain demanding further research, such as follows: Do these proteins play different roles in astrocytes than in neurons to regulate synapses? Do they interact with the same partners (such as receptors or cell adhesion molecules) as their neuronal counterparts?

3.2 *Astrocytes Express Many of the Same Voltage-Gated Ion Channels as Neurons*

Ion channels mediate a wide range of functions at the synapse. In neurons, ion channels are necessary for the generation and propagation of the action potential, neurotransmitter release, and postsynaptic neuronal function [121, 122]. In astrocytes, which do not fire action potentials, ion channel research has been primarily focused on potassium ion buffering at the synapse, which is mainly mediated by inward rectifying (such as Kir4.1) K^+ channels, which will not be discussed here (reviewed in [123–125]). Instead, we focus this section on ion channels that were classically studied in the context of neuronal activity, including the voltage-gated sodium, potassium, and calcium channels. Though these channels are expressed by astrocytes (Fig. 1e; Table 1), their roles at the synapse are largely unknown.

3.2.1 Voltage-Gated Tetrodotoxin-Sensitive Sodium (Na^+) Channels (Na_V1.X)

Na_Vs are critical for controlling neuronal excitability and have mostly been studied in this context [126]. In mammals, they are typically made up of a pore-forming α-subunit and regulatory β-subunits, which modulate Na^+ current amplitudes, the α-subunit expression levels [127], and contribute to cell-cell adhesion [128]. Of the different channel subtypes, we identified NaV1.6 (*Scn8a*) and NaVβ1 (*Scn1b*) expressed in astrocytes (Fig. 1e). Though these channels appear to have low expression in astrocytes in the healthy brain, there is evidence of their upregulation after stress or injury [129, 130]. Increased protein levels of both Nav1.6 [131] and Na_Vβ1 [132] were observed by IHC in hippocampal astrocytes in a rat model of kainic acid-induced temporal lobe epilepsy (TLE). The increase was observed in glial fibrillary acidic protein (GFAP; *Gfap*) [131] and vimentin [132]-positive reactive astrocytes as soon as a week after induction of status epilepticus (SE) and persisted for up to 4 months post-injury. Because β-subunits of sodium channels can act as cell adhesion molecules, post-SE astrocytic Na_Vβ1 upregulation may underlie remodeling after brain injury. These findings imply a potential role for astrocytes in the development of epileptogenic disorders, specifically after excitotoxic injury.

3.2.2 Voltage-Gated Potassium (K^+) Channels (K_V)

K_Vs are critical components of neuronal action potential repolarization [133]. While extensive work has characterized other types of potassium channels in astrocytes [124, 125, 134], the voltage-gated K_V channels are relatively unexplored. Our sequencing analysis and literature reports identify the presence of several of these channels in astrocytes, namely K_V1.6 (*Kcna6*), K_V1.3 (*Kcnab3*), and K_V4.3 (*Kcnd3*).

K_V1.6 was detected via immunocytochemistry (ICC) and patch clamp experiments showing dendrotoxin-sensitive K^+ currents in cultured rat astrocytes [135]. Additional work in rat cultures shows with ICC that K_V1.6 is largely retained in the endoplasmic reticulum, while $K_V1.3$ is highly enriched in the Golgi and may mediate membrane polarity of these organelles [136]. Moreover, a specific $K_V1.3$ inhibitor psoralen decreases the amplitude of astrocyte calcium signals in rat cultures leading to increased glutamate uptake, as indicated by altered D-[3H] aspartate levels used to measure glutamate transport. [137] These fascinating results potentially couple astrocytic voltage-gated channels to pH sensing. Further, $K_V4.3$ was shown to colocalize with GFAP in cultured rat astrocytes in both somata and astrocytic processes, and whole-cell patch clamp experiments performed in the presence of tetraethylammonium (TEA) or 4-aminopyridine elucidate that these channels contribute to about 70% of astrocytic K^+ A currents [138], though the biological functions that this imparts were not explored.

Another class of K^+ channels is the two-pore domain (K_2P) channels, a relatively novel class of K^+ leak channels identified in neurons [139]. Mammals express 15 K_2P channel variants, with Kcnk1 (TWIK1) expressed in both astrocytes and neurons in our cortical dataset (Fig. 1e), though other K_2P variants have been identified in both cell types in other studies [139]. Neuronal *Kcnk1* gates excitability by modulating membrane potential [140]. Dysregulation of *Kcnk1* is linked to Parkinson's disease (PD), though its mechanistic role in this pathology is largely unknown [141]. In astrocytes, Kcnk1 (TWIK1) and *Kcnk2* (TREK1) were shown by IHC to colocalize with astrocytic markers GFAP and glutamate-aspartate transporter (GLAST; EAAT1; *Slc1a3*) in rat hippocampus, while voltage clamp experiments demonstrate reduced astrocytic passive K^+ conductance in response to quinine and pH reduction, which selectively inhibit these channels [142]. Moreover, it has been shown using the sniffer patch technique that these channels facilitate glutamate release upon astrocytic G protein-coupled receptor (GPCR) stimulation [143].

3.2.3 Voltage-Gated Calcium (Ca^{2+}) Channels (Ca_V)

Ca_Vs that are present at the presynaptic terminal regulate the vesicular release required for synaptic transmission [144]. We identified *Cacna1a* mRNA, which encodes the P/Q type Ca^{2+} channel $Ca_V2.1$, in the astrocytic translatome (Fig. 1e). $Ca_V2.1$ dysregulation is linked to ataxia [145], neurodevelopmental [146], and neurodegenerative [147] disorders. In cultured astrocytes, $Ca_V2.1$ has been identified using ICC and calcium imaging using its specific pharmacological inhibitor agatoxin [148]. Given that extracellular Ca^{2+} entry has been shown to induce astrocytic stored calcium release [149], these findings suggest that $CA_V2.1$ may be part of a calcium toolkit involved in the regulation of astrocytic store-released calcium signaling.

In summary, these studies demonstrate that astrocytes express ion channels which were largely thought to be specific to maintaining neuronal membrane properties. Interestingly, some of these channels are expressed in subcellular organelles or are upregulated in astrocytes in response to injury. However, the biological role these channels play in astrocytes, and how this relates to synapse regulation is still largely unknown, raising some fundamental questions: Do these genes encode for functional ion-conducting channels? How do these ion channels contribute to astrocytic function, and in particular, with regard to its regulation of synapses?

3.3 *Astrocytes and Neurons Express Many of the Same Synaptic Receptors but Utilize Downstream Signaling Components in Unique Ways*

Converting the electrical action potential to a chemical signal is central to information processing at the synapse. Signaling molecules, e.g., neurotransmitters or neuropeptides, are released from presynaptic terminals, diffuse across the synaptic cleft, and bind their target receptors, which can be present not only on the neuron but also on neighboring astrocytes. In fact, astrocytes express many neurotransmitter and neuromodulator receptors and can also release an array of molecules (referred to as gliotransmitters) including glutamate, GABA, ATP, adenosine, and neuropeptides [150]. This bidirectional signaling relationship between astrocytes and neurons regulates many aspects of synaptic development, maintenance, and plasticity [28, 150–154]. Despite the known overlap of signaling components in astrocytes and neurons, the majority of astrocyte research is focused on neurotransmitter uptake and metabolism, a critically important function to ensure synaptic integrity, while thorough characterization of the astrocytic expression patterns and functions of neurotransmitter receptors is lacking. Here, we focus on neurotransmitter and neuromodulator receptors, reviewing evidence for their protein expression and cell-specific functions in astrocytes.

3.3.1 Glutamate Receptors

Glutamate is the primary excitatory neurotransmitter in the mammalian brain. Upon release to the synaptic cleft, it is taken up and metabolized by astrocytes where it is converted to glutamine and released to be utilized by neurons and other cells as a precursor for glutamate and, subsequently, GABA [155, 156]. Precise regulation of synaptic glutamate is critical for brain health, while disruptions to glutamate homeostasis are linked to cell death due to excitotoxicity, neurodegeneration, and other disorders [157, 158, 60, 61]. Glutamate influences many aspects of astrocytic function, including the release of calcium from intracellular stores [64, 159, 160], changes in morphology [161], metabolism [40, 155], secretion of signaling molecules [162–164], and gene expression [43], which in turn modulate astrocyte regulation of the synapse. However, while mechanisms of glutamate metabolism and uptake through specialized transporters have been thoroughly studied in astrocytes [165, 166], the function of both ionotropic and metabotropic glutamate receptors has been primarily investigated in neurons. Here, we highlight the astrocytic expression patterns and functions of some of these receptors.

Ionotropic glutamate receptors AMPARs, NMDARs, and kainate receptors (KARs) are ligand-gated ion channels that conduct cations upon glutamate stimulation.

AMPARs mediate the majority of fast glutamatergic signaling in the brain [167] and play a central role in synaptic plasticity [167, 168]. AMPARs form heterotetramers composed of subunits GLUA1-4 [169] (*Gria1-4*). The specific subunit combination confers distinct physiological characteristics, such as the channel's calcium permeability, which is determined by the presence of GLUA2 [170, 171]. Neurons and astrocytes express all four AMPAR subunits (Fig. 1e), though at different levels [106], implying that astrocytic AMPARs may have unique functional specializations. Electrophysiological studies using specific pharmacological inhibitors demonstrate that astrocytes in most areas of the CNS express functional AMPARs [106]. Interestingly, astrocytes, like neurons, are characterized by regionally specific AMPAR subunit compositions, where GLUA2-containing AMPARs dominate in the cortex, hippocampus, and olfactory bulb, while the cerebellum expresses mainly GLUA4-containing AMPARs [106, 167]. In neurons, C-terminal phosphorylation and trafficking of AMPARs are linked to long-term potentiation (LTP), long-term depression (LTD), and modulation of glutamatergic transmission [172]. While astrocytic roles of AMPAR-mediated plasticity are relatively unexplored, calcium imaging studies show that AMPAR stimulation induces calcium flux, the major intracellular signaling mechanism in astrocytes, influencing gliotransmitter release [173, 174]. Moreover, in white matter astrocytes, activation of AMPARs contributes to excitotoxic damage after spinal cord injury [175].

NMDARs mediate glutamatergic signaling with slower synaptic responses relative to AMPARs. At rest, NMDARs are blocked by Mg^{2+} ions within the channel pore and require both glutamate and depolarization to be activated [176]. Their main function is to mediate LTP and LTD through calcium-dependent membrane trafficking of AMPARs [177], as well as promote development and survival [178]. In most neurons, NMDARs are heterotetramers comprised of two obligatory glycine-binding subunits GLUN1 (*Grin1*) and glutamate-binding GLUN2A-D (*Grin2a-d*) subunits, which confer different levels of Mg^{2+} sensitivity and, specifically in development, GLUN3 (*Grin3a-b*) subunits [179]. Like with AMPARs, the subunit combination confers distinct physiological characteristics with prominent implications for neural excitability, plasticity, and synaptic function in neurons. Importantly, though it is well established that astrocytes modulate neuronal NMDAR activation through the release of glutamate and D serine [180–186], fewer studies have investigated the functional roles of NMDARs located within astrocytic membranes.

All seven NMDAR subunits have been identified in cultured rat and human astrocytes, albeit at lower abundance than those observed in neurons [187] (Fig. 1e), and with differential subunit compositions [188]. Electrophysiological studies in culture and acute mouse cortical brain slices indicate that, unlike neurons, astrocytes can express receptors composed of Mg^{2+} insensitive subunits based on their differential sensitivity to GLUN2C/D-selective antagonist UBP141 [188] and elicit functionally unique calcium signals in response to receptor activation [187].

Pivotally, these findings demonstrate that mature astrocytes, in contrast to neurons, express NMDARs characterized by tri-heteromeric structures which include GLUN1, GLUN2C/D, and GLUN3 subunits and can be active at resting membrane potential. Moreover, physiological NMDA stimulation enhances astrocytic GS synthesis, while overexposure compromises GS, aquaporin-4 (AQP4; *Aqp4*), and Kir 4.1 (*Kcnj10*) expression, thus likely contributing to the pathophysiology of excitotoxicity [179, 187, 189].

Kainate receptors (KARs) like NMDARs mediate slow kinetic cation currents in response to glutamate [190]. In neurons, these receptors are widely distributed throughout the brain and play roles in development and plasticity, with disruptions in KAR expression or function linked to epilepsy and anxiety [190–192]. Though our analysis detected very low mRNA levels of KAR *Grik1-4* subunits in cortical astrocytes (Fig. 1e), it was shown by IHC that astrocytic GRIK1-5 (*Grik1-5*) protein levels are increased in GFAP-positive reactive astrocytes one week after the induction of status epilepticus (SE) in adult rodents [193]. Interestingly, elevated astrocytic GRIK1 and 5 expression persisted at 8 weeks post-SE, which corresponds with the development of spontaneous epileptic activity that occurs as a consequence of SE [194]. These findings imply that astrocytic KAR upregulation may play a role in the development of adult idiopathic epilepsies and seizure disorders, which occur after CNS injury.

Metabotropic glutamate receptors (MGLURs) are G protein-coupled receptors (GPCRs) that activate intracellular signaling cascades in response to glutamate [195]. We identified the mRNA expression of MGLURs 2, 3, and 5 (*Grm2, 3, 5*) in astrocytes (Fig. 1e). In neurons, MGLUR5 is enriched in brain regions related to learning and memory, including the hippocampus, cortex, and amygdala, where it associates with NMDARs [196]. MGLUR5 stimulation induces phospholipase C (PLC) activation followed by intracellular calcium signaling cascades and the subsequent activation of protein kinase C (PKC), which facilitates synaptic plasticity, affecting learning and memory [196, 197]. In astrocytes, MGLUR5 expression is highest during development where it induces calcium signaling cascades, stimulating gliotransmission in regions including cortex, hippocampus, and striatum [198, 199]. Disruption of astrocytic MGLUR5 is linked to depression [200], and in human iPSC-derived astrocytes, MGLUR5 blockade enhances astrocytic phagocytosis [201]. Interestingly, astrocytic MGLUR5 is transiently re-expressed in adult brains after pathological events, where it mediates synapse remodeling [202].

Neuronal MGLUR2 and MGLUR3 are primarily presynaptic and promote inhibition of neuronal activity, including inhibition of adenylyl cyclase and calcium channels, and activation of potassium channels [195]. Surprisingly, stimulation of MGLUR2 and MGLUR3 in astrocytes facilitates the release of intracellular calcium inducing downstream signals that enhance excitatory neuronal network activity [160]. While the precise cellular pathways mediating these functions are unknown, recent evidence suggests that in astrocytes, MGLUR activity is cooperative. In the developing auditory cortex, sensory-evoked neuronal activity induces robust calcium signals in astrocytes, but only when both MGLUR3 and MGLUR5 are simultaneously stimulated [203].

3.3.2 GABA Receptors

GABA signaling mediates the majority of inhibition in the adult brain through its hyperpolarizing effects on the postsynaptic neuron [204]. GABAergic circuits are found across the entire brain and are critical for maintaining homeostasis of excitation state. Loss or reduction of GABAergic cells is associated with epilepsy, Huntington's disease, ADHD, and other conditions characterized by excitatory imbalance [205–207], rendering GABA receptors (GABARs) as prominent targets for therapeutics, such as anesthetics, analgesics, and antipsychotics [208]. Neurons and astrocytes are both GABAergic (GABA-releasing) and GABAceptive (GABA sensing) cells expressing both ionotropic ($GABA_A$) and metabotropic ($GABA_B$) GABARs (Fig. 1e). Much like with glutamate, research regarding GABA in astrocytes has been mainly focused on understanding its transport and metabolism, while the function of astrocytic GABARs is less understood. We highlight the roles of GABARs in astrocytes below.

$GABA_A$Rs are ligand-gated chloride channels with complex molecular compositions formed from a potential combination of 19 subunits (α-θ) encoded across seven different chromosomes in mammals [209], resulting in a large number of isoforms with distinct physiological properties, though their distributions in astrocytes have not been extensively characterized. Evidence from qPCR studies suggests that astrocytes may lack specific receptor subunits compared to neurons, influencing their localization in the membrane [210], and our analysis identified astrocytic expression of several subunits including *Gabra1-5*, *Gabrb 1-3*, *Gabrg1-2*, and *Gabrd* (Fig. 1e). Transient neuronal stimulation of $GABA_A$Rs results in phasic inhibition via calcium/calmodulin-stimulated protein kinase II (CAMKII), while sustained tonic inhibition is induced through protein kinase A (PKA) activation. In astrocytes, stimulation of $GABA_A$R also induces CAMKII activity, which is thought to modulate glutamatergic gliotransmission, and the expression of GABA transporters (GATs), which mediate GABA removal from the synaptic cleft, and is thus a critical mechanism for maintaining excitation/inhibition balance [211]. Notably, unlike in neurons, $GABA_A$R stimulation is depolarizing in astrocytes because their resting membrane potential is hyperpolarized relative to the equilibrium potential for chloride [211].

Much of the work detailing the roles of $GABA_A$R signaling in astrocytes has been focused on the effects of drugs such as general anesthetics. For example, it was shown that the anesthetics etomidate and sevoflurane enhance astrocyte chloride currents and prolong decay kinetics in hippocampal rodent brain slices [212]. Moreover, in cultured cells it is demonstrated that these drug interactions with astrocytic $GABA_A$Rs lead to the astrocytic release of paracrine factors, which underlie an increase in tonic inhibitory currents in neurons, purportedly contributing to cognitive deficits observed in postoperative patients [212].

$GABA_B$Rs are GPCRs that facilitate slow, persistent GABA signaling leading to inhibition of adenylyl cyclase and voltage-gated Ca^{2+} channels and activation of K^+ channels in neurons [213]. $GABA_B$Rs are composed of two subunits expressed in both astrocytes and neurons: $GABA_{B1}$ (*Gabbr1*) and $GABA_{B2}$ (*Gabbr2*; Fig. 1e).

Like other astrocytic GPCRs, and in contrast to their function in neurons, $GABA_BR$ signaling promotes the release of stored calcium, activating downstream signaling pathways that induce, among other things, changes in astrocytic gene expression. One important regulatory role of GABAergic signaling in astrocytes is the control of morphogenesis. Immunofluorescence experiments in juvenile mice show that targeted deletion of astrocytic $GABA_BR$ results in a loss of morphological complexity, corresponding with reduced activity in neuronal circuits [214]. This indicates that GABA-mediated crosstalk between astrocytes and neurons underlies an important, bidirectional signaling relationship necessary for proper circuit development.

3.3.3 Neuromodulatory Receptors and Ligands

Neuromodulators, such as D serine, adenosine, and sphingosine, regulate neuronal excitability and circuit function through GPCR-induced downstream signaling cascades. Astrocytes and neurons both express various neuromodulatory signaling molecules and receptors (Fig. 1e), the roles of which in astrocytes are highlighted below.

Sphingosine-1-phosphate receptor type 1 (S1PR1; *S1pr1*) is a GPCR activated by sphingosine-1-phosphate (SP1; *Sp1*) to regulate neurotransmitter release and neuronal excitability [215]. In astrocytes, S1PR1 expression is induced by neuronal contact at the synapse during development driving astrocyte morphological maturation and complexity [216]. It has been demonstrated that astrocytic S1PR1 in spinal cord synapses modulate neuropathic pain and that astrocyte-specific ablation of *S1pr1* leads to complete reduction of neuropathic pain following nerve injury, pointing to astrocytic S1PR1 as the primary target for S1P activity in this context. [217]

Adenosine receptors are GPCRs involved in the regulation of sleep, blood flow, LTP, LTD, and inflammation and are neuroprotective against excitotoxicity [218–221]. We show the presence of $A_{1A}R$ (*Adora1a*) and $A_{2A}R$ (*Adora2a*) mRNA in astrocytes (Fig. 1e). Though the effects of astrocytic release and subsequent hydrolysis of ATP have been characterized with regard to many of these functions [222–225], the activity of their adenosine receptors is less well described. In astrocytes, $A_{1A}R$ expression is considered protective, in part through the induction of phosphoinositide 3-kinase (PI3K) pathway [226, 227]. Specifically, astrocytic expression of $A_{1A}R$ is linked to a reduction in intracellular glutamate. After oxygen-glucose deprivation, $A_{1A}R$ dimerizes with $A_{2A}R$ to modulate GLT1 levels affecting glutamate levels and thus excitotoxicity [220]. Further, astrocytic $A_{2A}R$ alterations are closely associated with pathological phenotypes in Alzheimer's disease [228], whereupon its upregulation corresponds with disease severity in humans and contributes to memory loss in rodent models [229].

Somatostatin (SST; *Sst*) is a peptide released from inhibitory interneurons in the CNS acting on both the pre- and postsynaptic neurons to regulate excitability [230]. In the cortex, SST is co-released with GABA where it diffusely regulates the activity of the postsynaptic neuron and neighboring cells, including astrocytes, which, like neurons, express somatostatin receptor type 2 (SSTR2; *Sstr2*). In the rodent visual cortex, it was recently demonstrated that optogenetic stimulation of SST-releasing interneurons induces a robust increase in calcium signal amplitude and frequency in perisynaptic astrocyte processes through the coactivation of $GABA_BR$. This leads to the disinhibition of layer 2/3 pyramidal neuronal activity, measured by whole-cell patch clamp. This effect is ablated in inositol-1,4,5-trisphosphate receptor type 2 (IP3R2; *Itpr2*)-deficient mice, indicating that astrocytic SST stimulation modulates these responses in a calcium-dependent manner [231]. Further, upregulated SSTR2 has been identified in astrocyte-derived glioma, suggesting that the receptor plays a role in astrocytic tumorigenesis and disease progression, which may serve as a promising therapeutic target in translational studies [232].

To summarize, here we overviewed a diverse array of synaptic receptors expressed in astrocytes providing insight into their potential implications for normal physiology, pathology, and their viability as therapeutic targets for synaptic disorders. Importantly, we demonstrate that many of these receptors differ in their subunit composition between astrocytes and neurons, which holds prominent implications for their physiological functions across various contexts including development and pathology. Further, we reveal a distinction between the canonical function of GPCRs exhibiting either excitatory or inhibitory responses in neurons but mediating predominantly excitatory responses in astrocytes (at least with regard to calcium signals). This differentiation is critical for understanding the regulatory mechanisms of astrocytes, which influence synapse function. Overall, the stimulation of glutamatergic, GABAergic, and modulatory receptors in astrocytes is shown to facilitate robust crosstalk with neurons, emphasizing the need for studies to consider all synaptic components rather than neuronal constituents exclusively. Future work leveraging this comprehensive view is needed to answer the following questions: How do the physiological characteristics of the ionotropic neurotransmitter receptors (e.g., ligand-gated ion channels) in astrocytes differ from those in neurons? Does that influence astrocytic synapse regulation? Do neurotransmitters and neuromodulators exert the same regulatory effects in astrocytes as they do in neurons? If so, is this accomplished by similar intracellular signaling pathways?

3.4 The Matter of Vesicular Release from Astrocytes

One fundamental property of synaptic function is the release of neurotransmitters and neuromodulators, which form the basis of chemical communication between brain cells. This process involves the fusion of neurotransmitter-containing vesicles with the presynaptic membrane, release of their contents into the synaptic cleft, and the recycling of vesicles and subsequent replenishment with cargo [233]. Though this process has been extensively studied in neurons, the concept of vesicular release from astrocytes has been a highly debated topic in the field. Intriguingly, astrocytes express many of the same proteins related to vesicular release as neurons albeit for the most part, at lower levels (Fig. 1e). Here, we review studies that describe the presence, functions, and pathological implications of vesicular release components in astrocytes and compare their expression patterns and roles with those characterized in neurons.

3.4.1 Vesicular Neurotransmitter Transporters

These solute carriers located in the membranes of synaptic vesicles are categorized into three families, *Slc17*, *Slc18*, and *Slc32*, based on their affinity for negatively charged, positively charged, and neutral cargo, respectively [234–236]. Though vesicular transporters have been extensively characterized in neurons, the presence of astrocytic synaptic vesicles, and their related proteins, has been controversial [237]. A recent study leveraged HiPlex RNAscope experiments to demonstrate that *Slc17a7* and *Slc17a6*, which encode vesicular glutamate transporters (VGLUT) 1 and 2, respectively, are expressed in distinct subpopulations of astrocytes. While VGLUT1-positive astrocytes were primarily restricted to the hippocampus, VGLUT2-positive cells were identified in the substantia nigra. Using a combinatorial approach of designer receptors exclusively activated by designer drugs (DREADDs) and iGluSnFR glutamate sensor, it was demonstrated that stimulating these astrocytes induces fast glutamatergic gliotransmission in adult rodents. Further, targeted deletion of either VGLUT1 or VGLUT2 led to functional deficits in cortico-hippocampal and dopaminergic nigrostriatal circuits, respectively, suggesting an important role for astrocytic vesicular glutamate release in modulating these circuits [88].

3.4.2 Synaptic Vesicle Fusion and Recycling-Related Proteins

Vesicle-associated membrane protein 2 (VAMP2; *Vamp2*), also known as synaptobrevin-2, is a member of the SNARE complex, along with syntaxin and SNAP25 [238]. Changes in the neuronal levels and functions of VAMP2 have been associated with developmental pathophysiology [10], neurodegenerative disorders [239], and epilepsy [240]. Astrocytes show immunoreactivity to VAMP2 by ICC, and increases in cytosolic calcium levels are shown to induce VAMP2 mobilization and vesicle docking at the membrane of cultured rat astrocytes [241, 242]. VAMP2 is primarily located in perisynaptic astrocyte processes, where it is shown to modulate synaptic efficacy [243]. Targeted ablation of astrocytic VAMP2 leads to a decreased frequency of miniature and evoked excitatory postsynaptic currents (m/eEPSCs) in cultured neurons, indicating a role for VAMP2-mediated astrocytic glutamate release in modulating synaptic strength [243].

Synaptotagmin-11 (SYT11; *Syt11*) regulates vesicular fusion by binding calcium in presynaptic terminals [244]. Constitutive deletion of SYT11 leads to death shortly after birth in mice, while its conditional KO in glutamatergic forebrain neurons leads to deficits in synaptic plasticity and memory [245], and its mutation is associated with schizophrenia [246] and PD [247]. In cultured astrocytes, SYT11 is localized to caveolae, membrane microinvaginations that are not present in neurons, where it modulates responses to mechanical stimuli [248]. Because caveolae are distributed across the astrocytic membrane, SYT11-mediated functions in these structures could have implications for specialized vesicular release-mediated gliotransmission.

Complexin 2 (CPLX2; *Cplx2*) is a SNARE complex component that, in neurons, is specific to excitatory presynaptic terminals, where it mediates vesicular fusion through interactions with other SNARE proteins [249]. Astrocytic CPLX2, the sole complexin expressed in these cells, has been identified at the protein level in cell culture using ICC [250]. Though studies detailing its functions are mostly lacking, the astrocytic expression of CPLX2 points to a potential mechanism for the release of VGLUT1- and VGLUT2-containing vesicles.

Clathrin light-chain A (CLTA; *Clta*) is a major component of clathrin function, mediating the endocytic recycling of synaptic vesicles [251]. In astrocytes, this has mainly been studied in the context of Alzheimer's disease (AD), in which clathrin inhibition disrupts the endocytosis of Aβ proteins, suggesting it as a therapeutic target for mitigating AD pathogenesis [252].

Taken together, we provide evidence for the astrocytic expression of vesicular release-related proteins. Importantly, we discuss recent findings that identify VGLUT1 and VGLUT2 functions in distinct astrocytic subpopulations. This stresses the need for considering regional heterogeneity in the characterization of proteins in astrocytes (and other cells). Moreover, the presence of SNARE components in astrocytes suggests the occurrence of synaptic vesicle release processes in astrocytes. More work is needed to characterize the precise mechanisms of vesicular release and recycling in astrocytes and answer prominent questions surrounding their roles, such as follows: Which molecules are released from astrocytic vesicles? Do astrocytes exhibit specialized synaptic regions analogous to the neuronal presynaptic active zones?

4 Concluding Remarks and Outlook

Astrocytes are critically important for the development and maintenance of many synapse types across the CNS. In this chapter, we describe astrocyte-neuron interactions at the synapse by bringing attention to synaptic genes and their protein products expressed by both cell types. We overview these in the context of synapse organization, signaling receptors and channels, and vesicular release and, fascinatingly, show that astrocytes robustly express numerous synapse-related proteins known to play critical roles in synapse structure and function, which have been thus far studied primarily in neurons. Further, we show that the majority of the discussed astrocyte-focused studies provide evidence for their protein expression in astrocytes, while their functional roles are largely unknown, underscoring a need to characterize their astrocytic roles, particularly in vivo. For example, work describing the physiological characteristics of astrocytic proteins that in neurons comprise the postsynaptic density will be critical for understanding the ways that astrocytes organize their synapse-interacting compartment. This is especially important since no clear evidence of postsynaptic density, as it is present in neurons, has been observed thus far in astrocytes by electron microscopy (EM). While this may be due to limitations of sample preparation or imaging parameters, it could also reflect a different mode of astrocyte-specific organization at the tripartite synapse.

Furthermore, because astrocytes express proteins that, in neurons, are localized to either pre- or postsynaptic compartments, it is unclear how these structures are arranged within the synapse-contacting astrocyte process. Are both postsynaptic density and vesicular release-related proteins present within the same astrocytic process? Or are there "pre"- or "post"-synaptic specialized astrocytic contacts that, on one hand, sense the chemical signal released from the neuron, and on the other, mediate the astrocytic response, such as release of gliotransmitters? Thus, delineating the astrocyte-specific roles will require studies that leverage multifaceted approaches to determine whether they underlie identical or unique functions at the synapse compared to neurons. Finally, we show that for some of the discussed proteins, namely voltage- and ligand-gated ion channels, astrocytes have the remarkable capacity to upregulate their expression in the event of CNS injury or insult. What roles these upregulated channels play in astrocytes in the injured brain are unknown and need to be systematically addressed, particularly in the context of targeting astrocytes for therapeutics.

Altogether, we highlight the potential roles these uncharacterized astrocytic proteins may play in synapse regulation, and the importance of objectively considering traditionally neuron-centric questions surrounding them, such as: In diseases characterized by mutations in synaptic genes (such as ASD, epilepsy, and AD), what is the astrocytic contribution to pathophysiology? When we attempt to target these elements for therapeutic manipulation, how are we affecting cellular interactions between astrocytes and neurons and what are the consequences? Future studies focusing on the molecular mechanisms of astrocytic regulation at the synapse will be able to answer these questions, providing critical advancement to the field by not only increasing scientific knowledge, but also identifying novel targets for translational work.

Acknowledgments This work was supported by the NIH/NINDS grant (R01NS133047) to IFT. We thank Dr. Heath Blackmon for the critical reading of the manuscript and Dr. James Cai for advice on the meta-analysis. Components of Fig. 1a were created using Biorender. We apologize to those whose work was not cited due to our oversight or to space considerations.

References

1. Batool S, et al. Synapse formation: from cellular and molecular mechanisms to neurodevelopmental and neurodegenerative disorders. J Neurophysiol. 2019;121:1381–97. https://doi.org/10.1152/jn.00833.2018.
2. Cohen-Cory S. The developing synapse: construction and modulation of synaptic structures and circuits. Science. 2002;298:770–6. https://doi.org/10.1126/science.1075510.

3. Cagnetta R, Flanagan JG, Sonenberg N. Control of selective mRNA translation in neuronal subcellular compartments in health and disease. J Neurosci. 2023;43:7247–63. https://doi.org/10.1523/JNEUROSCI.2240-22.2023.
4. Sudhof TC. The presynaptic active zone. Neuron. 2012;75:11–25. https://doi.org/10.1016/j.neuron.2012.06.012.
5. Sheng M, Kim E. The postsynaptic organization of synapses. Cold Spring Harb Perspect Biol. 2011;3 https://doi.org/10.1101/cshperspect.a005678.
6. Meldolesi J. Post-synapses in the brain: role of dendritic and spine structures. Biomedicines. 2022;10 https://doi.org/10.3390/biomedicines10081859.
7. Kassabian B, et al. Developmental epileptic encephalopathy in DLG4-related synaptopathy. Epilepsia. 2023; https://doi.org/10.1111/epi.17876.
8. Meftah S, Gan J. Alzheimer's disease as a synaptopathy: evidence for dysfunction of synapses during disease progression. Front Synaptic Neurosci. 2023;15:1129036. https://doi.org/10.3389/fnsyn.2023.1129036.
9. Reece AS, Hulse GK. Perturbation of 3D nuclear architecture, epigenomic aging and dysregulation, and cannabinoid synaptopathy reconfigures conceptualization of cannabinoid pathophysiology: part 2-metabolome, immunome, synaptome. Front Psych. 2023;14:1182536. https://doi.org/10.3389/fpsyt.2023.1182536.
10. Salpietro V, et al. Mutations in the neuronal vesicular SNARE VAMP2 affect synaptic membrane fusion and impair human neurodevelopment. Am J Hum Genet. 2019;104:721–30. https://doi.org/10.1016/j.ajhg.2019.02.016.
11. Antoine MW, Langberg T, Schnepel P, Feldman DE. Increased excitation-inhibition ratio stabilizes synapse and circuit excitability in four autism mouse models. Neuron. 2019;101:648–661.e644. https://doi.org/10.1016/j.neuron.2018.12.026.
12. Galineau L, et al. Glutamatergic synapse in autism: a complex story for a complex disorder. Mol Psychiatry. 2023;28:801–9. https://doi.org/10.1038/s41380-022-01860-9.
13. Ford TJL, Jeon BT, Lee H, Kim WY. Dendritic spine and synapse pathology in chromatin modifier-associated autism spectrum disorders and intellectual disability. Front Mol Neurosci. 2022;15:1048713. https://doi.org/10.3389/fnmol.2022.1048713.
14. Anderson MP. Arrested glutamatergic synapse development in human partial epilepsy. Epilepsy Curr. 2010;10:153–8. https://doi.org/10.1111/j.1535-7511.2010.01386.x.
15. Loucks CM, et al. EFHC1, implicated in juvenile myoclonic epilepsy, functions at the cilium and synapse to modulate dopamine signaling. eLife. 2019;8 https://doi.org/10.7554/eLife.37271.
16. Chen X, Durisic N, Lynch JW, Keramidas A. Inhibitory synapse deficits caused by familial alpha1 GABA(a) receptor mutations in epilepsy. Neurobiol Dis. 2017;108:213–24. https://doi.org/10.1016/j.nbd.2017.08.020.
17. Andersen JV, et al. Glutamate metabolism and recycling at the excitatory synapse in health and neurodegeneration. Neuropharmacology. 2021;196:108719. https://doi.org/10.1016/j.neuropharm.2021.108719.
18. Handley EE, et al. Synapse dysfunction of layer V pyramidal neurons precedes neurodegeneration in a mouse model of TDP-43 Proteinopathies. Cereb Cortex. 2017;27:3630–47. https://doi.org/10.1093/cercor/bhw185.
19. Henstridge CM, Tzioras M, Paolicelli RC. Glial contribution to excitatory and inhibitory synapse loss in neurodegeneration. Front Cell Neurosci. 2019;13:63. https://doi.org/10.3389/fncel.2019.00063.
20. Zhou J, et al. The neuronal pentraxin Nptx2 regulates complement activity and restrains microglia-mediated synapse loss in neurodegeneration. Sci Transl Med. 2023;15:eadf0141. https://doi.org/10.1126/scitranslmed.adf0141.
21. Araque A, et al. Tripartite synapses: glia, the unacknowledged partner. Trends Neurosci. 1999;22:208–15. https://doi.org/10.1016/S0166-2236(98)01349-6.
22. Halassa MM, Fellin T, Haydon PG. The tripartite synapse: roles for gliotransmission in health and disease. Trends Mol Med. 2007;13:54–63. https://doi.org/10.1016/j.molmed.2006.12.005.

23. Semyanov A, Verkhratsky A. Astrocytic processes: from tripartite synapses to the active milieu. Trends Neurosci. 2021;44:781–92. https://doi.org/10.1016/j.tins.2021.07.006.
24. Perea G, Navarrete M, Araque A. Tripartite synapses: astrocytes process and control synaptic information. Trends Neurosci. 2009;32:421–31. https://doi.org/10.1016/j.tins.2009.05.001.
25. Schober AL, Wicki-Stordeur LE, Murai KK, Swayne LA. Foundations and implications of astrocyte heterogeneity during brain development and disease. Trends Neurosci. 2022;45:692–703. https://doi.org/10.1016/j.tins.2022.06.009.
26. Bastian C, et al. Aging astrocytes metabolically support aging axon function by proficiently regulating astrocyte-neuron lactate shuttle. Exp Neurol. 2022;357:114173. https://doi.org/10.1016/j.expneurol.2022.114173.
27. Boisvert MM, Erikson GA, Shokhirev MN, Allen NJ. The aging astrocyte transcriptome from multiple regions of the mouse brain. Cell Rep. 2018;22:269–85. https://doi.org/10.1016/j.celrep.2017.12.039.
28. Farhy-Tselnicker I, Allen NJ. Astrocytes, neurons, synapses: a tripartite view on cortical circuit development. Neural Dev. 2018;13:7. https://doi.org/10.1186/s13064-018-0104-y.
29. Guerra-Gomes S, Sousa N, Pinto L, Oliveira JF. Functional roles of astrocyte calcium elevations: from synapses to behavior. Front Cell Neurosci. 2018;11:427. https://doi.org/10.3389/fncel.2017.00427.
30. Halassa MM, Fellin T, Takano H, Dong J-H, Haydon PG. Synaptic Islands defined by the territory of a single astrocyte. J Neurosci. 2007;27:6473–7. https://doi.org/10.1523/jneurosci.1419-07.2007.
31. Bushong EA, Martone ME, Jones YZ, Ellisman MH. Protoplasmic astrocytes in CA1 stratum Radiatum occupy separate anatomical domains. J Neurosci. 2002;22:183–92. https://doi.org/10.1523/JNEUROSCI.22-01-00183.2002.
32. Baldwin KT, Eroglu C. Molecular mechanisms of astrocyte-induced synaptogenesis. Curr Opin Neurobiol. 2017;45:113–20. https://doi.org/10.1016/j.conb.2017.05.006.
33. Walker CD, Risher WC, Risher M-L. Regulation of synaptic development by astrocyte signaling factors and their emerging roles in substance abuse. Cells. 2020;9:297. https://doi.org/10.3390/cells9020297.
34. Kadry H, Noorani B, Cucullo L. A blood–brain barrier overview on structure, function, impairment, and biomarkers of integrity. Fluids Barriers CNS. 2020;17:69. https://doi.org/10.1186/s12987-020-00230-3.
35. de Rus Jacquet A, et al. The contribution of inflammatory astrocytes to BBB impairments in a brain-chip model of Parkinson's disease. Nat Commun. 2023;14:3651. https://doi.org/10.1038/s41467-023-39038-8.
36. Bellot-Saez A, Kékesi O, Morley JW, Buskila Y. Astrocytic modulation of neuronal excitability through K+ spatial buffering. Neurosci Biobehav Rev. 2017;77:87–97. https://doi.org/10.1016/j.neubiorev.2017.03.002.
37. Boddum K, et al. Astrocytic GABA transporter activity modulates excitatory neurotransmission. Nat Commun. 2016;7:13572. https://doi.org/10.1038/ncomms13572.
38. Cuellar-Santoyo AO, et al. Revealing the contribution of astrocytes to glutamatergic neuronal transmission. Front Cell Neurosci. 2023;16:1037641. https://doi.org/10.3389/fncel.2022.1037641.
39. Schousboe A, Bak LK, Waagepetersen HS. Astrocytic control of biosynthesis and turnover of the neurotransmitters glutamate and GABA. Front Endocrinol (Lausanne). 2013;4(102) https://doi.org/10.3389/fendo.2013.00102.
40. Schousboe A, Scafidi S, Bak LK, Waagepetersen HS, McKenna MC. Glutamate metabolism in the brain focusing on astrocytes. Adv Neurobiol. 2014;11:13–30. https://doi.org/10.1007/978-3-319-08894-5_2.
41. Ghirardini E, et al. Expression of functional inhibitory neurotransmitter transporters GlyT1, GAT-1, and GAT-3 by astrocytes of inferior colliculus and hippocampus. Mol Brain. 2018;11:4. https://doi.org/10.1186/s13041-018-0346-y.

42. Holt MG. Astrocyte heterogeneity and interactions with local neural circuits. Essays Biochem. 2023;67:93–106. https://doi.org/10.1042/EBC20220136.
43. Farhy-Tselnicker I, et al. Activity-dependent modulation of synapse-regulating genes in astrocytes. elife. 2021;10 https://doi.org/10.7554/eLife.70514.
44. Bayraktar OA, et al. Astrocyte layers in the mammalian cerebral cortex revealed by a single-cell in situ transcriptomic map. Nat Neurosci. 2020;23:500–9. https://doi.org/10.1038/s41593-020-0602-1.
45. Karpf J, et al. Dentate gyrus astrocytes exhibit layer-specific molecular, morphological and physiological features. Nat Neurosci. 2022;25:1626–38. https://doi.org/10.1038/s41593-022-01192-5.
46. Teschemacher AG, Gourine AV, Kasparov S. A role for astrocytes in sensing the brain microenvironment and neuro-metabolic integration. Neurochem Res. 2015;40:2386–93. https://doi.org/10.1007/s11064-015-1562-9.
47. Adermark L, et al. Astrocytes modulate extracellular neurotransmitter levels and excitatory neurotransmission in dorsolateral striatum via dopamine D2 receptor signaling. Neuropsychopharmacology. 2022;47:1493–502. https://doi.org/10.1038/s41386-021-01232-x.
48. Liu X, et al. Astrocytes in neural circuits: key factors in synaptic regulation and potential targets for neurodevelopmental disorders. Front Mol Neurosci. 2021;14:729273. https://doi.org/10.3389/fnmol.2021.729273.
49. Phatnani H, Maniatis T. Astrocytes in neurodegenerative disease. Cold Spring Harb Perspect Biol. 2015;7 https://doi.org/10.1101/cshperspect.a020628.
50. Hu C, Tao L, Cao X, Chen L. The solute carrier transporters and the brain: physiological and pharmacological implications. Asian J Pharm Sci. 2020;15:131–44. https://doi.org/10.1016/j.ajps.2019.09.002.
51. Chipman PH, et al. Astrocyte GluN2C NMDA receptors control basal synaptic strengths of hippocampal CA1 pyramidal neurons in the stratum radiatum. eLife. 2021;10 https://doi.org/10.7554/eLife.70818.
52. Lia A, et al. Ion channels and ionotropic receptors in astrocytes: physiological functions and alterations in Alzheimer's disease and glioblastoma. Life (Basel). 2023;13 https://doi.org/10.3390/life13102038.
53. Wang D-S, et al. Crosstalk between GABAA receptors in astrocytes and neurons triggered by general anesthetic drugs. Transl Res. 2023; https://doi.org/10.1016/j.trsl.2023.11.007.
54. Buscemi L, et al. Homer1 scaffold proteins govern Ca2+ dynamics in Normal and reactive astrocytes. Cereb Cortex. 2017;27:2365–84. https://doi.org/10.1093/cercor/bhw078.
55. Stogsdill JA, et al. Astrocytic neuroligins control astrocyte morphogenesis and synaptogenesis. Nature. 2017;551:192–7. https://doi.org/10.1038/nature24638.
56. Collins SM, Belagodu AP, Reed SL, Galvez R. SHANK1 is differentially expressed during development in CA1 hippocampal neurons and astrocytes. Dev Neurobiol. 2018;78:363–73. https://doi.org/10.1002/dneu.22564.
57. Asher RA, et al. Neurocan is upregulated in injured brain and in cytokine-treated astrocytes. J Neurosci. 2000;20:2427–38. https://doi.org/10.1523/JNEUROSCI.20-07-02427.2000.
58. Zieger HL, Choquet D. Nanoscale synapse organization and dysfunction in neurodevelopmental disorders. Neurobiol Dis. 2021;158:105453. https://doi.org/10.1016/j.nbd.2021.105453.
59. Zoghbi HY, Bear MF. Synaptic dysfunction in neurodevelopmental disorders associated with autism and intellectual disabilities. Cold Spring Harb Perspect Biol. 2012;4 https://doi.org/10.1101/cshperspect.a009886.
60. Jackson J, et al. Targeting the synapse in Alzheimer's disease. Front Neurosci. 2019;13. https://doi.org/10.3389/fnins.2019.00735.
61. Herrero-Navarro A, et al. Astrocytes and neurons share region-specific transcriptional signatures that confer regional identity to neuronal reprogramming. Sci Adv. 2021;7 https://doi.org/10.1126/sciadv.abe8978.

62. Furlanis E, Traunmuller L, Fucile G, Scheiffele P. Landscape of ribosome-engaged transcript isoforms reveals extensive neuronal-cell-class-specific alternative splicing programs. Nat Neurosci. 2019;22:1709–17. https://doi.org/10.1038/s41593-019-0465-5.
63. Glock C, et al. The translatome of neuronal cell bodies, dendrites, and axons. Proc Natl Acad Sci USA. 2021;118 https://doi.org/10.1073/pnas.2113929118.
64. Chai H, et al. Neural circuit-specialized astrocytes: transcriptomic, proteomic, morphological, and functional evidence. Neuron. 2017;95:531–549.e539. https://doi.org/10.1016/j.neuron.2017.06.029.
65. Clarke, L. E. et al. Normal aging induces A1-like astrocyte reactivity. Proceedings of the National Academy of Sciences. 2018;115:E1896–E1905. https://doi.org/10.1073/pnas.1800165115.
66. Zhang Y, et al. An RNA-sequencing transcriptome and splicing database of glia, neurons, and vascular cells of the cerebral cortex. J Neurosci. 2014;34:11929. https://doi.org/10.1523/JNEUROSCI.1860-14.2014.
67. Bakken TE, et al. Single-nucleus and single-cell transcriptomes compared in matched cortical cell types. PLoS One. 2018;13:e0209648. https://doi.org/10.1371/journal.pone.0209648.
68. Batiuk MY, et al. Identification of region-specific astrocyte subtypes at single cell resolution. Nat Commun. 2020;11:1220. https://doi.org/10.1038/s41467-019-14198-8.
69. Tasic B, et al. Shared and distinct transcriptomic cell types across neocortical areas. Nature. 2018;563:72–8. https://doi.org/10.1038/s41586-018-0654-5.
70. Jin Y, Garner CC. Molecular mechanisms of presynaptic differentiation. Annu Rev Cell Dev Biol. 2008;24:237–62. https://doi.org/10.1146/annurev.cellbio.23.090506.123417.
71. Sorokina, O. Synaptic proteome SQLite database, 2000–2020. 2021. https://doi.org/10.7488/ds/3017.
72. Sorokina O, et al. A unified resource and configurable model of the synapse proteome and its role in disease. Sci Rep. 2021;11:9967. https://doi.org/10.1038/s41598-021-88945-7.
73. Sanz E, et al. Cell-type-specific isolation of ribosome-associated mRNA from complex tissues. Proc Natl Acad Sci USA. 2009;106:13939–44. https://doi.org/10.1073/pnas.0907143106.
74. Heiman M, Kulicke R, Fenster RJ, Greengard P, Heintz N. Cell type-specific mRNA purification by translating ribosome affinity purification (TRAP). Nat Protoc. 2014;9:1282–91. https://doi.org/10.1038/nprot.2014.085.
75. Durinck S, Spellman PT, Birney E, Huber W. Mapping identifiers for the integration of genomic datasets with the R/Bioconductor package biomaRt. Nat Protoc. 2009;4:1184–91. https://doi.org/10.1038/nprot.2009.97.
76. Müller K, Wickham H, James DA, Falcon S. RSQLite: SQLite Interface for R. R package version 2.3.7. 2024. https://github.com/r-dbi/RSQLite, https://rsqlite.r-dbi.org.
77. Heberle H, Meirelles GV, da Silva FR, Telles GP, Minghim R. InteractiVenn: a web-based tool for the analysis of sets through Venn diagrams. BMC Bioinformatics. 2015;16:169. https://doi.org/10.1186/s12859-015-0611-3.
78. Gulyassy P, et al. Proteomic comparison of different synaptosome preparation procedures. Amino Acids. 2020;52:1529–43. https://doi.org/10.1007/s00726-020-02912-6.
79. Chen Y, Strickland MR, Soranno A, Holtzman DM. Apolipoprotein E: structural insights and links to Alzheimer disease pathogenesis. Neuron. 2021;109:205–21. https://doi.org/10.1016/j.neuron.2020.10.008.
80. Defilippi P, Di Stefano P, Cabodi S. p130Cas: a versatile scaffold in signaling networks. Trends Cell Biol. 2006;16:257–63. https://doi.org/10.1016/j.tcb.2006.03.003.
81. Schoch S, et al. SNARE function analyzed in synaptobrevin/VAMP knockout mice. Science. 2001;294:1117–22. https://doi.org/10.1126/science.1064335.
82. Sudhof TC, Rizo J. Synaptic vesicle exocytosis. Cold Spring Harb Perspect Biol. 2011;3 https://doi.org/10.1101/cshperspect.a005637.
83. Blasi J, et al. Botulinum neurotoxin a selectively cleaves the synaptic protein SNAP-25. Nature. 1993;365:160–3. https://doi.org/10.1038/365160a0.

84. Pirovich DB, Da'dara AA, Skelly PJ. Multifunctional fructose 1,6-bisphosphate aldolase as a therapeutic target. Front Mol Biosci. 2021;8:719678. https://doi.org/10.3389/fmolb.2021.719678.
85. Kucukdereli H, et al. Control of excitatory CNS synaptogenesis by astrocyte-secreted proteins Hevin and SPARC. Proc Natl Acad Sci USA. 2011;108:E440–9. https://doi.org/10.1073/pnas.1104977108.
86. Xiao L, Yang X, Loh YP. Neurotrophic, gene regulation, and cognitive functions of carboxypeptidase E-neurotrophic factor-alpha1 and its variants. Front Neurosci. 2019;13:243. https://doi.org/10.3389/fnins.2019.00243.
87. Szklarczyk D, et al. The STRING database in 2023: protein-protein association networks and functional enrichment analyses for any sequenced genome of interest. Nucleic Acids Res. 2023;51:D638–46. https://doi.org/10.1093/nar/gkac1000.
88. de Ceglia R, et al. Specialized astrocytes mediate glutamatergic gliotransmission in the CNS. Nature. 2023;622:120–9. https://doi.org/10.1038/s41586-023-06502-w.
89. Blanco-Suarez E, Liu T-F, Kopelevich A, Allen NJ. Astrocyte-secreted chordin-like 1 drives synapse maturation and limits plasticity by increasing synaptic GluA2 AMPA receptors. Neuron. 2018;100:1116–1132.e1113. https://doi.org/10.1016/j.neuron.2018.09.043.
90. Dalva MB, McClelland AC, Kayser MS. Cell adhesion molecules: signalling functions at the synapse. Nat Rev Neurosci. 2007;8:206–20. https://doi.org/10.1038/nrn2075.
91. Ziv NE, Fisher-Lavie A. Presynaptic and postsynaptic scaffolds: dynamics fast and slow. Neuroscientist. 2014;20:439–52. https://doi.org/10.1177/1073858414523321.
92. Baldwin KT, et al. HepaCAM controls astrocyte self-organization and coupling. Neuron. 2021;109:2427–2442.e2410. https://doi.org/10.1016/j.neuron.2021.05.025.
93. Garrett AM, Weiner JA. Control of CNS synapse development by γ-protocadherin-mediated astrocyte–neuron contact. J Neurosci. 2009;29:11723–31. https://doi.org/10.1523/jneurosci.2818-09.2009.
94. Nishida H, Okabe S. Direct astrocytic contacts regulate local maturation of dendritic spines. J Neurosci. 2007;27:331–40. https://doi.org/10.1523/JNEUROSCI.4466-06.2007.
95. Saint-Martin M, Goda Y. Astrocyte–synapse interactions and cell adhesion molecules. FEBS J. 2023;290:3512–26. https://doi.org/10.1111/febs.16540.
96. Tan CX, Eroglu C. Cell adhesion molecules regulating astrocyte-neuron interactions. Curr Opin Neurobiol. 2021;69:170–7. https://doi.org/10.1016/j.conb.2021.03.015.
97. Chen X, et al. Organization of the core structure of the postsynaptic density. Proc Natl Acad Sci USA. 2008;105:4453–8. https://doi.org/10.1073/pnas.0800897105.
98. de Bartolomeis A, Fiore G. Postsynaptic density scaffolding proteins at excitatory synapse and disorders of synaptic plasticity: implications for human behavior pathologies. Int Rev Neurobiol. 2004;59:221–54. https://doi.org/10.1016/S0074-7742(04)59009-8.
99. Chen X, et al. PSD-95 family MAGUKs are essential for anchoring AMPA and NMDA receptor complexes at the postsynaptic density. Proc Natl Acad Sci. 2015;112:E6983–92. https://doi.org/10.1073/pnas.1517045112.
100. El-Husseini AE-D, et al. Synaptic strength regulated by palmitate cycling on PSD-95. Cell. 2002;108:849–63. https://doi.org/10.1016/S0092-8674(02)00683-9.
101. Keith D, El-Husseini A. Excitation control: balancing PSD-95 function at the synapse. Front Mol Neurosci. 2008;1:4. https://doi.org/10.3389/neuro.02.004.2008.
102. Ehrlich I, Malinow R. Postsynaptic density 95 controls AMPA receptor incorporation during long-term potentiation and experience-driven synaptic plasticity. J Neurosci. 2004;24:916–27. https://doi.org/10.1523/JNEUROSCI.4733-03.2004.
103. Sheng M, Cummings J, Roldan LA, Jan YN, Jan LY. Changing subunit composition of heteromeric NMDA receptors during development of rat cortex. Nature. 1994;368:144–7. https://doi.org/10.1038/368144a0.
104. Kivisäkk P, et al. Increased levels of the synaptic proteins PSD-95, SNAP-25, and neurogranin in the cerebrospinal fluid of patients with Alzheimer's disease. Alzheimers Res Ther. 2022;14:58. https://doi.org/10.1186/s13195-022-01002-x.

105. Astillero-Lopez V, et al. Neurodegeneration and astrogliosis in the entorhinal cortex in Alzheimer's disease: stereological layer-specific assessment and proteomic analysis. Alzheimers Dement. 2022;18:2468–80. https://doi.org/10.1002/alz.12580.
106. Ceprian M, Fulton D. Glial cell AMPA receptors in nervous system health, injury and disease. Int J Mol Sci. 2019;20:2450. https://doi.org/10.3390/ijms20102450.
107. Skowrońska K, Obara-Michlewska M, Zielińska M, Albrecht J. NMDA receptors in astrocytes: in search for roles in neurotransmission and astrocytic homeostasis. Int J Mol Sci. 2019;20:309. https://doi.org/10.3390/ijms20020309.
108. May HJ, et al. Truncating variants in the SHANK1 gene are associated with a spectrum of neurodevelopmental disorders. Genet Med. 2021;23:1912–21. https://doi.org/10.1038/s41436-021-01222-w.
109. Qin Y, et al. A recurrent SHANK1 mutation implicated in autism spectrum disorder causes autistic-like core behaviors in mice via downregulation of mGluR1-IP3R1-calcium signaling. Mol Psychiatry. 2022;27:2985–98. https://doi.org/10.1038/s41380-022-01539-1.
110. Sungur AÖ, Schwarting RKW, Wöhr M. Early communication deficits in the Shank1 knockout mouse model for autism spectrum disorder: developmental aspects and effects of social context. Autism Res. 2016;9:696–709. https://doi.org/10.1002/aur.1564.
111. Hung AY, et al. Smaller dendritic spines, weaker synaptic transmission, but enhanced spatial learning in mice lacking Shank1. J Neurosci. 2008;28:1697–708. https://doi.org/10.1523/JNEUROSCI.3032-07.2008.
112. Naisbitt S, et al. Shank, a novel family of postsynaptic density proteins that binds to the NMDA receptor/PSD-95/GKAP complex and cortactin. Neuron. 1999;23:569–82. https://doi.org/10.1016/s0896-6273(00)80809-0.
113. Südhof TC. Neuroligins and Neurexins link synaptic function to cognitive disease. Nature. 2008;455:903–11. https://doi.org/10.1038/nature07456.
114. Craig AM, Kang Y. Neurexin–neuroligin signaling in synapse development. Curr Opin Neurobiol. 2007;17:43–52. https://doi.org/10.1016/j.conb.2007.01.011.
115. Xia Q-Q, et al. Neuroligins differentially mediate subtype-specific synapse formation in pyramidal neurons and interneurons. Neurosci Bull. 2019;35:497–506. https://doi.org/10.1007/s12264-019-00347-y.
116. Liu X, et al. Roles of neuroligins in central nervous system development: focus on glial neuroligins and neuron neuroligins. J Transl Med. 2022;20:418. https://doi.org/10.1186/s12967-022-03625-y.
117. Golf SR, Trotter JH, Nakahara G, Sudhof TC. Astrocytic neuroligins are not required for synapse formation or a normal astrocyte cytoarchitecture. bioRxiv. 2023. https://doi.org/10.1101/2023.04.10.536254.
118. Medina MA, et al. Wnt/β-catenin signaling stimulates the expression and synaptic clustering of the autism-associated Neuroligin 3 gene. Transl Psychiatry. 2018;8:1–12. https://doi.org/10.1038/s41398-018-0093-y.
119. Bamji SX, et al. Role of β-catenin in synaptic vesicle localization and presynaptic assembly. Neuron. 2003;40:719–31.
120. Lutgen V, Narasipura SD, Sharma A, Min S, Al-Harthi L. β-Catenin signaling positively regulates glutamate uptake and metabolism in astrocytes. J Neuroinflammation. 2016;13:242. https://doi.org/10.1186/s12974-016-0691-7.
121. Voglis G, Tavernarakis N. The role of synaptic ion channels in synaptic plasticity. EMBO Rep. 2006;7:1104–10. https://doi.org/10.1038/sj.embor.7400830.
122. Burke KJ Jr, Bender KJ. Modulation of ion channels in the axon: mechanisms and function. Front Cell Neurosci. 2019;13:221. https://doi.org/10.3389/fncel.2019.00221.
123. McNeill J, Rudyk C, Hildebrand ME, Salmaso N. Ion channels and electrophysiological properties of astrocytes: implications for emergent stimulation technologies. Front Cell Neurosci. 2021;15:644126. https://doi.org/10.3389/fncel.2021.644126.
124. Seifert G, Henneberger C, Steinhauser C. Diversity of astrocyte potassium channels: an update. Brain Res Bull. 2018;136:26–36. https://doi.org/10.1016/j.brainresbull.2016.12.002.

125. Zhou M, Du Y, Aten S, Terman D. On the electrical passivity of astrocyte potassium conductance. J Neurophysiol. 2021;126:1403–19. https://doi.org/10.1152/jn.00330.2021.
126. Catterall WA, Goldin AL, Waxman SG. International Union of Pharmacology. XLVII. Nomenclature and structure-function relationships of voltage-gated sodium channels. Pharmacol Rev. 2005;57:397–409. https://doi.org/10.1124/pr.57.4.4.
127. de Lera Ruiz M, Kraus RL. Voltage-gated sodium channels: structure, function, pharmacology, and clinical indications. J Med Chem. 2015;58:7093–118. https://doi.org/10.1021/jm501981g.
128. Brackenbury WJ, Isom LL. Na channel beta subunits: overachievers of the ion channel family. Front Pharmacol. 2011;2:53. https://doi.org/10.3389/fphar.2011.00053.
129. England JD, Gamboni F, Ferguson MA, Levinson SR. Sodium channels accumulate at the tips of injured axons. Muscle Nerve. 1994;17:593–8. https://doi.org/10.1002/mus.880170605.
130. Henry MA, Freking AR, Johnson LR, Levinson SR. Sodium channel Nav1.6 accumulates at the site of infraorbital nerve injury. BMC Neurosci. 2007;8:56. https://doi.org/10.1186/1471-2202-8-56.
131. Zhu H, et al. Remarkable alterations of Nav1.6 in reactive astrogliosis during epileptogenesis. Sci Rep. 2016;6:38108. https://doi.org/10.1038/srep38108.
132. Gorter JA, van Vliet EA, Lopes da Silva FH, Isom LL, Aronica E. Sodium channel beta1-subunit expression is increased in reactive astrocytes in a rat model for mesial temporal lobe epilepsy. Eur J Neurosci. 2002;16:360–4. https://doi.org/10.1046/j.1460-9568.2002.02078.x.
133. Kim DM, Nimigean CM. Voltage-gated potassium channels: a structural examination of selectivity and gating. Cold Spring Harb Perspect Biol. 2016;8 https://doi.org/10.1101/cshperspect.a029231.
134. Zhang X, Wan JQ, Tong XP. Potassium channel dysfunction in neurons and astrocytes in Huntington's disease. CNS Neurosci Ther. 2018;24:311–8. https://doi.org/10.1111/cns.12804.
135. Smart SL, Bosma MM, Tempel BL. Identification of the delayed rectifier potassium channel, Kv1.6, in cultured astrocytes. Glia. 1997;20:127–34. https://doi.org/10.1002/(sici)1098-1136(199706)20:2<127::aid-glia4>3.0.co;2-6.
136. Zhu J, Yan J, Thornhill WB. The Kv1.3 potassium channel is localized to the cis-Golgi and Kv1.6 is localized to the endoplasmic reticulum in rat astrocytes. FEBS J. 2014;281:3433–45. https://doi.org/10.1111/febs.12871.
137. Grimaldi A, et al. Kv1.3 activity perturbs the homeostatic properties of astrocytes in glioma. Sci Rep. 2018;8:7654. https://doi.org/10.1038/s41598-018-25940-5.
138. Bekar LK, et al. Complex expression and localization of inactivating Kv channels in cultured hippocampal astrocytes. J Neurophysiol. 2005;93:1699–709. https://doi.org/10.1152/jn.00850.2004.
139. Luo Y, Huang L, Liao P, Jiang R. Contribution of neuronal and glial two-pore-domain potassium channels in health and neurological disorders. Neural Plast. 2021;2021:8643129. https://doi.org/10.1155/2021/8643129.
140. Choi JH, et al. TWIK-1/TASK-3 heterodimeric channels contribute to the neurotensin-mediated excitation of hippocampal dentate gyrus granule cells. Exp Mol Med. 2018;50:1–13. https://doi.org/10.1038/s12276-018-0172-4.
141. Meurers BH, et al. Dopamine depletion induces distinct compensatory gene expression changes in DARPP-32 signal transduction cascades of striatonigral and striatopallidal neurons. J Neurosci. 2009;29:6828–39. https://doi.org/10.1523/JNEUROSCI.5310-08.2009.
142. Zhou M, et al. TWIK-1 and TREK-1 are potassium channels contributing significantly to astrocyte passive conductance in rat hippocampal slices. J Neurosci. 2009;29:8551–64. https://doi.org/10.1523/JNEUROSCI.5784-08.2009.
143. Woo DH, et al. TREK-1 and Best1 channels mediate fast and slow glutamate release in astrocytes upon GPCR activation. Cell. 2012;151:25–40. https://doi.org/10.1016/j.cell.2012.09.005.
144. Dolphin AC. Functions of presynaptic voltage-gated calcium channels. Function (Oxf). 2021;2:zqaa027. https://doi.org/10.1093/function/zqaa027.

145. Bavassano C, et al. Bicistronic CACNA1A gene expression in neurons derived from spinocerebellar ataxia type 6 patient-induced pluripotent stem cells. Stem Cells Dev. 2017;26:1612–25. https://doi.org/10.1089/scd.2017.0085.
146. Folacci M, et al. Functional characterization of four known Cav2.1 variants associated with neurodevelopmental disorders. Membranes (Basel). 2023;13 https://doi.org/10.3390/membranes13010096.
147. Luo X, et al. Clinically severe CACNA1A alleles affect synaptic function and neurodegeneration differentially. PLoS Genet. 2017;13:e1006905. https://doi.org/10.1371/journal.pgen.1006905.
148. Schulte A, et al. Homeostatic calcium fluxes, ER calcium release, SOCE, and calcium oscillations in cultured astrocytes are interlinked by a small calcium toolkit. Cell Calcium. 2022;101:102515. https://doi.org/10.1016/j.ceca.2021.102515.
149. Lemos FO, Bultynck G, Parys JB. A comprehensive overview of the complex world of the endo- and sarcoplasmic reticulum Ca(2+)-leak channels. Biochim Biophys Acta, Mol Cell Res. 2021;1868:119020. https://doi.org/10.1016/j.bbamcr.2021.119020.
150. Sahlender DA, Savtchouk I, Volterra A. What do we know about gliotransmitter release from astrocytes? Philos Trans R Soc B: Biol Sci. 2014;369:20130592. https://doi.org/10.1098/rstb.2013.0592.
151. Shan L, Zhang T, Fan K, Cai W, Liu H. Astrocyte-neuron signaling in synaptogenesis. Front Cell Dev Biol. 2021;9. https://doi.org/10.3389/fcell.2021.680301.
152. Chung W-S, Allen NJ, Eroglu C. Astrocytes control synapse formation, function, and elimination. Cold Spring Harb Perspect Biol. 2015;7:a020370. https://doi.org/10.1101/cshperspect.a020370.
153. Stevens B. Neuron-astrocyte signaling in the development and plasticity of neural circuits. Neurosignals. 2008;16:278–88. https://doi.org/10.1159/000123038.
154. Perez-Catalan NA, Doe CQ, Ackerman SD. The role of astrocyte-mediated plasticity in neural circuit development and function. Neural Dev. 2021;16:1. https://doi.org/10.1186/s13064-020-00151-9.
155. Mahmoud S, Gharagozloo M, Simard C, Gris D. Astrocytes maintain glutamate homeostasis in the CNS by controlling the balance between glutamate uptake and release. Cells. 2019;8:184. https://doi.org/10.3390/cells8020184.
156. O'Donovan SM, Sullivan CR, McCullumsmith RE. The role of glutamate transporters in the pathophysiology of neuropsychiatric disorders. NPJ Schizophr. 2017;3:32. https://doi.org/10.1038/s41537-017-0037-1.
157. Olloquequi J, Cornejo-Córdova E, Verdaguer E, et al. Excitotoxicity in the pathogenesis of neurological and psychiatric disorders: Therapeutic implications. Journal of Psychopharmacology. 2018;32(3):265–275. https://doi.org/10.1177/0269881118754680
158. Dong X-x, Wang Y, Qin Z-h. Molecular mechanisms of excitotoxicity and their relevance to pathogenesis of neurodegenerative diseases Acta Pharmacologica Sinica. 2009;30(4):379–387. https://doi.org/10.1038/aps.2009.24
159. Durkee CA, Araque A. Diversity and specificity of astrocyte-neuron communication. Neuroscience. 2019;396:73–8. https://doi.org/10.1016/j.neuroscience.2018.11.010.
160. Durkee CA, et al. Gi/o protein-coupled receptors inhibit neurons but activate astrocytes and stimulate gliotransmission. Glia. 2019;67:1076–93. https://doi.org/10.1002/glia.23589.
161. Morel L, Higashimori H, Tolman M, Yang Y. VGluT1+ neuronal glutamatergic signaling regulates postnatal developmental maturation of cortical protoplasmic Astroglia. J Neurosci. 2014;34:10950–62. https://doi.org/10.1523/JNEUROSCI.1167-14.2014.
162. Sun W, et al. Glutamate-dependent neuroglial calcium signaling differs between young and adult brain. Science. 2013;339:197–200. https://doi.org/10.1126/science.1226740.
163. Covelo A, Araque A. Neuronal activity determines distinct gliotransmitter release from a single astrocyte. eLife. 2018;7 https://doi.org/10.7554/eLife.32237.
164. Hatashita Y, et al. Spontaneous and multifaceted ATP release from astrocytes at the scale of hundreds of synapses. Glia. 2023;71:2250–65. https://doi.org/10.1002/glia.24392.

165. Chaudhry FA, et al. Glutamate transporters in glial plasma membranes: highly differentiated localizations revealed by quantitative ultrastructural immunocytochemistry. Neuron. 1995;15:711–20. https://doi.org/10.1016/0896-6273(95)90158-2.
166. Genoud C, et al. Plasticity of astrocytic coverage and glutamate transporter expression in adult mouse cortex. PLoS Biol. 2006;4:e343. https://doi.org/10.1371/journal.pbio.0040343.
167. Henley JM, Wilkinson KA. Synaptic AMPA receptor composition in development, plasticity and disease. Nat Rev Neurosci. 2016;17:337–50. https://doi.org/10.1038/nrn.2016.37.
168. Greger IH, Watson JF, Cull-Candy SG. Structural and functional architecture of AMPA-type glutamate receptors and their auxiliary proteins. Neuron. 2017;94:713–30. https://doi.org/10.1016/j.neuron.2017.04.009.
169. Kamalova A, Nakagawa T. AMPA receptor structure and auxiliary subunits. J Physiol. 2021;599:453–69. https://doi.org/10.1113/JP278701.
170. Cull-Candy SG, Farrant M. Ca(2+) -permeable AMPA receptors and their auxiliary subunits in synaptic plasticity and disease. J Physiol. 2021;599:2655–71. https://doi.org/10.1113/JP279029.
171. Gaidin SG, Kosenkov AM. Calcium-permeable AMPA receptors: an underestimated pharmacological target for the therapy of brain pathologies. Neural Regen Res. 2023;18:2669–70. https://doi.org/10.4103/1673-5374.373714.
172. Chater TE, Goda Y. The role of AMPA receptors in postsynaptic mechanisms of synaptic plasticity. Front Cell Neurosci. 2014;8. https://doi.org/10.3389/fncel.2014.00401.
173. Droste D, et al. Ca2+-permeable AMPA receptors in mouse olfactory bulb astrocytes. Sci Rep. 2017;7:44817. https://doi.org/10.1038/srep44817.
174. Iino M, et al. Glia-synapse interaction through Ca2+-permeable AMPA receptors in Bergmann glia. Science. 2001;292:926–9. https://doi.org/10.1126/science.1058827.
175. Li S, Stys PK. Mechanisms of ionotropic glutamate receptor-mediated Excitotoxicity in isolated spinal cord White matter. J Neurosci. 2000;20:1190–8. https://doi.org/10.1523/JNEUROSCI.20-03-01190.2000.
176. Sherwood MW, Oliet SHR, Panatier A. NMDARs, coincidence detectors of astrocytic and neuronal activities. Int J Mol Sci. 2021;22:7258. https://doi.org/10.3390/ijms22147258.
177. Yang X, et al. Trafficking of NMDA receptors is essential for hippocampal synaptic plasticity and memory consolidation. Cell Rep. 2022;40:111217. https://doi.org/10.1016/j.celrep.2022.111217.
178. Mu Y, Zhao C, Toni N, Yao J, Gage FH. Distinct roles of NMDA receptors at different stages of granule cell development in the adult brain. eLife. 2015;4:e07871. https://doi.org/10.7554/eLife.07871.
179. Lee C-H, et al. NMDA receptor structures reveal subunit arrangement and pore architecture. Nature. 2014;511:191–7. https://doi.org/10.1038/nature13548.
180. Fellin T, et al. Neuronal synchrony mediated by astrocytic glutamate through activation of extrasynaptic NMDA receptors. Neuron. 2004;43:729–43. https://doi.org/10.1016/j.neuron.2004.08.011.
181. Han KS, et al. Channel-mediated astrocytic glutamate release via Bestrophin-1 targets synaptic NMDARs. Mol Brain. 2013;6:4. https://doi.org/10.1186/1756-6606-6-4.
182. Henneberger C, Papouin T, Oliet SH, Rusakov DA. Long-term potentiation depends on release of D-serine from astrocytes. Nature. 2010;463:232–6. https://doi.org/10.1038/nature08673.
183. Lee CJ, et al. Astrocytic control of synaptic NMDA receptors. J Physiol. 2007;581:1057–81. https://doi.org/10.1113/jphysiol.2007.130377.
184. Potapenko ES, Biancardi VC, Zhou Y, Stern JE. Astrocytes modulate a postsynaptic NMDA-GABAA-receptor crosstalk in hypothalamic neurosecretory neurons. J Neurosci. 2013;33:631–40. https://doi.org/10.1523/JNEUROSCI.3936-12.2013.
185. Koh W, et al. Astrocytes render memory flexible by releasing D-serine and regulating NMDA receptor tone in the hippocampus. Biol Psychiatry. 2022;91:740–52. https://doi.org/10.1016/j.biopsych.2021.10.012.

186. Neame S, et al. The NMDA receptor activation by d-serine and glycine is controlled by an astrocytic Phgdh-dependent serine shuttle. Proc Natl Acad Sci USA. 2019;116:20736–42. https://doi.org/10.1073/pnas.1909458116.
187. de Oca M, Balderas P, Aguilera P. A metabotropic-like flux-independent NMDA receptor regulates Ca2+ exit from endoplasmic reticulum and mitochondrial membrane potential in cultured astrocytes. PLoS One. 2015;10:e0126314. https://doi.org/10.1371/journal.pone.0126314.
188. Palygin O, Lalo U, Pankratov Y. Distinct pharmacological and functional properties of NMDA receptors in mouse cortical astrocytes. Br J Pharmacol. 2011;163:1755–66. https://doi.org/10.1111/j.1476-5381.2011.01374.x.
189. Balu DT, et al. Neurotoxic astrocytes express the D-serine synthesizing enzyme, serine racemase, in Alzheimer's disease. Neurobiol Dis. 2019;130:104511. https://doi.org/10.1016/j.nbd.2019.104511.
190. Vignes M, Collingridge GL. The synaptic activation of kainate receptors. Nature. 1997;388:179–82. https://doi.org/10.1038/40639.
191. Falcón-Moya R, Sihra TS, Rodríguez-Moreno A. Kainate receptors: role in epilepsy. Front Mol Neurosci. 2018;11:217. https://doi.org/10.3389/fnmol.2018.00217.
192. Zhuo M. Cortical kainate receptors and behavioral anxiety. Mol Brain. 2017;10:16. https://doi.org/10.1186/s13041-017-0297-8.
193. Vargas JR, Takahashi DK, Thomson KE, Wilcox KS. The expression of Kainate receptor subunits in hippocampal astrocytes following experimentally induced status epilepticus. J Neuropathol Exp Neurol. 2013;72. https://doi.org/10.1097/NEN.1090b1013e3182a1094b1266, 10.1097/NEN.0b013e3182a4b266.
194. Williams PA, White AM, Clark S, Ferraro DJ, Swiercz W, Staley KJ, Dudek FE. Development of Spontaneous Recurrent Seizures after Kainate-Induced Status Epilepticus. J Neuroscience. 2009;29(7):2103–12. https://doi.org/10.1523/JNEUROSCI.0980-08.2009.
195. Niswender CM, Conn PJ. Metabotropic glutamate receptors: physiology, pharmacology, and disease. Annu Rev Pharmacol Toxicol. 2010;50:295–322. https://doi.org/10.1146/annurev.pharmtox.011008.145533.
196. Rodrigues SM, Bauer EP, Farb CR, Schafe GE, LeDoux JE. The group I metabotropic glutamate receptor mGluR5 is required for fear memory formation and long-term potentiation in the lateral amygdala. J Neurosci. 2002;22:5219–29. https://doi.org/10.1523/JNEUROSCI.22-12-05219.2002.
197. de Souza JM, et al. mGluR5 ablation leads to age-related synaptic plasticity impairments and does not improve Huntington's disease phenotype. Sci Rep. 2022;12:8982. https://doi.org/10.1038/s41598-022-13029-z.
198. D'Ascenzo M, et al. mGluR5 stimulates gliotransmission in the nucleus accumbens. Proc Natl Acad Sci. 2007;104:1995–2000. https://doi.org/10.1073/pnas.0609408104.
199. Biber K, et al. Expression and signaling of group I metabotropic glutamate receptors in astrocytes and microglia. J Neurochem. 1999;72:1671–80. https://doi.org/10.1046/j.1471-4159.1999.721671.x.
200. Liu C, Yang T-Q, Zhou Y-D, Shen Y. Reduced astrocytic mGluR5 in the hippocampus is associated with stress-induced depressive-like behaviors in mice. Neurosci Lett. 2022;784:136766. https://doi.org/10.1016/j.neulet.2022.136766.
201. de Lima IBQ, et al. Blockade of mGluR5 in astrocytes derived from human iPSCs modulates astrocytic function and increases phagocytosis. Front Immunol. 2023;14. https://doi.org/10.3389/fimmu.2023.1283331.
202. Danjo Y, et al. Transient astrocytic mGluR5 expression drives synaptic plasticity and subsequent chronic pain in mice. J Exp Med. 2022;219:e20210989. https://doi.org/10.1084/jem.20210989.
203. Kellner V, et al. Dual metabotropic glutamate receptor signaling enables coordination of astrocyte and neuron activity in developing sensory domains. Neuron. 2021;109:2545–2555.e2547. https://doi.org/10.1016/j.neuron.2021.06.010.

204. Ganguly K, Schinder AF, Wong ST, Poo M-M. GABA itself promotes the developmental switch of neuronal GABAergic responses from excitation to inhibition. Cell. 2001;105:521–32. https://doi.org/10.1016/S0092-8674(01)00341-5.
205. Hsu Y-T, Chang Y-G, Chern Y. Insights into GABAAergic system alteration in Huntington's disease. Open Biol. 2018;8:180165. https://doi.org/10.1098/rsob.180165.
206. Edden RAE, Crocetti D, Zhu H, Gilbert DL, Mostofsky SH. Reduced GABA concentration in attention-deficit/hyperactivity disorder. Arch Gen Psychiatry. 2012;69:750–3. https://doi.org/10.1001/archgenpsychiatry.2011.2280.
207. Treiman DM. GABAergic mechanisms in epilepsy. Epilepsia. 2001;42(Suppl 3):8–12. https://doi.org/10.1046/j.1528-1157.2001.042suppl.3008.x.
208. Khayenko V, Maric HM. Targeting GABAAR-associated proteins: new modulators, labels and concepts. Front Mol Neurosci. 2019;12:1662–5099. https://doi.org/10.3389/fnmol.2019.00162.
209. Ghit A, Assal D, Al-Shami AS, Hussein DEE. GABAA receptors: structure, function, pharmacology, and related disorders. J Genetic Eng Biotechnol. 2021;19:123. https://doi.org/10.1186/s43141-021-00224-0.
210. Höft S, Griemsmann S, Seifert G, Steinhäuser C. Heterogeneity in expression of functional ionotropic glutamate and GABA receptors in astrocytes across brain regions: insights from the thalamus. Philosophical Transactions of the Royal Society B: Biological Sciences 2014;369(1654). https://doi.org/10.1098/rstb.2013.0602.
211. Liu J, Feng X, Wang Y, Xia X, Zheng JC. Astrocytes: GABAceptive and GABAergic cells in the brain. Front Cell Neurosci. 2022;16:892497. https://doi.org/10.3389/fncel.2022.892497.
212. Chung W, Wang D-S, Khodaei S, Pinguelo A, Orser BA. GABAA receptors in astrocytes are targets for commonly used intravenous and inhalational general anesthetic drugs. Front Aging Neurosci. 2022;13:802582. https://doi.org/10.3389/fnagi.2021.802582.
213. Terunuma M. Diversity of structure and function of GABAB receptors: a complexity of GABAB-mediated signaling. Proc Jpn Acad Ser B Phys Biol Sci. 2018;94:390–411. https://doi.org/10.2183/pjab.94.026.
214. Cheng Y-T, et al. Inhibitory input directs astrocyte morphogenesis through glial GABABR. Nature. 2023;617:369–76. https://doi.org/10.1038/s41586-023-06010-x.
215. Skoug C, Martinsson I, Gouras GK, Meissner A, Duarte JMN. Sphingosine 1-Phoshpate receptors are located in synapses and control spontaneous activity of mouse neurons in culture. Neurochem Res. 2022;47:3114–25. https://doi.org/10.1007/s11064-022-03664-3.
216. Singh SK, Kordula T, Spiegel S. Neuronal contact upregulates astrocytic sphingosine-1-phosphate receptor 1 to coordinate astrocyte-neuron cross communication. Glia. 2022;70:712–27. https://doi.org/10.1002/glia.24135.
217. Chen Z, et al. Sphingosine-1-phosphate receptor 1 activation in astrocytes contributes to neuropathic pain. Proc Natl Acad Sci USA. 2019;116:10557–62. https://doi.org/10.1073/pnas.1820466116.
218. Garcia-Gil M, Camici M, Allegrini S, Pesi R, Tozzi MG. Metabolic aspects of adenosine functions in the brain. Front Pharmacol. 2021;12:672182. https://doi.org/10.3389/fphar.2021.672182.
219. Dunwiddie TV, Masino SA. The role and regulation of adenosine in the central nervous system. Annu Rev Neurosci. 2001;24:31–55. https://doi.org/10.1146/annurev.neuro.24.1.31.
220. Hou X, Li Y, Huang Y, Zhao H, Gui L. Adenosine receptor A1-A2a Heteromers regulate EAAT2 expression and glutamate uptake via YY1-induced repression of PPARgamma transcription. PPAR Res. 2020;2020:2410264. https://doi.org/10.1155/2020/2410264.
221. Stockwell J, Jakova E, Cayabyab FS. Adenosine A1 and A2A receptors in the brain: current research and their role in neurodegeneration. Molecules. 2017;22:676. https://doi.org/10.3390/molecules22040676.
222. Haydon PG. Astrocytes and the modulation of sleep. Curr Opin Neurobiol. 2017;44:28–33. https://doi.org/10.1016/j.conb.2017.02.008.

223. Lezmy J, et al. Astrocyte Ca(2+)-evoked ATP release regulates myelinated axon excitability and conduction speed. Science. 2021;374:eabh2858. https://doi.org/10.1126/science.abh2858.
224. Tan Z, et al. Glia-derived ATP inversely regulates excitability of pyramidal and CCK-positive neurons. Nat Commun. 2017;8:13772. https://doi.org/10.1038/ncomms13772.
225. Zhang JM, et al. ATP released by astrocytes mediates glutamatergic activity-dependent heterosynaptic suppression. Neuron. 2003;40:971–82. https://doi.org/10.1016/s0896-6273(03)00717-7.
226. Bjorklund O, Shang M, Tonazzini I, Dare E, Fredholm BB. Adenosine A1 and A3 receptors protect astrocytes from hypoxic damage. Eur J Pharmacol. 2008;596:6–13. https://doi.org/10.1016/j.ejphar.2008.08.002.
227. Ciccarelli R, et al. Molecular signalling mediating the protective effect of A1 adenosine and mGlu3 metabotropic glutamate receptor activation against apoptosis by oxygen/glucose deprivation in cultured astrocytes. Mol Pharmacol. 2007;71:1369–80. https://doi.org/10.1124/mol.106.031617.
228. Lopes CR, Cunha RA, Agostinho P. Astrocytes and adenosine A(2A) receptors: active players in Alzheimer's disease. Front Neurosci. 2021;15:666710. https://doi.org/10.3389/fnins.2021.666710.
229. Orr AG, et al. Astrocytic adenosine receptor A2A and Gs-coupled signaling regulate memory. Nat Neurosci. 2015;18:423–34. https://doi.org/10.1038/nn.3930.
230. Liguz-Lecznar M, Urban-Ciecko J, Kossut M. Somatostatin and somatostatin-containing neurons in shaping neuronal activity and plasticity. Front Neural Circuits. 2016;10:48. https://doi.org/10.3389/fncir.2016.00048.
231. Henriques VJ, Chiavegato A, Carmignoto G, Gomez-Gonzalo M. Astrocytes modulate somatostatin interneuron signaling in the visual cortex. Cells. 2022;11 https://doi.org/10.3390/cells11091400.
232. He J-H, et al. SSTR2 is a prognostic factor and a promising therapeutic target in glioma. Am J Transl Res. 2021;13:11223–34.
233. Sudhof TC. The synaptic vesicle cycle. Annu Rev Neurosci. 2004;27:509–47. https://doi.org/10.1146/annurev.neuro.26.041002.131412. PMID:15217342.
234. Reimer RJ. SLC17: A functionally diverse family of organic anion transporters. Molecular Aspects of Medicine. 2013;34(2-3):350–359. https://doi.org/10.1016/j.mam.2012.05.004.
235. Lawal HO, Krantz DE. SLC18: Vesicular neurotransmitter transporters for monoamines and acetylcholine. Molecular Aspects of Medicine. 2013;34(2–3):360–372. https://doi.org/10.1016/j.mam.2012.07.005.
236. Gasnier B. The SLC32 transporter, a key protein for the synaptic release of inhibitory amino acids. Pflugers Arch. 2004 Feb;447(5):756–9. https://doi.org/10.1007/s00424-003-1091-2. Epub 2003 May 16. PMID:12750892.
237. Bazargani N, Attwell D. Astrocyte calcium signaling: the third wave. Nat Neurosci. 2016;19:182–9. https://doi.org/10.1038/nn.4201.
238. Yan C, Jiang J, Yang Y, Geng X, Dong W. The function of VAMP2 in mediating membrane fusion: an overview. Front Mol Neurosci. 2022;15. https://doi.org/10.3389/fnmol.2022.948160.
239. Costa AS, et al. VAMP2 expression and genotype are possible discriminators in different forms of dementia. Front Aging Neurosci. 2022;14:858162. https://doi.org/10.3389/fnagi.2022.858162.
240. Cali E, Rocca C, Salpietro V, Houlden H. Epileptic phenotypes associated with SNAREs and related synaptic vesicle exocytosis machinery. Front Neurol. 2022;12:806506. https://doi.org/10.3389/fneur.2021.806506.
241. Mielnicka A, Michaluk P. Exocytosis in astrocytes. Biomolecules. 2021;11:1367. https://doi.org/10.3390/biom11091367.

242. Crippa D, et al. Synaptobrevin2-expressing vesicles in rat astrocytes: insights into molecular characterization, dynamics and exocytosis. J Physiol. 2006;570:567–82. https://doi.org/10.1113/jphysiol.2005.094052.
243. Schwarz Y, Zhao N, Kirchhoff F, Bruns D. Astrocytes control synaptic strength by two distinct v-SNARE-dependent release pathways. Nat Neurosci. 2017;20:1529–39. https://doi.org/10.1038/nn.4647.
244. Moghaddam HS, Arabi MH. Synaptotagmin-11 is a novel hotspot in the pathogenesis of Parkin-linked Parkinson's disease: new implications for clinical targeting. Mov Disord. 2018;33:582. https://doi.org/10.1002/mds.27369.
245. Shimojo M, et al. Synaptotagmin-11 mediates a vesicle trafficking pathway that is essential for development and synaptic plasticity. Genes Dev. 2019;33:365–76. https://doi.org/10.1101/gad.320077.118.
246. Inoue S, et al. Synaptotagmin XI as a candidate gene for susceptibility to schizophrenia. Am J Med Genet B Neuropsychiatr Genet. 2007;144B:332–40. https://doi.org/10.1002/ajmg.b.30465.
247. Ho GPH, Wilkie EC, White AJ, Selkoe DJ. Palmitoylation of the Parkinson's disease–associated protein synaptotagmin-11 links its turnover to α-synuclein homeostasis. Sci Signal. 2023;16:eadd7220. https://doi.org/10.1126/scisignal.add7220.
248. Yan S, et al. Synaptotagmin-11 regulates the functions of caveolae and responds to mechanical stimuli in astrocytes. FASEB J. 2020;34:2609–24. https://doi.org/10.1096/fj.201901715R.
249. Ramos-Miguel A, et al. Presynaptic proteins complexin-I and complexin-II differentially influence cognitive function in early and late stages of Alzheimer's disease. Acta Neuropathol. 2017;133:395–407. https://doi.org/10.1007/s00401-016-1647-9.
250. Hazell AS, Wang D. Identification of complexin II in astrocytes: a possible regulator of glutamate release in these cells. Biochem Biophys Res Commun. 2011;404:228–32. https://doi.org/10.1016/j.bbrc.2010.11.098.
251. Redlingshöfer L, McLeod F, Chen Y, Camus MD, Burden JJ, Palomer E, Briant K, Dannhauser PN, Salinas PC, Brodsky FM. Clathrin light chain diversity regulates membrane deformation in vitro and synaptic vesicle formation in vivo. Significance Proceedings of the National Academy of Sciences 2020;117(38):23527–538. https://doi.org/10.1073/pnas.2003662117.
252. Lee S-J, Seo B-R, Koh J-Y. Metallothionein-3 modulates the amyloid β endocytosis of astrocytes through its effects on actin polymerization. Mol Brain. 2015;8:84. https://doi.org/10.1186/s13041-015-0173-3.
253. Li Y, et al. The role of microRNA-148a and downstream DLGAP1 on the molecular regulation and tumor progression on human glioblastoma. Oncogene. 2019;38:7234–48. https://doi.org/10.1038/s41388-019-0922-3.
254. Liu L, Li X, Shi Y, Chen H. Long noncoding RNA DLGAP1-AS1 promotes the progression of glioma by regulating the miR-1297/EZH2 axis. Aging. 2021;13:12129–42. https://doi.org/10.18632/aging.202923.
255. Yang R, et al. ANK2 autism mutation targeting giant ankyrin-B promotes axon branching and ectopic connectivity. Proc Natl Acad Sci. 2019;116:15262–71. https://doi.org/10.1073/pnas.1904348116.
256. Chang K-J, et al. Glial ankyrins facilitate paranodal axoglial junction assembly. Nat Neurosci. 2014;17:1673–81. https://doi.org/10.1038/nn.3858.
257. Szewczyk LM, et al. Astrocytic β-catenin signaling via TCF7L2 regulates synapse development and social behavior. Mol Psychiatry. 2023:1–17. https://doi.org/10.1038/s41380-023-02281-y.
258. Kim SS, et al. Beta-COP regulates TWIK1/TREK1 heterodimeric channel-mediated passive conductance in astrocytes. Cells. 2022;11 https://doi.org/10.3390/cells11203322.
259. Wang W, et al. The contribution of TWIK-1 channels to astrocyte K(+) current is limited by retention in intracellular compartments. Front Cell Neurosci. 2013;7:246. https://doi.org/10.3389/fncel.2013.00246.

260. Mölders A, Koch A, Menke R, Klöcker N. Heterogeneity of the astrocytic AMPA-receptor transcriptome. Glia. 2018;66:2604–16. https://doi.org/10.1002/glia.23514.
261. Gérard F, Hansson E. Inflammatory activation enhances NMDA-triggered Ca2+ signalling and IL-1β secretion in primary cultures of rat astrocytes. Brain Res. 2012;1473:1–8. https://doi.org/10.1016/j.brainres.2012.07.032.
262. Lee M-C, et al. Characterisation of the expression of NMDA receptors in human astrocytes. PLoS One. 2010;5:e14123. https://doi.org/10.1371/journal.pone.0014123.
263. Jin LE, et al. mGluR2/3 mechanisms in primate dorsolateral prefrontal cortex: evidence for both presynaptic and postsynaptic actions. Mol Psychiatry. 2017;22:1615–25. https://doi.org/10.1038/mp.2016.129.
264. Mederos S, et al. GABAergic signaling to astrocytes in the prefrontal cortex sustains goal-directed behaviors. Nat Neurosci. 2021;24:82–92. https://doi.org/10.1038/s41593-020-00752-x.
265. Meier SD, Kafitz KW, Rose CR. Developmental profile and mechanisms of GABA-induced calcium signaling in hippocampal astrocytes. Glia. 2008;56:1127–37. https://doi.org/10.1002/glia.20684.
266. Sheth S, Brito R, Mukherjea D, Rybak LP, Ramkumar V. Adenosine receptors: expression, function and regulation. Int J Mol Sci. 2014;15:2024–52. https://doi.org/10.3390/ijms15022024.
267. Kitano T, Eguchi R, Okamatsu-Ogura Y, Yamaguchi S, Otsuguro K-I. Opposing functions of α- and β-adrenoceptors in the formation of processes by cultured astrocytes. J Pharmacol Sci. 2021;145:228–40. https://doi.org/10.1016/j.jphs.2020.12.005.
268. Gaidin SG, et al. Activation of alpha-2 adrenergic receptors stimulates GABA release by astrocytes. Glia. 2020;68:1114–30. https://doi.org/10.1002/glia.23763.
269. Franke H, et al. P2Y receptor expression on astrocytes in the nucleus accumbens of rats. Neuroscience. 2004;127:431–41. https://doi.org/10.1016/j.neuroscience.2004.05.003.
270. Ma W, et al. Astrocytic alpha4-containing nAChR signaling in the hippocampus governs the formation of temporal association memory. Cell Rep. 2023;42:112674. https://doi.org/10.1016/j.celrep.2023.112674.
271. Cong J, et al. Astroglial CB1 cannabinoid receptors mediate CP 55,940-induced conditioned place aversion through cyclooxygenase-2 signaling in mice. Front Cell Neurosci. 2021;15:772549. https://doi.org/10.3389/fncel.2021.772549.
272. Hegyi Z, et al. CB(1) receptor activation induces intracellular Ca(2+) mobilization and 2-arachidonoylglycerol release in rodent spinal cord astrocytes. Sci Rep. 2018;8:10562. https://doi.org/10.1038/s41598-018-28763-6.
273. Ramamoorthy P, Whim MD. Trafficking and fusion of neuropeptide Y-containing dense-Core granules in astrocytes. J Neurosci. 2008;28:13815–27. https://doi.org/10.1523/JNEUROSCI.5361-07.2008.
274. Singh P, et al. Single-vesicle architecture of synaptobrevin2 in astrocytes. Nat Commun. 2014;5:3780. https://doi.org/10.1038/ncomms4780.
275. Fong LK, et al. Full-length amyloid precursor protein regulates lipoprotein metabolism and amyloid-β clearance in human astrocytes. J Biol Chem. 2018;293:11341–57. https://doi.org/10.1074/jbc.RA117.000441.
276. Miñana R, Duran JM, Tomas M, Renau-Piqueras J, Guerri C. Neural cell adhesion molecule is endocytosed via a clathrin-dependent pathway. Eur J Neurosci. 2001;13:749–56. https://doi.org/10.1046/j.0953-816x.2000.01439.x.
277. He J, et al. Expression of fasciculation and elongation protein zeta-1 (FEZ1) in cultured rat neonatal astrocytes. Mol Cell Biochem. 2009;325:159–67. https://doi.org/10.1007/s11010-009-0030-8.
278. Sun YY, et al. Fasciculation and elongation protein zeta-1 (FEZ1) expression in reactive astrocytes in a rat model of Parkinson's disease. Neuropathol Appl Neurobiol. 2014;40:164–76. https://doi.org/10.1111/nan.12077.

279. Yu Z, et al. Four mood stabilizers commonly induce FEZ1 expression in human astrocytes. Bipolar Disord. 2011;13:486–99. https://doi.org/10.1111/j.1399-5618.2011.00946.x.
280. Leites EP, Morais VA. The PINK1-mediated crosstalk between neural cells and the underlying link to Parkinson's disease. Cells. 2021;10:1395. https://doi.org/10.3390/cells10061395.
281. Barodia SK, et al. PINK1 phosphorylates ubiquitin predominantly in astrocytes. npj Parkinsons Dis. 2019;5:1–9. https://doi.org/10.1038/s41531-019-0101-9.
282. Ko HM, et al. Tyrosine kinase Fyn regulates iNOS expression in LPS-stimulated astrocytes via modulation of ERK phosphorylation. Biochem Biophys Res Commun. 2018;495:1214–20. https://doi.org/10.1016/j.bbrc.2017.11.143.
283. Lee C, Low CYB, Wong SY, Lai MKP, Tan MGK. Selective induction of alternatively spliced FynT isoform by TNF facilitates persistent inflammatory responses in astrocytes. Sci Rep. 2017;7:43651. https://doi.org/10.1038/srep43651.
284. Chun JT, Crispino M, Tocco G. The dual response of protein kinase Fyn to neural trauma: early induction in neurons and delayed induction in reactive astrocytes. Exp Neurol. 2004;185:109–19. https://doi.org/10.1016/j.expneurol.2003.09.019.

Homeostasis to Allostasis: Prefrontal Astrocyte Roles in Cognitive Flexibility and Stress Biology

Bolati Wulaer, Mika A. Holtz, and Jun Nagai

Abstract In the intricate landscape of neurophysiology, astrocytes have been traditionally cast as homeostatic cells; however, their mechanistic involvement in allostasis—particularly how they modulate the adaptive response to stress and its accumulative impact that disrupts cognitive functions and precipitates psychiatric disorders—is now starting to be unraveled. Here, we address the gap by positing astrocytes as crucial allostatic players whose molecular adaptations underlie cognitive flexibility in stress-related neuropsychiatric conditions. We review how astrocytes, responding to stress mediators such as glucocorticoid and epinephrine/norepinephrine, undergo morphological and functional transformations that parallel the maladaptive changes. Our synthesis of recent findings reveals that these glial changes, especially in the metabolically demanding prefrontal cortex, may underlie some of the neuropsychiatric mechanisms characterized by the disruption of energy metabolism and astrocytic networks, compromised glutamate clearance, and diminished synaptic support. We argue that astrocytes extend beyond their homeostatic role, actively participating in the brain's allostatic response, especially by modulating energy substrates critical for cognitive functions.

Keywords Glia · Astrocytes · Cognition · Stress · Psychiatry

Abbreviations

5-HT	Serotonin
ACh	Acetylcholine
ACC	Anterior cingulate cortex
ACTH	Adrenocorticotropic hormone
ATP	Adenosine triphosphate

B. Wulaer · M. A. Holtz · J. Nagai (✉)
Laboratory for Glia-Neuron Circuit Dynamics, RIKEN Center for Brain Science, Wako, Saitama, Japan
e-mail: jun.nagai@riken.jp

E. Blanco-Suarez, I. Farhy-Tselnicker (eds.), *Astrocyte-Neuron Interactions in Health and Disease*, Advances in Neurobiology 39,
https://doi.org/10.1007/978-3-031-64839-7_6

AR	Adrenergic receptor
cAMP	Cyclic adenosine monophosphate
CRH	Corticotropin-releasing hormone
CSDS	Chronic social defeat stress
CUS	Chronic unpredictable stress
CVS	Chronic variable stress
DA	Dopamine
DREADD	Designer receptors exclusively activated by designer drugs
DRN	Dorsal raphe nuclei
eCB	Endocannabinoid
ESD	Early social deprivation
EAAT1	Excitatory amino acid transporter 1
FDG	2-deoxy-2-[^{18}F]fluoro-D-glucose
GABA	Gamma-aminobutyric acid
GFAP	Glial fibrillary acidic protein
GFP	Green fluorescent protein
GLUT	Glucose transporter
GPCR	G-protein-coupled receptor
GR	Glucocorticoid receptor
GSK3β	Glycogen synthase kinase 3β
HA	Histamine
HPA	Hypothalamic-pituitary-adrenal
HPC	Hippocampus
hPMCA2w/b	Human plasma membrane Ca^{2+} pump 2w/b
L-AAA	L-α-aminoadipate
LC	Locus coeruleus
LDH	Lactate dehydrogenase
MCT	Monocarboxylate transporter
MDD	Major depressive disorder
MERTK	Mer receptor tyrosine kinase
MRI	Magnetic resonance imaging
mPFC	Medial prefrontal cortex
mTOR	Mammalian target of rapamycin
NAc	Nucleus accumbens
NAD	Nicotinamide adenine dinucleotide
NADH	Nicotinamide adenine dinucleotide (reduced form)
NE	Norepinephrine
NMDA	N-Methyl-D-aspartate
OFC	Orbitofrontal cortex
PAG	Periaqueductal gray
PDH	Pyruvate dehydrogenase
PET	Positron emission tomography
PFKFB3	Phosphofructokinase/fructose-bisphosphatase 3
PFC	Prefrontal cortex
PPP	Pentose phosphate pathway

PTSD	Posttraumatic stress disorder
SNARE	Soluble NSF attachment protein receptor
SNc	Substantia nigra pars compacta
TCA	Tricarboxylic acid
TMN	Tuberomammillary nucleus
TXNIP	Thioredoxin interacting protein
VIP	Vasoactive intestinal peptide
VTA	Ventral tegmental area

1 Introduction

1.1 Homeostasis, Allostasis, and Allostatic Overload

The concepts of homeostasis, allostasis, and allostatic overload are central to understanding the body's response to stress and its implications for health, particularly in neurophysiology and neuropsychiatric disorders [1–4]. Homeostasis refers to the body's ability to maintain a stable internal environment, which is crucial for survival. The autonomic nervous system and the hypothalamic-pituitary-adrenal (HPA) axis are essential in this process [5]. When confronted with acute threats, the hypothalamus releases corticotropin-releasing hormone (CRH), which triggers the production of corticosterone. Alongside, the sympathetic nervous system is activated, leading to the release of epinephrine, facilitating the "fight, freeze, or flight" reaction necessary for immediate adaptation to environmental changes [6].

Allostasis extends beyond homeostasis, describing the active processes that maintain stability in and adapt to a different state. It involves robust, energy-intensive responses to significant challenges, whether physical, chemical, psychosocial, or environmental. Allostasis supports homeostasis by emphasizing flexible adaptation [3, 7]. However, when these adaptive responses fail to cease after a stressor has subsided, the cumulative burden on the body can lead to allostatic overload. This "wear and tear" reflects the cost of chronic engagement of the body's adaptive systems, leading to potential dysregulation of physiological systems, which may contribute to the development of disease [1, 3, 8]. In short, while allostasis is essential for adaptation and survival, extreme or prolonged stress can be detrimental, serving as a significant risk factor for diseases including major depressive disorder (MDD).

1.2 Neurobiology of Brain Allostasis

Allostasis, which involves the integration of multiple central and peripheral systems, including hormonal signals and neural circuits, is foundational for understanding stress responses. When overwhelmed by stress, resulting in allostatic overload, stress resilience may transition to stress susceptibility that accompanies the shift from cognitive flexibility to inflexibility [1].

The HPA axis and the locus coeruleus (LC), a norepinephrine (NE)-producing nucleus in the brainstem, modulate arousal and attention, collaborate to regulate the body's response to stressors [9, 10], and influence the forebrain regions [11] including the prefrontal cortex (PFC). The PFC is essential for cognitive flexibility, the adaptive behavior necessary for successfully navigating changing environments [12–16], whose functionality is critical for the process of allostatic adaptation in the face of adversity. In an allostatic response, the increased NE release from LC neurons projecting to the PFC, hippocampus (HPC), and ventral tegmental area (VTA) [11, 17] is indicative of NE's role in facilitating an adaptive response. Glucocorticoids released by the kidney via the HPA axis lead to structural and functional changes in neural circuits including the PFC, which can affect an individual's balance between cognitive adaptation and maladaptation [18]. Molecularly, the PFC's response to stress is shaped by transcriptional and epigenetic modifications in cells, such as the differential expression of transcription factors such as Zfp189 and changes in histone acetylation and DNA methylation [19]. These cellular and molecular changes reflect long-term adaptations in the PFC to stress exposure and have significant implications for cognition and emotional regulation, underpinning the development of cognitive inflexibility and stress-related neuropsychiatric disorders.

1.3 A Window into Brain Allostasis: Astrocytes

Astrocytes, with their remarkable heterogeneity in form and function, are traditionally recognized for their role in maintaining central nervous system homeostasis and defense [20–22]. These multifunctional cells regulate neurotransmitters, ions, and metabolites; facilitate neurogenesis; and modulate synapses in ways that are behaviorally consequential [23, 24]. They form syncytia through gap junctions, allowing for the transport of second messengers, ions, and bioactive molecules that are critical to intercellular communication, and they are vital in preserving the integrity of the blood-brain barrier and in regulating cerebral blood flow, thus bridging central and peripheral systems [25, 26].

In terms of intercellular communication, astrocytes express an array of receptors for various neurotransmitters and neurohormones, such as glutamate, gamma-aminobutyric acid (GABA), ATP, adenosine, glycine, acetylcholine (ACh), endocannabinoid (eCB), dopamine (DA), norepinephrine (NE), and glucocorticoids [27–29]. This diversity equips astrocytes to respond adaptively to neural activity fluctuations and environmental changes, highlighting their capacity to adjust molecularly to variations in both brain tissue environments and external stressors. The traditional view of astrocytes as passive responders is being reevaluated. Now, they are seen as active participants in neural circuit modulations, capable of context-specific adaptations to environmental challenges [30–32]. This suggests a dynamic astrocytic involvement in the brain's response to stress, holding an allostatic role where they not only respond to but also adapt to physiological demands.

However, when the stressor to the adaptive brain cells persists or is too severe, leading to allostatic overload, these once adaptive astrocytic changes may become maladaptive, contributing to the development and progression of disorders and injury [32–36]. This underscores the importance of understanding astrocytes' role in both homeostasis and the maladaptive processes that underpin disease states, offering insight into the nuanced interplay between protective and detrimental aspects of astrocytic function.

1.4 Focus of the Chapter

This chapter is in four sections, dedicated to an examination of PFC astrocytes in the brain's adaptive and maladaptive responses that lead to cognitive flexibility and inflexibility, respectively. Initially, we delve into experimental approaches to dissect cognitive flexibility and inflexibility, highlighting the functional significance of the PFC. This includes a foundational understanding of its anatomy in both primates and rodents. Subsequently, our review pivots to scrutinize the latest research on PFC astrocytes. This analysis concentrates on their physiological and pathophysiological roles in regulating behaviors during allostatic adaptation. We cover a spectrum of astrocyte interactions, ranging from with local neurons and long-range projections, to peripheral neurohormones. The third segment of our chapter presents an extensive literature review. It describes the persistent molecular dysfunctions observed in PFC astrocytes in stress-induced conditions representative of MDD, across human, nonhuman primates, and rodent studies. In the final section, we cast light on the significance of astrocyte metabolism in understanding the prevalence and implications of PFC astrocyte dysfunction under allostatic load and overload, through the lens of neuroenergetics.

Our review does not encompass the general roles of astrocytes in learning and memory, which have been well-reviewed elsewhere [23, 24, 29, 37–39]. Here, we propose novel insights into the pivotal role of PFC astrocytes in the nervous system's adaptation to environmental changes. Our synthesis of evidence from various species and the latest scientific findings points toward the potential of targeting astrocyte molecular adaptation under stress for therapeutic innovations.

2 Cognitive Flexibility and Prefrontal Cortex

2.1 Cognitive Flexibility and Behavioral Tests

Animals possess an innate ability to adapt their behavior in response to environmental changes for survival [40]. Take a mouse in a crop field as an example: When the crops change, the mouse's prior strategies to find food become obsolete. It must then relearn, using sensory inputs—such as vision and smell—to create a new cognitive map for locating food. This adaptive behavior is critical, involving several

Table 1 Some behavioral tests to assess cognition in rodents

Test type	Description	Key processes involved	Main brain Regions Implicated
Object- or location-based memory tests	Assess recognition memory by leveraging a rodent's tendency to explore novel objects. Rodents are familiarized with objects and then presented with a new or relocated familiar object. The exploration time of these objects indicates memory and cognitive processing	Basic perceptual abilities and attention	Cerebral cortex, HPC, thalamus, striatum
Pavlovian conditioning tests	Evaluate associative learning where rodents form associations between a neutral stimulus and a salient event (e.g., sound and food delivery), leading to anticipatory behavior	Memory, outcome prediction based on cues, and the integration of associative cues and outcomes	PFC, HPC, amygdala, thalamus, NAc
Instrumental conditioning tests	Focus on goal-directed behavior, requiring rodents to learn that certain actions lead to rewards. This includes tasks such as the T-maze or lever pressing for food. It also involves reversal learning tests for cognitive flexibility assessment	Planning, decision-making, error monitoring, inhibition of non-rewarding behaviors, and adaptation to changes in reward conditions	PFC, striatum (including the NAc)

cognitive domains for executive functions: planning new strategies, shifting attention, applying working memory of past experiences, making informed decisions, learning from errors, and exercising inhibitory control to avoid old, ineffective patterns [41, 42]. These steps are vital for the mouse to successfully forage in the newly altered environment, demonstrating the animal's cognitive flexibility and problem-solving skills in real-time scenarios [16, 43]. Not just in rodents, but this flexibility is innate and necessary in humans as well—we constantly adapt our internal processes to changing physical and social environments [15, 44].

For experimentally evaluating cognitive flexibility and executive functions, rodents are a useful model (Table 1). Here, we list behavioral tests that are routinely used mainly for rodents [16, 43, 45–47]. Most of behavioral tests require attention, which often involves NE signals from the LC, and spatial/contextual encoding by the HPC, and each progressively demands more PFC engagement. These cognitive tests help reveal the PFC's crucial role in integrating sensory input, past experiences, and anticipated outcomes to guide behavior, especially in tasks that demand adapting to new rules or situations, indicative of cognitive flexibility and executive control.

We also highlight the following stress-related behavioral models [48] for their utility in studying molecular, cellular, and circuit consequences of allostatic overload (Table 2). A prolonged stress often induces depressive-like symptoms. Research in both animal models and human psychology indicates that individuals who are

Table 2 Some stress models in rodent research

Model name	Description
Chronic unpredictable stress (CUS) model	Involves subjecting animals to a series of diverse and unpredictable stressors over a prolonged period, typically several days to weeks, to induce symptoms resembling human stress-related disorders. Stressors include physical, psychological, and environmental challenges, aiming to produce responses such as anhedonia, anxiety-like behavior, cognitive impairments, and decreased social interaction
Chronic social defeat stress (CSDS) model	Involves repeated subordination by a more aggressive animal (encounters) over 10 days, leading to individual differences among genetically similar animals. Some exhibit depressive-like behaviors such as social avoidance and anhedonia, while others show resilience. Considered to have greater etiological validity
Early-life stress models	Focus on the impact of stress during developmental stages on later-life outcomes. Common paradigms include maternal separation and reduced bedding, demonstrating that the extent of stress exposure and stress response are not linearly correlated. Long periods of maternal deprivation can lead to increased susceptibility to stress, HPA axis hyperactivity, and altered glucocorticoid responses in adulthood, whereas moderate early-life stress may promote resilience. Early behavioral traits can predict lifespan outcomes

susceptible to stress often exhibit diminished cognitive flexibility, struggling to adapt their thinking to meet changing situational demands [49–52] This observation suggests that resilience to stress is crucial for maintaining cognitive flexibility. Since the PFC plays a vital role in cognitive flexibility, understanding the cellular and molecular changes in the PFC that occur in both resilient and susceptible individuals can offer significant insights.

2.2 *Anatomy of Prefrontal Cortex*

The PFC is universally recognized as an essential brain region for higher cognitive processing, with considerable research in both primates and rodents enhancing our understanding of its functions. Since the 1990s, there has been a significant increase in PFC studies, with early work in monkeys highlighting the dorsolateral PFC's role in spatial working memory, followed by human neuroimaging studies that further reinforce the PFC's involvement in attention and working memory that are required for cognitive flexibility [12, 14, 15].

2.2.1 Definition in Rodents and Primates

Discrepancies in the anatomical definition of the PFC between species have posed challenges [15, 16]. In primates, the PFC commonly refers to the granular areas of the lateral PFC and includes the orbitofrontal cortex (OFC), which is understood without the need to specify "prefrontal." In contrast, rodent studies initially defined the PFC based on connections with the mediodorsal nucleus of the thalamus, which led to debates about whether regions such as the medial agranular cortex (premotor area M2) should be classified as part of the PFC. Primate researchers often use the term anterior cingulate cortex (ACC) to refer to agranular areas in the medial frontal cortex, whereas comparable regions in rodents are classified as the medial prefrontal cortex (mPFC), including prelimbic and infralimbic subregions.

Despite these differences, there is a growing consensus that the focus should be on common functions rather than anatomical structures when comparing the PFC across species. This is exemplified by research showing that rodent orbital areas, such as the OFC and its subregions, are involved in encoding expected outcomes, like the primate PFC [53–55].

In this review, we embrace a comprehensive approach, incorporating findings from primate studies of the granular lateral PFC, ventromedial PFC, OFC, and agranular ACC, as well as rodent studies of the mPFC, including the prelimbic and infralimbic subregions, ACC, and OFC. By focusing on functional parallels, we aim to bridge the gap between primate and rodent research, offering a more cohesive understanding of the PFC's role in cognitive processes.

2.2.2 Connectivity

Rather than functioning in isolation, the PFC operates as an integral component within the interconnected brain regions [16, 56]. In rodents, the PFC receives projections from various brain regions, such as glutamatergic projections from the hippocampus, amygdala, and thalamus, serotonergic (5-HT) projections from the dorsal raphe nuclei (DRN), NE projections from the LC, ACh from the basal forebrain, DA from the VTA and substantia nigra (SNc), and histamine (HA) from the tuberomammillary nucleus (TMN). In return, PFC neurons send glutamatergic projections to the NAc, HPC, amygdala, lateral habenula, thalamus and hypothalamus, DRN, VTA, and periaqueductal gray (PAG). Neurons within the PFC communicate with each other with glutamate, GABA, and eCB [16, 57]. The PFC's role in cognition and behavior is not limited to a single localized function but is intricately involved in a larger neural network.

2.2.3 Cells

The rodent PFC's neuronal composition predominantly includes excitatory pyramidal neurons (~80–90%) organized across all layers and inhibitory GABAergic interneurons (~10–20%) that mainly populate in layer II/III and V, which are further classified based on the expression of markers such as parvalbumin, somatostatin, or VIP [16]. Astrocytes, microglia, oligodendrocytes and their precursors, the glial constituents of the PFC, interact intensively with neurons within the region [58, 59]. Astrocytes express a palette of receptors to sense the transmitters mentioned above and exhibit dynamic Ca^{2+} signaling in response to neurotransmitters, influencing neuronal activity and, consequently, broader brain circuits. These astrocytic signals occur spontaneously and in response to neuronal activity, enabling their involvement in the PFC's response to environmental challenges.

3 Prefrontal Astrocytes in Homeostasis and Allostasis

We here focus on recent rodent studies that illustrate the role of PFC astrocytes in brain allostasis, highlighting the adaptive and maladaptive alterations of astrocyte signaling and their contributions to neural functionality, particularly in the context of cognitive flexibility and stress response.

3.1 *Local Astrocyte-Neuron Interaction*

Astrocytes directly interact with GABAergic interneurons in the mouse mPFC through $GABA_B$ receptors to influence goal-directed behaviors in the T-maze [60]. Genetic deletion of $GABA_B$ receptors selectively in mPFC astrocytes disrupts neuronal firing pattern, which are believed to regulate the preservation of working memory. Although direct optical activation of broadly GABAergic neurons did not restore cognitive dysfunctions induced by the absence of $GABA_BRs$ in astrocytes, optical stimulation of astrocyte GPCR signaling by melanopsin restored behavioral deficits. Synaptically, astrocyte GPCR signaling stimulation mediated the potentiation of inhibitory synaptic transmission. As mentioned above, astrocytes respond to various neurotransmitters, including GABA, to evoke GPCR/Ca^{2+} signaling. Selective perturbation of astrocytes using the Gq-DREADD hM3Dq to increase the human plasma membrane Ca^{2+} pump (hPMCA2w/b) and CalEx to attenuate Ca^{2+} signaling increased and decreased ethanol consumption, respectively [61]. These manipulations influenced ethanol-induced locomotor activity and sedative-hypnotic effects, indicating astrocyte roles in regulating the acute stimulatory effects of low-dose ethanol. RNA-seq of PFC astrocytes from mice after chronic ethanol consumption revealed changes in gene expressions including calcium-related signaling [62]. Overall, interactions between astrocytes and neurons within the PFC via GABA and GPCR/ Ca^{2+} signaling play in the synaptic adaptations underpinning goal-directed actions and decision-making processes.

Decision-making processes related to eating rewards are intricately linked to allostasis, as the brain continuously adapts to internal and external cues to maintain energy balance. The contexts of obesity, eating disorders, and alcohol use disorders exemplify how maladaptive allostatic responses can lead to impaired decision-making, where the reward systems override satiety signals. Obesity and alcohol intake affect brain circuits including hypothalamic, mesolimbic, and prefrontal pathways that govern eating and drinking behaviors. In rats that had an unrestricted access to a high-calorie diet over 40 days, inhibitory synaptic transmission onto excitatory pyramidal neurons in the OFC layer II/III was suppressed by postsynaptic eCB release [57]. The excess eCB was induced by a glutamate spillover, due to diminished efficiency of astrocyte glutamate transporters GLT-1. This astrocytic dysfunction contributed to an altered heterosynaptic plasticity—a process crucial for adapting synaptic strength—which was disrupted by the obesogenic diet. Glutamate homeostasis could be restored with dietary supplementation of N-acetylcysteine, a cystine prodrug reported to enhance GLT-1 activity. While the study stops short of extending these findings to behavioral causality, astrocyte-neuron interactions that distort synaptic plasticity may regulate food-related decision-making and inhibitory control.

3.2 *Interregional Interaction*

Despite their cellular territory being local, astrocytes exert a regulatory influence that extends beyond their immediate environment, impacting interregional communication through the brain's interconnected network. Disruptions in PFC astrocyte activity, such as those caused by expressing a dominant-negative SNARE [63], can desynchronize theta oscillations between HPC and PFC, impairing memory performance (but see [64]). Modifications such as overexpressing MHC-I in the mPFC astrocytes impact synaptic connections to the dorsal striatum [65]. IP_3R2 deletion mice show depressive-like behavior, which is reversed by channelrhodopsin-2 (ChR2) optogenetic stimulation of mPFC astrocytes, altering interregional connectivity including PFC-to-striatum pathway, as observed through functional MRI [66]. ChR2 stimulation in astrocytes is known to cause increases in extracellular K^+ levels to globally affect neuronal firing rates [67]. Careful considerations are required to elucidate the physiological relevance of any studies using ChR2 in astrocytes, because astrocytes do not fire action potentials and propagate electrical signals.

Astrocytes are responsive to projections from monoamine nuclei, such as the VTA for dopamine and LC for NE [29]. Notably, PFC astrocytes can respond to dopamine via α1-adrenergic receptors (α1-AR), usually linked to NE [68]. Using two-photon microscopy imaging *in vivo* and *ex vivo*, the study revealed that dopamine activates astrocytic calcium signaling through α1-ARs without engaging the classical dopamine receptors and cAMP pathways. In addition, the uniquely prolonged Ca^{2+} signaling, unlinked to locomotion unlike in primary sensory cortex [68], in PFC astrocytes may highlight its involvement in higher-order cognitive processes.

3.3 *Brain-Body Interaction*

Stress involves a complex interplay between the central and peripheral systems in maintaining allostasis. The key neurohormone molecules in this response are glucocorticoids, steroid hormones released by the adrenal cortex in the kidney in response to stress, which mobilize energy resources by increasing blood glucose levels primarily released by the liver to ensure that the body can cope with the demands of stress [5, 69]. This interaction is mediated by the hypothalamus that releases CRH, which stimulates the pituitary gland to secrete adrenocorticotropic hormone (ACTH). ACTH then acts on the adrenal cortex to release glucocorticoids into the bloodstream. Communication between the central nervous system and peripheral organs is critical in managing and adapting to stressors through synaptic mechanisms, particularly focusing on the role of astrocytes and glucocorticoid receptors (GRs) in the brain [18, 70].

The vulnerability to social defeat stress was found to correlate with a decrease in the expression of GR in astrocytes. Genetic deletion of astrocytic GRs leads to depressive-like behaviors, while restoring these receptors in the mPFC reverses these effects [70]. This study shows that astrocytic GRs influence Ca^{2+} activity and ATP release in response to stress. A recent study found that stress hormones increased the expression of the MERTK phagocytic receptor in astrocytes through GR *in vitro*, *in vivo* mouse brains, and in human brain organoids [18]. GR induces phagocytosis in astrocytes via the GR-MERTK pathway in mouse models of early social deprivation (ESD). There was a notable increase in astrocytic phagocytosis and a reduction in excitatory synaptic connections onto GABAergic neurons that may induce E/I imbalance, selectively observed in the OFC. These changes are not evident in other brain regions such as the mPFC, HPC, and striatum.

4 Prefrontal Astrocytes with Allostatic Overload in Neuropsychiatric Disorders

In instances where adaptive responses are inadequately activated or persist beyond the cessation of a stressor, there can be a cumulative impact on the body, resulting in allostatic overload, potentially leading to the dysregulation of physiological systems and thus increasing the likelihood of disease development altering PFC function and accompanying cognitive inflexibility, notably MDD. MDD represents the most prevalent psychiatric condition, exhibiting a lifetime prevalence of approximately 16% in the United States [71]. It manifests heterogeneously, though can be driven by dramatic or repeated stressful life events. Common symptoms include a persistently low or depressed mood, anhedonia, feelings of guilt or worthlessness, reduced energy and concentration, altered appetite, psychomotor abnormalities, sleep disturbances, and suicidal ideation. Patients with eating disorders such as anorexia nervosa, which affects about 0.8% of the population in the United States [72], show heightened stress response and co-morbidity with MDD and thus may share similar causes [73]. Research on such stress-related conditions accompanies quite similar astrocyte alterations, characterized by disrupted astrocytic networks, impaired glutamate clearance, and reduced synaptic support, particularly within the PFC.

4.1 Observational Evidence from Cross-Species Studies

4.1.1 Morphology

Astrocytes come into close contact with blood vessels, synapses, and other glia. Changes in their morphologies can indicate changes in their functions and interactions with surrounding cells that then may reflect changes in cognitive function. Postmortem studies from the past two decades on patients with MDD and/or suicidality indicate a decrease in the astrocytic marker GFAP in prefrontal regions [74–76]. Furthermore, GFAP-positive prefrontal astrocytes show decreased cell density and size in these patients [74, 77, 78]. However, not all astrocytes express GFAP, or to the same level, and it cannot be a universal indicator for astrocyte morphology and pathology [79]. Vimentin was seen to increase in rhesus monkeys displaying self-injurious behavior, and vimentin-positive prefrontal astrocytes showed similar morphological changes in suicidal human patients, particularly in the PFC white matter [80]. AQP4-positive astrocyte contact with blood vessels also decreased in MDD patients, particularly in the OFC gray matter [81].

Rodent CUS models show similar patterns in prefrontal astrocytes. Regarding GFAP, expression levels and immunoreactive cell density decreased [78, 82–84]. Expressing GFP under astrocytic promoters revealed astrocyte atrophy, where cell size decreases and processes shorten and increase in proximal complexity [85–88], which was reversed with the antidepressant fluoxetine [87]. Anorexia mouse models show decreases in astrocyte density and ramification as well [89, 90]. Overall, both human and rodent studies indicate that decreases in astrocyte number, size, and distal complexity accompany depressive symptoms.

4.1.2 Gap Junctions

Gap junctions are pivotal in the establishment of astrocytic networks, which in turn play a crucial role in modulating neuronal function beyond the synaptic and microcircuit levels. Postmortem prefrontal cortex samples from suicidal patients show decreased expression of both Connexin 30 (Cx30) and Cx43, the major gap junction proteins expressed in astrocytes, and increased methylation of their genes [75, 91]. This pattern of diminished expression and weakened gap junction coupling is also evident in rodent models subjected to CUS [85, 92, 93].

4.1.3 Glutamate Transporters

As mentioned above, one of the most well-known functions of astrocytes is glutamate homeostasis at the synapse, which controls neuronal excitability and synaptic efficacy. In the context of MDD, human patient samples showed decreased expression of GLT1 (Slc1a2) and GLAST (Slc1a3) in the cortex—two glutamate transporters expressed in astrocytes [94]. Glutamine synthase (GS) expression also decreased in the prefrontal cortex of suicide patients and GS-positive astrocyte density decreased in MDD patients, indicating a disruption in astrocyte glutamate metabolism [95, 96]. CUS rats also show decreased GLT1 and GLAST expression in prefrontal astrocytes, as well as decreases in ^{13}C-acetate metabolism that indicates slowed glutamate metabolism [83, 84]. Anorexia model rats exhibit decreased endogenous glutamate and glutamine levels and decreased GLT1, GLAST, and GS expression in the PFC [89]. Administering drugs known to stimulate glutamate uptake and metabolism—ceftriaxone and riluzole—were observed to have antidepressant effects in mice [83, 97].

4.2 Causal Evidence from Rodent Studies

Rodent models allow for a more detailed look at the causal role of astrocyte responses to stress on behavioral changes. Ablation of PFC astrocytes in rats using an astrocyte-specific toxin l-α-aminoadipate (L-AAA) resulted in depressive-like behaviors, which another study reversed by L-glutamine infusions [98]. The same ablation in the mPFC specifically disrupted cognitive flexibility, memory, and attention [99, 100].

While restoring Cx30 and Cx43 expression in CUS mice rescued depressive-like behaviors, depleting Cx30 and Cx43 led to opposite symptoms. Inhibiting GLT-1 in the PFC led to depressive-like behaviors [101]. Blocking the astrocytic release of ATP, D-serine, and glutamate in the PFC decreased spatial and long-term memory ability, and IP injections of D-serine restored the changes [63]. Though there has yet to be more focus on prefrontal astrocyte changes in morphology, gap junctions, and glutamate transport, the current evidence strongly suggests the direct involvement of astrocytes in behavioral responses to stress.

5 Astrocyte-Neuron Metabolic Coupling for Allostasis

Building upon the discussion of astrocyte-neuron interactions characterized by dynamic neurotransmitter activities and morphological changes, we turn our attention to the critical aspect of astrocyte metabolism in the neuroenergetically demanding PFC, extending beyond the typical astrocyte-neuron interactions involving neural circuit dynamics and morphological changes. We provide evidence on the metabolic intensity of the PFC in adult brains and explore in depth how the metabolic functions of astrocytes play a key regulatory role. Additionally, we examine the significant implications of astrocyte metabolic dysfunctions in the PFC, particularly in relation to neuropsychiatric disorders characterized by cognitive inflexibility and prefrontal cortex impairments.

5.1 *Metabolically Demanding Prefrontal Cortex*

The energy budget of the brain is largely consumed by neurons. Neuronal energy demand (ATP use) is due to the ion flux required for generating action potentials and sustaining resting potentials [102], as well as plasticity processes such as cytoskeletal rearrangements, protein trafficking, or increased gene expression [103]. This demand is met through metabolic support mechanisms, which include glial glucose uptake and the shuttling of serine and lactate from glia to neurons to fuel cell respiration via mitochondria [26, 104].

Disruptions in this local energy supply can lead to a breakdown of brain-wide network coding, as evidenced by various human studies. The concept of an "overloaded" or "failed hub" has been proposed [105], where hubs with high energy demands become compromised, as seen in many diseased conditions [106–112]. With aging, the brain becomes increasingly reliant on hub regions [113, 114], and the PFC, especially its lateral and ventromedial regions, is identified as key players among hubs, connecting multiple brain networks [115, 116].

A transcriptomic analysis revealed the higher energy demand in hub regions is supported by more frequent neuronal firing and higher gene expressions related to ATP production and the tricarboxylic acid (TCA) cycle [117, 118]. These findings underscore the PFC's role in energy-intensive cognitive processes. The human PFC is also characterized by its remarkable "neoteny"—the retention of developmental features into adulthood [119, 120]. This neoteny is associated with higher levels of aerobic glycolysis, a metabolic process that correlates with synaptic growth gene expression and is particularly pronounced during childhood. In the adult brain, regions exhibiting neoteny maintain relatively high levels of aerobic glycolysis, suggesting a sustained capacity for synaptic plasticity [121].

Together, the PFC's extensive energy demands, the necessity of metabolic support for maintaining neural activity and plasticity for adaptation, and the implications of metabolic disruptions provide crucial insights into the neurobiological underpinnings of cognitive function and the etiology of various neuropsychiatric conditions.

5.2 Astrocyte Metabolic Support of Neurons

Pioneering organ-level studies by Kety and Schmidt [122] and Sokoloff [123] in the mid-twentieth century established glucose as the brain's essential energy substrate, metabolized to carbon dioxide and water to yield 32–36 ATP molecules per glucose molecule under oxidative conditions. Despite neurons accounting for most of the brain's total energy consumption [102], they take up an amount of glucose from the circulation, at the most, roughly equal to that of glia under basal conditions [124, 125]. Glial cells—notably, astrocytes, and oligodendrocytes—are predominantly glycolytic, processing glucose to lactate. Neurons, however, mainly process lactate to pyruvate for the TCA cycle in the mitochondria and for the associated oxidative phosphorylation, the most efficient ATP-producing pathway. When brain function is evoked, glia are the primary cells that increase glucose uptake and transfer lactate to neurons [104].

Distinct metabolic profiles are characteristic of various cells and tissues within the body. In skeletal muscles, for instance, the rapidly contracting "white" fibers primarily engage in glycolysis for energy, in contrast to the "red" fibers, which rely on oxidative processes [126]. Similarly, several molecular factors support the specific metabolic properties of astrocytes and neurons and their coupling [104]:

1. Enzymatic Activity: Astrocytes exhibit a unique profile of enzyme expression favoring the conversion of pyruvate to lactate. Pyruvate dehydrogenase (PDH), which provides carbon atoms for the TCA cycle, has lower activity in astrocytes due to high phosphorylation levels driven by a high expression of pyruvate dehydrogenase kinase isoform 4 (PDK4).
2. Mitochondrial Organization: Neurons have a highly efficient organization of the mitochondrial respiratory chain complex (numbered I-IV). Conversely, astrocytes have a particular organization with the most complex I coupled from supercomplexes. The low dependence of astrocytes on mitochondria is further suggested by mice with reduced cytochrome oxidase activity in astrocytes from lacking Cox10 in GLAST-Cre-targeted astrocytes which resulted in no neurodegenerative signs over a period exceeding one year.
3. Protein Expression: Astrocytes predominantly express LDH5, suited for lactate production, while neurons express LDH1, indicative of lactate consumption. This differential expression supports the directionality of lactate from astrocytes to neurons.

4. Transporter Distribution: Monocarboxylate transporters (MCTs) that operate transmembrane lactate transport are distributed cell specifically, with astrocytes expressing the low-affinity MCT1 and MCT4 and neurons expressing the high-affinity MCT2, which is optimal for astrocyte release and neuron influx of lactate.
5. Redox State: Astrocytes have a higher NADH:NAD+ ratio, indicating a more reducing state that thermodynamically favors the conversion of pyruvate to lactate. Although neurons and glia take up glucose at similar rates under basal conditions, neurons channel glucose primarily through the pentose phosphate pathway (PPP) to produce NADPH, which combats oxidative stress produced by their intense oxidative activity. Indeed, inhibiting PFKFB3 degradation, pivotal for the PPP, leads to neuronal oxidative damage and death.

5.3 Astrocyte Metabolic Support of Cognition

5.3.1 Memory Consolidation

Studies initiated in the 1990s by Gibbs and Hertz [127] revealed the necessity of glycolysis, glycogenolysis, and lactate shuttling for memory consolidation during passive avoidance learning in a neonatal chick model. Similarly, in the rat hippocampus, it has been observed that learning significantly elevates extracellular lactate levels, originating from glycogen primarily in astrocytes [128]. This glycogen breakdown and subsequent lactate release are crucial for the formation and maintenance of long-term memory and potentiation of synaptic strength.

Interference with astrocytic lactate transporters, such as MCT1 and MCT4, induces amnesia, which is reversible by L-lactate but not glucose, while disruption of the neuronal lactate transporter MCT2 leads to irreversible amnesia, underlining the necessity of lactate import into neurons for long-term memory formation [128, 129]. This process also underpins critical molecular changes, including the induction of phospho-CREB, Arc, and phospho-cofilin. Further supporting the energy-independent action of lactate on neuronal plasticity, lactate (but not glucose or pyruvate) induces the expression of plasticity-related genes such as Arc, Egr1, and Bdnf in cultured mouse cortical neurons [130]. This occurs at concentrations as low as 2.5 and 5 mM, indicating lactate's role in amplifying NMDA receptor-mediated currents. Collectively, this series of studies suggests a broader role of astrocyte-neuron lactate shuttling in long-term memory formation through supporting energy demand from ion flux during action potential firing and gene expression changes for synaptic plasticity.

5.3.2 Cognitive Flexibility

The role of astrocyte lactate shuttling in cognitive flexibility, particularly in the PFC, has been explored by recent findings in various rodent models. For instance, CUS in rodent models resulted in decreased levels of glycolysis, lactate dehydrogenase A (LDHA), and L-lactate in the dorsolateral part of the PFC, signifying reduced lactate production [131]. L-lactate has been shown to enhance synchrony between the amygdala and ACC, which in turn supports decision-making processes [132]. The optogenetic Gi activation of mouse ACC astrocytes led to an impairment in cognitive flexibility that accompanies decreased L-lactate levels, which were rescued upon administration of L-lactate [133]. This evidence supports that astrocyte-mediated metabolic processes are necessary for cognitive functions.

5.4 Signaling Regulation of Astrocyte Metabolism in Allostasis

Despite the considerable variability in stressors encountered by both humans and animals across studies, a unifying hypothesis emerges: Alterations in astrocytic functions—specifically metabolic flux, cellular morphology, gap junctions, and glutamate transport—are intricately interconnected through complex molecular signaling cascades. Stress alters astrocyte morphology and network connectivity, particularly via gap junctions in the PFC and other brain regions [92, 134], potentially impairing the cells' ability to shuttle key energy substrates such as lactate. A causal link between GLT1 and glucose influx was indicated by an *in vivo* rodent study using 2-deoxy-2-[^{18}F]fluoro-D-glucose (FDG) positron emission tomography (PET) imaging [135]. Postmortem studies of stress-related disorders show disruptions in all these astrocytic functions (see above), implicating compromised metabolic support. However, the precise relationships among these molecular cascades and the impact of excessive stress on inducing maladaptive changes remain largely unknown. We explore plausible signaling interplays between stressors and astrocytes in allostasis, allostatic overload, and the potential mechanisms of therapeutic interventions.

5.4.1 In Allostasis

Astrocytes, serving as the primary glycogen storage in the brain, support neuronal function through glycogenolysis and the astrocyte-neuron lactate shuttle. Stress influences astrocytic adrenergic receptor signaling and intracellular pathways in the mouse neocortex [136]. Air-puff acute stress induces transient astrocytic Ca^{2+}

signaling via α1-AR, while prolonged stress escalates Ca^{2+} signaling and cAMP levels through β-adrenoceptors (β-AR), triggering glycogenolysis. Astrocytic β-ARs modulate cAMP, influencing both glycogenolysis and lactate shuttling, crucial for memory consolidation in the HPC [137]. Notably, β-ARs, rather than cortisol synthesis, mediate shifts in brain state and network connectivity under fear-related acute stress [9, 10].

5.4.2 With Allostatic Overload

In a rat model of corticosterone-induced depression, Txnip, whose protein product is a negative regulator of GLUT1 transmembrane localization, is upregulated in the PFC, with its overexpression in mPFC astrocytes leading to depressive-like behaviors [138]. However, the mechanism by which corticosterone induces Txnip expression remains unclear. Acute stress to mice, a 20-minute forced swim, induces astrocytic hypertrophy in the mouse neocortex, potentially rupturing gap junction channels and the astrocyte metabolic network, thereby limiting neuronal access to energy reserves in a glucocorticoid-dependent manner [139]. This process, accompanied by impaired synaptic plasticity, can be mitigated by L-lactate supplementation. Astrocyte-specific transcriptomic analysis under such conditions reveals altered Wnt/β-catenin signaling which is known to induce resilience [140] and anxiolytic effects [141], and influence connexin expression [142], highlighting the astrocyte syncytium's role as a distributed energy source, which is disrupted by stress and glucocorticoid signaling. This is in accord with other findings: Restoring the lactate supply rescues stress-related depressive-like behaviors [131, 143]. The linkage between glucocorticoids and Wnt signaling remains an area for further investigation. Rapid channel closure of Cx43, demonstrated with a synthetic glucocorticoid receptor agonist dexamethasone, suggests a pathway by which glucocorticoids alter functional coupling under stress [144].

5.4.3 Adaptation vs Maladaptation

Glucocorticoid-induced mechanisms in the neocortex and HPC appear deleterious; however, it is possible that stress favors energy supply and plasticity in regions critical for threat assessment and allostatic regulation. Such differences may involve diversity in astrocyte states [20, 30–32, 35]. Innate interregional and intraregional diversity in astrocytes and adaptive astrocyte state changes in allostasis, including intracellular molecular programs, epigenomic profiles, receptor expression levels, and morphological interactions with other brain cells can significantly affect how astrocytes respond to stressors. As discussed above, β-ARs mediate NE-induced astrocyte glycogenolysis that could assumingly be classified as allostatic responses. However, extensive studies have suggested that α1-ARs may mediate detrimental effects of stress under allostatic overload. For example, in the monkey dorsolateral

PFC (dlPFC), α1-ARs in layer III astrocytes are implicated in the treatment of under-aroused subjects, whereas α1-AR blockade is central to treating stress-related disorders such as posttraumatic stress disorders (PTSD) [145]. Astrocyte pathophysiological changes observed in PTSD patients are quite like those observed in MDD [146]. PTSD involves excessive NE signaling [147–149], and brains of suicide patients show increased α1-ARs in the dorsal PFC [150], highlighting the specific effects of α1-AR under high NE conditions that impair PFC's cognitive abilities [151]. Clinically, α1-AR antagonists are widely used in disorders associated with stress and excessive NE signaling [152–154].

5.4.4 In Stress Disorders and Therapeutics

Ketamine, recognized for its rapid antidepressant effects, enhances glucose uptake in PFC astrocytes via the glucose transporter 3 signaling pathway [155]. In culture, ketamine increases astrocyte cytoplasmic cAMP that accompanies the rapid release of neurotrophic factors and redistribution of plasmalemmal cholesterol, suggesting a potential role of ketamine for enhancing astrocyte cAMP and astrocytic support of neuronal activity and plasticity as well as myelin maintenance by oligodendrocytes for depression treatment [156].

Despite its longstanding position as a first-line treatment for mood disorders and its superior efficacy in suicide prevention compared to other mood stabilizers and antidepressants, the mechanisms underlying lithium's therapeutic actions remain partially elusive. Lithium acts on glycogen synthase kinase 3β (GSK3β), inhibiting glycogen synthase. This is thus modulating glycogenolysis and glucose availability, potentially facilitating lactate shuttling in response to stressor-induced NE signals. This mechanism may play a part in how astrocytes energetically support neurons. Inhibition of GSK3β by lithium correlates with a notable augmentation in both the density and size of astrocytes [157]. However, the inhibitory effect of lithium on GSK3β activity occurs at doses higher than those used therapeutically, suggesting that this may not be its physiological mechanism of action. At therapeutic concentrations (1–2 mM), lithium is reported to affect GPCRs and the G protein cycle, indicating an alternative mechanism by which lithium may impact astrocyte function [158]. The mechanistic action of lithium remains an area for extensive further research.

The diversity in astrocyte responses to stress underscores the necessity for comprehensive research for characterizing astrocyte allostatic states. In a broader prospect, future studies should aim to tease apart what constitutes context-specific astrocyte states [20, 30–32, 35] at a single-cell level through transcriptomics or proteomics, as well as investigate "astrocyte state-specific" manipulations for causal studies. Humanized research models, including noninvasive astrocyte signaling imaging and AI-based brain function inference from peripheral data, are also crucial for understanding the metabolic states of astrocytes and their links to biomarker developments. A multivariate analytical approach may reveal the complex interplay of molecules and pathways that shape astrocyte phenotypes across different conditions and stages of disease. These findings could lead to predictive models for the functional outcomes of mixed protective and deleterious pathways and identify key regulatory hub signaling pathways.

6 Outlook

This chapter critically examines the complex roles of astrocytes in maintaining the balance of brain homeostasis and allostasis, with a focused lens on the PFC as a pivotal site for cognitive flexibility in responses to environmental changes, stressors, and metabolic challenges therein. This review also explores the intricate link between astrocyte functionality under allostatic overload and the development of neuropsychiatric disorders, particularly MDD. We highlight how astrocyte metabolism may underlie the pathophysiology of these conditions. This understanding will be enhanced and detailed through the use of innovative tools offering unprecedented spatial and temporal resolution, such as genetically encoded fluorescent biosensors for imaging intracellular glucose [159] and intracellular and extracellular L-lactate [160, 161]. Allostatic astrocytes may apply to a context-specific state of astrocytes called reactive astrocytes, once considered a uniform response to brain pathology, which actually exhibit diverse, disease-specific phenotypes [32, 35, 36]. Their documented involvement in neurological, neurodevelopmental, and neurodegenerative diseases suggests that these astrocytic adaptive changes are not merely reactive but are part of an allostatic adaptation aimed at protecting and repairing neural tissue. Drawing upon empirical evidence from various species, we shed light on the emerging potential of astrocyte molecular adaptations and maladaptations to serve as both biomarkers and viable targets for therapeutic intervention.

References

1. Cathomas F, Murrough JW, Nestler EJ, Han M-H, Russo SJ. Neurobiology of resilience: interface between mind and body. Biol Psychiatry. 2019;86:410–20.
2. Mason S. Lactate shuttles in neuroenergetics-homeostasis, allostasis and beyond. Front Neurosci. 2017;11:43.
3. McEwen BS. Stress, adaptation, and disease. Allostasis and allostatic load. Ann N Y Acad Sci. 1998;840:33–44.
4. Charney DS. Psychobiological mechanisms of resilience and vulnerability: implications for successful adaptation to extreme stress. Am J Psychiatry. 2004;161:195–216.
5. Herman JP, et al. Regulation of the hypothalamic-pituitary-adrenocortical stress response. Compr Physiol. 2016;6:603–21.

6. Ulrich-Lai YM, Herman JP. Neural regulation of endocrine and autonomic stress responses. Nat Rev Neurosci. 2009;10:397–409.
7. Logan JG, Barksdale DJ. Allostasis and allostatic load: expanding the discourse on stress and cardiovascular disease. J Clin Nurs. 2008;17:201–8.
8. Yehuda R, LeDoux J. Response variation following trauma: a translational neuroscience approach to understanding PTSD. Neuron. 2007;56:19–32.
9. Hermans EJ, et al. Stress-related noradrenergic activity prompts large-scale neural network reconfiguration. Science. 2011;334:1151–3.
10. Krystal JH, Neumeister A. Noradrenergic and serotonergic mechanisms in the neurobiology of posttraumatic stress disorder and resilience. Brain Res. 2009;1293:13–23.
11. Poe GR, et al. Locus coeruleus: a new look at the blue spot. Nat Rev Neurosci. 2020;21:644–59.
12. Carlén M. What constitutes the prefrontal cortex? Science. 2017;358:478–82.
13. Anastasiades PG, Carter AG. Circuit organization of the rodent medial prefrontal cortex. Trends Neurosci. 2021;44:550–63.
14. Friedman NP, Robbins TW. The role of prefrontal cortex in cognitive control and executive function. Neuropsychopharmacol Off Publ Am Coll Neuropsychopharmacol. 2022;47:72–89.
15. Laubach M, Amarante LM, Swanson K, White SR. What, if anything, is rodent prefrontal cortex? eNeuro 2018;5. ENEURO.0315-18.2018
16. Le Merre P, Ährlund-Richter S, Carlén M. The mouse prefrontal cortex: unity in diversity. Neuron. 2021;109:1925–44.
17. Isingrini E, et al. Resilience to chronic stress is mediated by noradrenergic regulation of dopamine neurons. Nat Neurosci. 2016;19:560–3.
18. Byun YG, et al. Stress induces behavioral abnormalities by increasing expression of phagocytic receptor MERTK in astrocytes to promote synapse phagocytosis. Immunity. 2023;56:2105–2120.e13.
19. Lorsch ZS, et al. Stress resilience is promoted by a Zfp189-driven transcriptional network in prefrontal cortex. Nat Neurosci. 2019;22:1413–23.
20. Khakh BS, Sofroniew MV. Diversity of astrocyte functions and phenotypes in neural circuits. Nat Neurosci. 2015;18:942–52.
21. Allen NJ, Lyons DA. Glia as architects of central nervous system formation and function. Science. 2018;362:181–5.
22. Verkhratsky A, Nedergaard M. Physiology of astroglia. Physiol Rev. 2018;98(1):239–389. https://pubmed.ncbi.nlm.nih.gov/29351512/
23. Nagai J, et al. Behaviorally consequential astrocytic regulation of neural circuits. Neuron. 2021;109:576–96.
24. Oliveira JF, Sardinha VM, Guerra-Gomes S, Araque A, Sousa N. Do stars govern our actions? Astrocyte involvement in rodent behavior. Trends Neurosci. 2015;38:535–49.
25. Attwell D, et al. Glial and neuronal control of brain blood flow. Nature. 2010;468:232–43.
26. Nortley R, Attwell D. Control of brain energy supply by astrocytes. Curr Opin Neurobiol. 2017;47:80–5.
27. Porter JT, McCarthy KD. Astrocytic neurotransmitter receptors in situ and in vivo. Prog Neurobiol. 1997;51:439–55.
28. Chai H, et al. Neural circuit-specialized astrocytes: transcriptomic, proteomic, morphological, and functional evidence. Neuron. 2017;95:531–549.e9.
29. Akther S, Hirase H. Assessment of astrocytes as a mediator of memory and learning in rodents. Glia. 2022;70:1484–505.
30. Yu X, et al. Context-specific striatal astrocyte molecular responses are phenotypically exploitable. Neuron. 2020;108:1146–1162.e10.
31. Lee H-G, Wheeler MA, Quintana FJ. Function and therapeutic value of astrocytes in neurological diseases. Nat Rev Drug Discov. 2022;21:339–58.
32. Escartin C, et al. Reactive astrocyte nomenclature, definitions, and future directions. Nat Neurosci. 2021;24:312–25.

33. Brandebura AN, Paumier A, Onur TS, Allen NJ. Astrocyte contribution to dysfunction, risk and progression in neurodegenerative disorders. Nat Rev Neurosci. 2023;24:23–39.
34. Linnerbauer M, Wheeler MA, Quintana FJ. Astrocyte crosstalk in CNS inflammation. Neuron. 2020;108:608–22.
35. Burda JE, et al. Divergent transcriptional regulation of astrocyte reactivity across disorders. Nature. 2022;606:557–64.
36. Sofroniew MV. Astrocyte reactivity: subtypes, states, and functions in CNS innate immunity. Trends Immunol. 2020;41:758–70.
37. Yu X, Nagai J, Khakh BS. Improved tools to study astrocytes. Nat Rev Neurosci. 2020;21:121–38.
38. Santello M, Toni N, Volterra A. Astrocyte function from information processing to cognition and cognitive impairment. Nat Neurosci. 2019;22:154–66.
39. Koob AO. Astrocytes imagined. J Integr Neurosci. 2022;21:112.
40. Mansouri FA, Koechlin E, Rosa MGP, Buckley MJ. Managing competing goals – a key role for the frontopolar cortex. Nat Rev Neurosci. 2017;18:645–57.
41. Ragozzino ME. The contribution of the medial prefrontal cortex, orbitofrontal cortex, and dorsomedial striatum to behavioral flexibility. Ann N Y Acad Sci. 2007;1121:355–75.
42. Tanji J, Hoshi E. Behavioral planning in the prefrontal cortex. Curr Opin Neurobiol. 2001;11:164–70.
43. Hamilton DA, Brigman JL. Behavioral flexibility in rats and mice: contributions of distinct frontocortical regions. Genes Brain Behav. 2015;14:4–21.
44. Klein-Flügge MC, Bongioanni A, Rushworth MFS. Medial and orbital frontal cortex in decision-making and flexible behavior. Neuron. 2022;110:2743–70.
45. Antunes M, Biala G. The novel object recognition memory: neurobiology, test procedure, and its modifications. Cogn Process. 2012;13:93–110.
46. Fanselow MS, Wassum KM. The origins and organization of vertebrate Pavlovian conditioning. Cold Spring Harb Perspect Biol. 2016;8:a021717.
47. LeDoux JE, Moscarello J, Sears R, Campese V. The birth, death and resurrection of avoidance: a reconceptualization of a troubled paradigm. Mol Psychiatry. 2017;22:24–36.
48. Gyles TM, Nestler EJ, Parise EM. Advancing preclinical chronic stress models to promote therapeutic discovery for human stress disorders. Neuropsychopharmacology. 2024;49:215–26.
49. Akil H, Nestler EJ. The neurobiology of stress: vulnerability, resilience, and major depression. Proc Natl Acad Sci. 2023;120:e2312662120.
50. Rosado AF, Bevilacqua LM, Moreira ELG, Kaster MP. Behavioral flexibility impacts on coping and emotional responses in male mice submitted to social defeat stress. Prog Neuro-Psychopharmacol Biol Psychiatry. 2023;123:110696.
51. Gabrys RL, Tabri N, Anisman H, Matheson K. Cognitive control and flexibility in the context of stress and depressive symptoms: the cognitive control and flexibility questionnaire. Front Psychol. 2018;9:2219.
52. Bryan CJ, Ray-Sannerud B, Heron EA. Psychological flexibility as a dimension of resilience for posttraumatic stress, depression, and risk for suicidal ideation among Air Force personnel. J Context Behav Sci. 2015;4:263–8.
53. Schoenbaum G, Chiba AA, Gallagher M. Orbitofrontal cortex and basolateral amygdala encode expected outcomes during learning. Nat Neurosci. 1998;1:155–9.
54. Bissonette GB, Powell EM. Reversal learning and attentional set-shifting in mice. Neuropharmacology. 2012;62:1168–74.
55. Graybeal C, et al. Paradoxical reversal learning enhancement by stress or prefrontal cortical damage: rescue with BDNF. Nat Neurosci. 2011;14:1507–9.
56. Zimmermann KS, Yamin JA, Rainnie DG, Ressler KJ, Gourley SL. Connections of the mouse orbitofrontal cortex and regulation of goal-directed action selection by brain-derived neurotrophic factor. Biol Psychiatry. 2017;81:366–77.

57. Lau BK, et al. Obesity-induced astrocyte dysfunction impairs heterosynaptic plasticity in the orbitofrontal cortex. Cell Rep. 2021;36:109563.
58. Ortinski P, Reissner K, Turner J, Anderson TL, Scimemi A. Control of complex behavior by astrocytes and microglia. Neurosci Biobehav Rev. 2022;137:104651.
59. Shimizu T, et al. Oligodendrocyte dynamics dictate cognitive performance outcomes of working memory training in mice. Nat Commun. 2023;14:6499.
60. Mederos S, et al. GABAergic signaling to astrocytes in the prefrontal cortex sustains goal-directed behaviors. Nat Neurosci. 2021;24:82–92.
61. Erickson EK, et al. Cortical astrocytes regulate ethanol consumption and intoxication in mice. Neuropsychopharmacology. 2021;46:500–8.
62. Erickson EK, Farris SP, Blednov YA, Mayfield RD, Harris RA. Astrocyte-specific transcriptome responses to chronic ethanol consumption. Pharmacogenomics J. 2018;18:578–89.
63. Sardinha VM, et al. Astrocytic signaling supports hippocampal–prefrontal theta synchronization and cognitive function. Glia. 2017;65:1944–60.
64. Fujita T, et al. Neuronal transgene expression in dominant-negative SNARE mice. J Neurosci. 2014;34:16594–604.
65. Wulaer B, et al. Overexpression of astroglial major histocompatibility complex class I in the medial prefrontal cortex impairs visual discrimination learning in mice. Mol Brain. 2020;13:170.
66. Liu J, et al. Astrocyte dysfunction drives abnormal resting-state functional connectivity in depression. Sci Adv. 2022;8:eabo2098.
67. Octeau JC, et al. Transient, consequential increases in extracellular potassium ions accompany channelrhodopsin2 excitation. Cell Rep. 2019;27:2249–2261.e7.
68. Pittolo S, et al. Dopamine activates astrocytes in prefrontal cortex via α1-adrenergic receptors. Cell Rep. 2022;40:111426.
69. Evanson NK, Tasker JG, Hill MN, Hillard CJ, Herman JP. Fast feedback inhibition of the HPA axis by glucocorticoids is mediated by endocannabinoid signaling. Endocrinology. 2010;151:4811–9.
70. Lu C-L, et al. Glucocorticoid receptor–dependent astrocytes mediate stress vulnerability. Biol Psychiatry. 2022;92:204–15.
71. Kessler RC, et al. The epidemiology of major depressive disorder: results from the National Comorbidity Survey Replication (NCS-R). JAMA. 2003;289:3095–105.
72. Grilo CM. Treatment of eating disorders: current status, challenges, and future directions. Annu Rev Clin Psychol. 2024. https://doi.org/10.1146/annurev-clinpsy-080822-043256.
73. Calvo-Rivera MP, Navarrete-Páez MI, Bodoano I, Gutiérrez-Rojas L. Comorbidity between anorexia nervosa and depressive disorder: a narrative review. Psychiatry Investig. 2022;19:155–63.
74. Cotter DR, Pariante CM, Everall IP. Glial cell abnormalities in major psychiatric disorders: the evidence and implications. Brain Res Bull. 2001;55:585–95.
75. Nagy C, Torres-Platas SG, Mechawar N, Turecki G. Repression of astrocytic connexins in cortical and subcortical brain regions and prefrontal enrichment of H3K9me3 in depression and suicide. Int J Neuropsychopharmacol. 2017;20:50–7.
76. Torres-Platas SG, Nagy C, Wakid M, Turecki G, Mechawar N. Glial fibrillary acidic protein is differentially expressed across cortical and subcortical regions in healthy brains and downregulated in the thalamus and caudate nucleus of depressed suicides. Mol Psychiatry. 2016;21:509–15.
77. Cotter D, et al. Reduced neuronal size and glial cell density in area 9 of the dorsolateral prefrontal cortex in subjects with major depressive disorder. Cereb Cortex. 2002;12:386–94.
78. Rajkowska G, et al. Astrocyte pathology in the ventral prefrontal white matter in depression. J Psychiatr Res. 2018;102:150–8.
79. Baldwin KT, Murai KK, Khakh BS. Astrocyte morphology. Trends Cell Biol. 2023:S0962-8924(23)00204-0. https://doi.org/10.1016/j.tcb.2023.09.006.

80. Ramsey J, Martin EC, Purcell OM, Lee KM, MacLean AG. Self-injurious behaviours in rhesus macaques: potential glial mechanisms. J Intellect Disabil Res. 2018;62:1008–17.
81. Rajkowska G, Hughes J, Stockmeier CA, Javier Miguel-Hidalgo J, Maciag D. Coverage of blood vessels by astrocytic endfeet is reduced in major depressive disorder. Biol Psychiatry. 2013;73:613–21.
82. Woodburn SC, Bollinger JL, Wohleb ES. Synaptic and behavioral effects of chronic stress are linked to dynamic and sex-specific changes in microglia function and astrocyte dystrophy. Neurobiol Stress. 2021;14:100312.
83. Banasr M, et al. Glial pathology in an animal model of depression: reversal of stress-induced cellular, metabolic and behavioral deficits by the glutamate-modulating drug riluzole. Mol Psychiatry. 2010;15:501–11.
84. Huang D, et al. Positive effects of running exercise on astrocytes in the medial prefrontal cortex in an animal model of depression. J Comp Neurol. 2022;530:3056–71.
85. Aten S, et al. Chronic stress impairs the structure and function of astrocyte networks in an animal model of depression. Neurochem Res. 2023;48:1191–210.
86. Codeluppi SA, et al. Chronic stress alters astrocyte morphology in mouse prefrontal cortex. Int J Neuropsychopharmacol. 2021;24:842–53.
87. Lin S-S, et al. Electroacupuncture prevents astrocyte atrophy to alleviate depression. Cell Death Dis. 2023;14:1–10.
88. Tynan RJ, et al. Chronic stress-induced disruption of the astrocyte network is driven by structural atrophy and not loss of astrocytes. Acta Neuropathol (Berl). 2013;126:75–91.
89. Reyes-Ortega P, et al. Anorexia disrupts glutamate-glutamine homeostasis associated with astroglia in the prefrontal cortex of young female rats. Behav Brain Res. 2022;420:113715.
90. Hurley MM, et al. Adolescent female rats prone to the activity based anorexia (ABA) paradigm have altered hedonic responses and cortical astrocyte density compared to resistant animals. Appetite. 2022;168:105666.
91. Ernst C, et al. Dysfunction of astrocyte connexins 30 and 43 in dorsal lateral prefrontal cortex of suicide completers. Biol Psychiatry. 2011;70:312–9.
92. Huang D, et al. Dysfunction of astrocytic connexins 30 and 43 in the medial prefrontal cortex and hippocampus mediates depressive-like behaviours. Behav Brain Res. 2019;372:111950.
93. Zheng Q-L, et al. Korean red ginseng alleviate depressive disorder by improving astrocyte gap junction function. J Ethnopharmacol. 2021;281:114466.
94. Choudary PV, et al. Altered cortical glutamatergic and GABAergic signal transmission with glial involvement in depression. Proc Natl Acad Sci. 2005;102:15653–8.
95. Sequeira A, et al. Global brain gene expression analysis links glutamatergic and GABAergic alterations to suicide and major depression. PLoS One. 2009;4:e6585.
96. Bernstein H-G, et al. Reduced density of glutamine synthetase immunoreactive astrocytes in different cortical areas in major depression but not in bipolar I disorder. Front Cell Neurosci. 2015;9.
97. Mineur YS, Picciotto MR, Sanacora G. Antidepressant-like effects of ceftriaxone in male C57BL/6J mice. Biol Psychiatry. 2007;61:250–2.
98. Lee Y, et al. Glutamine deficiency in the prefrontal cortex increases depressive-like behaviours in male mice. J Psychiatry Neurosci. 2013;38:183–91.
99. Lima A, et al. Astrocyte pathology in the prefrontal cortex impairs the cognitive function of rats. Mol Psychiatry. 2014;19:834–41.
100. Brockett AT, et al. Evidence supporting a role for astrocytes in the regulation of cognitive flexibility and neuronal oscillations through the Ca2+ binding protein S100β. PLoS One. 2018;13:e0195726.
101. John CS, et al. Blockade of astrocytic glutamate uptake in the prefrontal cortex induces anhedonia. Neuropsychopharmacology. 2012;37:2467–75.
102. Attwell D, Laughlin SB. An energy budget for signaling in the grey matter of the brain. J Cereb Blood Flow Metab Off J Int Soc Cereb Blood Flow Metab. 2001;21:1133–45.
103. Engl E, Attwell D. Non-signalling energy use in the brain. J Physiol. 2015;593:3417–29.

104. Magistretti PJ, Allaman I. Lactate in the brain: from metabolic end-product to signalling molecule. Nat Rev Neurosci. 2018;19:235–49.
105. Stam CJ. Modern network science of neurological disorders. Nat Rev Neurosci. 2014;15:683–95.
106. Kliemann D, et al. Intrinsic functional connectivity of the brain in adults with a single cerebral hemisphere. Cell Rep. 2019;29:2398–2407.e4.
107. Imai K, et al. The neural network basis of altered decision-making in patients with amyotrophic lateral sclerosis. Ann Clin Transl Neurol. 2020;7:2115–26.
108. Kawabata K, et al. Distinct manifestation of cognitive deficits associate with different resting-state network disruptions in non-demented patients with Parkinson's disease. J Neurol. 2018;265:688–700.
109. Ogura A, et al. Semantic deficits in ALS related to right lingual/fusiform gyrus network involvement. EBioMedicine. 2019;47:506–17.
110. Maesawa S, et al. Evaluation of resting state networks in patients with gliomas: connectivity changes in the unaffected side and its relation to cognitive function. PLoS One. 2015;10:e0118072.
111. Yamamoto M, Bagarinao E, Shimamoto M, Iidaka T, Ozaki N. Involvement of cerebellar and subcortical connector hubs in schizophrenia. NeuroImage Clin. 2022;35:103140.
112. Cunnane SC, et al. Brain energy rescue: an emerging therapeutic concept for neurodegenerative disorders of ageing. Nat Rev Drug Discov. 2020;19:609–33.
113. Goyal MS, et al. Loss of brain aerobic glycolysis in normal human aging. Cell Metab. 2017;26:353–360.e3.
114. Petralia RS, Mattson MP, Yao PJ. Communication breakdown: the impact of ageing on synapse structure. Ageing Res Rev. 2014;14:31–42.
115. Bagarinao E, et al. Identifying the brain's connector hubs at the voxel level using functional connectivity overlap ratio. NeuroImage. 2020;222:117241.
116. Gordon EM, et al. Three distinct sets of connector hubs integrate human brain function. Cell Rep. 2018;24:1687–1695.e4.
117. Richiardi J, et al. Brain networks. Correlated gene expression supports synchronous activity in brain networks. Science. 2015;348:1241–4.
118. Fulcher BD, Fornito A. A transcriptional signature of hub connectivity in the mouse connectome. Proc Natl Acad Sci USA. 2016;113:1435–40.
119. Petanjek Z, et al. Extraordinary neoteny of synaptic spines in the human prefrontal cortex. Proc Natl Acad Sci USA. 2011;108:13281–6.
120. Kolb B, et al. Experience and the developing prefrontal cortex. Proc Natl Acad Sci USA. 2012;109:17186–93.
121. Goyal MS, Hawrylycz M, Miller JA, Snyder AZ, Raichle ME. Aerobic glycolysis in the human brain is associated with development and neotenous gene expression. Cell Metab. 2014;19:49–57.
122. Kety SS, Schmidt CF. The nitrous oxide method for the quantitative determination of cerebral blood flow in man: theory, procedure and normal values. J Clin Invest. 1948;27:476–83.
123. Sokoloff L. Localization of functional activity in the central nervous system by measurement of glucose utilization with radioactive deoxyglucose. J Cereb Blood Flow Metab. 1981;1:7–36.
124. Nehlig A, Wittendorp-Rechenmann E, Lam CD. Selective uptake of [14C]2-deoxyglucose by neurons and astrocytes: high-resolution microautoradiographic imaging by cellular 14C-trajectography combined with immunohistochemistry. J Cereb Blood Flow Metab Off J Int Soc Cereb Blood Flow Metab. 2004;24:1004–14.
125. Chuquet J, Quilichini P, Nimchinsky EA, Buzsáki G. Predominant enhancement of glucose uptake in astrocytes versus neurons during activation of the somatosensory cortex. J Neurosci. 2010;30:15298–303.
126. Brooks GA. Cell-cell and intracellular lactate shuttles. J Physiol. 2009;587:5591–600.

127. Gibbs ME, Anderson DG, Hertz L. Inhibition of glycogenolysis in astrocytes interrupts memory consolidation in young chickens. Glia. 2006;54:214–22.
128. Suzuki A, et al. Astrocyte-neuron lactate transport is required for long-term memory formation. Cell. 2011;144:810–23.
129. Tadi M, Allaman I, Lengacher S, Grenningloh G, Magistretti PJ. Learning-induced gene expression in the hippocampus reveals a role of neuron -astrocyte metabolic coupling in long term memory. PLoS One. 2015;10:e0141568.
130. Yang J, et al. Lactate promotes plasticity gene expression by potentiating NMDA signaling in neurons. Proc Natl Acad Sci. 2014;111:12228–33.
131. Yao S, et al. Astrocytic lactate dehydrogenase A regulates neuronal excitability and depressive-like behaviors through lactate homeostasis in mice. Nat Commun. 2023;14:729.
132. Wang J, et al. Astrocytic l-lactate signaling facilitates amygdala-anterior cingulate cortex synchrony and decision making in rats. Cell Rep. 2017;21:2407–18.
133. Akter M, et al. Astrocyte and L-lactate in the anterior cingulate cortex modulate schema memory and neuronal mitochondrial biogenesis. elife. 2023;12:e85751.
134. Xia C-Y, et al. Corticosterone impairs gap junctions in the prefrontal cortical and hippocampal astrocytes via different mechanisms. Neuropharmacology. 2018;131:20–30.
135. Zimmer ER, et al. [18F]FDG PET signal is driven by astroglial glutamate transport. Nat Neurosci. 2017;20:393–5.
136. Oe Y, et al. Distinct temporal integration of noradrenaline signaling by astrocytic second messengers during vigilance. Nat Commun. 2020;11:471.
137. Gao V, et al. Astrocytic β2-adrenergic receptors mediate hippocampal long-term memory consolidation. Proc Natl Acad Sci USA. 2016;113:8526–31.
138. Pan S-M, et al. Thioredoxin interacting protein drives astrocytic glucose hypometabolism in corticosterone-induced depressive state. J Neurochem. 2022;161:84–100.
139. Murphy-Royal C, et al. Stress gates an astrocytic energy reservoir to impair synaptic plasticity. Nat Commun. 2020;11:2014.
140. Dias C, et al. β-catenin mediates stress resilience through Dicer1/microRNA regulation. Nature. 2014;516:51–5.
141. Vidal R, et al. Targeting β-Catenin in GLAST-expressing cells: impact on anxiety and depression-related behavior and hippocampal proliferation. Mol Neurobiol. 2019;56:553–66.
142. Ai Z, Fischer A, Spray DC, Brown AM, Fishman GI. Wnt-1 regulation of connexin43 in cardiac myocytes. J Clin Invest. 2000;105:161–71.
143. Carrard A, et al. Peripheral administration of lactate produces antidepressant-like effects. Mol Psychiatry. 2018;23:392–9.
144. Samarasinghe RA, et al. Nongenomic glucocorticoid receptor action regulates gap junction intercellular communication and neural progenitor cell proliferation. Proc Natl Acad Sci USA. 2011;108:16657–62.
145. Datta D, et al. Noradrenergic α1-adrenoceptor actions in the primate dorsolateral prefrontal cortex. J Neurosci. 2019;39:2722–34.
146. Li B, Zhang D, Verkhratsky A. Astrocytes in post-traumatic stress disorder. Neurosci Bull. 2022;38:953–65.
147. O'Donnell T, Hegadoren KM, Coupland NC. Noradrenergic mechanisms in the pathophysiology of post-traumatic stress disorder. Neuropsychobiology. 2004;50:273–83.
148. Southwick SM, et al. Role of norepinephrine in the pathophysiology and treatment of post-traumatic stress disorder. Biol Psychiatry. 1999;46:1192–204.
149. Southwick SM, et al. Abnormal noradrenergic function in posttraumatic stress disorder. Arch Gen Psychiatry. 1993;50:266–74.
150. Arango V, Ernsberger P, Sved AF, Mann JJ. Quantitative autoradiography of alpha 1- and alpha 2-adrenergic receptors in the cerebral cortex of controls and suicide victims. Brain Res. 1993;630:271–82.
151. Birnbaum S, Gobeske KT, Auerbach J, Taylor JR, Arnsten AFT. A role for norepinephrine in stress-induced cognitive deficits: α-1-adrenoceptor mediation in the prefrontal cortex. Biol Psychiatry. 1999;46:1266–74.

152. George KC, Kebejian L, Ruth LJ, Miller CWT, Himelhoch S. Meta-analysis of the efficacy and safety of prazosin versus placebo for the treatment of nightmares and sleep disturbances in adults with posttraumatic stress disorder. J Trauma Dissociation Off J Int Soc Study Dissociation ISSD. 2016;17:494–510.
153. De Berardis D, et al. Targeting the noradrenergic system in posttraumatic stress disorder: a systematic review and meta-analysis of Prazosin trials. Curr Drug Targets. 2015;16:1094–106.
154. Singh B, Hughes AJ, Mehta G, Erwin PJ, Parsaik AK. Efficacy of Prazosin in posttraumatic stress disorder: a systematic review and meta-analysis. Prim Care Companion CNS Disord. 2016;18(4).
155. Ouyang X, et al. Ketamine ameliorates depressive-like behaviors in mice through increasing glucose uptake regulated by the ERK/GLUT3 signaling pathway. Sci Rep. 2021;11:18181.
156. Lasič E, et al. Astrocyte specific remodeling of plasmalemmal cholesterol composition by ketamine indicates a new mechanism of antidepressant action. Sci Rep. 2019;9:10957.
157. Rivera AD, Butt AM. Astrocytes are direct cellular targets of lithium treatment: novel roles for lysyl oxidase and peroxisome-proliferator activated receptor-γ as astroglial targets of lithium. Transl Psychiatry. 2019;9:211.
158. Farhy-Tselnicker, et al. Dual regulation of G proteins and the G-protein-activated K+ channels by lithium. Proc Natl Acad Sci USA. 2014;111:5018–23.
159. Keller JP, et al. In vivo glucose imaging in multiple model organisms with an engineered single-wavelength sensor. Cell Rep. 2021;35:109284.
160. Nasu Y, et al. Lactate biosensors for spectrally and spatially multiplexed fluorescence imaging. Nat Commun. 2023;14:6598.
161. Nasu Y, et al. A genetically encoded fluorescent biosensor for extracellular L-lactate. Nat Commun. 2021;12:7058.

Astrocyte-Neuron Interactions in Substance Use Disorders

Eden V. Harder, Janay P. Franklin, Jonathan W. VanRyzin, and Kathryn J. Reissner

Abstract Engagement of astrocytes within the brain's reward circuitry has been apparent for approximately 30 years, when noncontingent drug administration was observed to lead to cytological markers of reactive astrocytes. Since that time, advanced approaches in rodent behavior and astrocyte monitoring have revealed complex interactions between astrocytes with drug type, animal sex, brain region, and dose and duration of drug administration. A number of studies now collectively reveal that rodent drug self-administration followed by prolonged abstinence results in decreased features of structure and synaptic colocalization of astrocytes. In addition, stimulation of astrocytes in the nucleus accumbens with DREADD receptors or pharmacological compounds opposes drug-seeking behavior. These findings provide a clear path for ongoing investigation into astrocytes as mediators of drug action in the brain and underscore the potential therapeutic utility of astrocytes in the regulation of drug craving and relapse vulnerability.

Keywords Astrocyte · Synapse · Cocaine · Heroin · Ethanol

Abbreviations

AMPAR	α-amino-3-hydroxy-5-methyl-4-isoxazolepropionic acid receptor
ATP	Adenosine triphosphate
BBB	Blood-brain barrier
CB1	Cannabinoid receptor 1

Eden V. Harder and Janay P. Franklin contributed equally with all other contributors.

E. V. Harder · J. P. Franklin · J. W. VanRyzin · K. J. Reissner (✉)
Department of Psychology & Neuroscience, Neuroscience Center, UNC Chapel Hill, Chapel Hill, NC, USA
e-mail: edenvh@live.unc.edu; jphunt@med.unc.edu; jvanr@unc.edu; reissner@unc.edu

E. Blanco-Suarez, I. Farhy-Tselnicker (eds.), *Astrocyte-Neuron Interactions in Health and Disease*, Advances in Neurobiology 39,
https://doi.org/10.1007/978-3-031-64839-7_7

CNS	Central nervous system
D1	Dopamine receptor 1
DEGs	Differentially expressed genes
DREADD	Designer receptors exclusively activated by designer drugs
EAAT1/2	Excitatory amino acid transporters 1 and 2
GABA	Gamma-aminobutyric acid
GFAP	Glial fibrillary acidic protein
GLAST	High-affinity glutamate/aspartate transporter
GLT-1	Glutamate transporter-1
GPCR	G-protein-coupled receptors
IL-6	Interleukin-6
IP	Intraperitoneal
IP3R2	Inositol 1,4,5-trisphosphate (IP3) receptor type 2
GFP	Green fluorescent protein
mGluR2/3	Metabotropic glutamate receptors 2 and 3
MSN	Medium spiny neurons
NAc	Nucleus accumbens
NAC	N-acetylcysteine
NAcc	Nucleus accumbens core
NAs	Nucleus accumbens shell
NMDAR	N-methyl-d-aspartate receptors
PAPs	Peripheral astrocyte processes
PFC	Prefrontal cortex
PND	Postnatal day
PTSD	Posttraumatic stress disorder
SNARE	Soluble N-ethylmaleimide-sensitive factor attachment
SUD	Substance use disorder
TSP1/2	Thrombospondin 1 and 2
VTA	Ventral tegmental area
xCT	Cystine-glutamate antiporter
nAChR	Nicotinic acetylcholine receptors
kg	Kilogram
mg	Milligram
hr	Hour

1 Introduction

1.1 Animal Models of Substance Use Disorders

The use of animal models has been critical to understanding both the brain's reward neurocircuitry in adaptive reward learning, as well as effects of drug use on neurons and glia. The earliest studies in the 1950s utilized rodent intracranial self-stimulation procedures to identify primary nodes in the reward circuitry. Subsequently, microdialysis and, later, voltammetry were used to monitor dopamine release, and more recently, studies have employed in vivo calcium imaging and chemogenetics and optogenetics for recording and control of this circuitry, respectively [1–4]. Nonhuman primate and human imaging studies have revealed a high degree of cross-species conservation among seminal regions mediating reward processing, highlighting the critical role of reward learning in survival and evolution [5, 6].

The brain's reward circuitry is a multinodal circuit with bidirectional synaptic communication including dopaminergic output from the ventral tegmental area (VTA) to the nucleus accumbens (NAc), dorsal striatum, hippocampus, amygdala, and prefrontal cortex [7] (Fig. 1). Glutamatergic and GABAergic mesocorticolimbic projections descend from the prefrontal cortex back to the VTA, as well as to the NAc, which serves as a central hub in reward processing and a limbic-motor integrator [8]. Complex interactions within this circuitry carry information regarding salience, hedonic value, and contextual associations of drug use. Responses within these pathways mediate both the acute reinforcing effects of drug exposure, as well as chronic adaptations which lead to dependence, compulsive drug seeking, and relapse [9, 10].

Because of the strengths of rodent models for studying cellular changes in response to stimuli, these studies have been integral in the investigation of the effects of drug use on astrocyte structure, function, and engagement with synapses. Preclinical animal models of drug use are largely compartmentalized into contingent versus noncontingent, reflecting whether drug is administered contingent on a behavior of the animal (i.e., self-administration) or not (noncontingent, experimenter-administered) [11, 12]. While self-administration is largely considered the gold standard of drug administration, because it necessitates and encapsulates the motivation of the subject, limitations include increased expense and duration of studies, as well as technical challenges including considerations regarding catheter patency and animal learning. A number of variations of the self-administration approach exist, based on duration of drug access, negative consequences of drug use, and mode of abstinence (home cage versus extinction training). Strengths of noncontingent models include simplicity of design and drug administration and can be effective for investigating cellular effects of drug exposure and mechanisms of drug reward memory. As in all animal models, each approach affords unique benefits and caveats, and varied constraints with regard to validity toward human experience. In Sect. 4 below, we will consider evidence for effects of drug use on astrocytes in human studies and how these align with what has been learned from rodent studies [13].

1.2 Evidence for Engagement of Astrocyte in Drug Effects

Early evidence for effects of drugs of abuse on astrocytes began in the mid-1990s, with monitoring of changes in GFAP expression within and beyond the brain's reward circuitry following noncontingent drug administration in rodents [14–17].

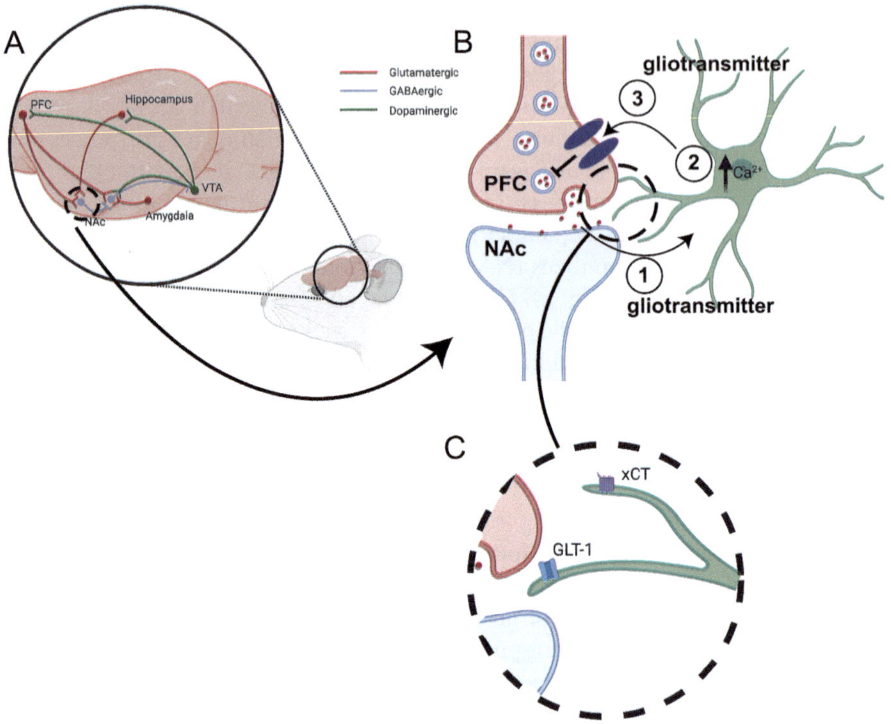

Fig. 1 Circuit and synaptic view of astrocytes in the reward circuitry. (**a**) Major nodes in the reward neurocircuitry include the ventral tegmental area (VTA), hippocampus, prefrontal cortex (PFC), amygdala, and nucleus accumbens (NAc) which communicate via reciprocal projections releasing neurotransmitters glutamate (red), dopamine (green), and GABA (blue). The NAc receives inputs from all of these regions and serves as an integrator of limbic signals. Many of the astrocyte adaptations in response to drug use have been reported in the NAc, in large part due to its salient role in reward and motivation. Dotted circle in **a** is provided in greater detail in (**b**), depicting bidirectional communication between astrocytes and synapses. Astrocytes respond to neurotransmitter release (1) with complex intracellular calcium responses (2). Often in response to calcium signaling, factors are released from astrocytes (3) including glutamate, adenosine, serine/D-serine, and more, which can act both presynaptically and postsynaptically to modulate synaptic function. (**c**) Enlargement of two mediators of glutamate homeostasis, xCT (catalytic subunit of glutamate antiporter xC-) and GLT-1. Both xCT and GLT-1 are down persistently downregulated by cocaine self-administration, leading to reduced glutamatergic tone on presynaptic inhibitory mGluR receptors, as well as impaired uptake of synaptically released glutamate, driving maladaptive synaptic processing and contributing to increased synaptic drive for reward and drug seeking. (Figure generated with BioRender)

Since that time, enduring effects of drug administration and drug use on astrocytes in both animal models and human populations have been reported, and long-lasting changes in neuron-astrocyte communication are now recognized as a hallmark cellular feature associated with rodent self-administration of multiple drug classes (for review, see [18–21]).

Astrocytes are well-positioned mediators of drug effects within the brain's reward circuitry. First, astrocytes express a broad range of neurotransmitter receptors and transporters that mediate drug effects, albeit at lower levels than in neurons [22]; see also http://astrocyternaseq.org/. Notably, however, the GluN2C NMDA receptor subunit is exclusively expressed on astrocytes in the striatum and mediates cocaine reward memory [23]. Of particular relevance, regulators of glutamate homeostasis, including the cystine-glutamate exchange system and the high-affinity glutamate transporters EAAT2/GLT-1 and EAAT1/GLAST, are enriched on astrocytes and are chronically downregulated following rodent self-administration of multiple drug classes [24, 25]. Dysregulation of glutamate homeostasis in the nucleus accumbens contributes to cellular mechanisms of drug-seeking behavior and has provided a target for the generation of tools to countermand evidence of drug craving. The effects of cocaine and other drugs on this system and the progress in normalizing glutamate homeostasis as a treatment intervention for substance use disorders (SUDs) are discussed in more detail throughout this chapter.

This chapter will first focus on the effects of substance use on astrocyte structure and function, with emphasis on communication and regulation of neurons and synapses. Following analysis of evidence determined from preclinical animal models of drug exposure, we will separately consider the translational implications of these observations and evidence collected from human studies.

2 Effects of Drugs on Astrocyte Structure and Synaptic Colocalization

Astrocytes are structurally complex cells, with a characteristic branched morphology, often described as "star-shaped" due to their extensive branch network. Astrocytes possess GFAP-enriched intermediate filament processes, as well as peripheral astrocyte processes (PAPs) which broker structural and functional interactions with synapses. Historically, changes in GFAP expression have been used to identify structural changes in astrocytes, particularly those that reflect a reactive phenotype. However, recognition of the fact that GFAP comprises a minority of astrocyte territory and is largely absent from actin-containing peripheral processes has led to more comprehensive approaches to understanding astrocyte structure [26].

The physical interaction between astrocyte PAPs and neuronal synapses, and the bidirectional communication between these components which regulate synaptic function, is referenced as the "tripartite synapse" [27]. The tripartite synapse reflects the role of astrocytes in the maintenance of the CNS through neurotransmitter uptake and synaptic transmission and plasticity, among other critical roles. Specifically, the PAPs play a significant role in glutamate clearance from the synapse via abundant expression of EAAT-2/GLT-1, the primary high-affinity glutamate transporter in the brain [28]. As tools to study astrocytes become more refined, more sophisticated analyses have expanded to look not just at protein expression and gross measures of astrocyte reactivity, but to study more closely the effects of drug use on both structural and functional interactions with synapses. Below, we will examine the state of knowledge regarding different drug classes on all of these measures, with particular consideration for how factors such as route of administration, time, and sex influence observed results.

2.1 Alcohol

Considerable studies have revealed evidence of effects of alcohol on astrocyte function and structure (for review, see [29–31]). Both in vitro and in vivo studies have demonstrated varied effects on astrocyte morphology, dependent on the region, sex, and method of alcohol administration. For example, early studies using cultured astrocytes harvested from rat fetuses of ethanol-treated female rats revealed decreased GFAP expression [32]. In another study investigating effects of prenatal alcohol exposure on astrocytes, female rats were fed an alcohol or isocaloric diet before and during gestation; decreased GFAP expression was observed in primary cultured astrocytes prepared from 21-day-old neonates of alcohol-fed female rats, in comparison to those harvested from control rats [33]. Complementary to in vitro studies, in vivo ethanol vapor exposure has demonstrated bidirectional changes in GFAP-positive astrocytes, varying based on brain region, sex, and administration protocol [34, 35]. Adolescent ethanol gavage administration results in decreased synaptic colocalization of astrocytes in both the hippocampus and the prefrontal cortex in adulthood, underscoring persistent, long-lasting effects of drug exposure on astrocytes, as discussed in detail below [36–38].

2.2 Psychostimulants

Analogous to ethanol-induced morphological changes in astrocytes, the use of psychostimulants, including amphetamine, methamphetamine, and cocaine, produces changes in astrocyte structural features. These changes include but are not limited to changes in GFAP expression, which, similarly to alcohol, vary based on administration paradigms. Amphetamines and methamphetamines in vitro or in vivo generally lead to astrogliosis as revealed by an increase in GFAP expression, across various administration paradigms [39–41]. More recently, a study was conducted using human cerebral organoids—human pluripotent stem cells that are capable of mimicking the biological function of organs and cells—to examine the effect of prenatal methamphetamine exposure on fetal brain development [42]. This study corroborated findings with previous studies, in that organoids treated with methamphetamine showed increased GFAP expression and broader filaments in GFAP-positive astrocytes. Additionally, these findings further supported the occurrence of astrogliosis by measuring IL-6, which is a pro-inflammatory cytokine typically altered during inflammatory responses, such as drug exposure. Increased IL-6 expression following methamphetamine exposure was parallel to elevated GFAP expression in the organoids.

In vivo studies on the effects of methamphetamine exposure on astrocyte structure have also been conducted utilizing both experimenter-administered, noncontingent, and contingent self-administration paradigms. However, studies employing noncontingent administration versus self-administration of methamphetamine have revealed opposing findings. For example, Castellano et al. [39] showed that varying concentrations (2.5, 5, and 10 mg/kg) of intraperitoneal (IP) methamphetamine injections led to increased GFAP expression in methamphetamine-treated versus saline-treated female mice. However, another study using a self-administration of methamphetamine (2 h/10–14 days, 20 μg/50 μL bolus) protocol followed by extinction reported no significant differences in GFAP-positive astrocyte quantity, astrocyte process diameter, or changes in other structural features in the nucleus accumbens core (NAcc) of adult male rats [43]. Even so, methamphetamine-extinguished rats displayed decreased NAcc astrocyte-synaptic contact in comparison to yoked saline controls, which was quantified by colocalization of synapsin I and membrane-associated astrocyte label, Lck-GFP. Leitão and colleagues found that repeated noncontingent IP administration of methamphetamine (4–10 mg/kg; 2 h apart) led to no significant differences in total GFAP protein levels and immunoreactivity in the CA3 subregion of the hippocampus between saline- and methamphetamine-treated male mice [44]. However, there were increased number and length of astrocyte processes, suggesting that the branching complexity of the astrocytes was altered postinjection [44]. Together, these studies with others, show that methamphetamine-induced changes in astrocyte structure are dependent on region, sex, and route of drug administration.

Like methamphetamine, studies have demonstrated that amphetamine impacts astrocyte structure, by inducing astrocyte activation which is typically indicated by increased GFAP expression [18, 45]. Most studies have found that GFAP expression is increased following amphetamine exposure. For example, multiple in vivo studies that employed noncontingent amphetamine administration led to significant increases in GFAP expression in astrocytes in varying regions including the ventral caudate putamen [46] and prelimbic cortex [47]. Although both studies observed increased GFAP expression following an abstinence period (10 and 21 days, respectively), the former study (10 days) implemented a challenge dose of amphetamine (0.5 mg/kg) or saline following the abstinence period. Franke et al. [48] observed similar amphetamine-induced increases in GFAP immunoreactivity in the striatum and NAc, but also observed increased branching and astrocyte process length and increased GFAP-positive astrocytes as Basmadjian et al. did at 21 days [47]. The aforementioned studies found similar alterations in astrocyte structure in various regions, suggesting that a hypertrophic astrocyte phenotype is produced by amphetamine exposure.

In contrast, other studies that used noncontingent amphetamine injections did not find any significant differences in GFAP expression after acute administration of amphetamine at increasing doses (1, 2, and 4 mg/kg) in the hippocampus [40].

However, repeated administration of amphetamine (2 mg/kg) for one week produced an increase in GFAP expression in the hippocampus [40]. Taken together, these studies suggest that changes in GFAP expression are region- and regimen-dependent, as alterations in GFAP were generally elicited following repeated administration of amphetamine.

In recent decades, studies investigating cocaine's effects on astrocytes and astrocyte involvement in pathophysiological effects of cocaine in the brain have grown significantly [21]. Like other drugs of abuse, cocaine exposure leads to varied effects on astrocyte structure and function, which has been shown to be dependent on route, dose, and time course. As addressed above, earlier studies monitored changes in GFAP expression as the determining factor for changes in astrocyte structure or number. Such studies have produced variable results of cocaine-induced effects on astrocyte morphology. Early studies reported increased GFAP expression in the hippocampus, prefrontal cortex, and NAc following noncontingent cocaine administration [15, 16]. Both of these studies utilized repeated intraperitoneal injections of cocaine (7 days), which led to increased GFAP expression in the NAc, PFC, dentate gyrus, and striatum. Fattore et al. [16] also observed increased GFAP expression after 1 day of cocaine injections, in addition to after 7 days of injections. However, between day 8 and 14, there was a decrease in GFAP expression to saline control levels. Complementary to these studies, others who documented increased GFAP expression also utilized chronic cocaine administration paradigms where repeated cocaine injections were given and/or paired with a period of withdrawal [49, 50].

Conversely, using similar behavioral models, some studies showed that both acute and chronic cocaine exposure did not induce any changes in GFAP expression. Namely, Bowers and Kalivas did not observe changes in GFAP immunoreactivity in the NAc, PFC, or NAs 2 h after acute noncontingent administration of cocaine [15]. Others have corroborated these findings using similar administration models and in diverse regions including striatum and anterior cingulate. In contrast, fewer studies have confirmed a decrease in GFAP expression, such as Scofield et al. who found this observation following cocaine self-administration (2 h/10 day, 0.2 mg per infusion) paired with 14–16 days of extinction [51]. Notably, these opposing findings may suggest a role of administration paradigm, as both Scofield et al. and Jarvis et al. used cocaine self-administration as opposed to noncontingent administration methods [51, 52]. Further, Jarvis et al. observed a trending, but not significant, reversal of decreased GFAP expression in cocaine-administering rats following extinction (20 days).

Because the complete complex, branching astrocyte morphology cannot be fully appreciated, visualized, and analyzed using only GFAP immunoreactivity, more recent studies have utilized a modified GFP viral construct, Lck-GFP. Lck-GFP is a fusion of a membrane-associated, myristoylated Src kinase peptide with GFP, allowing enhanced visualization of the fine peripheral astrocyte processes in comparison to GFAP [53]. Lck-GFP has been used to quantify astrocyte structure following cocaine self-administration and extinction as well as home cage abstinence, revealing that NAc astrocytes are significantly decreased in structure, which supports prior observations of decreased GFAP expression following cocaine self-administration [51, 54, 55]. Importantly, 10 days of cocaine self-administration resulted in enduring hypotrophy of astrocyte structure and synaptic colocalization when measured 45 days (but not 24 h) following the last cocaine self-administration session [54]. Notably, Kim et al. showed these changes in NAc astrocyte structure to be sex-specific, as these effects of cocaine self-administration on astrocyte structure were not observed in female rats [54]. Overall, the studies discussed previously show that astrocytes are structurally affected by various exposure models of cocaine and that these changes are time- and sex-dependent, as well as long-lasting. Although more studies have been conducted to decipher underlying mechanisms of cocaine-induced astrocyte structural changes, additional studies are needed to further understand the functional implications of these structural alterations and downstream relation to cocaine-seeking behaviors.

2.3 Opioids

The majority of studies conducted to date regarding effects of opioid drugs on astrocytes have been performed with either morphine or heroin. Studies with morphine have largely been performed using noncontingent experimenter-administered morphine, with observation of increased GFAP expression in various regions of the brain. For example, Song and Zhao [56] and Garrido et al. [57] showed that intraperitoneal injections of morphine increased GFAP immunoreactivity in male Sprague Dawley rats. While Garrido and colleagues did not observe changes in GFAP immunoreactivity following chronic morphine administration (20 mg/kg/13 days), increased GFAP was observed 12 h after acute morphine administration in the NAc shell, PFC, and VTA. Relatedly, Song and Zhao [56] observed similar effects of morphine on GFAP expression, but these changes occurred following chronic morphine administration (50 mg/kg daily for 9 days) and were observed in different regions including the spinal cord, posterior cingulate cortex, and hippocampus. A follow-up study determined that the increase in GFAP expression was due to astrocyte hypertrophy rather than increased astrocyte numbers, as no proliferation was found. An earlier study [14] employed chronic administration of morphine via subcutaneous pellets of morphine (75 mg/day) for 5 days. Chronic, but not acute, morphine administration induced increased GFAP immunoreactivity in the VTA, with no effect in the substantia nigra or locus coeruleus, or other regions. Additionally, Lazriev et al. [58] corroborated earlier studies that found no change in astrocyte proliferation in the NAc, caudate nucleus, and lateral septal nucleus following intraperitoneal morphine injections, but found region-dependent contrasting effects on astrocyte branching complexity. Interestingly, morphine led to decreased astrocyte process length in the caudate nucleus, but increased process length in the NAc and lateral septal nucleus. In contrast, Narita et al. [59] observed increased GFAP expression in primary cortical astrocytes, who were harvested from newborn mice, following noncontingent morphine injections.

Heroin studies using noncontingent administration and self-administration have shown drug-induced morphological plasticity of astrocytes in multiple regions. Following daily increasing doses of noncontingent heroin injections (1, 2.5, 5, 7.5, 10, mg/mL), increased GFAP immunoreactivity has been observed in the dentate gyrus 24 h after the last injection. However, this change was not observed immediately following the last injection, suggesting a role of withdrawal for this increase in GFAP immunoreactivity [60]. Other studies using heroin self-administration (14 days) followed by extinction (12–13 days) showed heroin self-administration led to increased branch points and filament length of GFAP-positive cells in the prelimbic region of the prefrontal cortex [61]. Although overall GFAP expression was not quantified, morphological alterations of individual GFAP-positive astrocytes were observed in heroin-administering rats. Additionally, this study and others [62] have found that heroin-administering rats exhibit altered astrocyte-synaptic contact following heroin self-administration and extinction, through analysis of colocalization of Lck-GFP-labeled astrocyte processes with synapsin I as a presynaptic marker. However, Siemsen et al. [61] found that astrocyte-synaptic contact was increased in the prelimbic cortex following heroin self-administration and extinction, whereas Kruyer et al. [62] observed reduced astrocyte-synaptic contact in the NAc. Although opposing results were observed, both studies validate the dynamic effects of heroin on astrocyte structure, which may alter astrocyte function under heroin exposure and underscore that drug-induced effects on astrocytes vary based on brain region. Very few studies to date have investigated effects of other opioids on astrocyte structure or activity, including other prescription opiates.

2.4 Nicotine

In addition to psychostimulants and opioids, astrocytes are influenced by nicotine exposure and affect nicotine-seeking behavior [63]. Astrocytes express nicotinic acetylcholine receptors, which contribute to their response to nicotine exposure [64, 65]. While self-administration of nicotine leads to decreased GFAP expression in nicotine-extinguished rats compared to yoked saline controls [63], noncontingent administration with osmotic minipumps leads to an increase in GFAP immunoreactivity, either when implanted into pregnant females for observations in offspring [66] or following 2 weeks of administration [67]. Additionally, in vitro studies using primary astrocytes cultured from mice at PND2-4 showed that dose-dependent nicotine incubation significantly increased GFAP expression, while in vivo nicotine administration via osmotic minipumps increased astrocyte volume and territory volume in the CA1 of the hippocampus, substantia nigra, and PFC [64]. Here again, studies show that the resultant effects of nicotine are region-, paradigm-, and concentration-dependent.

3 Effects of Drug Use on Astrocyte Function

Astrocyte function in the brain broadly encompasses the maintenance of neurotransmitter and ion homeostasis [68], formation and regulation of synaptic structure and function [69], and maintenance of the blood-brain barrier (BBB) [70]. In the developed brain, astrocytes can contact up to 140,000 synapses, sending peripheral astrocyte processes (PAPs) into the synaptic cleft [71, 72]. Astrocyte PAPs express several proteins that allow for the maintenance of glutamate homeostasis [73] and the formation of synapses [74]. Beyond their roles in supporting synaptic function, astrocytes possess the machinery to bidirectionally communicate with neuronal synapses. More specifically, astrocyte receptors bind neurotransmitters released at the synapse, respond with elevations in intracellular calcium, and in turn release gliotransmitters that can impact signaling occurring at the synapse. Finally, through the extension of end feet, astrocytes together with endothelial cells and pericytes form and maintain the BBB [75]. Astrocytes specifically secrete factors that promote the formation of tight junctions between endothelial cells, which restrict the entry of pathogens and toxins into the brain [76, 77]. All of these functions are critical for normal brain function and can be disrupted by drugs of abuse.

3.1 Astrocytes Acutely Respond to Drugs of Abuse and the Neurotransmitter Systems with Which They Interact

Nearly all drugs of abuse engage the mesolimbic dopaminergic system, resulting in the elevation of extracellular levels of dopamine [1, 78, 79]. A number of studies have demonstrated that astrocytes in culture will respond to either dopamine or psychostimulant, opiate, ethanol, or nicotine application [80–86]. Notably, one relatively recent and salient study investigated astrocyte Ca^{2+} responses to dopamine within the mesolimbic dopaminergic system in vivo [87]. Corkrum and colleagues demonstrated that astrocytes within the reward circuitry respond with increases in calcium event probability to dopamine release, through the D1 receptor [87]. Within the nucleus accumbens, astrocytes respond with Ca^{2+} transients directly to locally administered dopamine, and approximately 50% of nucleus accumbens astrocytes respond subsequent to stimulation of the terminals projecting from the ventral tegmental area to the nucleus accumbens [87]. Beyond their ability to interact with astrocytes by promoting an elevation in dopaminergic signaling, several categories of drugs, including opiates, nicotine, and cannabinoid drugs, may also interact with astrocytes directly through their interaction with other receptors expressed on astrocytes.

As described briefly above, astrocyte calcium signaling is fundamental to astrocyte function. Stimulation of astrocytes by neurotransmitters results in calcium mobilization, primarily through release from internal stores [88–94]. These calcium dynamics can lead to the release of gliotransmitters, which can, in turn, influence neuronal activity [95, 96]. While beyond the scope of this discussion, it is important to note that astrocyte calcium dynamics may encode information via many properties of their signals, including amplitude, frequency, duration, spatial location, and spread [97]. While much work remains to fully elucidate the role of astrocyte calcium in astrocyte function, as well as brain function more broadly, it has been postulated that astrocyte calcium may function to impact synaptic transmission and plasticity, synapse formation and removal, vascular tone, and inflammation [98].

Astrocyte calcium is evoked by broad categories of drugs. For example, psychostimulant drugs including amphetamine and methamphetamine increase astrocyte calcium signaling both in culture and in vivo [81, 83, 87]. Ethanol similarly induces intracellular calcium transients in a subset of astrocytes in primary culture [29, 80, 82]. Opiate drugs elevate astrocyte calcium through direct interaction with the astrocyte mu-opioid receptor [99]. More specifically, in acute brain slices, application of morphine increases astrocyte calcium responses in an IP_3R2-mediated manner [99]. Similar data have been collected in cultured astrocytes [86, 100]. Nicotine can directly act on the $\alpha7$ nACh receptor, which is expressed in astrocytes [101], and activity of this receptor induces robust calcium activity in cultured astrocytes [85, 102]. Lastly, cannabinoid exposure can also evoke astrocyte Ca^{2+} responses. Astrocytes in the hippocampus express endocannabinoid CB1 receptors, and their activation either by endocannabinoids released by neurons or by exogenous cannabinoid drugs results in elevations in intracellular calcium and consequent glutamate release from astrocytes [103, 104].

Astrocytes possess machinery required for signaling, including receptors, elements involved in the vesicular release, and gliotransmitters [105]. Specifically, astrocytes express GABA [106], glutamate [107], endocannabinoid [103], and dopamine receptors [87]. They also express synaptic-like micro-vesicles and SNARE proteins, which are known to mediate vesicular fusion [108] and gliotransmitters including glutamate, GABA, ATP/adenosine, and D-serine [109]. This allows astrocytes to detect neurotransmitter release at the synapse and respond via signaling which can affect both presynaptic and postsynaptic receptors. In this way, astrocytes may impact signaling and plasticity at the synapse [105]. While the mechanism by which astrocytes release gliotransmitters has been debated, it is clear that though astrocytes are not electrically excitable, they demonstrate stimulus-induced and spontaneous calcium signaling [110]. This signaling can be mediated by G-protein coupled receptors, which activate IP_3R2 signaling on the endoplasmic reticulum, resulting in an increase of intracellular calcium [111]. While the mechanisms involved in downstream release of gliotransmitters are still being fully elucidated, it is generally well-accepted that the recruitment of internal calcium stores subsequent to GPCR stimulation on astrocytes is required for gliotransmission [105]. Through these mechanisms of supporting and actively modulating synaptic function, astrocytes affect the neuronal signaling in circuits underlying a variety of behaviors including memory processes, emotional and motivational behavior, fear learning, circadian rhythms, feeding behaviors, sensory-motor responses, and decision-making [112].

By these mechanisms, drugs of abuse can both acutely (described above) and chronically impact astrocyte responses to dopamine or drug stimuli, as well as neuron-astrocyte communication. While relatively little is known about the chronic effects of drugs of abuse on astrocyte-neuron communication, recent literature indicates that it is significantly affected. For instance, after a history of chronic cocaine, neuron-astrocyte communication is significantly altered. More specifically, in the nucleus accumbens shell, spontaneous astrocyte calcium transients are decreased in duration and amplitude, and astrocyte network connectivity is changed, with greater connection density and network efficiency but reduced modularity [113]. Interestingly, slow-inward currents on medium spiny neurons (MSNs), which are indicative of astrocyte-released glutamate, were increased in duration and frequency in a subset of MSNs. In contrast, after 24-hour nicotine treatment, cultured astrocytes show increased calcium event frequency in both the soma and the processes [64].

These changes in astrocyte signaling can promote significant changes in synaptic activity through the aforementioned downstream release of gliotransmitters, which, in turn, can impact synaptic transmission. Perhaps most relevant to this discussion are findings collected by Corkrum and colleagues demonstrating this point. These demonstrate that amphetamine acts through the D1 receptor expressed on astrocytes to elevate astrocyte calcium, release ATP from astrocytes, and depress excitatory synaptic transmission in the NAc via presynaptic Gi/o-coupled adenosine A1 receptors [87]. Furthermore, these changes in neuron-astrocyte communication were found to be behaviorally relevant, in that preventing the amphetamine-induced changes in astrocyte calcium dynamics by astrocyte dopamine D1-receptor knockout or IP_3R2 knockout prevented corticostriatal synaptic depression and mitigated amphetamine-induced locomotor enhancement. These effects mirror those described in Scofield et al. [114], which demonstrated that hM3Dq-DREADD stimulation of astrocytes promotes glial glutamate release and decreases cocaine-seeking behavior through action at presynaptic mGluR2/3.

3.2 Roles of Astrocytes in Drug-Induced Neuroplasticity and Drug Seeking

Some of the earliest observations regarding the functional effects of drug exposure in vivo on astrocyte function were in regard to cocaine effects on glutamate homeostasis [73]. Under physiological conditions, astrocytes regulate and maintain glutamate homeostasis, the balance between synaptic and extrasynaptic glutamate, through the action of coordinated glutamate uptake and export [73]. Astrocyte PAPs robustly express the high-affinity glutamate transporter, GLT-1, which is responsible for 90% of synaptic glutamate uptake [115]. Extrasynaptically, astrocytes express the cystine-glutamate exchanger, xCT, which maintains ambient levels of extrasynaptic glutamate [116] by exchanging extracellular cystine for intracellular glutamate. This maintains extracellular glutamate tone on presynaptic mGluR2/3 receptors, thereby inhibiting synaptic glutamate release [116]. Expression and activity of GLT-1, xCT, and mGluR2/3 are all altered by chronic exposure to drugs of abuse, including opiates and psychostimulants [117–119]. Importantly, regulation of all of these targets, and normalization of astrocyte-mediated glutamate homeostasis, is one of the leading mechanisms in the search for pharmacotherapeutic treatment for psychostimulant use disorders. For example, the antioxidant and cystine prodrug N- acetylcysteine (NAC) normalizes nucleus accumbens glutamate levels, restores drug-induced supporting of GLT-1 and xCT, and reduces drug seeking for cocaine, heroin, and nicotine [63, 120, 121]. Accordingly, NAC and related compounds targeting glutamate homeostasis have been intensely studied as anti-relapse medications, discussed in more detail below [20, 25, 122–124].

In addition to decreased expression and activity of astrocyte regulators of glutamate homeostasis, self-administration of cocaine, heroin, and methamphetamine results in decreased synaptic interaction of astrocyte membranes, as evidenced by their significantly reduced colocalization with both presynaptic and postsynaptic markers [43, 51, 55, 62]. Collectively, these drug-induced changes contribute to reduced glutamate clearance from the synapse [125] as well as reduced extracellular inhibitory glutamatergic tone [126]. This may exacerbate the pathophysiological upregulation of synaptic transmission at corticostriatal synapses induced by active drug seeking following a history of self-administration [127]. Additionally, this excess glutamate may stimulate mGluRs and NMDARs at the synapse [128], driving relapse and drug-seeking behavior.

Pharmacologically mitigating these changes in astrocyte protein expression opposes drug-seeking behavior. Specifically, the administration of cysteine prodrug, N-acetylcysteine, which restores xCT levels [129] and ameliorates deficits in GLT-1 [130] opposes cue-induced relapse to cocaine-seeking [129]. Similarly, hM3Dq-DREADD stimulation of astrocytes promotes glial glutamate release and decreases cocaine-seeking behavior in an mGluR2/3-dependent manner [114]. Importantly however, while evidence indicates that nucleus accumbens synapses may oppose drug seeking for both cocaine, methamphetamine, and heroin [43, 62, 114], cortical astrocytes may promote ethanol consumption [131].

Early stages of cocaine self-administration lead to adaptations in both astrocytes and neurons, which recapitulate mechanisms of development, contributing to drug reinforcement. During development, astrocytes promote synapse formation by releasing thrombospondins, TSP 1 and TSP 2 [74]. TSP 1 and TSP 2 interact with their corresponding neuronal receptor $\alpha 2\delta - 1$ to induce differentiation of the presynapse and postsynapse and promote clustering of synaptic proteins, thereby generating synapses that are enriched with NMDARs, but are AMPAR-silent [132, 133]. After developmental stages have passed, TSPs are lowly expressed in the nucleus accumbens, but neuronal expression of $\alpha 2\delta - 1$ expression remains intact [134].

In acute brain slices of the nucleus accumbens collected from adult mice, perfusion of cocaine increases astrocyte calcium [135] and chemogenetic stimulation of astrocytes using astrocyte-specific hM4Di-DREADDs in the dorsal striatum similarly increases astrocyte calcium and downstream release of TSP1 [136]. These data combined with the hypothesis that developmental mechanisms may contribute to learning gave rise to the hypothesis that cocaine may rejuvenate astrocyte ability to perform this developmental synaptogenic function in adulthood, providing a means by which astrocytes may regulate dynamic changes in nucleus accumbens circuitry, leading to cue-induced cocaine seeking [137].

Independent of astrocyte involvement, it is known that the administration of cocaine results in the generation of AMPAR-silent synapses [138, 139]. In vivo, following repeated intraperitoneal injections of cocaine, excitatory silent synapses appear in the nucleus accumbens. These synapses express GluN2B-NMDARs, but are AMPAR-silent. These new silent synapses mature over the course of withdrawal, resulting in the strengthening of cocaine-generated excitatory synapses. More specifically, after drug withdrawal, these synapses recruit AMPARs and, in doing so, stabilize and consolidate cocaine memories. These synapses remain dynamic even after maturation, re-silencing, and re-recruiting AMPARs as a function of reintroduction to cocaine cues [140, 141].

Importantly, cocaine-induced synaptogenesis occurs via a TSP-dependent mechanism [135]. Administration of gabapentin, which disrupts $\alpha 2\delta - 1$ and TSP signaling [133, 142] blocks cocaine-induced generation of silent synapses and spinogenesis, as well as cue-primed cocaine-seeking behavior [135]. Separate studies indicate that $\alpha 2\delta - 1$ is upregulated by cocaine self-administration and extinction, and intra-NAc microinjection of gabapentin opposes cue-induced reinstatement to cocaine seeking following extinction [143]. Moreover, genetic deletion or pharmacological inhibition of $\alpha 2\delta - 1$ blocks effects of morphine on synaptic transmission and conditioned place preference; however, it is unclear if this is through a TSP-dependent mechanism [144].

4 Translational Implications: Human Studies and Astrocyte-Based Candidate Therapeutics

4.1 *Postmortem Studies in Humans: Strengths and Limitations*

Thus far in the chapter, we have highlighted evidence from preclinical studies using cell culture and animal models to demonstrate significant roles for astrocytes in the neuropathology of drug use. Despite the wealth of knowledge established in animal models, far less is known about how these pathologies manifest and contribute to features of substance use disorders (SUDs) in humans.

Studying SUDs in humans is complicated for a number of reasons: First, scientists are often limited to post hoc analyses using postmortem tissue, which is limited in access (most people do not donate their body to science) and can be confounded by a number of factors including manner of death, postmortem interval, and comorbid psychiatric or other medical conditions [145]. Moreover, studies using postmortem human tissue are often quite limited by experimental techniques, largely relying on histological analyses and only more recently, single-nucleus sequencing [146, 147].

Therefore, because live imaging methods are not yet advanced enough to provide evidence of astrocyte engagement, studying astrocytes in the human brain in the context of SUDs relies on determining correlations between an individual's life history of drug use and gross structural or gene expression changes in specific regions of the brain. As mentioned previously, astrocytes express most receptors and transporters which mediate drug-induced neural signaling, suggesting that astrocytes may be affected directly by exposure. However, astrocyte changes may also be secondary to drug use, that is, responding via morphological or functional change as a result of alterations in neuron number or function as a consequence of drug use.

Second, studies of substance use in humans are complicated due to the complexities of the individual. Each person is a compilation of their own unique genetic makeup (which may influence susceptibility to addiction or complications from using drugs of abuse), their own genetics, history of stress, life circumstances, etc., all of which will influence one's likelihood to use drugs or their susceptibility to addiction. SUDs confer a high degree of heritability [148], and genetic risks and vulnerabilities may influence glial responses to drug use. Importantly, life history and particularly adverse experiences create a strong influence on risk for SUDs [149–151]. Additionally, drug users may often engage in different patterns of drug use (recreational or occasional use versus chronic and habitual use), durations of drug use, and/or polydrug use, which further complicate the ability to attribute any scientific findings to a singular drug. Third and finally, human studies are often confounded by comorbid mental health or other medical conditions that may also contribute to observed postmortem pathology. Co-occurrence of SUDs with other psychiatric diagnoses is high; among individuals with diagnosed SUD, nearly 40% will have an additional mental health challenge [152, 153].

Given the limitations of postmortem analyses of human tissue and the low number of studies that directly investigate astrocyte pathology in the postmortem human brain, the true extent to which chronic drug use leads to astrocyte dysregulation in humans is unclear. However, postmortem studies have been useful in shedding light on astrocyte pathology and corroborating preclinical evidence. Some reports find no significant alterations in GFAP immunoreactivity (indicative of reactive astrocytes), for example, in the prefrontal cortex of individuals who used heroin [154]. Yet others find shifts in astrocyte morphological phenotype toward reactivity in various brain regions such as the hippocampus following heroin use [155] and the hippocampus, striatum, and cortex of individuals using methamphetamine [156–158]. Finally, others have found that astrocyte numbers and densities significantly decrease, suggesting possible atrophy and degeneration of astrocyte populations across several regions of the brain following polydrug use [159]. A recent study which employed single-nucleus RNA-seq on postmortem prefrontal cortical samples from alcohol-dependent and control individuals revealed that the greatest number of identified differentially expressed genes (DEGs) were from astrocytes [160].

Observations of decreases in astrocyte number and density are most commonly reported in postmortem studies of individuals with alcohol use disorders. For example, there is a reduction in the size of astrocyte nuclei and decreased astrocyte density in the prefrontal cortex of alcohol-dependent individuals [161, 162], which corroborates other findings in the hippocampus which also indicate a reduction in astrocyte number [163]. Thus, it seems that chronic or long-term alcohol use may lead to a degenerative phenotype in astrocytes, which may contribute to the neurological deficits seen in individuals with alcohol use disorder [31].

These human postmortem studies highlight two critical ideas with respect to how drugs of abuse impact the brain, particularly astrocytes. First, the studies that indicate increases in astrocytic markers, either by protein measurement or histological analysis, seem to be related to cases of significant drug exposure (i.e., overdose) or chronic usage, but likely in the early phase. Second, studies of chronic, but more long-term drug usage indicate that measures of astrocytes actually decrease. That is, as drug usage prolongs, the degenerative effects of the drugs on neuron and astrocyte physiology increase. Thus, it is likely that the reactive phenotypes of astrocytes (hypertrophic morphology, increases in GFAP protein expression, etc.) observed postmortem may be due to acute toxicity and/or indirect effects on astrocytes, whereas prolonged usage or abstinence from drug exposure may lead to the more hypotrophic/atrophic phenotype observed in preclinical models.

4.2 Clinical Trials

Given the abundance of preclinical research discussed in previous sections, there is ample evidence to suggest that the fundamental and common pathology underlying many drug use disorders is disruption of glutamate signaling and its regulation by astrocytes. As such, the most commonly studied and, perhaps most promising, drugs in human clinical trials have focused on restoring glutamate balance during drug taking or relapse, as these time windows offer the most optimal point of intervention for drug users [164–166].

Clinical trials examining the efficacy of N-acetylcysteine (NAC) find that at higher doses, study participants often have decreased craving during the extent of the study across a range of drug classes and mental health comorbidities. For example, a double-blind study of military veterans diagnosed with both substance use disorder and PTSD found that 2400 mg/day treatment with NAC across 8 weeks significantly decreased craving and depression while improving other symptoms of PTSD [167]. Similar decreases in craving have been in a double-blind study of methamphetamine-dependent users (1200 mg/day for 4 weeks; [168]) and improved odds of maintaining abstinence in a study of cannabis-dependent adolescents ((2400 mg/day for 8 weeks; [169]). Studies on cocaine craving have been a bit more mixed, however, with one trial finding that NAC treatment (1200-2400 mg/day for 4 days) reduced cocaine craving when administered with an experimenter-delivered IV dose of cocaine [170], while another found no overall effect of NAC on cocaine craving in cocaine-dependent users ((1200 or 2400 mg/day for 8 weeks) [171]). Somewhat encouraging, the latter study did find that participants who had already achieved abstinence from cocaine had longer relapse times and lower reported craving [171].

While astrocyte modulators certainly hold promise for therapeutic potential for substance use disorders, the study of their application in the clinic is still in the early stages. Furthermore, questions remain as to the long-term efficacy of drugs such as NAC to produce lasting reductions in craving beyond the initial treatment window. Importantly, the greatest efficacy of pharmacological treatments against craving and relapse vulnerability is almost certainly found through combined treatment with cognitive and/or behavioral therapies and attention to environmental and community support.

5 Conclusions

In summary, knowledge regarding effects of drug use on astrocytes has developed tremendously since the earliest observations regarding effects of experimenter-administered drugs on GFAP immunoreactivity in the 1990s. We now have a broad appreciation of the influence of varied paradigms and protocols of drug administration on a myriad of measures of astrocyte gene expression and function. Additionally, we have increased appreciation that directionality, magnitude, and nature of astrocyte effects depend heavily on factors including sex, dose, duration of exposure, and time since exposure. In particular, lasting disruptions in the synaptic regulation of astrocytes are increasingly appreciated, particularly given the enduring disruptions in glutamate homeostasis observed following rodent self-administration [73].

Given the current knowledge regarding the engagement of astrocytes and astrocyte-neuron communication by drug use, specific priorities can be identified. Of note, no studies have investigated effects of polydrug use on astrocyte-neuron communication, which is of concern given the widespread nature polydrug use. In addition, other than GLT-1 and xCT, few specific genes regulated by drug use have been identified in astrocytes. Identification of genetic adaptations will reveal mechanisms for changes in astrocyte calcium dynamics and synaptic regulation and will reveal more targets for candidate intervention beyond mediators of glutamate homeostasis. Translational utility for treatment of substance use disorders and investigation of how drug use in animal models engages astrocytes have revealed important principles regarding experience-dependent astrocyte plasticity and underscore the fundament roles of astrocytes in synaptic function and regulation of behavior.

References

1. Di Chiara G, Imperato A. Drugs abused by humans preferentially increase synaptic dopamine concentrations in the mesolimbic system of freely moving rats. Proc Natl Acad Sci USA. 1988;85:5274–8.
2. Olds J, Milner P. Positive reinforcement produced by electrical stimulation of septal area and other regions of rat brain. J Comp Physiol Psychol. 1954;47:419–27.
3. Saunders BT, Richard JM, Janak PH. Contemporary approaches to neural circuit manipulation and mapping: focus on reward and addiction. Philos Trans R Soc Lond Ser B Biol Sci. 2015;370:20140210.
4. Wise RA. Dopamine, learning and motivation. Nat Rev Neurosci. 2004;5:483–94.
5. Lewis RG, Florio E, Punzo D, Borrelli E. The brain's reward system in health and disease. Adv Exp Med Biol. 2021;1344:57–69.
6. Volkow ND, Wang GJ, Fowler JS, Tomasi D. Addiction circuitry in the human brain. Annu Rev Pharmacol Toxicol. 2012;52:321–36.
7. Russo SJ, Nestler EJ. The brain reward circuitry in mood disorders. Nat Rev Neurosci. 2013;14:609–25.
8. Mogenson GJ, Jones DL, Yim CY. From motivation to action: functional interface between the limbic system and the motor system. Prog Neurobiol. 1980;14:69–97.
9. Koob GF, Volkow ND. Neurocircuitry of addiction. Neuropsychopharmacology. 2010;35:217–38.
10. Koob GF, Volkow ND. Neurobiology of addiction: a neurocircuitry analysis. Lancet Psychiatry. 2016;3:760–73.
11. Kuhn BN, Kalivas PW, Bobadilla AC. Understanding addiction using animal models. Front Behav Neurosci. 2019;13:262.
12. Spanagel R. Animal models of addiction. Dialogues Clin Neurosci. 2017;19:247–58.
13. Venniro M, Banks ML, Heilig M, Epstein DH, Shaham Y. Improving translation of animal models of addiction and relapse by reverse translation. Nat Rev Neurosci. 2020;21:625–43.
14. Beitner-Johnson D, Guitart X, Nestler EJ. Glial fibrillary acidic protein and the mesolimbic dopamine system: regulation by chronic morphine and Lewis-Fischer strain differences in the rat ventral tegmental area. J Neurochem. 1993;61:1766–73.
15. Bowers MS, Kalivas PW. Forebrain astroglial plasticity is induced following withdrawal from repeated cocaine administration. Eur J Neurosci. 2003;17:1273–8.
16. Fattore L, Puddu MC, Picciau S, Cappai A, Fratta W, Serra GP, Spiga S. Astroglial in vivo response to cocaine in mouse dentate gyrus: a quantitative and qualitative analysis by confocal microscopy. Neuroscience. 2002;110:1–6.
17. Silva-Araujo AL, Tavares MA. Expression of glial fibrillary acidic protein in the rat retina after exposure to psychostimulants. Retina. 1995;15:241–7.
18. Kim R, Healey KL, Sepulveda-Orengo MT, Reissner KJ. Astroglial correlates of neuropsychiatric disease: from astrocytopathy to astrogliosis. Prog Neuro-Psychopharmacol Biol Psychiatry. 2018;87:126–46.
19. Kruyer A, Scofield MD. Astrocytes in addictive disorders. Adv Neurobiol. 2021;26:231–54.
20. Scofield MD, Heinsbroek JA, Gipson CD, Kupchik YM, Spencer S, Smith AC, Roberts-Wolfe D, Kalivas PW. The nucleus accumbens: mechanisms of addiction across drug classes reflect the importance of glutamate homeostasis. Pharmacol Rev. 2016a;68:816–71.
21. Wang J, Holt LM, Huang HH, Sesack SR, Nestler EJ, Dong Y. Astrocytes in cocaine addiction and beyond. Mol Psychiatry. 2022;27:652–68.
22. Porter JT, McCarthy KD. Astrocytic neurotransmitter receptors in situ and in vivo. Prog Neurobiol. 1997;51:439–55.
23. Shelkar GP, Gandhi PJ, Liu J, Dravid SM. Cocaine preference and neuroadaptations are maintained by astrocytic NMDA receptors in the nucleus accumbens. Sci Adv. 2022;8:eabo6574.

24. Reissner KJ, Uys JD, Schwacke JH, Comte-Walters S, Rutherford-Bethard JL, Dunn TE, Blumer JB, Schey KL, Kalivas PW. AKAP signaling in reinstated cocaine seeking revealed by iTRAQ proteomic analysis. J Neurosci. 2011;31:5648–58.
25. Spencer S, Kalivas PW. Glutamate transport: a new bench to bedside mechanism for treating drug abuse. Int J Neuropsychopharmacol. 2017;20:797–812.
26. Bushong EA, Martone ME, Jones YZ, Ellisman MH. Protoplasmic astrocytes in CA1 stratum radiatum occupy separate anatomical domains. J Neurosci. 2002;22:183–92.
27. Araque A, Parpura V, Sanzgiri RP, Haydon PG. Tripartite synapses: glia, the unacknowledged partner. Trends Neurosci. 1999;22:208–15.
28. Rimmele TS, Rosenberg PA. GLT-1: the elusive presynaptic glutamate transporter. Neurochem Int. 2016;98:19–28.
29. Adermark L, Bowers MS. Disentangling the role of astrocytes in alcohol use disorder. Alcohol Clin Exp Res. 2016;40:1802–16.
30. Coulter OR, Walker CD, Risher ML. Astrocyte-specific Ca(2+) activity: mechanisms of action, experimental tools, and roles in ethanol-induced dysfunction. Biochem Cell Biol. 2023;101:410–21.
31. Miguel-Hidalgo JJ. Molecular neuropathology of astrocytes and oligodendrocytes in alcohol use disorders. Front Mol Neurosci. 2018;11:78.
32. Renau-Piqueras J, Zaragoza R, De Paz P, Baguena-Cervellera R, Megias L, Guerri C. Effects of prolonged ethanol exposure on the glial fibrillary acidic protein-containing intermediate filaments of astrocytes in primary culture: a quantitative immunofluorescence and immunogold electron microscopic study. J Histochem Cytochem. 1989;37:229–40.
33. Saez R, Burgal M, Renau-Piqueras J, Marques A, Guerri C. Evolution of several cytoskeletal proteins of astrocytes in primary culture: effect of prenatal alcohol exposure. Neurochem Res. 1991;16:737–47.
34. Franke H. Influence of chronic alcohol treatment on the GFAP-immunoreactivity in astrocytes of the hippocampus in rats. Acta Histochem. 1995;97:263–71.
35. Wilhelm CJ, Hashimoto JG, Roberts ML, Bloom SH, Andrew MR, Wiren KM. Astrocyte dysfunction induced by alcohol in females but not males. Brain Pathol. 2016;26:433–51.
36. Healey KL, Bell A, Scofield MD, Swartzwelder HS. Adolescent intermittent ethanol exposure reduces astrocyte-synaptic proximity in the adult medial prefrontal cortex in rats: reversal by gabapentin. Addict Neurosci. 2022;4:100047.
37. Healey KL, Kibble S, Hodges S, Reissner KJ, Testen A, Wills TA, Acheson SK, Siemsen BM, McFaddin JA, Scofield MD, Swartzwelder HS. Enduring alterations in hippocampal astrocytesynaptic proximity following adolescent alcohol exposure: reversal by gabapentin. Neural Regen Res. 2020;15:1496–501.
38. Walker CD, Sexton HG, Hyde J, Greene B, Risher ML. Diverging effects of adolescent ethanol exposure on tripartite synaptic development across prefrontal cortex subregions. Cells. 2022;11
39. Castellano P, Nwagbo C, Martinez LR, Eugenin EA. Methamphetamine compromises gap junctional communication in astrocytes and neurons. J Neurochem. 2016;137:561–75.
40. Frey BN, Andreazza AC, Cereser KM, Martins MR, Petronilho FC, de Souza DF, Tramontina F, Goncalves CA, Quevedo J, Kapczinski F. Evidence of astrogliosis in rat hippocampus after d-amphetamine exposure. Prog Neuro-Psychopharmacol Biol Psychiatry. 2006;30:1231–4.
41. Zhang Y, Zhu T, Zhang X, Chao J, Hu G, Yao H. Role of high-mobility group box 1 in methamphetamine-induced activation and migration of astrocytes. J Neuroinflammation. 2015;12:156.
42. Dang J, Tiwari SK, Agrawal K, Hui H, Qin Y, Rana TM. Glial cell diversity and methamphetamine-induced neuroinflammation in human cerebral organoids. Mol Psychiatry. 2021;26:1194–207.
43. Siemsen BM, Reichel CM, Leong KC, Garcia-Keller C, Gipson CD, Spencer S, McFaddin JA, Hooker KN, Kalivas PW, Scofield MD. Effects of methamphetamine self-administration and extinction on astrocyte structure and function in the nucleus accumbens core. Neuroscience. 2019;406:528–41.

44. Leitao RA, Fontes-Ribeiro CA, Silva AP. The effect of parthenolide on methamphetamine-induced blood-brain barrier and astrocyte alterations. Eur J Clin Investig. 2022;52:e13694.
45. Hol EM, Pekny M. Glial fibrillary acidic protein (GFAP) and the astrocyte intermediate filament system in diseases of the central nervous system. Curr Opin Cell Biol. 2015;32:121–30.
46. Armstrong V, Reichel CM, Doti JF, Crawford CA, McDougall SA. Repeated amphetamine treatment causes a persistent elevation of glial fibrillary acidic protein in the caudate-putamen. Eur J Pharmacol. 2004;488:111–5.
47. Basmadjian OM, Occhieppo VB, Marchese NA, Silvero CM, Becerra MC, Baiardi G, Bregonzio C. Amphetamine induces oxidative stress, glial activation and transient angiogenesis in prefrontal cortex via AT(1)-R. Front Pharmacol. 2021;12:647747.
48. Franke H, Kittner H, Grosche J, Illes P. Enhanced P2Y1 receptor expression in the brain after sensitisation with d-amphetamine. Psychopharmacology. 2003;167:187–94.
49. Blanco-Calvo E, Rivera P, Arrabal S, Vargas A, Pavon FJ, Serrano A, Castilla-Ortega E, Galeano P, Rubio L, Suarez J, Rodriguez de Fonseca F. Pharmacological blockade of either cannabinoid CB1 or CB2 receptors prevents both cocaine-induced conditioned locomotion and cocaine-induced reduction of cell proliferation in the hippocampus of adult male rat. Front Integr Neurosci. 2014;7:106.
50. Zhu W, Mao Z, Zhu C, Li M, Cao C, Guan Y, Yuan J, Xie G, Guan X. Adolescent exposure to cocaine increases anxiety-like behavior and induces morphologic and neurochemical changes in the hippocampus of adult rats. Neuroscience. 2016;313:174–83.
51. Scofield MD, Li H, Siemsen BM, Healey KL, Tran PK, Woronoff N, Boger HA, Kalivas PW, Reissner KJ. Cocaine self-administration and extinction leads to reduced glial fibrillary acidic protein expression and morphometric features of astrocytes in the nucleus accumbens core. Biol Psychiatry. 2016b;80:207–15.
52. Jarvis R, Tamashiro-Orrego A, Promes V, Tu L, Shi J, Yang Y. Cocaine self-administration and extinction inversely alter neuron to glia exosomal dynamics in the nucleus accumbens. Front Cell Neurosci. 2019;13:581.
53. Benediktsson AM, Schachtele SJ, Green SH, Dailey ME. Ballistic labeling and dynamic imaging of astrocytes in organotypic hippocampal slice cultures. J Neurosci Methods. 2005;141:41–53.
54. Kim R, Testen A, Harder EV, Brown NE, Witt EA, Bellinger TJ, Franklin JP, Reissner KJ. Abstinence-Dependent Effects of Long-Access Cocaine Self-Administration on Nucleus Accumbens Astrocytes Are Observed in Male, But Not Female, Rats. eNeuro. 2022b;9
55. Testen A, Sepulveda-Orengo MT, Gaines CH, Reissner KJ. Region-specific reductions in morphometric properties and synaptic colocalization of astrocytes following cocaine self-administration and extinction. Front Cell Neurosci. 2018;12:246.
56. Song P, Zhao ZQ. The involvement of glial cells in the development of morphine tolerance. Neurosci Res. 2001;39:281–6.
57. Garrido E, Perez-Garcia C, Alguacil LF, Diez-Fernandez C. The alpha2-adrenoceptor antagonist yohimbine reduces glial fibrillary acidic protein upregulation induced by chronic morphine administration. Neurosci Lett. 2005;383:141–4.
58. Lazriev IL, Kiknadze GI, Kutateladze II, Nebieridze MI. Effect of morphine on the number and branching of astrocytes in various regions of rat brain. Bull Exp Biol Med. 2001;131:248–50.
59. Narita M, Miyatake M, Narita M, Shibasaki M, Shindo K, Nakamura A, Kuzumaki N, Nagumo Y, Suzuki T. Direct evidence of astrocytic modulation in the development of rewarding effects induced by drugs of abuse. Neuropsychopharmacology. 2006;31:2476–88.
60. Parekh SV, Paniccia JE, Lebonville CL, Lysle DT. Dorsal hippocampal interleukin-1 signaling mediates heroin withdrawal-enhanced fear learning. Psychopharmacology. 2020;237:3653–64.
61. Siemsen BM, Denton AR, Parrila-Carrero J, Hooker KN, Carpenter EA, Prescot ME, Brock AG, Westphal AM, Leath MN, McFaddin JA, Jhou TC, McGinty JF, Scofield MD. Heroin self-administration and extinction increase prelimbic cortical astrocyte-synapse proximity and alter dendritic spine morphometrics that are reversed by N-acetylcysteine. Cells. 2023;12

62. Kruyer A, Scofield MD, Wood D, Reissner KJ, Kalivas PW. Heroin cue-evoked astrocytic structural plasticity at nucleus accumbens synapses inhibits heroin seeking. Biol Psychiatry. 2019;86:811–9.
63. Namba MD, Kupchik YM, Spencer SM, Garcia-Keller C, Goenaga JG, Powell GL, Vicino IA, Hogue IB, Gipson CD. Accumbens neuroimmune signaling and dysregulation of astrocytic glutamate transport underlie conditioned nicotine-seeking behavior. Addict Biol. 2020;25:e12797.
64. Aryal SP, Fu X, Sandin JN, Neupane KR, Lakes JE, Grady ME, Richards CI. Nicotine induces morphological and functional changes in astrocytes via nicotinic receptor activity. Glia. 2021;69:2037–53.
65. Hernandez-Morales M, Garcia-Colunga J. Effects of nicotine on K+ currents and nicotinic receptors in astrocytes of the hippocampal CA1 region. Neuropharmacology. 2009;56:975–83.
66. Abdel-Rahman A, Dechkovskaia AM, Sutton JM, Chen WC, Guan X, Khan WA, Abou-Donia MB. Maternal exposure of rats to nicotine via infusion during gestation produces neurobehavioral deficits and elevated expression of glial fibrillary acidic protein in the cerebellum and CA1 subfield in the offspring at puberty. Toxicology. 2005;209:245–61.
67. Hawkins JL, Denson JE, Miley DR, Durham PL. Nicotine stimulates expression of proteins implicated in peripheral and central sensitization. Neuroscience. 2015;290:115–25.
68. Parpura V, Verkhratsky A. Homeostatic function of astrocytes: Ca(2+) and Na(+) signalling. Transl Neurosci. 2012;3:334–44.
69. Volterra A, Meldolesi J. Astrocytes, from brain glue to communication elements: the revolution continues. Nat Rev Neurosci. 2005;6:626–40.
70. Abbott NJ, Patabendige AA, Dolman DE, Yusof SR, Begley DJ. Structure and function of the blood-brain barrier. Neurobiol Dis. 2010;37:13–25.
71. Oberheim NA, Wang X, Goldman S, Nedergaard M. Astrocytic complexity distinguishes the human brain. Trends Neurosci. 2006;29:547–53.
72. Santello M, Cali C, Bezzi P. Gliotransmission and the tripartite synapse. Adv Exp Med Biol. 2012;970:307–31.
73. Kalivas PW. The glutamate homeostasis hypothesis of addiction. Nat Rev Neurosci. 2009;10:561–72.
74. Clarke LE, Barres BA. Emerging roles of astrocytes in neural circuit development. Nat Rev Neurosci. 2013;14:311–21.
75. Cabezas R, Avila M, Gonzalez J, El-Bacha RS, Baez E, Garcia-Segura LM, Jurado Coronel JC, Capani F, Cardona-Gomez GP, Barreto GE. Astrocytic modulation of blood brain barrier: perspectives on Parkinson's disease. Front Cell Neurosci. 2014;8:211.
76. Kubotera H, Ikeshima-Kataoka H, Hatashita Y, Allegra Mascaro AL, Pavone FS, Inoue T. Astrocytic endfeet re-cover blood vessels after removal by laser ablation. Sci Rep. 2019;9:1263.
77. Zhao Z, Nelson AR, Betsholtz C, Zlokovic BV. Establishment and dysfunction of the blood-brain barrier. Cell. 2015;163:1064–78.
78. Pontieri FE, Tanda G, Di Chiara G. Intravenous cocaine, morphine, and amphetamine preferentially increase extracellular dopamine in the "shell" as compared with the "core" of the rat nucleus accumbens. Proc Natl Acad Sci USA. 1995;92:12304–8.
79. Volkow ND, Fowler JS, Wang GJ, Swanson JM, Telang F. Dopamine in drug abuse and addiction: results of imaging studies and treatment implications. Arch Neurol. 2007;64:1575–9.
80. Allansson L, Khatibi S, Olsson T, Hansson E. Acute ethanol exposure induces [Ca2+]i transients, cell swelling and transformation of actin cytoskeleton in astroglial primary cultures. J Neurochem. 2001;76:472–9.
81. Canedo T, Portugal CC, Socodato R, Almeida TO, Terceiro AF, Bravo J, Silva AI, Magalhaes JD, Guerra-Gomes S, Oliveira JF, Sousa N, Magalhaes A, Relvas JB, Summavielle T. Astrocyte-derived TNF and glutamate critically modulate microglia activation by methamphetamine. Neuropsychopharmacology. 2021;46:2358–70.

82. Kim HB, Lu Y, Oh SC, Morris J, Miyashiro K, Kim J, Eberwine J, Sul JY. Astrocyte ethanol exposure reveals persistent and defined calcium response subtypes and associated gene signatures. J Biol Chem. 2022a;298:102147.
83. Miyatake M, Narita M, Shibasaki M, Nakamura A, Suzuki T. Glutamatergic neurotransmission and protein kinase C play a role in neuron-glia communication during the development of methamphetamine-induced psychological dependence. Eur J Neurosci. 2005;22:1476–88.
84. Salazar M, Pariente JA, Salido GM, Gonzalez A. Ethanol induces glutamate secretion by Ca2+ mobilization and ROS generation in rat hippocampal astrocytes. Neurochem Int. 2008;52:1061–7.
85. Sharma G, Vijayaraghavan S. Nicotinic cholinergic signaling in hippocampal astrocytes involves calcium-induced calcium release from intracellular stores. Proc Natl Acad Sci USA. 2001;98:4148–53.
86. Stiene-Martin A, Hauser KF. Morphine suppresses DNA synthesis in cultured murine astrocytes from cortex, hippocampus and striatum. Neurosci Lett. 1993;157:1–3.
87. Corkrum M, Covelo A, Lines J, Bellocchio L, Pisansky M, Loke K, Quintana R, Rothwell PE, Lujan R, Marsicano G, Martin ED, Thomas MJ, Kofuji P, Araque A. Dopamine-evoked synaptic regulation in the nucleus accumbens requires astrocyte activity. Neuron. 2020;105:1036–1047.e5.
88. Dani JW, Chernjavsky A, Smith SJ. Neuronal activity triggers calcium waves in hippocampal astrocyte networks. Neuron. 1992;8:429–40.
89. Jourdain P, Bergersen LH, Bhaukaurally K, Bezzi P, Santello M, Domercq M, Matute C, Tonello F, Gundersen V, Volterra A. Glutamate exocytosis from astrocytes controls synaptic strength. Nat Neurosci. 2007;10:331–9.
90. Latour I, Gee CE, Robitaille R, Lacaille JC. Differential mechanisms of Ca2+ responses in glial cells evoked by exogenous and endogenous glutamate in rat hippocampus. Hippocampus. 2001;11:132–45.
91. Murphy TH, Blatter LA, Wier WG, Baraban JM. Rapid communication between neurons and astrocytes in primary cortical cultures. J Neurosci. 1993;13:2672–9.
92. Pasti L, Volterra A, Pozzan T, Carmignoto G. Intracellular calcium oscillations in astrocytes: a highly plastic, bidirectional form of communication between neurons and astrocytes in situ. J Neurosci. 1997;17:7817–30.
93. Porter JT, McCarthy KD. GFAP-positive hippocampal astrocytes in situ respond to glutamatergic neuroligands with increases in [Ca2+]i. Glia. 1995a;13:101–12.
94. Porter JT, McCarthy KD. Adenosine receptors modulate [Ca2+]i in hippocampal astrocytes in situ. J Neurochem. 1995b;65:1515–23.
95. Fellin T, Pascual O, Gobbo S, Pozzan T, Haydon PG, Carmignoto G. Neuronal synchrony mediated by astrocytic glutamate through activation of extrasynaptic NMDA receptors. Neuron. 2004;43:729–43.
96. Fiacco TA, McCarthy KD. Intracellular astrocyte calcium waves in situ increase the frequency of spontaneous AMPA receptor currents in CA1 pyramidal neurons. J Neurosci. 2004;24:722–32.
97. Rusakov DA, Zheng K, Henneberger C. Astrocytes as regulators of synaptic function: a quest for the Ca2+ master key. Neuroscientist. 2011;17:513–23.
98. Khakh BS, McCarthy KD. Astrocyte calcium signaling: from observations to functions and the challenges therein. Cold Spring Harb Perspect Biol. 2015;7:a020404.
99. Corkrum M, Rothwell PE, Thomas MJ, Kofuji P, Araque A. Opioid-mediated astrocyte-neuron signaling in the nucleus accumbens. Cells. 2019;8:586.
100. Stiene-Martin A, Zhou R, Hauser KF. Regional, developmental, and cell cycle-dependent differences in mu, delta, and kappa-opioid receptor expression among cultured mouse astrocytes. Glia. 1998;22:249–59.
101. Xu ZQ, Zhang WJ, Su DF, Zhang GQ, Miao CY. Cellular responses and functions of alpha7 nicotinic acetylcholine receptor activation in the brain: a narrative review. Ann Transl Med. 2021;9:509.

102. Sharma G, Vijayaraghavan S. Nicotinic receptor signaling in nonexcitable cells. J Neurobiol. 2002;53:524–34.
103. Navarrete M, Araque A. Endocannabinoids mediate neuron-astrocyte communication. Neuron. 2008;57:883–93.
104. Navarrete M, Diez A, Araque A. Astrocytes in endocannabinoid signalling. Philos Trans R Soc Lond Ser B Biol Sci. 2014;369:20130599.
105. Papouin T, Dunphy J, Tolman M, Foley JC, Haydon PG. Astrocytic control of synaptic function. Philos Trans R Soc Lond Ser B Biol Sci. 2017;372:20160154.
106. Fraser DD, Mudrick-Donnon LA, MacVicar BA. Astrocytic GABA receptors. Glia. 1994;11:83–93.
107. Verkhratsky A, Kirchhoff F. NMDA Receptors in glia. Neuroscientist. 2007;13:28–37.
108. Bohmbach K, Schwarz MK, Schoch S, Henneberger C. The structural and functional evidence for vesicular release from astrocytes in situ. Brain Res Bull. 2018;136:65–75.
109. Harada K, Kamiya T, Tsuboi T. Gliotransmitter release from astrocytes: functional, developmental, and pathological implications in the brain. Front Neurosci. 2015;9:499.
110. Cornell-Bell AH, Finkbeiner SM, Cooper MS, Smith SJ. Glutamate induces calcium waves in cultured astrocytes: long-range glial signaling. Science. 1990;247:470–3.
111. Sperlagh B, Zsilla G, Baranyi M, Illes P, Vizi ES. Purinergic modulation of glutamate release under ischemic-like conditions in the hippocampus. Neuroscience. 2007;149:99–111.
112. Kofuji P, Araque A. Astrocytes and behavior. Annu Rev Neurosci. 2021;44:49–67.
113. O'Donovan B, Neugornet A, Neogi R, Xia M, Ortinski P. Cocaine experience induces functional adaptations in astrocytes: implications for synaptic plasticity in the nucleus accumbens shell. Addict Biol. 2021;26:e13042.
114. Scofield MD, Boger HA, Smith RJ, Li H, Haydon PG, Kalivas PW. Gq-DREADD selectively initiates glial glutamate release and inhibits cue-induced cocaine seeking. Biol Psychiatry. 2015;78:441–51.
115. Schousboe A. Transport and metabolism of glutamate and GABA in neurons are glial cells. Int Rev Neurobiol. 1981;22:1–45.
116. Baker DA, Xi ZX, Shen H, Swanson CJ, Kalivas PW. The origin and neuronal function of in vivo nonsynaptic glutamate. J Neurosci. 2002;22:9134–41.
117. Fischer-Smith KD, Houston AC, Rebec GV. Differential effects of cocaine access and withdrawal on glutamate type 1 transporter expression in rat nucleus accumbens core and shell. Neuroscience. 2012;210:333–9.
118. Knackstedt LA, LaRowe S, Mardikian P, Malcolm R, Upadhyaya H, Hedden S, Markou A, Kalivas PW. The role of cystine-glutamate exchange in nicotine dependence in rats and humans. Biol Psychiatry. 2009;65:841–5.
119. Pomierny-Chamiolo L, Miszkiel J, Frankowska M, Mizera J, Filip M. Neuroadaptive changes in metabotropic glutamate mGlu2/3R expression during different phases of cocaine addiction in rats. Pharmacol Rep. 2017;69:1073–81.
120. Baker DA, McFarland K, Lake RW, Shen H, Tang XC, Toda S, Kalivas PW. Neuroadaptations in cystine-glutamate exchange underlie cocaine relapse. Nat Neurosci. 2003;6:743–9.
121. Zhou W, Kalivas PW. N-acetylcysteine reduces extinction responding and induces enduring reductions in cue- and heroin-induced drug-seeking. Biol Psychiatry. 2008;63:338–40.
122. Moussawi K, Kalivas PW. Group II metabotropic glutamate receptors (mGlu2/3) in drug addiction. Eur J Pharmacol. 2010;639:115–22.
123. Roberts-Wolfe DJ, Kalivas PW. Glutamate transporter GLT-1 as a therapeutic target for substance use disorders. CNS Neurol Disord Drug Targets. 2015;14:745–56.
124. Zhang LY, Zhou YQ, Yu ZP, Zhang XQ, Shi J, Shen HW. Restoring glutamate homeostasis in the nucleus accumbens via endocannabinoid-mimetic drug prevents relapse to cocaine seeking behavior in rats. Neuropsychopharmacology. 2021;46:970–81.
125. Piet R, Vargova L, Sykova E, Poulain DA, Oliet SH. Physiological contribution of the astrocytic environment of neurons to intersynaptic crosstalk. Proc Natl Acad Sci USA. 2004;101:2151–5.

126. Moran MM, McFarland K, Melendez RI, Kalivas PW, Seamans JK. Cystine/glutamate exchange regulates metabotropic glutamate receptor presynaptic inhibition of excitatory transmission and vulnerability to cocaine seeking. J Neurosci. 2005;25:6389–93.
127. McFarland K, Lapish CC, Kalivas PW. Prefrontal glutamate release into the core of the nucleus accumbens mediates cocaine-induced reinstatement of drug-seeking behavior. J Neurosci. 2003;23:3531–7.
128. Henneberger C, et al. LTP induction boosts glutamate spillover by driving withdrawal of perisynaptic astroglia. Neuron. 2020;108:919–936.e911.
129. Madayag A, Lobner D, Kau KS, Mantsch JR, Abdulhameed O, Hearing M, Grier MD, Baker DA. Repeated N-acetylcysteine administration alters plasticity-dependent effects of cocaine. J Neurosci. 2007;27:13968–76.
130. Reissner KJ, Gipson CD, Tran PK, Knackstedt LA, Scofield MD, Kalivas PW. Glutamate transporter GLT-1 mediates N-acetylcysteine inhibition of cocaine reinstatement. Addict Biol. 2015;20:316–23.
131. Erickson EK, DaCosta AJ, Mason SC, Blednov YA, Mayfield RD, Harris RA. Cortical astrocytes regulate ethanol consumption and intoxication in mice. Neuropsychopharmacology. 2021;46:500–8.
132. Christopherson KS, Ullian EM, Stokes CC, Mullowney CE, Hell JW, Agah A, Lawler J, Mosher DF, Bornstein P, Barres BA. Thrombospondins are astrocyte-secreted proteins that promote CNS synaptogenesis. Cell. 2005;120:421–33.
133. Eroglu C, Allen NJ, Susman MW, O'Rourke NA, Park CY, Ozkan E, Chakraborty C, Mulinyawe SB, Annis DS, Huberman AD, Green EM, Lawler J, Dolmetsch R, Garcia KC, Smith SJ, Luo ZD, Rosenthal A, Mosher DF, Barres BA. Gabapentin receptor alpha2delta-1 is a neuronal thrombospondin receptor responsible for excitatory CNS synaptogenesis. Cell. 2009;139:380–92.
134. Hoffman JR, Dixit VM, O'Shea KS. Expression of thrombospondin in the adult nervous system. J Comp Neurol. 1994;340:126–39.
135. Wang J, Li KL, Shukla A, Beroun A, Ishikawa M, Huang X, Wang Y, Wang YQ, Yang Y, Bastola ND, Huang HH, Kramer LE, Chao T, Huang YH, Sesack SR, Nestler EJ, Schluter OM, Dong Y. Cocaine triggers astrocyte-mediated synaptogenesis. Biol Psychiatry. 2021;89:386–97.
136. Nagai J, Rajbhandari AK, Gangwani MR, Hachisuka A, Coppola G, Masmanidis SC, Fanselow MS, Khakh BS. Hyperactivity with disrupted attention by activation of an astrocyte synaptogenic cue. Cell. 2019;177:1280–1292 e1220.
137. Dong Y, Nestler EJ. The neural rejuvenation hypothesis of cocaine addiction. Trends Pharmacol Sci. 2014;35:374–83.
138. Brown TE, Lee BR, Mu P, Ferguson D, Dietz D, Ohnishi YN, Lin Y, Suska A, Ishikawa M, Huang YH, Shen H, Kalivas PW, Sorg BA, Zukin RS, Nestler EJ, Dong Y, Schluter OM. A silent synapse-based mechanism for cocaine-induced locomotor sensitization. J Neurosci. 2011;31:8163–74.
139. Huang YH, Lin Y, Mu P, Lee BR, Brown TE, Wayman G, Marie H, Liu W, Yan Z, Sorg BA, Schluter OM, Zukin RS, Dong Y. In vivo cocaine experience generates silent synapses. Neuron. 2009;63:40–7.
140. Wright WJ, Dong Y. Silent synapses in cocaine-associated memory and beyond. J Neurosci. 2021;41:9275–85.
141. Wright WJ, Graziane NM, Neumann PA, Hamilton PJ, Cates HM, Fuerst L, Spenceley A, MacKinnon-Booth N, Iyer K, Huang YH, Shaham Y, Schluter OM, Nestler EJ, Dong Y. Silent synapses dictate cocaine memory destabilization and reconsolidation. Nat Neurosci. 2020;23:32–46.
142. Risher WC, Kim N, Koh S, Choi JE, Mitev P, Spence EF, Pilaz LJ, Wang D, Feng G, Silver DL, Soderling SH, Yin HH, Eroglu C. Thrombospondin receptor alpha2delta-1 promotes synaptogenesis and spinogenesis via postsynaptic Rac1. J Cell Biol. 2018;217:3747–65.

143. Spencer S, Brown RM, Quintero GC, Kupchik YM, Thomas CA, Reissner KJ, Kalivas PW. alpha2delta-1 signaling in nucleus accumbens is necessary for cocaine-induced relapse. J Neurosci. 2014;34:8605–11.
144. Jin D, Chen H, Chen SR, Pan HL. alpha2delta-1 protein drives opioid-induced conditioned reward and synaptic NMDA receptor hyperactivity in the nucleus accumbens. J Neurochem. 2023;164:143–57.
145. Stan AD, Ghose S, Gao XM, Roberts RC, Lewis-Amezcua K, Hatanpaa KJ, Tamminga CA. Human postmortem tissue: what quality markers matter? Brain Res. 2006;1123:1–11.
146. Chehimi SN, Crist RC, Reiner BC. Unraveling psychiatric disorders through neural single-cell transcriptomics approaches. Genes (Basel). 2023;14:771.
147. Lehrmann E, Freed WJ. Transcriptional correlates of human substance use. Ann N Y Acad Sci. 2008;1139:34–42.
148. Goldman D, Oroszi G, Ducci F. The genetics of addictions: uncovering the genes. Nat Rev Genet. 2005;6:521–32.
149. Brady KT, Sinha R. Co-occurring mental and substance use disorders: the neurobiological effects of chronic stress. Am J Psychiatry. 2005;162:1483–93.
150. Dube SR, Felitti VJ, Dong M, Chapman DP, Giles WH, Anda RF. Childhood abuse, neglect, and household dysfunction and the risk of illicit drug use: the adverse childhood experiences study. Pediatrics. 2003;111:564–72.
151. Felitti VJ, Anda RF, Nordenberg D, Williamson DF, Spitz AM, Edwards V, Koss MP, Marks JS. Relationship of childhood abuse and household dysfunction to many of the leading causes of death in adults. The Adverse Childhood Experiences (ACE) Study. Am J Prev Med. 1998;14:245–58.
152. Han B, Compton WM, Blanco C, Colpe LJ. Prevalence, treatment, and unmet treatment needs of US adults with mental health and substance use disorders. Health Aff (Millwood). 2017;36:1739–47.
153. Regier DA, Farmer ME, Rae DS, Locke BZ, Keith SJ, Judd LL, Goodwin FK. Comorbidity of mental disorders with alcohol and other drug abuse. Results from the Epidemiologic Catchment Area (ECA) Study. JAMA. 1990;264:2511–8.
154. Ferrer-Alcon M, Garcia-Sevilla JA, Jaquet PE, La Harpe R, Riederer BM, Walzer C, Guimon J. Regulation of nonphosphorylated and phosphorylated forms of neurofilament proteins in the prefrontal cortex of human opioid addicts. J Neurosci Res. 2000;61:338–49.
155. Weber M, Scherf N, Kahl T, Braumann UD, Scheibe P, Kuska JP, Bayer R, Buttner A, Franke H. Quantitative analysis of astrogliosis in drug-dependent humans. Brain Res. 2013;1500:72–87.
156. Kitamura O, Takeichi T, Wang EL, Tokunaga I, Ishigami A, Kubo S. Microglial and astrocytic changes in the striatum of methamphetamine abusers. Leg Med (Tokyo). 2010;12:57–62.
157. Mahmoudiasl GR, Abbaszadeh HA, Rezaei-Tavirani M, Abdollahifar MA, Sadeghi Y, Khoramgah MS, Niknazar S, Darabi S. Postmortem study of molecular and histological changes in the CA1 hippocampal region of chronic methamphetamine user. Iran J Pharm Res. 2019;18:2067–82.
158. Zhang Z, Gong Q, Feng X, Zhang D, Quan L. Astrocytic clasmatodendrosis in the cerebral cortex of methamphetamine abusers. Forensic Sci Res. 2017;2:139–44.
159. Buttner A, Weis S. Neuropathological alterations in drug abusers: the involvement of neurons, glial, and vascular systems. Forensic Sci Med Pathol. 2006;2:115–26.
160. Brenner E, Tiwari GR, Kapoor M, Liu Y, Brock A, Mayfield RD. Single cell transcriptome profiling of the human alcohol-dependent brain. Hum Mol Genet. 2020;29:1144–53.
161. Miguel-Hidalgo JJ, Rajkowska G. Comparison of prefrontal cell pathology between depression and alcohol dependence. J Psychiatr Res. 2003;37:411–20.
162. Miguel-Hidalgo JJ, Overholser JC, Meltzer HY, Stockmeier CA, Rajkowska G. Reduced glial and neuronal packing density in the orbitofrontal cortex in alcohol dependence and its relationship with suicide and duration of alcohol dependence. Alcohol Clin Exp Res. 2006;30:1845–55.

163. Korbo L. Glial cell loss in the hippocampus of alcoholics. Alcohol Clin Exp Res. 1999;23:164–8.
164. Berk M, Malhi GS, Gray LJ, Dean OM. The promise of N-acetylcysteine in neuropsychiatry. Trends Pharmacol Sci. 2013;34:167–77.
165. LaRowe SD, Myrick H, Hedden S, Mardikian P, Saladin M, McRae A, Brady K, Kalivas PW, Malcolm R. Is cocaine desire reduced by N-acetylcysteine? Am J Psychiatry. 2007;164:1115–7.
166. Vocci F, Ling W. Medications development: successes and challenges. Pharmacol Ther. 2005;108:94–108.
167. Back SE, McCauley JL, Korte KJ, Gros DF, Leavitt V, Gray KM, Hamner MB, DeSantis SM, Malcolm R, Brady KT, Kalivas PW. A double-blind, randomized, controlled pilot trial of N-acetylcysteine in veterans with posttraumatic stress disorder and substance use disorders. J Clin Psychiatry. 2016;77:e1439–46.
168. Mousavi SG, Sharbafchi MR, Salehi M, Peykanpour M, Karimian Sichani N, Maracy M. The efficacy of N-acetylcysteine in the treatment of methamphetamine dependence: a double-blind controlled, crossover study. Arch Iran Med. 2015;18:28–33.
169. Gray KM, Carpenter MJ, Baker NL, DeSantis SM, Kryway E, Hartwell KJ, McRae-Clark AL, Brady KT. A double-blind randomized controlled trial of N-acetylcysteine in cannabis-dependent adolescents. Am J Psychiatry. 2012;169:805–12.
170. Amen SL, Piacentine LB, Ahmad ME, Li SJ, Mantsch JR, Risinger RC, Baker DA. Repeated N-acetyl cysteine reduces cocaine seeking in rodents and craving in cocaine-dependent humans. Neuropsychopharmacology. 2011;36:871–8.
171. LaRowe SD, Kalivas PW, Nicholas JS, Randall PK, Mardikian PN, Malcolm RJ. A double-blind placebo-controlled trial of N-acetylcysteine in the treatment of cocaine dependence. Am J Addict. 2013;22:443–52.

Astrocytes in Pain Perception: A Systems Neuroscience Approach

Angélica Rivera-Villaseñor, Frida Higinio-Rodríguez, and Mónica López-Hidalgo

Abstract Astrocytes play an active role in the function of the brain integrating neuronal activity and regulating back neuronal dynamic. They have recently emerged as active contributors of brain's emergent properties such as perceptions. Here, we analyzed the role of astrocytes in pain perception from the lens of systems neuroscience, and we do this by analyzing how astrocytes encode nociceptive information within brain processing areas and how they are key regulators of the internal state that determines pain perception. Specifically, we discuss the dynamic interactions between astrocytes and neuromodulators, such as noradrenaline, highlighting their role in shaping the level of activation of the neuronal ensemble, thereby influencing the experience of pain. Also, we will discuss the possible implications of an "Astro-NeuroMatrix" in the integration of pain across sensory, affective, and cognitive dimensions of pain perception.

Keywords Astrocytes · Pain · Perception · Astro-NeuroMatrix · Systems neuroscience

Abbreviations

ACC	Anterior cingulate cortex
AMY	Amygdala
ATP	Adenosine triphosphate

A. Rivera-Villaseñor · F. Higinio-Rodríguez
Escuela Nacional de Estudios Superiores, Universidad Nacional Autónoma de México, Queretaro, Qro., Mexico

Instituto de Neurobiología, Universidad Nacional Autónoma de México, Queretaro, Qro., Mexico

M. López-Hidalgo (✉)
Escuela Nacional de Estudios Superiores, Universidad Nacional Autónoma de México, Queretaro, Qro., Mexico
e-mail: lopezhidalgo@unam.mx

E. Blanco-Suarez, I. Farhy-Tselnicker (eds.), *Astrocyte-Neuron Interactions in Health and Disease*, Advances in Neurobiology 39,
https://doi.org/10.1007/978-3-031-64839-7_8

Cav1.2	Calcium channel, voltage-dependent, L-type, alpha 1C subunit
CeM	Center-medial nuclei of the amygdala
Cx	Connexin
DST	Dynamic system theory
GABA	γ-Aminobutyric acid
Hip	Hippocampus
IC	Insular cortex
PFC	Prefrontal cortex
PO	Posterior nuclei
S1	Primary somatosensory cortex
S2	Secondary somatosensory cortex
Thal	Thalamus
TRPA1	Transient receptor potential cation channel, subfamily A, member 1
VPL	Ventral posterolateral nuclei

1 Introduction

Pain is a highly variable and complex experience that has aroused the interest of researchers across disciplines for centuries. It can serve as an alarming signal to protect us from something harmful when acute but can be a pathological feature of the brain in chronic conditions [1]. There are many classifications based on etiology, duration, and neurophysiological mechanisms that provide valuable insights into the diverse manifestations of pain and guide personalized treatment [2]. Despite this, pain affects millions worldwide, leading to significant morbidity, reduced quality of life, and high economic costs [3].

We are used to considering pain as a purely sensory phenomenon; however, pain is a big disrupter of brain homeostasis, thereby initiating neuronal, hormonal, and behavioral changes to reinstate it. Also, pain has a reciprocal relationship with affective states, and emotional experiences can either decrease or exacerbate pain, depending in part on the personal and social contexts. On the other side, pain can evoke complex emotional responses that can extend into the domain of suffering characterized by distress and existential threat [4, 5]. Rather than viewing pain as a mere summation of its components, it requires conceptualizing it as a dynamic system, shaped by the simultaneous activity of various neural units across cortical and subcortical structures [6]. This dynamic interplay involves the integration of sensory inputs with individual internal states, which evolve and manifest through observable behaviors.

Central to this understanding is the role of astrocytes, non-neuronal cells traditionally relegated to supportive roles in brain function, that are now recognized as orchestrators of several brain functions. Astrocytes encode nociceptive information in several sensory structures, and they modulate neural activity and regulate the brain's internal state [7, 8]. Astrocytes contribute to the functional integration of pain by forming interconnected networks across brain regions implicated in pain processing. These astrocytic networks synchronize neural activity, shaping neuronal oscillations [9–11] and facilitating the binding of diverse sensory, emotional, and cognitive features of pain. Such dynamic interactions underscore the role of astrocytes as integral mediators of pain perception and processing and urge the necessity to take them into account to have a comprehensive view of pain perception.

2 Pain: An Emergent Property of the Brain Dynamics

Sherrington defined nociception as the sensation produced by the activation of receptors (nociceptors) by noxious stimuli (mechanical, thermal, or chemical) that can potentially damage tissue [12]. The conscious perception of nociception is pain, and although both processes can relate, they are also independent. Nociception can lead to brain responses without causing pain [13], and pain can occur without a nociceptive stimulus [14]. Pain emerges from integrating sensory features with memories and emotional and cognitive context that results in a coherent unpleasant experience [15]. For example, many of us will agree that someone introducing a needle into our skin would be painful; however, if we ask people if getting a tattoo (that uses needles) is painful, you would have a less unanimous answer, and in fact, many people would agree it is something pleasant, although, the skin broke, an inflammatory response was triggered, nociceptors were activated, and many pain processing brain areas responded. This means that the behavioral response to a nociceptive stimulus depends on an individual experience that varies in a particular situation.

If the consciousness of an experience determines the perception, then the question is how and at what point the pain is created [16]. Although these remain open questions, pain is an emergent property of the brain that depends on the simultaneous activity of several units of processing (neuronal ensembles, networks, etc.), located in cortical and subcortical structures. Then, pain cannot be explained by the sum of its components; instead, it is required to understand it as a dynamic system. Then, the perception of pain depends on the integration of a stimulus (U) with an individual internal state (X). This is not static, but changes over time ($\mathrm{d}X/\mathrm{d}t$) and is manifested through behavior [17].

$$\frac{\mathrm{d}X}{\mathrm{d}t} = f(X) + U$$

To understand how this dynamic system (the brain) creates pain, it is necessary to analyze the contribution, intermingling and interconnections of all the elements, including glial cells such as astrocytes.

2.1 Astrocytes Encode Nociceptive Information

Cutaneous and nociceptive stimuli are encoded in the somatosensory system. A-delta and C-fibers transmit nociceptive information [18] from the periphery to the spinal cord. Here, they contact superficial nociceptive specific (laminae I–II) or wide dynamic range neurons located in the deeper lamina (III–VI) where non-nociceptive and nociceptive stimulations converge [19, 20]. Spinal neurons encode noxious stimuli with fast calcium transients, whose amplitude and frequency correlate with the intensity and duration of the stimulus [21, 22]. C-fibers respond to repetitive activity with a short-lasting increase in the discharge [23, 24] that induces neuroplasticity, leading to persistent pain typical in chronic cases. Nociceptive information is sent then to the thalamus (ventral posterolateral nuclei, VPL; posterior nuclei, PO; intralaminar nuclei, IN) [25–27] and from there, in parallel, to cortical areas including but not limited to primary (S1) and secondary (S2) somatosensory cortex, insular, anterior cingulate cortex (ACC), prefrontal cortex (PFC), amygdala, and hippocampus [28–31]. All these are involved in encoding sensorial, cognitive, and affective components of pain [32]. The emergence of pain depends on the coordinated activity of this large-scale network described as a pain matrix or neuromatrix [33, 34].

Nociceptive-evoked calcium activity in astrocytes has not been evaluated; instead, high-intensity mechanical stimulation or acute pain models (noxious stimuli) have been used as an approach to understand how astrocytes encode nociceptive information. This is relevant because there is high variability in the nociceptive threshold between subjects, a nociceptive stimulus for someone could not be nociceptive for someone else. Also, high-intensity mechanical stimulation activates not only nociceptive fibers but also non-nociceptive (cutaneous) fibers (if you are pinched by someone you would maybe feel pain, at the same time, you would also notice the texture or the temperature of the object). Then, to understand how astrocytes encode nociceptive information, selective activation of nociceptors would be required. Now that this point is clear; here, we would use noxious stimulus to infer how astrocytes encode nociceptive responses.

In the spinal cord, noxious but not cutaneous stimulation of the tail elicited astrocyte calcium responses. In an acute mouse pain model, dorsal horn astrocytes responded to formalin and capsaicin administration with calcium responses [35, 36], showing that astrocytes respond to a nociceptive stimulus (Fig. 1). In the somatosensory cortex, electrical and mechanical stimulation of the hindlimb evoked calcium increases in astrocytes [26, 37, 40, 41]. Astrocyte activation has also been observed in other pain processing areas such as the hippocampus, amygdala, and prefrontal cortex (Fig. 1) [7, 38, 42]. Astrocyte calcium signals evoked by noxious stimuli encode the frequency and intensity of the stimulus [26, 40]. However, the onset and peak of the astrocyte calcium response are delayed in comparison with neurons (0.5 s neurons and 2.5 s to astrocytes in the spinal cord; [22]), limiting the representation of temporal patterns of neuronal activity in individual calcium events [43]. Instead, they are summed and integrated over time (and with the individual's internal state) to provide a coherent response. Astrocytes also encode nociceptive stimulus at a population level, by recruiting astrocytes in non-stimulated areas eliciting large-scale calcium responses. For example, in the spinal cord, astrocyte

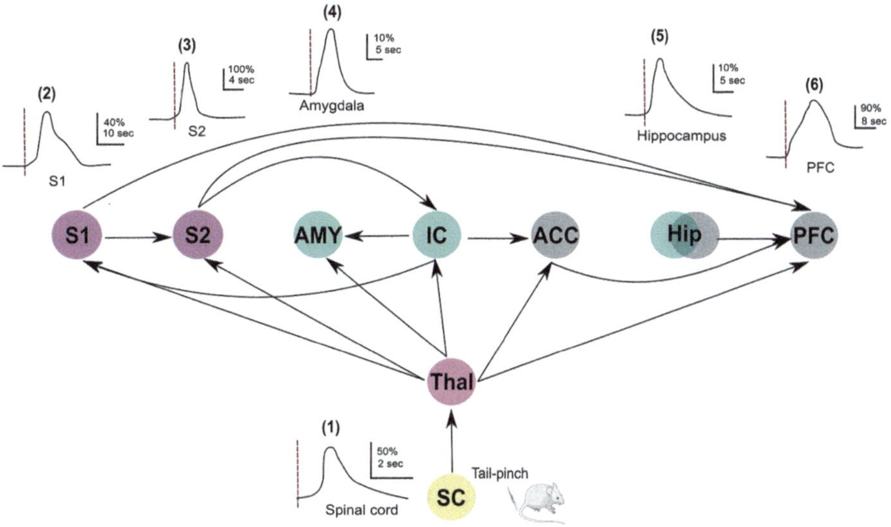

Fig. 1 Astrocyte calcium activity induced by noxious stimuli is observed in diverse pain-related neural circuits. Parallel processing of pain components in the spinal cord (SC), thalamus (Thal), primary somatosensory cortex (S1); secondary somatosensory cortex (S2), amygdala (AMY), insular cortex (IC), anterior cingulate cortex (ACC), hippocampus (Hip), and prefrontal cortex (PFC). The inserts showed astrocyte calcium activity following nociceptive stimulation. The onset of the stimuli is indicated by dashed red lines. (Traces modified from (1) Sekiguchi et al. [22], (2) Bekar et al. [37], (3, 6) Higinio-Rodríguez et al. [7], (4) Suthard et al. [38], (5) Qin et al. [39])

activity elicited simultaneous calcium activity in the ipsilateral and contralateral sides of the dorsal horn, which extends rostral for at least three spinal segments (L2–L5) [22]. Similarly, in S1, calcium activity was observed in topographic areas that were not directly affected by the stimulus [37].

The functional role of astrocyte calcium activity evoked by noxious stimulus on nociception is not clear [44]. On one side, astrocytes have an antinociceptive role in the spinal cord mediating the inhibition of nociceptive fibers due to cutaneous stimulation (non-noxious) (like when you rub yourself after you hit) [44]. On the other side, astrocytes control the dynamic range of cortical network activity optimizing the sensibility to sensory inputs [26, 41]. Similarly, thalamic astrocytes modulate tactile neuronal acuity through GABA release, inhibiting thalamic projecting neurons and adjusting their sensorial dynamic range [45, 46].

2.2 *Astrocytes: Orchestrators of Internal State Regulation*

The internal state refers to the basal condition of an organism that is dynamically established by the communication between the brain and the body [47]. Depending on this state, sensory information can be represented as pain and manifested through behavior. For example, if you step on a nail, you probably will stop and take it out as soon as possible. However, if you were running away, you will likely continue running until you are safe before taking out the nail. It is even possible that you may not realize you have the nail until your internal state (now in stress) is back to a resting state (safe situation). It is important to understand that the perception of pain intensity and its manifestations through behavior depends on the level of the internal state. Several situations in daily life illustrate this. For example, when high levels of stress and arousal occur in wars, during sports activities, or in street fights, the pain perception decreases favoring survival.

Neurotransmitters like glutamate and GABA regulate local circuits, while neuromodulator systems use long-range projections to coordinate distributed brain-wide neural activity setting the level of the internal state of local circuits [48, 49]. Neuromodulation systems respond to noxious stimuli (e.g., footshock or irritant injection). Thereby, nociception activates dopaminergic neurons from the ventral tegmental area [50, 51], cholinergic neurons from the medial septal region [52], and noradrenergic neurons from the locus coeruleus [53]. This last system has a dual role in pain perception. It decreases the nociception at the spinal cord through descending pathways and regulates the response to pain such as aversion and anxiety-like behavior through the activity of the ascending projections (to cortical areas) [54].

The responsiveness of astrocytes to sensory stimuli varies depending on the up/down states or the activity to locus coeruleus as occurs to cortical astrocytes from S1 and visual cortex [37, 55–57]. This situated astrocytes as local connector hubs [58]. According to dynamic system theory (DST), a connector hub has connections diversely distributed across communities, they are wired to integrate and process information and are usually implicated in diverse cognitive tasks [59]. In this sense, astrocytes can integrate local inputs and top-down and bottom-up afferents along with the activity of neuromodulator systems [60–63], pointing to a central role in perception processes such as pain.

Maintaining a loop between the brain and the body relies on interoception (the perception of the internal body and brain signals) and the pathways (cranial nerves, hormones, etc.) to communicate both systems. The interoception requires internal references that act as attractors (an attractor in DST refers to a state toward the system moves). The process that keeps these references by adjusting the cellular or systems activity in response to changing external conditions is homeostasis [64, 65]. Astrocytes are key homeostatic cells, and they regulate brain interstitial pH, the permeability of the blood-brain barrier, and metabolite transport [66]. They also maintain adequate neurotransmitter concentrations [67] and rigorously control the extracellular levels of ions such as K^+, Na^+, and Ca^{+2}. When these mechanisms are dysregulated, they can produce changes in the excitability of individual neurons or neural circuits [68]. For example, without the proper function of astrocytes, sustained neuronal activity could lead to a local increase of extracellular K^+ and neuronal hyperexcitability as occurs in chronic pain. Abnormal K^+ channels dynamic in astrocytes increase the excitability of the neurons in the somatosensory system [8]. This could lead to central sensitization due to repetitive stimulation of C-fibers.

2.3 *Astrocyte Participation in the Functional Integration of Pain*

The emergence of pain is dynamically adjusted to the changing demands of everyday life. This flexibility requires to sum up and integration of the information processed by the units located across the pain matrix [6]. These include binding the different characteristics of an object and its associations with the context; for example, in the tattoo example, the information of the needle crossing the skin is bound with visual information that provides spatial context and memories. All this generates an experience that, depending on the affective connotation, if it is unpleasant, will be (probably) experimented with and described as pain. The functional connections and interaction between all these brain networks occur during the "binding process" [6].

The mechanisms underlying the binding process are still unclear. However, it requires the synchronic activity of large-scale networks [69–72]. A network can be defined as spatially distributed, but functionally linked brain regions that are continuously communicating with each other [73]. Since the establishment of the "neuronal theory" in the 1880s by Ramon y Cajal, individual neurons were defined as the anatomical and functional units of the nervous system [74]. The connectivity between brain areas is made by their projections (axons). Based on this, first-order networks refer to the direct axonal connection between two regions (A–B) (e.g., the network between the thalamus and S1) (Fig. 2b), whereas second-order networks do not have a direct connection between A and B; instead, it is indirectly connected through a third region (C) (A–C–B) (Fig. 2c).

In parallel to the "neuron theory," Camillo Golgi proposed the "reticular theory" [78] that viewed the nervous system as a singular continuing network system (syncytium) [79]. Although the neuronal theory prevailed and the reticulum idea vanished, the discovery of gap junctions between astrocytes led to a reconsideration of the reticular theory, demonstrating that these cells can also connected by continuity through gap junctions [80]. It is then possible a third model of networks as an alternative for communication between A and B. This is to be through a reticular network (a path) of astrocytes coupled by gap junctions forming a first-order neuron-astrocyte network (Fig. 2d).

Astrocyte networks are formed by connexins, primarily by Cx43 and Cx30 [10, 11, 81]. These connexins are expressed in many brain regions of the pain neuromatrix where they have been associated with the coordination of neuronal circuits [11, 82–84] and in which expression and activity increase during pain states [85]. Within the astrocyte networks, calcium can diffuse and recruit local and distant astrocytes [86]. Calcium waves have been reported within the neuromatrix, in the spinal cord [87, 88], thalamus, somatosensory cortex [89], cingulate cortex [90], prefrontal cortex [91], and amygdala [92]. Calcium diffusion in response to nociceptive stimuli has also been reported. Putting it all together, it is feasible to propose that astrocyte paths are important for the "binding process" and are involved in pain perception by synchronizing large-scale networks.

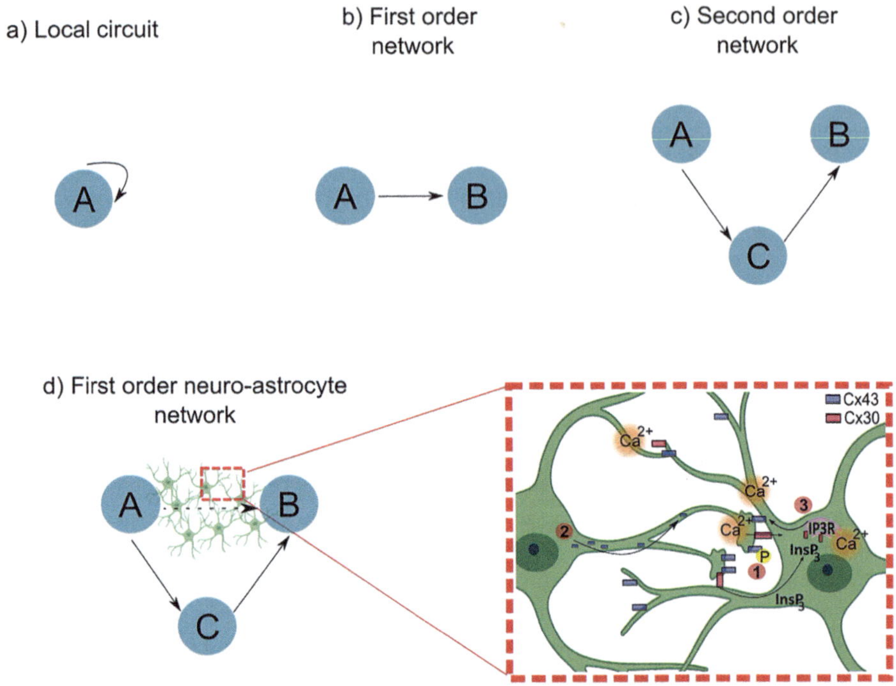

Fig. 2 Functional organization of neuronal and astrocyte networks. (**a**) The communication of neighboring neurons from the same region is a local circuit. (**b**) First-order networks, direct communication between regions A and B; (**c**) second-order networks, an indirect connection between A and B through a direct connection with C. (**d**) First-order neuro-astrocyte networks by which A and B can be directly connected by an astrocyte network coupled by gap junctions (small green cells). Zoom of panel **d**, the phosphorylation of connexin [75], increases its expression in the membrane of the astrocytes [76], favoring the communication between astrocytes through calcium ions, and inositol 3 phosphate ($InsP_3$) establishes a functional network of the astrocytes [77]

The synchronous activity of large-scale networks induces neuronal oscillations that can be divided into bands according to their frequency in alpha (8–13 Hz), gamma (30–100 Hz), and theta (4–8 Hz) oscillations. In rodents, noxious stimulus intensity is linearly associated with an increase in gamma- and theta-band oscillations in the somatosensory cortex. This precedes the activation of other pain-related areas such as the cingulate cortex, the insula, the amygdala, and the prefrontal cortex [93–97], suggesting a recruiting of the neuromatrix for binding the painful information. Gamma-band activity is also important for maintaining the information (working memory) to quickly associate sensory, emotional, and cognitive features of pain in the prefrontal cortex [98, 99].

Astrocytes contribute to shaping gamma oscillations by enhancing inhibitory transmission [100, 101]. Sensory stimuli elicit a surge of neuronal network activity in the gamma range (30–50 Hz) followed by astrocyte activity that dampens the steady-state gamma activity [41]. Another mechanism of neural integration is the coupling of different frequency bands; in this sense, astrocyte activity enhances the coupling between gamma and theta [102].

2.4 *Pain, Suffering, and Astrocytes*

Pain can lead to suffering, and it occurs when a human recognizes that their wholeness as a person is under threat or injured [4, 5, 103]. Suffering is a multidimensional experience characterized by severe or specific distress, anger, fear, frustration, anxiety, and depression. It requires the cognitive (insula), emotional (cingulate cortex), and autonomic (cingulate cortex and insula) components of pain. We cannot sustain that suffering is characteristic of humans; however, the analysis of this complex experience cannot be directly evaluated in animals as it requires verbal reporting limiting a full understanding of the mechanisms underlying this process.

Pain experiences must be stored (fear memory) to retrieve them in potential risk situations to protect us [104]. By this, a painful situation can lead to a life-long change in how pain is experienced. Fear can both suppress and enhance pain. When in danger, fear suppresses pain perception as a survival mechanism. On the other side, fearful anticipation of imminent pain could enhance its perception. Local circuits in the prefrontal cortex [105] and hippocampus [71] representing nociception have been associated with fear memory and with maintaining chronic pain in states such as those occur in hyperalgesia and allodynia.

One important aspect of fear memory consolidation is attention and arousal, which improves signal-to-noise ratio through signal amplification, filtering, and enhanced gain control [106–108]. Pain is by nature a salient stimulus to pay attention to it and to increase the level of arousal and awareness [109]. Astrocytes play also a role in this by adjusting the signal-to-noise ratio through the regulation of extracellular K^+ levels [110]. They can also release gliotransmitters like GABA, ATP, or adenosine [26, 41, 46]. Furthermore, astrocytes from the prefrontal cortex and sensory cortex [37, 111–113] respond to noradrenaline release from the locus coeruleus with large-scale astrocyte activity, which suggests that they could be mediating noradrenaline effects on arousal and attention [114, 115].

Astrocytes are involved in several behavioral expressions of suffering such as fear memory. The activation of hippocampal astrocytes decreases fear-related anxiety behavior and contextual fear memory process that is dependent on cholinergic transmission [116, 117]. In the anterior cingulate cortex, aversive behavior is linearly dependent on the activation of the astrocytes [118–120]. This depends on the adrenergic signaling acting on astrocytes in response to nociceptive stimuli and plays a key role in mediating the consolidation of pain-related aversion and memory formation [121–123]. Aligned with this, Martin-Fernandez et al. [124] showed that astrocytic activation decreased the firing rate of neurons in the amygdala (CeM) reducing fear expression in a fear-conditioning paradigm [124].

Not all suffering is caused by pain, but in most cultures, emotional suffering is described using a language of pain. This applies to their suffering such as a "heartbroken" or seeing someone else suffering in grief by losing a loved one. In this situation, you still can experience pain for them (emotional empathy) "I feel what you feel." Is this language used just by chance? The answer is no. The paradox of defining emotions is challenging in objective scientific terms. Most researchers agree that emotions include certain expressive behaviors that are associated with internal brain states that we, as humans, subjectively experience as "feelings" [125] as Millan Kundera points: "For there is nothing heavier than compassion. Not even one's pain weighs so heavy as the pain one feels with someone, for someone, a pain intensified by the imagination and prolonged by a hundred echoes" (Milan Kundera, *The Unbearable Lightness of Being*). The concept of empathy involves both emotional and cognitive components [75, 126]. Emotional empathy, characterized by "feeling what others feel," is believed to be mediated by the parafascicular/mediodorsal thalamic nuclei and the anterior cingulate cortex [127]. During observational fear learning (a prosocial behavior) [128], neuronal activity of the anterior cingulate cortex was increased and synchronized (in the theta band) with the lateral amygdala. The interference of neuronal activity in the cingulate cortex (by blocking Cav1.2) impairs fear learning demonstrating its involvement in affective pain [129].

Cognitive empathy, involving "understanding what others feel," is associated with the activity of the anterior cingulate cortex, prelimbic, and infralimbic cortex [130, 131]. In 2004, Singer and colleagues showed that empathy for pain activates the bilateral anterior insula, the anterior cingulate cortex, and other areas associated with the affective component of pain [132, 133]. In particular, the anterior cingulate cortex encodes information about the affective state of others (essential for empathy), which is a function highly preserved in different species such as humans and rodents [134, 135]. Aligned with this, the cingulate cortex acts along with the nucleus accumbens in the social transfer of pain and analgesia in mice [136, 137], which reinforces this prosocial behavior, favoring the community instead of the individual [138]. Up to date, there is no information regarding the role of astrocytes in empathy; however, recently, it was shown that cerebellar astrocytes were involved in social interaction in adjusting Purkinje cell excitability and setting the tone of aggression [77]. Also, in a paradigm of social deprivation, astrocytes showed an increase in calcium activity compared to control mice, which was associated with an upregulation of astrocytic Ca^{2+} channel TRPA1. This increase was maladaptive because inhibition of TRPA1 restores hippocampal circuit function and rescues memory deficits after social deprivation [139].

3 Concluding Remarks

Considering astrocytes in pain perception is essential to understanding how the brain processes sensory information and generates subjective experiences. Astrocytes play a critical role in encoding sensory signals, integrating information across neuronal networks, and regulating the excitability of sensory circuits. They also regulate the internal state, which determines the magnitude of pain and the context of the experience. Astrocytes establish networks coupled by gap junctions that can extend through several brain areas and be involved in integrating sensory, affective, and cognitive components. Astrocytes are also important regulators in the expression of pain in behaviors such as stress, fear, and memory that are associated with a subjective experience like suffering. Furthermore, they could also be involved in social behaviors such as empathy that is closely related to pain and helps us to live in communities, feeling the pain of others.

Modern neuroscience now urges us to consider astrocytes in systems neuroscience to have a better understanding of the interplay between cellular and network-level processes involved in emergent properties like pain perception.

Acknowledgments ARV and FHR are doctoral students from the Programa de Doctorado en Ciencias Biomédicas, Universidad Nacional Autónoma de México (UNAM), and received fellowships from CONAHCYT (785184 and 1086257), supported by UNAM, Dirección General de Asuntos del Personal Académico, Universidad Nacional Autónoma de México (DGAPA, IA208022). We thank SP11 for the motivation and Dr. Rafael Olivares for technical assistance. Dr. Gerardo Rojas Piloni and Dr. Oswaldo Pérez provide critical feedback on this manuscript.

References

1. Swieboda P, Filip R, Prystupa A, Drozd M. Assessment of pain: types, mechanism and treatment. Ann Agric Environ Med. 2013;Spec no. 1:2–7.
2. Dowell D, Ragan KR, Jones CM, Baldwin GT, Chou R. CDC clinical practice guideline for prescribing opioids for pain — United States, 2022. MMWR Recomm Rep. 2022;71(RR-3):1–95. https://doi.org/10.15585/mmwr.rr7103a1.
3. Smith TJ, Hillner BE. The cost of pain. JAMA Netw Open. 2019;2(4):e191532. https://doi.org/10.1001/jamanetworkopen.2019.1532.
4. Cassell EJ. Teaching the fundamentals of primary care: a point of view. Milbank Q. 1995;73(3):373–405. PMID: 7659045
5. Cassell EJ. The phenomenon of suffering and its relationship to pain. In: Toombs SK, editor. Handbook of phenomenology and medicine, Philosophy and Medicine, vol. 68. Dordrecht: Springer; 2001. https://doi.org/10.1007/978-94-010-0536-4_20.
6. Feldman M. The value of uncertainty. Psychoanal Q. 2013;82(1):51–61. https://doi.org/10.1002/j.2167-4086.2013.00005.x.
7. Higinio-Rodríguez F, Rivera-Villaseñor A, Calero-Vargas I, López-Hidalgo M. From nociception to pain perception, possible implications of astrocytes. Front Cell Neurosci. 2022;16:972827. https://doi.org/10.3389/fncel.2022.972827. PMID: 36159392; PMCID: PMC9492445
8. Tsantoulas C, McMahon SB. Opening paths to novel analgesics: the role of potassium channels in chronic pain. Trends Neurosci. 2014;37(3):146–58. https://doi.org/10.1016/j.tins.2013.12.002. Epub 2014 Jan 21. PMID: 24461875; PMCID: PMC3945816
9. Oliveira JF, Araque A. Astrocyte regulation of neural circuit activity and network states. Glia. 2022;70(8):1455–66. https://doi.org/10.1002/glia.24178. Epub 2022 Apr 22. PMID: 35460131; PMCID: PMC9232995
10. Pannasch U, Rouach N. Emerging role for astroglial networks in information processing: from synapse to behavior. Trends Neurosci. 2013;36(7):405–17. https://doi.org/10.1016/j.tins.2013.04.004. Epub 2013 May 7. PMID: 23659852
11. Rouach N, Koulakoff A, Abudara V, Willecke K, Giaume C. Astroglial metabolic networks sustain hippocampal synaptic transmission. Science. 2008;322(5907):1551–5. https://doi.org/10.1126/science.1164022. PMID: 19056987

12. Sherrington CS. Textbook of physiology. Edinburgh: Churchill Livingstone; 1900.
13. Lee MC, Mouraux A, Iannetti GD. Characterizing the cortical activity through which pain emerges from nociception. J Neurosci. 2009;29(24):7909–16. https://doi.org/10.1523/JNEUROSCI.0014-09.2009. PMID: 19535602; PMCID: PMC6665631
14. Flor H, Nikolajsen L, Staehelin Jensen T. Phantom limb pain: a case of maladaptive CNS plasticity? Nat Rev Neurosci. 2006;7(11):873–81. https://doi.org/10.1038/nrn1991.
15. Peláez I, Martínez-Iñigo D, Barjola P, Cardoso S, Mercado F. Decreased pain perception by unconscious emotional pictures. Front Psychol. 2016;7:1636. https://doi.org/10.3389/fpsyg.2016.01636. Erratum in: Front Psychol 2018;9:1310. PMID: 27818642; PMCID: PMC5073127
16. Párraga JP, Castellanos A. A manifesto in defense of pain complexity: a critical review of essential insights in pain neuroscience. J Clin Med. 2023;12(22):7080. https://doi.org/10.3390/jcm12227080. PMID: 38002692; PMCID: PMC10672144
17. Betancourt A, Pérez O, Gámez J, Mendoza G, Merchant H. Amodal population clock in the primate medial premotor system for rhythmic tapping. Cell Rep. 2023;42(10):113234. https://doi.org/10.1016/j.celrep.2023.113234. Epub 2023 Oct 14. PMID: 37838944
18. Zimmerman A, Bai L, Ginty DD. The gentle touch receptors of mammalian skin. Science. 2014;346(6212):950–4. https://doi.org/10.1126/science.1254229. PMID: 25414303; PMCID: PMC4450345
19. Coghill R. Encoding of noxious information in the spinal cord. In: Gebhart GF, Schmidt RF, editors. Encyclopedia of pain. Berlin, Heidelberg: Springer; 2013. https://doi.org/10.1007/978-3-642-28753-4_1297.
20. Price D. Spinothalamic tract neurons, in deep dorsal horn. In: Schmidt R, Willis W, editors. Encyclopedia of pain. Berlin/Heidelberg: Springer; 2007. https://doi.org/10.1007/978-3-540-29805-2_4173.
21. Chisholm KI, Lo Re L, Polgár E, Gutierrez-Mecinas M, Todd AJ, McMahon SB. Encoding of cutaneous stimuli by lamina I projection neurons. Pain. 2021;162(9):2405–17. https://doi.org/10.1097/j.pain.0000000000002226. PMID: 33769365; PMCID: PMC8374708
22. Sekiguchi K, Shekhtmeyster P, Merten K, et al. Imaging large-scale cellular activity in spinal cord of freely behaving mice. Nat Commun. 2016;7:11450. https://doi.org/10.1038/ncomms11450.
23. Coghill RC. The distributed nociceptive system: a framework for understanding pain. Trends Neurosci. 2020;43(10):780–94. https://doi.org/10.1016/j.tins.2020.07.004. Epub 2020 Aug 13. PMID: 32800534; PMCID: PMC7530033
24. Hanai F, Yokota T. C-fiber responses of wide dynamic range neurons in the spinal dorsal horn of the cat. Pain Res. 1993;8(2):61–9). Japanese Association for the Study of Pain. https://doi.org/10.11154/pain.8.61.
25. Cover KK, Mathur BN. Rostral Intralaminar thalamus engagement in cognition and behavior. Front Behav Neurosci. 2021;15:652764. https://doi.org/10.3389/fnbeh.2021.652764. PMID: 33935663; PMCID: PMC8082140
26. Miguel-Quesada C, Zaforas M, Herrera-Pérez S, Lines J, Fernández-López E, Alonso-Calviño E, Ardaya M, Soria FN, Araque A, Aguilar J, Rosa JM. Astrocytes adjust the dynamic range of cortical network activity to control modality-specific sensory information processing. Cell Rep. 2023;42(8):112950. https://doi.org/10.1016/j.celrep.2023.112950. Epub 2023 Aug 5. PMID: 37543946
27. Yen CT, Lu PL. Thalamus and pain. Acta Anaesthesiol Taiwanica. 2013;51(2):73–80. https://doi.org/10.1016/j.aat.2013.06.011. Epub 2013 Aug 2. PMID: 23968658
28. Bushnell MC, Ceko M, Low LA. Cognitive and emotional control of pain and its disruption in chronic pain. Nat Rev Neurosci. 2013;14(7):502–11. https://doi.org/10.1038/nrn3516. Epub 30 May 2013. PMID: 23719569; PMCID: PMC4465351
29. Ong WY, Stohler CS, Herr DR. Role of the prefrontal cortex in pain processing. Mol Neurobiol. 2019;56(2):1137–66. https://doi.org/10.1007/s12035-018-1130-9. Epub 2018 Jun 6. PMID: 29876878; PMCID: PMC6400876

30. Oshiro Y, Quevedo AS, McHaffie JG, Kraft RA, Coghill RC. Brain mechanisms supporting discrimination of sensory features of pain: a new model. J Neurosci. 2009;29(47):14924–31. https://doi.org/10.1523/JNEUROSCI.5538-08.2009. PMID: 19940188; PMCID: PMC2823474
31. Xie Y, Huo F, Tang J. Cerebral cortex modulation of pain. Acta Pharmacol Sin. 2009;30:31–41. https://doi.org/10.1038/aps.2008.14.
32. Bornhövd K, Quante M, Glauche V, Bromm B, Weiller C, Büchel C. Painful stimuli evoke different stimulus-response functions in the amygdala, prefrontal, insula and somatosensory cortex: a single-trial fMRI study. Brain. 2002;125(Pt 6):1326–36. https://doi.org/10.1093/brain/awf137. PMID: 12023321
33. Melzack R. Pain – an overview. Acta Anaesthesiol Scand. 1999;43(9):880–4. https://doi.org/10.1034/j.1399-6576.1999.430903.x. PMID: 10522733
34. Tracey I, Mantyh PW. The cerebral signature for pain perception and its modulation. Neuron. 2007;55(3):377–91. https://doi.org/10.1016/j.neuron.2007.07.012.
35. Kohro Y, Matsuda T, Yoshihara K, Kohno K, Koga K, Katsuragi R, Oka T, Tashima R, Muneta S, Yamane T, Okada S, Momokino K, Furusho A, Hamase K, Oti T, Sakamoto H, Hayashida K, Kobayashi R, Horii T, Hatada I, Tozaki-Saitoh H, Mikoshiba K, Taylor V, Inoue K, Tsuda M. Spinal astrocytes in superficial laminae gate brainstem descending control of mechanosensory hypersensitivity. Nat Neurosci. 2020;23(11):1376–87. https://doi.org/10.1038/s41593-020-00713-4. Epub 2020 Oct 5. PMID: 33020652
36. Yoshihara K, Matsuda T, Kohro Y, Tozaki-Saitoh H, Inoue K, Tsuda M. Astrocytic Ca2+ responses in the spinal dorsal horn by noxious stimuli to the skin. J Pharmacol Sci. 2018;137(1):101–4. https://doi.org/10.1016/j.jphs.2018.04.007. Epub 2018 May 3. PMID: 29773517
37. Bekar LK, He W, Nedergaard M. Locus coeruleus alpha-adrenergic-mediated activation of cortical astrocytes in vivo. Cereb Cortex. 2008;18(12):2789–95. https://doi.org/10.1093/cercor/bhn040. Epub 2008 Mar 27. PMID: 18372288; PMCID: PMC2583159
38. Suthard RL, Senne RA, Buzharsky MD, Pyo AY, Dorst KE, Diep AH, Cole RH, Ramirez S. Basolateral amygdala astrocytes are engaged by the acquisition and expression of a contextual fear memory. J Neurosci. 2023;43(27):4997–5013. https://doi.org/10.1523/JNEUROSCI.1775-22.2023. Epub 2023 Jun 2. PMID: 37268419; PMCID: PMC10324998
39. Qin H, He W, Yang C, Li J, Jian T, Liang S, Chen T, Feng H, Chen X, Liao X, Zhang K. Monitoring astrocytic Ca2+ activity in freely behaving mice. Front Cell Neurosci. 2020;14:603095. https://doi.org/10.3389/fncel.2020.603095. PMID: 33343304; PMCID: PMC7744696
40. Gu X, Chen W, Volkow ND, Koretsky AP, Du C, Pan Y. Synchronized astrocytic Ca2+ responses in neurovascular coupling during somatosensory stimulation and for the resting state. Cell Rep. 2018;23(13):3878–90. https://doi.org/10.1016/j.celrep.2018.05.091. PMID: 29949771; PMCID: PMC7469112
41. Lines J, Martin ED, Kofuji P, Aguilar J, Araque A. Astrocytes modulate sensory-evoked neuronal network activity. Nat Commun. 2020;11(1):3689. https://doi.org/10.1038/s41467-020-17536-3. PMID: 32704144; PMCID: PMC7378834
42. Kanemaru K, Sekiya H, Xu M, Satoh K, Kitajima N, Yoshida K, Okubo Y, Sasaki T, Moritoh S, Hasuwa H, Mimura M, Horikawa K, Matsui K, Nagai T, Iino M, Tanaka KF. In vivo visualization of subtle, transient, and local activity of astrocytes using an ultrasensitive Ca(2+) indicator. Cell Rep. 2014;8(1):311–8. https://doi.org/10.1016/j.celrep.2014.05.056. Epub 2014 Jun 26. PMID: 24981861
43. Brockett AT, Kane GA, Monari PK, Briones BA, Vigneron PA, Barber GA, Bermudez A, Dieffenbach U, Kloth AD, Buschman TJ, Gould E. Evidence supporting a role for astrocytes in the regulation of cognitive flexibility and neuronal oscillations through the Ca2+ binding protein S100β. PLoS One. 2018;13(4):e0195726. https://doi.org/10.1371/journal.pone.0195726. PMID: 29664924; PMCID: PMC5903631

44. Xu Q, Ford NC, He S, Huang Q, Anderson M, Chen Z, Yang F, Crawford LK, Caterina MJ, Guan Y, Dong X. Astrocytes contribute to pain gating in the spinal cord. Sci Adv. 2021;7(45):eabi6287. https://doi.org/10.1126/sciadv.abi6287. Epub 2021 Nov 3. PMID: 34730998; PMCID: PMC8565904
45. Copeland CS, Wall TM, Sims RE, Neale SA, Nisenbaum E, Parri HR, Salt TE. Astrocytes modulate thalamic sensory processing via mGlu2 receptor activation. Neuropharmacology. 2017;121:100–10. https://doi.org/10.1016/j.neuropharm.2017.04.019. Epub 2017 Apr 14. PMID: 28416443; PMCID: PMC5480778
46. Kwak H, Koh W, Kim S, Song K, Shin JI, Lee JM, Lee EH, Bae JY, Ha GE, Oh JE, Park YM, Kim S, Feng J, Lee SE, Choi JW, Kim KH, Kim YS, Woo J, Lee D, Son T, Kwon SW, Park KD, Yoon BE, Lee J, Li Y, Lee H, Bae YC, Lee CJ, Cheong E. Astrocytes control sensory acuity via tonic inhibition in the thalamus. Neuron. 2020;108(4):691–706.e10. https://doi.org/10.1016/j.neuron.2020.08.013. Epub 2020 Sep 8. PMID: 32905785
47. Kanwal JK, Coddington E, Frazer R, Limbania D, Turner G, Davila KJ, Givens MA, Williams V, Datta SR, Wasserman S. Internal State: Dynamic, Interconnected Communication Loops Distributed Across Body, Brain, and Time. Integr Comp Biol. 2021;61(3):867–86. https://doi.org/10.1093/icb/icab101. PMID: 34115114; PMCID: PMC8623242
48. Brzosko Z, Mierau SB, Paulsen O. Neuromodulation of spike-timing-dependent plasticity: past, present, and future. Neuron. 2019;103(4):563–81. https://doi.org/10.1016/j.neuron.2019.05.041. PMID: 31437453
49. Nadim F, Bucher D. Neuromodulation of neurons and synapses. Curr Opin Neurobiol. 2014;29:48–56. https://doi.org/10.1016/j.conb.2014.05.003. Epub 2014 Jun 5. PMID: 24907657; PMCID: PMC4252488
50. Brischoux F, Chakraborty S, Brierley DI, Ungless MA. Phasic excitation of dopamine neurons in ventral VTA by noxious stimuli. Proc Natl Acad Sci USA. 2009;106(12):4894–9. https://doi.org/10.1073/pnas.0811507106. Epub 4 Mar 2009. PMID: 19261850; PMCID: PMC2660746
51. Gomez JA, Perkins JM, Beaudoin GM, Cook NB, Quraishi SA, Szoeke EA, Thangamani K, Tschumi CW, Wanat MJ, Maroof AM, Beckstead MJ, Rosenberg PA, Paladini CA. Ventral tegmental area astrocytes orchestrate avoidance and approach behavior. Nat Commun. 2019;10(1):1455. https://doi.org/10.1038/s41467-019-09131-y. PMID: 30926783; PMCID: PMC6440962
52. Dutar P, Lamour Y, Jobert A. Septohippocampal neurons in the rat. An in vivo. Intracellular study. Brain Res. 1985;340:135–42.
53. Hirata H, Aston-Jones G. A novel long-latency response of locus coeruleus neurons to noxious stimuli: mediation by peripheral C-fibers. J Neurophysiol. 1994;71(5):1752–61. https://doi.org/10.1152/jn.1994.71.5.1752. PMID: 8064346
54. Hirschberg S, Li Y, Randall A, Kremer EJ, Pickering AE. Functional dichotomy in spinal- vs prefrontal-projecting locus coeruleus modules splits descending noradrenergic analgesia from ascending aversion and anxiety in rats. elife. 2017;6:e29808. https://doi.org/10.7554/eLife.29808. PMID: 29027903; PMCID: PMC5653237
55. López-Hidalgo M, Kellner V, Schummers J. Astrocyte calcium responses to sensory input: influence of circuit organization and experimental factors. Front Neural Circuits. 2017;11:16. https://doi.org/10.3389/fncir.2017.00016. PMID: 28381991; PMCID: PMC5360724
56. Ma Z, Stork T, Bergles DE, Freeman MR. Neuromodulators signal through astrocytes to alter neural circuit activity and behaviour. Nature. 2016;539(7629):428–32. https://doi.org/10.1038/nature20145. Epub 2016 Nov 9. PMID: 27828941; PMCID: PMC5161596
57. Suárez-Pereira I, Llorca-Torralba M, Bravo L, Camarena-Delgado C, Soriano-Mas C, Berrocoso E. The role of the locus coeruleus in pain and associated stress-related disorders. Biol Psychiatry. 2022;91(9):786–97. https://doi.org/10.1016/j.biopsych.2021.11.023. Epub 2021 Dec 16. PMID: 35164940

58. Pacholko AG, Wotton CA, Bekar LK. Astrocytes-the ultimate effectors of long-range Neuromodulatory networks? Front Cell Neurosci. 2020;14:581075. https://doi.org/10.3389/fncel.2020.581075. PMID: 33192327; PMCID: PMC7554522
59. Bertolero M, Bassett DS, Studios MR. How matter becomes mind. Sci Am. 2019;321(1):26–33. https://www.jstor.org/stable/27265246
60. Perea G, Gómez R, Mederos S, Covelo A, Ballesteros JJ, Schlosser L, Hernández-Vivanco A, Martín-Fernández M, Quintana R, Rayan A, Díez A, Fuenzalida M, Agarwal A, Bergles DE, Bettler B, Manahan-Vaughan D, Martín ED, Kirchhoff F, Araque A. Activity-dependent switch of GABAergic inhibition into glutamatergic excitation in astrocyte-neuron networks. elife. 2016;5:e20362. https://doi.org/10.7554/eLife.20362. PMID: 28012274; PMCID: PMC5231406
61. Poskanzer KE, Yuste R. Astrocytes regulate cortical state switching in vivo. Proc Natl Acad Sci USA. 2016;113(19):E2675–84. https://doi.org/10.1073/pnas.1520759113. Epub 2016 Apr 27. PMID: 27122314; PMCID: PMC4868485
62. Reitman ME, Tse V, Mi X, Willoughby DD, Peinado A, Aivazidis A, Myagmar BE, Simpson PC, Bayraktar OA, Yu G, Poskanzer KE. Norepinephrine links astrocytic activity to regulation of cortical state. Nat Neurosci. 2023;26(4):579–93. https://doi.org/10.1038/s41593-023-01284-w. Epub 2023 Mar 30. PMID: 36997759; PMCID: PMC10089924
63. Requie LM, Gómez-Gonzalo M, Speggiorin M, Managò F, Melone M, Congiu M, Chiavegato A, Lia A, Zonta M, Losi G, Henriques VJ, Pugliese A, Pacinelli G, Marsicano G, Papaleo F, Muntoni AL, Conti F, Carmignoto G. Astrocytes mediate long-lasting synaptic regulation of ventral tegmental area dopamine neurons. Nat Neurosci. 2022;25(12):1639–50. https://doi.org/10.1038/s41593-022-01193-4. Epub 2022 Nov 17. Erratum in: Nat Neurosci. 2023;26(6):1129. PMID: 36396976
64. Billman GE. Homeostasis: the underappreciated and far too often ignored central organizing principle of physiology. Front Physiol. 2020;11:200.
65. Cannon WB, Rosenberg CE. Wisdom of the body. New York; 1932. W.W. Norton & Company, Inc
66. MacAulay N, Zeuthen T. Water transport between CNS compartments: contributions of aquaporins and cotransporters. Neuroscience. 2010;168(4):941–56. https://doi.org/10.1016/j.neuroscience.2009.09.016. Epub 2009 Sep 15. PMID: 19761815
67. Bringmann A, Grosche A, Pannicke T, Reichenbach A. GABA and glutamate uptake and metabolism in retinal glial (Müller) cells. Front Endocrinol (Lausanne). 2013;4:48. https://doi.org/10.3389/fendo.2013.00048. PMID: 23616782; PMCID: PMC3627989
68. Hertz L. Possible role of neuroglia: a potassium-mediated neuronal – neuroglial – neuronal impulse transmission system. Nature. 1965;206(989):1091–4. https://doi.org/10.1038/2061091a0. PMID: 5325441
69. Başar E. Brain oscillations in neuropsychiatric disease. Dialogues Clin Neurosci. 2013;15(3):291–300. https://doi.org/10.31887/DCNS.2013.15.3/ebasar. PMID: 24174901; PMCID: PMC3811101
70. Ghiani A, Maniglia M, Battaglini L, Melcher D, Ronconi L. Binding mechanisms in visual perception and their link with neural oscillations: a review of evidence from tACS. Front Psychol. 2021;12:643677. https://doi.org/10.3389/fpsyg.2021.643677. PMID: 33828509; PMCID: PMC8019716
71. Kim WB, Cho JH. Encoding of contextual fear memory in hippocampal-amygdala circuit. Nat Commun. 2020;11(1):1382. https://doi.org/10.1038/s41467-020-15121-2. PMID: 32170133; PMCID: PMC7069961
72. Sridharan D, Knudsen EI. Selective disinhibition: a unified neural mechanism for predictive and post hoc attentional selection. Vis Res. 2015;116(Pt B):194–209. https://doi.org/10.1016/j.visres.2014.12.010. Epub 2014 Dec 24. PMID: 25542276
73. van den Heuvel MP, Hulshoff Pol HE. Exploring the brain network: a review on resting-state fMRI functional connectivity. Eur Neuropsychopharmacol. 2010 Aug;20(8):519–34. https://doi.org/10.1016/j.euroneuro.2010.03.008. Epub 2010 May 14

74. Ramón y Cajal S. Sobre un nuevo proceder de impregnacion de la neuroglia y sus resultados en los centros nerviosos del hombre y animales. Trab Lab Invest Biol Univ Madrid. 1913;11:219–37.
75. Akyazi I, Eraslan E. Transmission of stress between cagemates: a study in rats. Physiol Behav. 2014;123:114–8. https://doi.org/10.1016/j.physbeh.2013.10.006. Epub 2013 Oct 23. PMID: 24161514
76. Araque A, Martín ED, Perea G, Arellano JI, Buño W. Synaptically released acetylcholine evokes Ca2+ elevations in astrocytes in hippocampal slices. J Neurosci. 2002;22(7):2443–50. https://doi.org/10.1523/JNEUROSCI.22-07-02443.2002. PMID: 11923408; PMCID: PMC6758296
77. Asano Y, Sasaki D, Ikoma Y, Matsui K. Glial tone of aggression. Neurosci Res. 2023;S0168-0102(23):00203-1. https://doi.org/10.1016/j.neures.2023.11.008. Epub ahead of print. PMID: 38007191
78. Cimino G. Reticular theory versus neuron theory in the work of Camillo Golgi. Physis Riv Int Stor Sci. 1999;36(2):431–72. PMID: 11640243
79. Kiyoshi CM, Zhou M. Astrocyte syncytium: a functional reticular system in the brain. Neural Regen Res. 2019 Apr;14(4):595–6. https://doi.org/10.4103/1673-5374.247462. PMID: 30632498; PMCID: PMC6352602
80. Shepherd, G. M. (2009). Neuron doctrine: historical background. En L. R. Squire (Ed.), Encyclopedia of neuroscience (pp. 691–695). Academic Press. https://doi.org/10.1016/B978-008045046-9.00988-8.
81. Pannasch U, Vargová L, Reingruber J, Ezan P, Holcman D, Giaume C, Syková E, Rouach N. Astroglial networks scale synaptic activity and plasticity. Proc Natl Acad Sci U S A. 2011;108(20):8467–72. https://doi.org/10.1073/pnas.1016650108. Epub 2011 May 2. PMID: 21536893; PMCID: PMC3100942
82. Giaume C, Koulakoff A, Roux L, Holcman D, Rouach N. Astroglial networks: a step further in neuroglial and gliovascular interactions. Nat Rev Neurosci. 2010;11(2):87–99. https://doi.org/10.1038/nrn2757. PMID: 20087359
83. Giaume C, Naus CC, Sáez JC, Leybaert L. Glial connexins and pannexins in the healthy and diseased brain. Physiol Rev. 2021;101(1):93–145. https://doi.org/10.1152/physrev.00043.2018. Epub 2020 Apr 23. PMID: 32326824
84. Mazaud D, Capano A, Rouach N. The many ways astroglial connexins regulate neurotransmission and behavior. Glia. 2021;69(11):2527–45. https://doi.org/10.1002/glia.24040. Epub 2021 Jun 8. PMID: 34101261
85. Solan JL, Lampe PD. Key connexin 43 phosphorylation events regulate the gap junction life cycle. J Membr Biol. 2007;217(1–3):35–41. https://doi.org/10.1007/s00232-007-9035-y. Epub 2007 Jul 15. PMID: 17629739; PMCID: PMC2596931
86. Zhang SS, Shaw RM. Trafficking highways to the intercalated disc: new insights unlocking the specificity of connexin 43 localization. Cell Commun Adhes. 2014;21(1):43–54. https://doi.org/10.3109/15419061.2013.876014. PMID: 24460200; PMCID: PMC4950511
87. Fujii Y, Maekawa S, Morita M. Astrocyte calcium waves propagate proximally by gap junction and distally by extracellular diffusion of ATP released from volume-regulated anion channels. Sci Rep. 2017;7:13115. https://doi.org/10.1038/s41598-017-13243-0.
88. Chang WP, Wu JJ, Shyu BC. Thalamic modulation of cingulate seizure activity via the regulation of gap junctions in mice thalamocingulate slice. PLoS One. 2013;8(5):e62952. https://doi.org/10.1371/journal.pone.0062952. PMID: 23690968; PMCID: PMC3653920
89. Hirase H, Qian L, Barthó P, Buzsáki G. Calcium dynamics of cortical astrocytic networks in vivo. PLoS Biol. 2004;2(4):e96. https://doi.org/10.1371/journal.pbio.0020096.
90. Chen FL, Dong YL, Zhang ZJ, Cao DL, Xu J, Hui J, Zhu L, Gao YJ. Activation of astrocytes in the anterior cingulate cortex contributes to the affective component of pain in an inflammatory pain model. Brain Res Bull. 2012;87(1):60–6. https://doi.org/10.1016/j.brainresbull.2011.09.022. Epub 2011 Oct 8

91. Tao XD, Liu ZR, Zhang YQ, Zhang XH. Connexin43 hemichannels contribute to working memory and excitatory synaptic transmission of pyramidal neurons in the prefrontal cortex of rats. Life Sci. 2021;286:120049. https://doi.org/10.1016/j.lfs.2021.120049. Epub 2021 Oct 15. PMID: 34662549
92. Linsambarth S, Carvajal FJ, Moraga-Amaro R, Mendez L, Tamburini G, Jimenez I, Verdugo DA, Gómez GI, Jury N, Martínez P, van Zundert B, Varela-Nallar L, Retamal MA, Martin C, Altenberg GA, Fiori MC, Cerpa W, Orellana JA, Stehberg J. Astroglial gliotransmitters released via Cx43 hemichannels regulate NMDAR-dependent transmission and short-term fear memory in the basolateral amygdala. FASEB J. 2022;36(2):e22134. https://doi.org/10.1096/fj.202100798RR. PMID: 35061296
93. Li G, Henriquez CS, Fröhlich F. Unified thalamic model generates multiple distinct oscillations with state-dependent entrainment by stimulation. PLoS Comput Biol. 2017;13(10):e1005797. https://doi.org/10.1371/journal.pcbi.1005797. PMID: 29073146; PMCID: PMC5675460
94. Peng W, Xia X, Yi M, Huang G, Zhang Z, Iannetti G, Hu L. Brain oscillations reflecting pain-related behavior in freely moving rats. Pain. 2018;159(1):106–18. https://doi.org/10.1097/j.pain.0000000000001069. PMID: 28953192; PMCID: PMC5737457
95. Ponsel S, Zhang J, Pilz M, Yanovsky Y, Brankačk J, Draguhn A. Alterations of distributed neuronal network oscillations during acute pain in freely-moving mice. IBRO Rep. 2020;9:195–206. https://doi.org/10.1016/j.ibror.2020.08.001. PMID: 32944670; PMCID: PMC7481812
96. Schulz E, May ES, Postorino M, Tiemann L, Nickel MM, Witkovsky V, Schmidt P, Gross J, Ploner M. Prefrontal gamma oscillations encode tonic pain in humans. Cereb Cortex. 2015;25(11):4407–14. https://doi.org/10.1093/cercor/bhv043. Epub 2015 Mar 8. PMID: 25754338; PMCID: PMC4816790
97. Tan LL, Oswald MJ, Heinl C, Retana Romero OA, Kaushalya SK, Monyer H, Kuner R. Gamma oscillations in somatosensory cortex recruit prefrontal and descending serotonergic pathways in aversion and nociception. Nat Commun. 2019;10(1):983. https://doi.org/10.1038/s41467-019-08873-z. PMID: 30816113; PMCID: PMC6395755
98. Lundqvist M, Rose J, Herman P, Brincat SL, Buschman TJ, Miller EK. Gamma and Beta bursts underlie working memory. Neuron. 2016;90(1):152–64. https://doi.org/10.1016/j.neuron.2016.02.028. Epub 2016 Mar 17. PMID: 26996084; PMCID: PMC5220584
99. May ES, Nickel MM, Ta Dinh S, Tiemann L, Heitmann H, Voth I, Tölle TR, Gross J, Ploner M. Prefrontal gamma oscillations reflect ongoing pain intensity in chronic back pain patients. Hum Brain Mapp. 2019;40(1):293–305. https://doi.org/10.1002/hbm.24373. Epub 2018 Sep 10. PMID: 30260531; PMCID: PMC6585682
100. Lee HS, Ghetti A, Pinto-Duarte A, Wang X, Dziewczapolski G, Galimi F, Huitron-Resendiz S, Piña-Crespo JC, Roberts AJ, Verma IM, Sejnowski TJ, Heinemann SF. Astrocytes contribute to gamma oscillations and recognition memory. Proc Natl Acad Sci U S A. 2014;111(32):E3343–52. https://doi.org/10.1073/pnas.1410893111. Epub 2014 Jul 28. PMID: 25071179; PMCID: PMC4136580
101. Makovkin S, Kozinov E, Ivanchenko M, Gordleeva S. Controlling synchronization of gamma oscillations by astrocytic modulation in a model hippocampal neural network. Sci Rep. 2022;12(1):6970. https://doi.org/10.1038/s41598-022-10649-3. PMID: 35484169; PMCID: PMC9050920
102. Sardinha VM, Guerra-Gomes S, Caetano I, Tavares G, Martins M, Reis JS, Correia JS, Teixeira-Castro A, Pinto L, Sousa N, Oliveira JF. Astrocytic signaling supports hippocampal-prefrontal theta synchronization and cognitive function. Glia. 2017;65(12):1944–60. https://doi.org/10.1002/glia.23205. Epub 2017 Sep 8. PMID: 28885722
103. Cassell EJ. Recognizing suffering. Hast Cent Rep. 1991;21(3):24–31. PMID: 1885294
104. Sandkühler J, Lee J. How to erase memory traces of pain and fear. Trends Neurosci. 2013 Jun;36(6):343–52. https://doi.org/10.1016/j.tins.2013.03.004. Epub 2013 Apr 18. PMID: 23602194; PMCID: PMC3679540

105. Stegemann A, Liu S, Retana Romero OA, Oswald MJ, Han Y, Beretta CA, Gan Z, Tan LL, Wisden W, Gräff J, Kuner R. Prefrontal engrams of long-term fear memory perpetuate pain perception. Nat Neurosci. 2023;26(5):820–9. https://doi.org/10.1038/s41593-023-01291-x. Epub 2023 Apr 6. PMID: 37024573; PMCID: PMC10166861
106. Eldar E, Cohen JD, Niv Y. The effects of neural gain on attention and learning. Nat Neurosci. 2013;16(8):1146–53. https://doi.org/10.1038/nn.3428. Epub 2013 Jun 16. PMID: 23770566; PMCID: PMC3725201
107. Ferguson KA, Cardin JA. Mechanisms underlying gain modulation in the cortex. Nat Rev Neurosci. 2020;21(2):80–92. https://doi.org/10.1038/s41583-019-0253-y. Epub 2020 Jan 7. PMID: 31911627; PMCID: PMC7408409
108. Lu ZL. Mechanisms of attention: psychophysics, cognitive psychology, and cognitive neuroscience. Jpn J Psychon Sci. 2008;27(1):38–45. PMID: 20523762; PMCID: PMC2879667
109. Lee IS, Necka EA, Atlas LY. Distinguishing pain from nociception, salience, and arousal: How autonomic nervous system activity can improve neuroimaging tests of specificity. Neuroimage. 2020 Jan 1;204:116254. doi: 10.1016/j.neuroimage.2019.116254. Epub 2019 Oct 8. PMID: 31604122; PMCID: PMC6911655.
110. Wang F, Smith NA, Xu Q, Fujita T, Baba A, Matsuda T, Takano T, Bekar L, Nedergaard M. Astrocytes modulate neural network activity by Ca2+-dependent uptake of extracellular K+. Sci Signal. 2012;5(218):ra26. https://doi.org/10.1126/scisignal.2002334. PMID: 22472648; PMCID: PMC3515082
111. Oe Y, Wang X, Patriarchi T, Konno A, Ozawa K, Yahagi K, Hirai H, Tsuboi T, Kitaguchi T, Tian L, McHugh TJ, Hirase H. Distinct temporal integration of noradrenaline signaling by astrocytic second messengers during vigilance. Nat Commun. 2020;11(1):471. https://doi.org/10.1038/s41467-020-14378-x. Erratum in: Nat Commun 2020;11(1):3447. PMID: 31980655; PMCID: PMC6981284.
112. Paukert M, Agarwal A, Cha J, Doze VA, Kang JU, Bergles DE. Norepinephrine controls astroglial responsiveness to local circuit activity. Neuron. 2014;82(6):1263–70. https://doi.org/10.1016/j.neuron.2014.04.038. PMID: 24945771; PMCID: PMC4080721
113. Pittolo S, Yokoyama S, Willoughby DD, Taylor CR, Reitman ME, Tse V, Wu Z, Etchenique R, Li Y, Poskanzer KE. Dopamine activates astrocytes in prefrontal cortex via α1-adrenergic receptors. Cell Rep. 2022;40(13):111426. https://doi.org/10.1016/j.celrep.2022.111426. PMID: 36170823; PMCID: PMC9555850
114. Calderon DP, Kilinc M, Maritan A, Banavar JR, Pfaff D. Generalized CNS arousal: an elementary force within the vertebrate nervous system. Neurosci Biobehav Rev. 2016;68:167–76. https://doi.org/10.1016/j.neubiorev.2016.05.014. Epub 2016 May 20. Erratum in: Neurosci Biobehav Rev. 2016;68:1021. PMID: 27216213; PMCID: PMC5003634
115. Wahis J, Holt MG. Astrocytes, noradrenaline, α1-adrenoreceptors, and neuromodulation: evidence and unanswered questions. Front Cell Neurosci. 2021;15:645691. https://doi.org/10.3389/fncel.2021.645691. PMID: 33716677; PMCID: PMC7947346
116. Li WP, Su XH, Hu NY, Hu J, Li XW, Yang JM, Gao TM. Astrocytes mediate cholinergic regulation of adult hippocampal neurogenesis and memory through M1 muscarinic receptor. Biol Psychiatry. 2022;92(12):984–98. https://doi.org/10.1016/j.biopsych.2022.04.019. Epub 2022 May 13. PMID: 35787318
117. Li Y, Li L, Wu J, Zhu Z, Feng X, Qin L, Zhu Y, Sun L, Liu Y, Qiu Z, Duan S, Yu YQ. Activation of astrocytes in hippocampus decreases fear memory through adenosine A1 receptors. elife. 2020;9:e57155. https://doi.org/10.7554/eLife.57155. PMID: 32869747; PMCID: PMC7505657
118. Chen FL, Dong YL, Zhang ZJ, Cao DL, Xu J, Hui J, Zhu L, Gao YJ. Activation of astrocytes in the anterior cingulate cortex contributes to the affective component of pain in an inflammatory pain model. Brain Res Bull. 2012a;87(1):60–6. https://doi.org/10.1016/j.brainresbull.2011.09.022. Epub 2011 Oct 8. PMID: 22004615
119. Chen MJ, Kress B, Han X, Moll K, Peng W, Ji RR, Nedergaard M. Astrocytic CX43 hemichannels and gap junctions play a crucial role in development of chronic neuropathic pain following spinal cord injury. Glia. 2012b;60:1660–70.

120. Iqbal Z, Liu S, Lei Z, Ramkrishnan AS, Akter M, Li Y. Astrocyte L-lactate signaling in the ACC regulates visceral pain aversive memory in rats. Cells. 2022;12(1):26. https://doi.org/10.3390/cells12010026. PMID: 36611820; PMCID: PMC9818423
121. Ding F, O'Donnell J, Thrane AS, Zeppenfeld D, Kang H, Xie L, Wang F, Nedergaard M. α1-Adrenergic receptors mediate coordinated Ca2+ signaling of cortical astrocytes in awake, behaving mice. Cell Calcium. 2013;54(6):387–94. https://doi.org/10.1016/j.ceca.2013.09.001. Epub 2013 Sep 24. PMID: 24138901; PMCID: PMC3858490
122. Kjaerby C, Rasmussen R, Andersen M, Nedergaard M. Does global astrocytic calcium signaling participate in awake brain state transitions and neuronal circuit function? Neurochem Res. 2017;42(6):1810–22. https://doi.org/10.1007/s11064-017-2195-y. Epub 2017 Feb 16. PMID: 28210958; PMCID: PMC5451293
123. Iqbal Z, Lei Z, Ramkrishnan AS, et al. Adrenergic signalling to astrocytes in anterior cingulate cortex contributes to pain-related aversive memory in rats. Commun Biol. 2023;6:10. https://doi.org/10.1038/s42003-022-04405-6.
124. Martin-Fernandez M, Jamison S, Robin LM, Zhao Z, Martin ED, Aguilar J, Benneyworth MA, Marsicano G, Araque A. Synapse-specific astrocyte gating of amygdala-related behavior. Nat Neurosci. 2017;20(11):1540–8. https://doi.org/10.1038/nn.4649. Epub 2017 Sep 25. PMID: 28945222; PMCID: PMC5903286
125. Dolan RJ. Emotion, cognition, and behavior. Science. 2002;298(5596):1191–4. https://doi.org/10.1126/science.1076358. PMID: 12424363
126. Gonzalez-Liencres C, Juckel G, Tas C, Friebe A, Brüne M. Emotional contagion in mice: the role of familiarity. Behav Brain Res. 2014;263:16–21. https://doi.org/10.1016/j.bbr.2014.01.020. Epub 2014 Jan 28. PMID: 24480421
127. Chen C, Martínez RM, Cheng Y. The developmental origins of the social brain: empathy, morality, and justice. Front Psychol. 2018;9:2584. https://doi.org/10.3389/fpsyg.2018.02584. PMID: 30618998; PMCID: PMC6302010
128. Prosociality JK. Curr Biol. 2016;26(16):R748–52. https://doi.org/10.1016/j.cub.2016.07.025. PMID: 27554648
129. Jeon D, Kim S, Chetana M, Jo D, Ruley HE, Lin SY, Rabah D, Kinet JP, Shin HS. Observational fear learning involves affective pain system and Cav1.2 Ca2+ channels in ACC. Nat Neurosci. 2010;13(4):482–8. https://doi.org/10.1038/nn.2504. Epub 2010 Feb 28. PMID: 20190743; PMCID: PMC2958925
130. Condé F, Audinat E, Maire-Lepoivre E, Crépel F. Afferent connections of the medial frontal cortex of the rat. A study using retrograde transport of fluorescent dyes. I. Thalamic afferents. Brain Res Bull. 1990;24:341–54.
131. Hoover WB, Vertes RP. Anatomical analysis of afferent projections to the medial prefrontal cortex in the rat. Brain Struct Funct. 2007;212(2):149–79. https://doi.org/10.1007/s00429-007-0150-4. Epub 2007 Jul 27. PMID: 17717690
132. Singer T, Seymour B, O'Doherty J, Kaube H, Dolan RJ, Frith CD. Empathy for pain involves the affective but not sensory components of pain. Science. 2004;303(5661):1157–62. https://doi.org/10.1126/science.1093535.
133. Wicker B, Keysers C, Plailly J, Royet JP, Gallese V, Rizzolatti G. Both of us disgusted in My insula: the common neural basis of seeing and feeling disgust. Neuron. 2003;40(3):655–64. https://doi.org/10.1016/s0896-6273(03)00679-2. PMID: 14642287
134. Langford DJ, Crager SE, Shehzad Z, Smith SB, Sotocinal SG, Levenstadt JS, Chanda ML, Levitin DJ, Mogil JS. Social modulation of pain as evidence for empathy in mice. Science. 2006;312(5782):1967–70. https://doi.org/10.1126/science.1128322.
135. Yang H, Jung S, Seo J, Khalid A, Yoo JS, Park J, Kim S, Moon J, Lee ST, Jung KH, Chu K, Lee SK, Jeon D. Altered behavior and neural activity in conspecific cagemates co-housed with mouse models of brain disorders. Physiol Behav. 2016;163:167–76. https://doi.org/10.1016/j.physbeh.2016.05.031. Epub 2016 May 18. PMID: 27211331
136. Kim A, Keum S, Shin HS. Observational fear behavior in rodents as a model for empathy. Genes Brain Behav. 2019;18(1):e12521. https://doi.org/10.1111/gbb.12521. Epub 2018 Oct 12. PMID: 30264490

137. Smith ML, Asada N, Malenka RC. Anterior cingulate inputs to nucleus accumbens control the social transfer of pain and analgesia. Science. 2021;371(6525):153–9. https://doi.org/10.1126/science.abe3040. Epub 2021 Jan 7. PMID: 33414216; PMCID: PMC7952019
138. Zhang M, Wu YE, Jiang M, Hong W. Cortical regulation of helping behaviour towards others in pain. Nature. 2024;626(7997):136–44. https://doi.org/10.1038/s41586-023-06973-x. Epub 2024 Jan 24. PMID: 38267578
139. Cheng YT, Woo J, Luna-Figueroa E, Maleki E, Harmanci AS, Deneen B. Social deprivation induces astrocytic TRPA1-GABA suppression of hippocampal circuits. Neuron. 2023;111(8):1301–1315.e5. https://doi.org/10.1016/j.neuron.2023.01.015. Epub 2023 Feb 13. PMID: 36787749; PMCID: PMC10121837

Astrocyte-Neuron Interactions in Spinal Cord Injury

Catrina Reyes and Mayssa H. Mokalled

Abstract Spinal cord injuries cause irreversible loss of sensory and motor functions. In mammals, intrinsic and extrinsic inhibitions of neuronal regeneration obstruct neural repair after spinal cord injury. Although astrocytes have been involved in a growing list of vital homeostatic functions in the nervous system, their roles after injury have fascinated and puzzled scientists for decades. Astrocytes undergo long-lasting morphological and functional changes after injury, referred to as reactive astrogliosis. Although reactive astrogliosis is required to contain spinal cord lesions and restore the blood-spinal cord barrier, reactive astrocytes have detrimental effects that inhibit neuronal repair and remyelination. Intriguingly, elevated regenerative capacity is preserved in some non-mammalian vertebrates, where astrocyte-like glial cells display exclusively pro-regenerative effects after injury. A detailed molecular and phenotypic catalog of the continuum of astrocyte reactivity states is an essential first step toward the development of glial cell manipulations for spinal cord repair.

Keywords Astrocytes · Spinal cord injury · Reactive astrogliosis · Glial scarring · Regeneration

Abbreviations

AQP	Aquaporin
CCL	CC Chemokine ligand
CSPG	Chondroitin sulfate proteoglycan
CTGF	Connective tissue growth factor

C. Reyes · M. H. Mokalled (✉)
Department of Developmental Biology, Washington University School of Medicine, Saint Louis, MO, USA
e-mail: c.reyes@wustl.edu; mmokalled@wustl.edu

E. Blanco-Suarez, I. Farhy-Tselnicker (eds.), *Astrocyte-Neuron Interactions in Health and Disease*, Advances in Neurobiology 39,
https://doi.org/10.1007/978-3-031-64839-7_9

FGF	Fibroblast growth factor
Gfap	Glial fibrillary acidic protein
IL	Interleukin
MAG	Myelin-associated glycoprotein
OMgp	Oligodendrocyte-myelin glycoprotein
OPC	Oligodendrocyte precursor cell
SCI	Spinal cord injury
Shh	Sonic hedgehog
TNF	Tumor necrosis factor

1 Introduction

This chapter provides an overview of the complex and fascinating roles of astrocytes after spinal cord injury. We first review the disease burden and multicellular complications associated with spinal cord injuries. With astrocytes displaying a range of reactivity states in response to different insults, we discuss the advantages and limitations of implementing different models and injury paradigms, highlighting the value of performing comparative studies to generate a comprehensive catalog of astrocyte reactivity states and functions. We then delve into the mechanisms that direct reactive astrogliosis and the dual consequences of reactive astrogliosis as a defense mechanism and anti-regenerative effector after spinal cord injury.

1.1 Spinal Cord Injury: Disease Burden and Multicellular Complications

1.1.1 Spinal Cord Injury Is a Devastating Condition and a Challenging Medical Priority

Spinal cord injuries (SCIs) comprise traumatic and non-traumatic damages that alter the anatomy and function of the spinal cord. Common causes of traumatic SCI include motor vehicle accidents, falls, sports-related injuries, or violence, whereas non-traumatic SCI results from acute or chronic diseases such as cancer, infection, or degenerative disk disease [1, 2]. SCI has devastating, long-term consequences for the physical, psychological, financial, and social well-being of patients and their caregivers. Over the past 30 years, the number of SCI cases has shown a 52.7% increase with over nine million cases worldwide and North America reporting the highest number of cases among developed regions [3]. With less than 1% of SCI patients experiencing complete neurological recovery by hospital discharge, over one million people are affected in North America alone with direct lifetime costs ranging between $1.1 and 4.6 million per patient [2, 4]. With the increased lifespan of the human population, the global burden of living with disability increased by

65.4% from 1990 to 2019 [3]. SCI is commonly viewed as a problem of motor control and sensory loss, resulting in at least partial paralysis. However, SCI drives systemic immune dysregulation across the body, negatively impacting major organ systems and general homeostatic mechanisms [5, 6]. Major concurrent SCI complications, such as neuropathic pain, increased risk of infection, bowel and bladder dysfunction, cardiovascular problems, and predisposition to obesity, diabetes, and liver disease, are challenging medical priorities [5, 7]. Despite advanced medical care, even SCI patients in developed regions have significantly reduced lifespan. SCI is therefore a medical priority that requires concerted research efforts to extend the lifespan and improve the health span of SCI patients.

1.1.2 Spinal Cord Injury Triggers Multicellular and Systemic Injury Complications

SCI leads to loss of motor, sensory, and autonomic functions. Understanding the role of astrocyte-neuron interactions in SCI requires a thorough understanding of various SCI complications, which cooperatively disrupt astrocyte functions after injury. SCI elicits complex multicellular responses that cause permanent neurological deficits and impede spontaneous regeneration in mammals [8–13]. Injury responses following SCI vary depending on the location, severity, and time of injury. SCI has been divided into four temporally distinct phases: acute (0 to 2 days), subacute (2 to 14 days), intermediate (14 days to 6 months), and chronic (>6 months) phases in humans. Acute SCI is marked by the primary physical damage that causes compression or transection to the spinal cord, disrupts spinal vascular flow, and compromises the blood-spinal cord barrier. Primary SCI damages trigger spontaneous cascades of inflammatory, glial, neuronal, and systemic complications that result in neurotoxic inflammation, overt glial cell reactivity, and cyclical cell loss [8, 9, 12, 13]. Even in the acute-to-subacute stages, multiple cellular complications result in secondary damages that are often more detrimental than the primary lesion. Shortly after injury, disrupted vascular flow and exaggerated inflammation cause swelling and additional compression that exacerbate the primary injury. Rapid disruption to calcium homeostasis and ischemic toxicity mediates neuronal and glial cell death, releasing high levels of glutamate into the extracellular milieu. Glutamate, which is cleared by astrocytes under homeostatic conditions, is poorly reabsorbed by surviving astrocytes [14, 15]. Glutamate cytotoxicity and necroptosis lead to cyclical propagation of secondary cell death. Excess glutamate activates NMDA and AMPA receptors, leading to cyclical propagation of glutamate cytotoxicity and necroptosis [14]. Intermediate-to-chronic SCI is marked by scarring, cystic cavitation, neural circuit remodeling, and vascular reorganization, which form additional barriers to regeneration [16–22]. At these stages, the extracellular milieu of the lesioned cord is flooded by reactive cell-derived or debris-derived molecules that are potent inhibitors of axon regeneration. Alterations in the composition of the extracellular matrix result in the accumulation of inhibitory proteoglycans in and around the lesion [20, 23–26]. Inhibitory molecules released by degenerating oligodendrocytes include neurite outgrowth inhibitor A (Nogo-A), oligodendrocyte-myelin glycoprotein (OMgp), and myelin-associated glycoprotein (MAG). Myelin-associated glycoproteins, upon binding to Nogo and TNF receptors, cause growth cone collapse and neurite retraction in a Rho/ROCK-dependent mechanism [25, 27–30]. Together, these injury complications exacerbate the inherently limited ability of the mammalian spinal cord to replenish lost neurons via adult neurogenesis or to regrow lesioned axon tracts. As primary SCI damages are irreversible, the major goals of SCI therapies have been to target and contain secondary damages. Considering the multicellular complications associated with secondary SCI damages, individual therapeutic approaches are unlikely to yield major effects. Thus, combinatorial strategies have been pursued as a more promising therapeutic avenue [31–34]. Comprehensive and simultaneous examination of neuronal and non-neuronal cells after SCI is therefore fundamental to understanding and manipulating the multicellular complexities of SCI injuries.

1.2 Models of Spinal Cord Injury

Animal models are indispensable for understanding the pathophysiology of SCI, investigating mechanisms of innate spinal cord repair, and testing regenerative therapies. An ideal SCI model should (a) mimic the anatomical and pathophysiological complications of human SCI, (b) possess an endogenous regenerative capacity, (c) be amenable to genetic and pharmacological manipulations, and (d) allow for high-powered and inexpensive studies and therapeutic testing. In the absence of a single animal model that fulfills all these criteria, multiple animal models and injury paradigms have been deployed.

1.2.1 Injury Paradigms

Animal models of SCI include contusion, compression, or transection injuries. These SCI models could be tweaked to target specific injury levels along the rosto-caudal axis of the injured animal and a range of injury severity from mild SCI to severe SCI [35]. Contusion models, which involve a weight-induced impact on the spinal cord, are thought to generate injuries that closely mimic the etiology and outcome of human SCI [36–38]. Compression or crush injuries are created by constricting the exposed cord with forceps or weight application [39–41]. Transection models can be complete transections or partial hemisections and are commonly used to study axon regrowth mechanisms.

1.2.2 Mammalian Models Mimic SCI Complications

Large animal models, such as non-human primate or swine models, are ideal to mimic the pathophysiology of SCI patients, including lesion size and the formation of cystic cavitation post-injury. However, substantial ethical, economical, and technical concerns are associated with large animal models including sample size, cost, and housing requirements. Therefore, large animal models can form an important intermediate model to validate preclinical data prior to initiating clinical studies. In contrast to large animal models, rat and mouse models are less expensive and are prominently used for SCI research and therapeutic testing. While the injury response in rats is anatomically similar to human lesions, including the formation of cystic cavities in the lesion core, mouse SCI lesions do not cavitate [42, 43]. Nonetheless, inherent differences in molecular signaling, regenerative capacity, and decreased amenability to genetic studies have challenged the throughput and translatability of rat SCI studies.

1.2.3 Regenerative Vertebrates Present Role Models for Innate Spinal Cord Repair

Unlike mammals, vertebrate species including teleost fish and amphibians have an innate ability to spontaneously recover from severe SCI. Following complete transection of spinal cord tissues, adult zebrafish are capable of reversing paralysis and regaining swim function within 6 to 8 weeks of injury [44, 45]. Pro-regenerative injury responses involving neurons, glia, and immune cells cooperate to achieve spontaneous and efficient repair in zebrafish [46–52]. Early after SCI, potent populations of progenitor cells, including central canal-surrounding ependymo-radial glial cells, are activated to replenish lost neurons and glia [46, 48, 50, 51]. Newly differentiated motor neurons and interneurons populate the regenerate tissue, as pre-existing neurons regrow axons across lesioned tissues [49–51]. Though less studied, astrocyte-like glial cells enact instrumental pro-regenerative responses throughout the course of regeneration [45, 48, 53]. However, a holistic understanding of the cellular interactions that coordinate the pro-regenerative responses to direct SC regeneration in zebrafish is to be acquired. Thus, how and why injury responses differ between zebrafish and mammalian astrocytes requires comprehensive molecular investigation and cross-species comparisons.

1.3 Astrocyte Responses after Mammalian Spinal Cord Injury

Astrocytes are the largest and most abundant glial cells in mammals (Freeman and Rowitch, 2013). They play developmental and physiological roles that are critical for the development, survival, and function of the central nervous system. Under physiological conditions, astrocytes provide trophic and metabolic support for neurons and play central roles in synapse development, pruning, and maturation [54, 55]. Astrocytes control the blood-spinal cord barrier and modulate spinal blood flow, fluid availability, and ion homeostasis [56, 57]. Spinal astrocytes are highly heterogeneous. Morphological, molecular, and functional differences between white and gray matter astrocytes have emerged in recent years [58, 59]. Accumulating evidence suggests that Gfap is predominantly expressed in white matter astrocytes, while S100β tends to be expressed in the astrocytes located in the gray matter [58, 59]. Even within the gray matter, astrocytes are regionally specialized into anatomical regions that correlate with the respective neurotransmitter properties of each region [60]. As astrocytes are shown to be involved in a growing list of vital homeostatic functions in the nervous system, their roles after injury have fascinated and puzzled scientists for decades.

1.3.1 The Glial Scar

Mammalian SCI is marked by scarring. The "glial scar," a historically used umbrella term that refers to SCI scarring, was thought to be detrimental for regeneration and blamed on astrocytes [12, 61, 62]. However, emerging technology and cell-specific manipulations have enabled a finer view and deeper understanding of SCI scarring. In response to SCI, multiple cell types rapidly migrate into the lesion site. Lesion-induced cells include immune cells (macrophages, neutrophils, and lymphocytes), fibroblasts, and pericytes in addition to glial cells (microglia, OPCs, and astrocytes). Microglia and macrophages are the first responders, with microglia proliferating and migrating toward the lesion, while monocyte-derived macrophages infiltrate from the periphery via the broken blood-spinal cord barrier. Lesion surrounding astrocytes proliferate, elongate, and form a cellular border that separates the necrotic lesion core from the spared neural tissue that surrounds the lesion [63–65]. With renewed appreciation for the anatomy and complexity of injury-induced scarring, the SCI scar has been more recently categorized into distinct fibrotic and glial compartments. While the fibrotic core is comprised of infiltrating immune cells, fibroblasts, and pericytes, the glial scar is comprised of microglia, NG2 glia, and astrocytes. Concomitantly, the concept that the glial scar is both detrimental and beneficial for functional recovery following SCI has emerged [12, 13, 21, 66]. Thus, the glial scar plays dual roles in pathophysiology of SCI.

In the acute stages of SCI, astrocytes proliferate and migrate toward the lesion site [53, 67, 68]. The border formed by reactive astrocytes isolates the fibrotic compartment of the scar from adjacent preserved neural tissues. Thus, scar-bordering astrocytes play a dual role by (a) limiting the expansion of the inflammatory core and further expansion of the fibrotic lesion core and (b) protecting spared neural tissues from exposure to scar-induced insults that trigger secondary damage [12, 69]. This initial astrocyte response is also necessary to reestablish the blood-spinal cord barrier [39, 70–72]. Subsequently, changes in the gene expression profiles of reactive astrocytes alter the microenvironmental landscape of the lesioned spinal cord, resulting in neurotoxicity and extrinsic inhibition of axon sprouting and functional recovery [22, 73–77]. Astrocytes increase the expression of intermediate filaments (Gfap, nestin, and vimentin) [71, 78–80] and release a variety of cytokines (TNF-α, IL-6, IL-10, and IL-1β) [81–83], chemokines (CCL2 and CCL3) [84, 85], growth factors (BDNF and GDNF) [86, 87], toxic amino acids (GABA and glutamate) [88, 89], and extracellular matrix molecules and proteoglycans (CSPGs, collagen I, fibronectin, and MMP-9) [78, 79, 90]. These molecules stimulate a variety of cellular responses by neurons, OPCs, and microglia, in addition to triggering cyclical reactivity in surrounding astrocytes [91]. In summary, mammalian astrocytes display multifarious injury responses that are overshadowed by anti-regenerative scar-forming cells and inhibitory extracellular molecules [12, 92, 93]. Thus, in light of our increased ability to examine the regional and functional heterogeneity of astrocyte subpopulations in response to SCI, it is important to move away from the umbrella term of the glial scar and to understand the concept and complexity of astrocyte reactivity in response to SCI.

1.3.2 Astrocyte Reactivity

In response to SCI, astrocytes undergo long-lasting morphological and functional changes referred to as "reactive astrogliosis" or "astrocyte reactivity" [94]. Various intrinsic and extrinsic factors stimulate reactive astrogliosis and play critical roles in the pathology and outcomes of SCI. Reactive astrogliosis results in a range of changes in the transcriptional profiles of reactive astrocytes and the adoption of newly acquired biochemical, morphological, metabolic, and physiological states in response to injury [95]. By causing reactive astrocytes to gain new functions and/or lose their ability to maintain homeostatic functions in the nervous system, reactive astrogliosis is an essential predictor and modulator of SCI outcome.

Cellular hypertrophy, thickening of astrocytic processes, and elevated expression of intermediate filament proteins such as Gfap are the minimum criteria used to define reactive astrocytes. Additional markers of astrocyte reactivity include Plaur, Mmp2, Mmp13, Axin2, Nes, and Ctnnb1, complement 3, type I collagen, Cdh2, Sox9, and chondroitin sulfate proteoglycan (CSPG)-related genes including Xylt1, Csgalnact1, Chst11, Pcan, Acan, and Slit2. While some of these genes have been associated with specific phenotypes such as A1/A2, neurotoxic/neuroprotective, and reactive/scar-forming [75, 96], multiple gene combinations are typically concomitantly expressed and inferring the functional phenotype of a reactive astrocyte based on a handful of markers proved to be misleading. Notably, astrocyte reactivity is a broad terminology that encompasses a continuum of transcriptional and functional states that astrocytes can adopt in response to diverse pathologies. Even within the pathology of SCI, astrocyte reactivity can vastly differ between injuries depending on the location of the lesion, time, and severity of the injury [91]. As the loss of some homeostatic functions can happen simultaneously with gains in protective and/or detrimental functions, computing the overall phenotype of a single reactive astrocyte into beneficial or detrimental is virtually impossible. Adding to the complexity of profiling a single astrocyte, the heterogeneity in astrocyte population prior to and after SCI adds to the challenges of determining astrocyte-induced phenotypes based on their specific reactivity states. In the absence of a prototypical reactive astrocyte and the wrong assumption of categorizing astrocytes into binary phenotypes, astrocyte reactivity has been a challenging research topic in SCI. An updated consensus among astrocyte biologists is to move beyond binary definitions to reflect the increasing complexity and functional heterogeneity of astrocytes [95].

1.3.3 Mechanisms of Astrocyte Activation

Following SCI, mechanical disruption of astrocyte membranes and processes activates mechanosensitive ion channels, resulting in rapid influx of extracellular sodium and calcium ions into astrocytes [97–100]. Further studies showed that membrane stretching is sufficient to release ions and signaling molecules including calcium, ATP, endothelin 1, and matrix metalloproteinases 9 [101–104]. In addition to mechanically induced reactivity, an array of chemokines, cytokines, transcription factors, and growth factors mediate astrocyte reactivity. These pathways include NF-κB, Stat3, Tgfβ, and MAPK. Proinflammatory cytokines such as TNF-α, interleukin-6, and interleukin-1β trigger the reactivity of astrocytes during the acute phase after SCI [81, 105, 106]. Early injury also triggers microglial activation, which induces the activation of astrocytes by secreting interleukin-1α, TNF, and C1q [107]. Reactive astrocytes release TNF-α, interleukin-6, and MMP9, which in turn activate more astrocytes in a feedback loop [108]. Stat3 signaling is a central pathway that directs astrocyte reactivity. Mice with astrocyte-specific deletion of Stat3 fail to upregulate Gfap, eliciting increased cell hypertrophy and exacerbated scar formation after SCI [109]. Stat3 deletion was also shown to limit the migration of astrocytes, resulting in widespread infiltration of inflammatory cells, degeneration of neurons, demyelination of axons, and severe motor deficits [71]. Conversely, Stat3 activation enhances the migration of reactive astrocytes toward the lesion, reduces the size of the fibrotic scar, and improves functional recovery [71]. These results provided a potential intervention strategy that aims to target Stat3 signaling pathway for SCI treatment. Additional astrocyte-derived signaling pathways that impact SCI include TGF-β signaling, which greatly contributes to the formation of reactive astrocytes and is a key upstream pathway in astrocyte reactivity. TGF-β could increase the expression of anti-regenerative molecules such as CSPGs, laminin, and fibronectin in reactive astrocytes [110]. These examples highlight multimodal modes by which SCI triggers astrocyte reactivity.

1.4 Reactive Astrocytes Play Dual Roles after Spinal Cord Injury

1.4.1 Reactive Astrocytes Have Detrimental Effects

Extrinsic Inhibition of Neuronal Repair Reactive astrocytes form a physical barrier that obstructs axonal growth. Reactive astrocytes secrete inhibitory proteins, such as CSPGs, which inhibit axon growth in vitro and in vivo. While CSPG-rich regions show inhibited axon regrowth, chondroitinase ABC, which removes CSPG glycosaminoglycan chains, was shown to counteract the inhibitory activity of CSPGs and facilitate axonal regeneration and functional recovery [16, 111, 112]. Besides, reactive astrocytes have a role in inhibiting the differentiation of neuronal progenitor cells by expressing insulin-like growth factor-binding proteins and CSPGs [113].

Inhibitory Roles on OPCs Oligodendrocyte precursor cells (OPCs) have extremely powerful remyelination ability as they can proliferate and differentiate to replenish oligodendrocytes that are lost after SCI. CSPGs play major inhibitory roles in neurons and OPCs. Astrocyte-derived CSPGs inhibit the migration and differentiation of OPCs in vitro, and the number of OPCs surrounding the lesion significantly increased when treated with the enzyme chondroitinase ABC [114, 115]. In addition to CSPGs, other astrocyte-derived molecules, such as BMP and ET1, can also inhibit the differentiation of OPCs and influence remyelination [116, 117].

Astrocytes Induce Edema and Secondary Injuries After SCI In the acute-to-subacute stages of SCI, edema develops rapidly at the lesion site and extends rostrocaudally beyond the lesion, leading to spinal fluid accumulation and increasing the overall volume of the spinal cord [118]. In contrast to vasogenic edema, which refers to the accumulation of interstitial fluid in the extracellular space following blood-spinal cord barrier disruption, cytotoxic edema refers to intracellular fluid accumulation in astrocytes [119]. Cytotoxic edema, a common pathology in compression injuries, causes secondary compression damages that exacerbate the primary injury and impact neurological outcome [118, 120–122]. As surgical decompressive interventions do not fully relieve edema or local pressure in the spinal cord, biological interventions are in great need. Water channels play central roles in the initiation and resolution of distinct forms of injury-induced edema. Aquaporin 4 (AQP4) is ubiquitously expressed in astrocytes, with prominent localization to the astrocytic end feet and glia limitans, which serve as direct contact points between the spinal cord and its surrounding cerebrospinal fluid [123]. Following thoracic compression injury, AQP4 mutant mice show reduced water content, enhanced neuronal survival, myelin sparing, and improved functional recovery [124, 125]. In contrast, in a model of thoracic contusion injury that causes vasogenic edema rather than cytotoxic edema, AQP4 deletion exacerbates injury outcomes, revealing a protective role for AQP4 in the clearance of vasogenic edema [126]. Thus, AQP4 water channels play a protective role in contusion or transection SCI (vasogenic edema), but a deleterious role in compression SCI (cytotoxic edema).

Astrocyte Reactivity and Neuropathic Pain Chronic pain affects >80% of SCI patients [127]. The pain may be nociceptive, neuropathic, or a combination of nociceptive and neuropathic. While musculoskeletal pain is the predominant source of nociceptive pain, neuropathic pain is the most common type of pain in SCI patients [127]. Reactive astrogliosis is associated with the development of neuropathic pain [128]. Multiple signaling pathways activated in reactive astrocytes have been shown to contribute to neuropathic pain. Glutamate transporters are downregulated in astrocytes in pain. Indeed, overexpression of GLT1 and GLAST in spinal astrocytes or reduction of extracellular glutamate levels is sufficient to prevent injury-induced allodynia [129–131]. These findings are consistent with a model in which decreased glutamate transporter in astrocytes leads to excessive activation of excitatory glutamatergic pathway, resulting in the development of neuropathic pain. JNK and ERK are major signaling pathways that have been implicated in the development of neuropathic pain [132]. In addition to JNK and glutamate signaling, various astrocytic molecules have been implicated in reactive astrogliosis-induced neuropathic pain, including the gap junction protein connexin 43 and the nuclear receptor protein PPAR$^{\gamma}$ [133, 134]. Connexin 43 is upregulated after SCI, suggesting increased connectivity between adjacent astrocytes [135]. Despite limited knowledge that implicates few molecules in neuropathic pain development after SCI, our understanding of the signaling pathways that trigger or limit pain remains limited.

1.4.2 Reactive Astrocytes Are a Defense Mechanism

Containing the Lesion and Impacting the Immune System At the acute stages of SCI, reactive astrocytes migrate rapidly around the lesion to seclude infiltrating immune cells and limit the extent of inflammation, which has markedly positive effects on functional recovery [71]. Supporting this beneficial role of astrocyte reactivity, selective and conditional ablation of reactive astrocytes in mouse SCI models results in significantly increased and prolonged infiltration of inflammatory cells around the lesion [39]. Various endogenous and exogenous factors trigger the release and accumulation of cellular debris and neurotoxic factors in the extracellular spaces after SCI. Moreover, reactive astrocytes can influence immune cell activation and mobilization by releasing various molecules such as TNF-α, TGF-β, and proteoglycans [136–140]. CSPGs have a close relationship with immune activity as they can recruit chemokines and growth factors that enhance the connection of immune cells [91].

Additional Beneficial Outputs of Astrocyte Reactivity in SCI SCI-induced disruption to the blood-spinal cord barrier exposes the lesioned cord to endogenous and exogenous blood-borne molecules that can result in detrimental effects on functional recovery. In response to these insults, reactive astrocytes upregulate Sonic hedgehog (Shh) as part of a reactive signaling cascade that restores tight junctions and directs the repair of the blood-spinal cord barrier [141]. Consequently, loss of reactive astrocytes leads to failure in repairing the damaged barrier. Reactive astrocytes can also reduce the impact of glutamate cytotoxicity on neurons and OPCs by clearing excess glutamate from blood or necrotic cells [142]. Reactive astrocytes were recently reported to play a crucial role in clearing cellular debris and neurotoxic factors from the lesion environment. Intriguingly, reactive astrocytes have the ability to phagocytose dead cells in vitro and in vivo via the upregulation of ABCA1 [143, 144]. Early astrocyte reactivity is therefore an ostensibly conserved response that evolved to protect the central nervous system against external injuries.

1.4.3 Species Differences in Astrocyte Reactivity

It is important to note that the timeline and outcomes of astrocyte reactivity differ among different species and SCI models. In rodents, microglia become activated before astrocytes, and both glial cell populations proliferate and migrate toward the injury site. In primates, glial cell activation is delayed relative to rodents. The response of astrocytes eventually leads to the formation of a dense astroglial border surrounding the fibrotic lesion core.

Though the complexities of glial cell responses to injury have been extensively studied in mammals, our understanding of how astrocytes respond to SCI in highly regenerative vertebrates such as zebrafish is relatively limited. In fact, a detailed description of zebrafish astrocytes was only reported in 2020 [145]. Following SCI in zebrafish, specialized glial cells form a bridge that is thought to provide a physical and signaling scaffold for cellular regrowth across the lesion [45, 53]. Previous studies implicated Fgf and Ctgf signaling in glial bridge formation in adult zebrafish. The glial bridging functions of Fgf signaling were established using transgenic dominant-negative Fgf receptor manipulations, *sprouty* mutants, and Fgf8 injection [53]. *ctgfa* genetic loss of function was shown to impair glial bridging, axon regrowth, and functional SC repair [45]. Genetic evidence indicates that *ctgfa* functions within a multi-nodal gene regulatory network that directs epithelial-to-mesenchymal transition and is necessary and sufficient to reprogram pro-regenerative glia after injury [48]. However, since Fgf and Ctgf are secreted extracellular proteins, the glial requirements of these molecules during bridging and of glial bridging during SC regeneration remained unclear. Previous studies suggested axon regrowth proceeds independently of the projection of glial processes across the lesion [46, 146, 147]. However, ablation of *ctgfa*$^+$ cells was shown to be sufficient to impair glial bridging, axon regrowth, and functional recovery after SCI, indicating *ctgfa*$^+$ cells are required during SC repair [148]. Regenerating axons often associate with elongated bipolar astrocytes across vertebrates, including zebrafish and mammals [45, 53, 149–151]. Bridging glia from adult zebrafish retain an elongated morphology [45, 53] and possess a mesenchymal signature that correlates with increased plasticity and immature astrocytic cell identity [48]. Similarly, in adult mice, a small proportion of elongated astrocytes, termed "astroglial bridges," correlate with increased axon regrowth under genetic manipulations such as PTEN deletion [151]. We thus propose that future comparative studies will shed light on the molecular similarities and differences between murine elongated astrocytes and zebrafish bridging glia. Indeed, zebrafish glia possess an astrocyte-like cell identity as transcriptomes of zebrafish bridging glia are thought to be most similar to scar-bordering astrocytes in mammals. These findings shed light on glial bridging as an effective, natural mechanism of spinal cord regeneration and established astrocyte reactivity as an indispensable component to achieve spinal cord repair.

2 Conclusion

In summary, spinal cord injuries are devastating conditions that require concerted efforts and combinatorial approaches. Astrocytes undergo a range of long-lasting morphological and functional changes in response to neural damages. It is likely that reactive astrogliosis evolved as an innate protective response to spinal cord injury. Yet, the anti-regenerative effects of reactive astrocytes present a major burden to spinal cord repair. Unraveling the molecular complexities of reactive astrogliosis using state-of-the-art technologies will offer unprecedented opportunities to modulate the phenotypic outputs of reactive astrocytes to promote spinal cord repair.

References

1. Ahuja CS, Nori S, Tetreault L, Wilson J, Kwon B, Harrop J, Choi D, Fehlings MG. Traumatic spinal cord injury-repair and regeneration. Neurosurgery. 2017a;80:S9–S22.
2. Ahuja CS, Wilson JR, Nori S, Kotter MRN, Druschel C, Curt A, Fehlings MG. Traumatic spinal cord injury. Nat Rev Dis Primers. 2017b;3:17018.
3. Liu Y, Yang X, He Z, Li J, Li Y, Wu Y, Manyande A, Feng M, Xiang H. Spinal cord injury: global burden from 1990 to 2019 and projections up to 2030 using Bayesian age-period-cohort analysis. Front Neurol. 2023;14:1304153.
4. SCI Facts and Figures. J Spinal Cord Med. 2017;40:872–3.
5. DiSabato DJ, Marion CM, Mifflin KA, Alfredo AN, Rodgers KA, Kigerl KA, Popovich PG, McTigue DM. System failure: systemic inflammation following spinal cord injury. Eur J Immunol. 2024;54:e2250274.
6. Morse LR, Field-Fote EC, Contreras-Vidal J, Noble-Haeusslein LJ, Rodreick M, Shields RK, Sofroniew M, Wudlick R, Zanca JM, Group, S.C.I.W. Meeting proceedings for SCI 2020: launching a decade of disruption in spinal cord injury research. J Neurotrauma. 2021;38:1251–66.
7. Adegeest CY, Ter Wengel PV, Peul WC. Traumatic spinal cord injury: acute phase treatment in critical care. Curr Opin Crit Care. 2023;29:659–65.
8. Brennan FH, Popovich PG. Emerging targets for reprograming the immune response to promote repair and recovery of function after spinal cord injury. Curr Opin Neurol. 2018;31:334–44.
9. Fouad K, Popovich PG, Kopp MA, Schwab JM. The neuroanatomical-functional paradox in spinal cord injury. Nat Rev Neurol. 2021;17:53–62.
10. He Z, Jin Y. Intrinsic control of axon regeneration. Neuron. 2016;90:437–51.
11. Milich LM, Ryan CB, Lee JK. The origin, fate, and contribution of macrophages to spinal cord injury pathology. Acta Neuropathol. 2019;137:785–97.
12. Sofroniew MV. Dissecting spinal cord regeneration. Nature. 2018;557:343–50.
13. Tran AP, Warren PM, Silver J. New insights into glial scar formation after spinal cord injury. Cell Tissue Res. 2022;387:319–36.
14. Li S, Stys PK. Mechanisms of ionotropic glutamate receptor-mediated excitotoxicity in isolated spinal cord white matter. J Neurosci. 2000;20:1190–8.
15. Liu M, Wu W, Li H, Li S, Huang LT, Yang YQ, Sun Q, Wang CX, Yu Z, Hang CH. Necroptosis, a novel type of programmed cell death, contributes to early neural cells damage after spinal cord injury in adult mice. J Spinal Cord Med. 2015;38:745–53.
16. Bradbury EJ, Moon LD, Popat RJ, King VR, Bennett GS, Patel PN, Fawcett JW, McMahon SB. Chondroitinase ABC promotes functional recovery after spinal cord injury. Nature. 2002;416:636–40.

17. Hata K, Fujitani M, Yasuda Y, Doya H, Saito T, Yamagishi S, Mueller BK, Yamashita T. RGMa inhibition promotes axonal growth and recovery after spinal cord injury. J Cell Biol. 2006;173:47–58.
18. Hollis ER 2nd, Ishiko N, Yu T, Lu CC, Haimovich A, Tolentino K, Richman A, Tury A, Wang SH, Pessian M, Jo E, Kolodkin A, Zou Y. Ryk controls remapping of motor cortex during functional recovery after spinal cord injury. Nat Neurosci. 2016;19:697–705.
19. Kaneko S, Iwanami A, Nakamura M, Kishino A, Kikuchi K, Shibata S, Okano HJ, Ikegami T, Moriya A, Konishi O, Nakayama C, Kumagai K, Kimura T, Sato Y, Goshima Y, Taniguchi M, Ito M, He Z, Toyama Y, Okano H. A selective Sema3A inhibitor enhances regenerative responses and functional recovery of the injured spinal cord. Nat Med. 2006;12:1380–9.
20. Schwab ME, Strittmatter SM. Nogo limits neural plasticity and recovery from injury. Curr Opin Neurobiol. 2014;27:53–60.
21. Shafqat A, Albalkhi I, Magableh HM, Saleh T, Alkattan K, Yaqinuddin A. Tackling the glial scar in spinal cord regeneration: new discoveries and future directions. Front Cell Neurosci. 2023;17:1180825.
22. Silver J, Miller JH. Regeneration beyond the glial scar. Nat Rev Neurosci. 2004;5:146–56.
23. Cafferty WB, Duffy P, Huebner E, Strittmatter SM. MAG and OMgp synergize with Nogo-A to restrict axonal growth and neurological recovery after spinal cord trauma. J Neurosci. 2010;30:6825–37.
24. Chio JCT, Punjani N, Hejrati N, Zavvarian MM, Hong J, Fehlings MG. Extracellular matrix and oxidative stress following traumatic spinal cord injury: physiological and pathophysiological roles and opportunities for therapeutic intervention. Antioxid Redox Signal. 2022;37:184–207.
25. Lee JK, Geoffroy CG, Chan AF, Tolentino KE, Crawford MJ, Leal MA, Kang B, Zheng B. Assessing spinal axon regeneration and sprouting in Nogo-, MAG-, and OMgp-deficient mice. Neuron. 2010b;66:663–70.
26. Li S, Liu BP, Budel S, Li M, Ji B, Walus L, Li W, Jirik A, Rabacchi S, Choi E, Worley D, Sah DW, Pepinsky B, Lee D, Relton J, Strittmatter SM. Blockade of Nogo-66, myelin-associated glycoprotein, and oligodendrocyte myelin glycoprotein by soluble Nogo-66 receptor promotes axonal sprouting and recovery after spinal injury. J Neurosci. 2004;24:10511–20.
27. Kubo T, Yamaguchi A, Iwata N, Yamashita T. The therapeutic effects of Rho-ROCK inhibitors on CNS disorders. Ther Clin Risk Manag. 2008;4:605–15.
28. Montani L, Gerrits B, Gehrig P, Kempf A, Dimou L, Wollscheid B, Schwab ME. Neuronal Nogo-A modulates growth cone motility via Rho-GTP/LIMK1/cofilin in the unlesioned adult nervous system. J Biol Chem. 2009;284:10793–807.
29. Naska S, Lin DC, Miller FD, Kaplan DR. p75NTR is an obligate signaling receptor required for cues that cause sympathetic neuron growth cone collapse. Mol Cell Neurosci. 2010;45:108–20.
30. Niederost B, Oertle T, Fritsche J, McKinney RA, Bandtlow CE. Nogo-A and myelin-associated glycoprotein mediate neurite growth inhibition by antagonistic regulation of RhoA and Rac1. J Neurosci. 2002;22:10368–76.
31. Anderson MA, O'Shea TM, Burda JE, Ao Y, Barlatey SL, Bernstein AM, Kim JH, James ND, Rogers A, Kato B, Wollenberg AL, Kawaguchi R, Coppola G, Wang C, Deming TJ, He Z, Courtine G, Sofroniew MV. Required growth facilitators propel axon regeneration across complete spinal cord injury. Nature. 2018;561:396–400.
32. DePaul MA, Lin CY, Silver J, Lee YS. Combinatory repair strategy to promote axon regeneration and functional recovery after chronic spinal cord injury. Sci Rep. 2017;7:9018.
33. Nakamura Y, Ueno M, Niehaus JK, Lang RA, Zheng Y, Yoshida Y. Modulation of both intrinsic and extrinsic factors additively promotes rewiring of corticospinal circuits after spinal cord injury. J Neurosci. 2021;41:10247–60.
34. Squair JW, Milano M, de Coucy A, Gautier M, Skinnider MA, James ND, Cho N, Lasne A, Kathe C, Hutson TH, Ceto S, Baud L, Galan K, Aureli V, Laskaratos A, Barraud Q, Deming TJ, Kohman RE, Schneider BL, He Z, Bloch J, Sofroniew MV, Courtine G, Anderson MA. Recovery of walking after paralysis by regenerating characterized neurons to their natural target region. Science. 2023;381:1338–45.

35. Zhang N, Fang M, Chen H, Gou F, Ding M. Evaluation of spinal cord injury animal models. Neural Regen Res. 2014;9:2008–12.
36. Kwon BK, Oxland TR, Tetzlaff W. Animal models used in spinal cord regeneration research. Spine (Phila Pa 1976). 2002;27:1504–10.
37. Rosenzweig ES, McDonald JW. Rodent models for treatment of spinal cord injury: research trends and progress toward useful repair. Curr Opin Neurol. 2004;17:121–31.
38. Stokes BT, Jakeman LB. Experimental modelling of human spinal cord injury: a model that crosses the species barrier and mimics the spectrum of human cytopathology. Spinal Cord. 2002;40:101–9.
39. Faulkner JR, Herrmann JE, Woo MJ, Tansey KE, Doan NB, Sofroniew MV. Reactive astrocytes protect tissue and preserve function after spinal cord injury. J Neurosci. 2004;24:2143–55.
40. Huang WL, George KJ, Ibba V, Liu MC, Averill S, Quartu M, Hamlyn PJ, Priestley JV. The characteristics of neuronal injury in a static compression model of spinal cord injury in adult rats. Eur J Neurosci. 2007;25:362–72.
41. Rivlin AS, Tator CH. Effect of duration of acute spinal cord compression in a new acute cord injury model in the rat. Surg Neurol. 1978;10:38–43.
42. Byrnes KR, Fricke ST, Faden AI. Neuropathological differences between rats and mice after spinal cord injury. J Magn Reson Imaging. 2010;32:836–46.
43. Lee DH, Lee JK. Animal models of axon regeneration after spinal cord injury. Neurosci Bull. 2013;29:436–44.
44. Burris B, Jensen N, Mokalled MH. Assessment of swim endurance and swim behavior in adult zebrafish. J Vis Exp; 2021.
45. Mokalled MH, Patra C, Dickson AL, Endo T, Stainier DY, Poss KD. Injury-induced ctgfa directs glial bridging and spinal cord regeneration in zebrafish. Science. 2016;354:630–4.
46. Briona LK, Poulain FE, Mosimann C, Dorsky RI. Wnt/ss-catenin signaling is required for radial glial neurogenesis following spinal cord injury. Dev Biol. 2015;403:15–21.
47. Cavone L, McCann T, Drake LK, Aguzzi EA, Oprisoreanu AM, Pedersen E, Sandi S, Selvarajah J, Tsarouchas TM, Wehner D, Keatinge M, Mysiak KS, Henderson BEP, Dobie R, Henderson NC, Becker T, Becker CG. A unique macrophage subpopulation signals directly to progenitor cells to promote regenerative neurogenesis in the zebrafish spinal cord. Dev Cell. 2021;56:1617–1630.e1616.
48. Klatt Shaw D, Saraswathy VM, Zhou L, McAdow AR, Burris B, Butka E, Morris SA, Dietmann S, Mokalled MH. Localized EMT reprograms glial progenitors to promote spinal cord repair. Dev Cell. 2021;56:613–626.e7.
49. Kuscha V, Frazer SL, Dias TB, Hibi M, Becker T, Becker CG. Lesion-induced generation of interneuron cell types in specific dorsoventral domains in the spinal cord of adult zebrafish. J Comp Neurol. 2012;520:3604–16.
50. Reimer MM, Sorensen I, Kuscha V, Frank RE, Liu C, Becker CG, Becker T. Motor neuron regeneration in adult zebrafish. J Neurosci. 2008;28:8510–6.
51. Saraswathy VM, Zhou L, McAdow AR, Burris B, Dogra D, Reischauer S, Mokalled MH. Myostatin is a negative regulator of adult neurogenesis after spinal cord injury in zebrafish. Cell Rep. 2022;41:111705.
52. Vandestadt C, Vanwalleghem GC, Khabooshan MA, Douek AM, Castillo HA, Li M, Schulze K, Don E, Stamatis SA, Ratnadiwakara M, Anko ML, Scott EK, Kaslin J. RNA-induced inflammation and migration of precursor neurons initiates neuronal circuit regeneration in zebrafish. Dev Cell. 2021;56:2364–2380.e68.
53. Goldshmit Y, Sztal TE, Jusuf PR, Hall TE, Nguyen-Chi M, Currie PD. Fgf-dependent glial cell bridges facilitate spinal cord regeneration in zebrafish. J Neurosci. 2012;32:7477–92.
54. Blanco-Suarez E, Caldwell AL, Allen NJ. Role of astrocyte-synapse interactions in CNS disorders. J Physiol. 2017;595:1903–16.
55. Paixao S, Klein R. Neuron-astrocyte communication and synaptic plasticity. Curr Opin Neurobiol. 2010;20:466–73.

56. Attwell D, Buchan AM, Charpak S, Lauritzen M, Macvicar BA, Newman EA. Glial and neuronal control of brain blood flow. Nature. 2010;468:232–43.
57. Takano T, Tian GF, Peng W, Lou N, Libionka W, Han X, Nedergaard M. Astrocyte-mediated control of cerebral blood flow. Nat Neurosci. 2006;9:260–7.
58. Ben Haim L, Rowitch DH. Functional diversity of astrocytes in neural circuit regulation. Nat Rev Neurosci. 2017;18:31–41.
59. Rusnakova V, Honsa P, Dzamba D, Stahlberg A, Kubista M, Anderova M. Heterogeneity of astrocytes: from development to injury – single cell gene expression. PLoS One. 2013;8:e69734.
60. Aldskogius H. Repairing CNS myelin--astrocytes have to do their jobs. Exp Neurol. 2005;192:7–10.
61. Ramón y Cajal S, Jones EG, Ramâon y Cajal S, DeFelipe J, Oxford Scholarship Online, N. Cajal's degeneration and regeneration of the nervous system. New York, Oxford: Oxford University Press; 1991.
62. Yang T, Dai Y, Chen G, Cui S. Dissecting the dual role of the glial scar and scar-forming astrocytes in spinal cord injury. Front Cell Neurosci. 2020;14:78.
63. Anjum A, Yazid MD, Fauzi Daud M, Idris J, Ng AMH, Selvi Naicker A, Ismail OHR, Athi Kumar RK, Lokanathan Y. Spinal cord injury: pathophysiology, multimolecular interactions, and underlying recovery mechanisms. Int J Mol Sci. 2020;21
64. Cooper JG, Jeong SJ, McGuire TL, Sharma S, Wang W, Bhattacharyya S, Varga J, Kessler JA. Fibronectin EDA forms the chronic fibrotic scar after contusive spinal cord injury. Neurobiol Dis. 2018;116:60–8.
65. Hawkins LA, Devitt A. Current understanding of the mechanisms for clearance of apoptotic cells-a fine balance. J Cell Death. 2013;6:57–68.
66. Pekny M, Wilhelmsson U, Pekna M. The dual role of astrocyte activation and reactive gliosis. Neurosci Lett. 2014;565:30–8.
67. Auguste KI, Jin S, Uchida K, Yan D, Manley GT, Papadopoulos MC, Verkman AS. Greatly impaired migration of implanted aquaporin-4-deficient astroglial cells in mouse brain toward a site of injury. FASEB J. 2007;21:108–16.
68. Dawley EM, Samson SO, Woodard KT, Matthias KA. Spinal cord regeneration in a tail autotomizing urodele. J Morphol. 2012;273:211–25.
69. Yuan YM, He C. The glial scar in spinal cord injury and repair. Neurosci Bull. 2013;29:421–35.
70. Myer DJ, Gurkoff GG, Lee SM, Hovda DA, Sofroniew MV. Essential protective roles of reactive astrocytes in traumatic brain injury. Brain. 2006;129:2761–72.
71. Okada S, Nakamura M, Katoh H, Miyao T, Shimazaki T, Ishii K, Yamane J, Yoshimura A, Iwamoto Y, Toyama Y, Okano H. Conditional ablation of Stat3 or Socs3 discloses a dual role for reactive astrocytes after spinal cord injury. Nat Med. 2006;12:829–34.
72. Sofroniew MV. Molecular dissection of reactive astrogliosis and glial scar formation. Trends Neurosci. 2009;32:638–47.
73. Andrews EM, Richards RJ, Yin FQ, Viapiano MS, Jakeman LB. Alterations in chondroitin sulfate proteoglycan expression occur both at and far from the site of spinal contusion injury. Exp Neurol. 2012;235:174–87.
74. Galtrey CM, Fawcett JW. The role of chondroitin sulfate proteoglycans in regeneration and plasticity in the central nervous system. Brain Res Rev. 2007;54:1–18.
75. Hara M, Kobayakawa K, Ohkawa Y, Kumamaru H, Yokota K, Saito T, Kijima K, Yoshizaki S, Harimaya K, Nakashima Y, Okada S. Interaction of reactive astrocytes with type I collagen induces astrocytic scar formation through the integrin-N-cadherin pathway after spinal cord injury. Nat Med. 2017;23:818–28.
76. Massey JM, Amps J, Viapiano MS, Matthews RT, Wagoner MR, Whitaker CM, Alilain W, Yonkof AL, Khalyfa A, Cooper NG, Silver J, Onifer SM. Increased chondroitin sulfate proteoglycan expression in denervated brainstem targets following spinal cord injury creates a barrier to axonal regeneration overcome by chondroitinase ABC and neurotrophin-3. Exp Neurol. 2008;209:426–45.

77. Massey JM, Hubscher CH, Wagoner MR, Decker JA, Amps J, Silver J, Onifer SM. Chondroitinase ABC digestion of the perineuronal net promotes functional collateral sprouting in the cuneate nucleus after cervical spinal cord injury. J Neurosci. 2006;26:4406–14.
78. Liddelow SA, Barres BA. Reactive astrocytes: production, function, and therapeutic potential. Immunity. 2017;46:957–67.
79. McKeon RJ, Schreiber RC, Rudge JS, Silver J. Reduction of neurite outgrowth in a model of glial scarring following CNS injury is correlated with the expression of inhibitory molecules on reactive astrocytes. J Neurosci. 1991;11:3398–411.
80. Wilhelmsson U, Bushong EA, Price DL, Smarr BL, Phung V, Terada M, Ellisman MH, Pekny M. Redefining the concept of reactive astrocytes as cells that remain within their unique domains upon reaction to injury. Proc Natl Acad Sci USA. 2006;103:17513–8.
81. Balasingam V, Tejada-Berges T, Wright E, Bouckova R, Yong VW. Reactive astrogliosis in the neonatal mouse brain and its modulation by cytokines. J Neurosci. 1994;14:846–56.
82. Rabchevsky AG, Weinitz JM, Coulpier M, Fages C, Tinel M, Junier MP. A role for transforming growth factor alpha as an inducer of astrogliosis. J Neurosci. 1998;18:10541–52.
83. Winter CG, Saotome Y, Levison SW, Hirsh D. A role for ciliary neurotrophic factor as an inducer of reactive gliosis, the glial response to central nervous system injury. Proc Natl Acad Sci USA. 1995;92:5865–9.
84. Ambrosini E, Remoli ME, Giacomini E, Rosicarelli B, Serafini B, Lande R, Aloisi F, Coccia EM. Astrocytes produce dendritic cell-attracting chemokines in vitro and in multiple sclerosis lesions. J Neuropathol Exp Neurol. 2005;64:706–15.
85. Lo U, Selvaraj V, Plane JM, Chechneva OV, Otsu K, Deng W. p38alpha (MAPK14) critically regulates the immunological response and the production of specific cytokines and chemokines in astrocytes. Sci Rep. 2014;4:7405.
86. Duarte Azevedo M, Sander S, Tenenbaum L. GDNF, A neuron-derived factor upregulated in glial cells during disease. J Clin Med. 2020;9
87. Saba J, Turati J, Ramirez D, Carniglia L, Durand D, Lasaga M, Caruso C. Astrocyte truncated tropomyosin receptor kinase B mediates brain-derived neurotrophic factor anti-apoptotic effect leading to neuroprotection. J Neurochem. 2018;146:686–702.
88. Jo S, Yarishkin O, Hwang YJ, Chun YE, Park M, Woo DH, Bae JY, Kim T, Lee J, Chun H, Park HJ, Lee DY, Hong J, Kim HY, Oh SJ, Park SJ, Lee H, Yoon BE, Kim Y, Jeong Y, Shim I, Bae YC, Cho J, Kowall NW, Ryu H, Hwang E, Kim D, Lee CJ. GABA from reactive astrocytes impairs memory in mouse models of Alzheimer's disease. Nat Med. 2014;20:886–96.
89. Oh SJ, Han KS, Park H, Woo DH, Kim HY, Traynelis SF, Lee CJ. Protease activated receptor 1-induced glutamate release in cultured astrocytes is mediated by Bestrophin-1 channel but not by vesicular exocytosis. Mol Brain. 2012;5:38.
90. Hsu JY, Bourguignon LY, Adams CM, Peyrollier K, Zhang H, Fandel T, Cun CL, Werb Z, Noble-Haeusslein LJ. Matrix metalloproteinase-9 facilitates glial scar formation in the injured spinal cord. J Neurosci. 2008;28:13467–77.
91. Li X, Li M, Tian L, Chen J, Liu R, Ning B. Reactive astrogliosis: implications in spinal cord injury progression and therapy. Oxidative Med Cell Longev. 2020;2020:9494352.
92. Dias DO, Goritz C. Fibrotic scarring following lesions to the central nervous system. Matrix Biol. 2018;68-69:561–70.
93. Grimpe B, Silver J. A novel DNA enzyme reduces glycosaminoglycan chains in the glial scar and allows microtransplanted dorsal root ganglia axons to regenerate beyond lesions in the spinal cord. J Neurosci. 2004;24:1393–7.
94. Sofroniew MV. Multiple roles for astrocytes as effectors of cytokines and inflammatory mediators. Neuroscientist. 2014;20:160–72.
95. Escartin C, Galea E, Lakatos A, O'Callaghan JP, Petzold GC, Serrano-Pozo A, Steinhauser C, Volterra A, Carmignoto G, Agarwal A, Allen NJ, Araque A, Barbeito L, Barzilai A, Bergles DE, Bonvento G, Butt AM, Chen WT, Cohen-Salmon M, Cunningham C, Deneen B, De Strooper B, Diaz-Castro B, Farina C, Freeman M, Gallo V, Goldman JE, Goldman SA, Gotz M, Gutierrez A, Haydon PG, Heiland DH, Hol EM, Holt MG, Iino M, Kastanenka KV, Kettenmann H, Khakh BS, Koizumi S, Lee CJ, Liddelow SA, MacVicar BA, Magistretti P,

Messing A, Mishra A, Molofsky AV, Murai KK, Norris CM, Okada S, Oliet SHR, Oliveira JF, Panatier A, Parpura V, Pekna M, Pekny M, Pellerin L, Perea G, Perez-Nievas BG, Pfrieger FW, Poskanzer KE, Quintana FJ, Ransohoff RM, Riquelme-Perez M, Robel S, Rose CR, Rothstein JD, Rouach N, Rowitch DH, Semyanov A, Sirko S, Sontheimer H, Swanson RA, Vitorica J, Wanner IB, Wood LB, Wu J, Zheng B, Zimmer ER, Zorec R, Sofroniew MV, Verkhratsky A. Reactive astrocyte nomenclature, definitions, and future directions. Nat Neurosci. 2021;24:312–25.
96. Zamanian JL, Xu L, Foo LC, Nouri N, Zhou L, Giffard RG, Barres BA. Genomic analysis of reactive astrogliosis. J Neurosci. 2012;32:6391–410.
97. Bowman CL, Ding JP, Sachs F, Sokabe M. Mechanotransducing ion channels in astrocytes. Brain Res. 1992;584:272–86.
98. Cullen DK, Vernekar VN, LaPlaca MC. Trauma-induced plasmalemma disruptions in three-dimensional neural cultures are dependent on strain modality and rate. J Neurotrauma. 2011;28:2219–33.
99. Floyd CL, Gorin FA, Lyeth BG. Mechanical strain injury increases intracellular sodium and reverses Na+/Ca2+ exchange in cortical astrocytes. Glia. 2005;51:35–46.
100. Islas L, Pasantes-Morales H, Sanchez JA. Characterization of stretch-activated ion channels in cultured astrocytes. Glia. 1993;8:87–96.
101. Hoffman SW, Rzigalinski BA, Willoughby KA, Ellis EF. Astrocytes generate isoprostanes in response to trauma or oxygen radicals. J Neurotrauma. 2000;17:415–20.
102. Neary JT, Kang Y, Tran M, Feld J. Traumatic injury activates protein kinase B/Akt in cultured astrocytes: role of extracellular ATP and P2 purinergic receptors. J Neurotrauma. 2005;22:491–500.
103. Neary JT, Kang Y, Willoughby KA, Ellis EF. Activation of extracellular signal-regulated kinase by stretch-induced injury in astrocytes involves extracellular ATP and P2 purinergic receptors. J Neurosci. 2003;23:2348–56.
104. Ostrow LW, Suchyna TM, Sachs F. Stretch induced endothelin-1 secretion by adult rat astrocytes involves calcium influx via stretch-activated ion channels (SACs). Biochem Biophys Res Commun. 2011;410:81–6.
105. Giulian D, Lachman LB. Interleukin-1 stimulation of astroglial proliferation after brain injury. Science. 1985;228:497–9.
106. Selmaj KW, Farooq M, Norton WT, Raine CS, Brosnan CF. Proliferation of astrocytes in vitro in response to cytokines. A primary role for tumor necrosis factor. J Immunol. 1990;144:129–35.
107. Liddelow SA, Guttenplan KA, Clarke LE, Bennett FC, Bohlen CJ, Schirmer L, Bennett ML, Munch AE, Chung WS, Peterson TC, Wilton DK, Frouin A, Napier BA, Panicker N, Kumar M, Buckwalter MS, Rowitch DH, Dawson VL, Dawson TM, Stevens B, Barres BA. Neurotoxic reactive astrocytes are induced by activated microglia. Nature. 2017;541:481–7.
108. Ridet JL, Malhotra SK, Privat A, Gage FH. Reactive astrocytes: cellular and molecular cues to biological function. Trends Neurosci. 1997;20:570–7.
109. Herrmann JE, Imura T, Song B, Qi J, Ao Y, Nguyen TK, Korsak RA, Takeda K, Akira S, Sofroniew MV. STAT3 is a critical regulator of astrogliosis and scar formation after spinal cord injury. J Neurosci. 2008;28:7231–43.
110. Gris P, Tighe A, Levin D, Sharma R, Brown A. Transcriptional regulation of scar gene expression in primary astrocytes. Glia. 2007;55:1145–55.
111. Lee H, McKeon RJ, Bellamkonda RV. Sustained delivery of thermostabilized chABC enhances axonal sprouting and functional recovery after spinal cord injury. Proc Natl Acad Sci USA. 2010a;107:3340–5.
112. Smith-Thomas LC, Stevens J, Fok-Seang J, Faissner A, Rogers JH, Fawcett JW. Increased axon regeneration in astrocytes grown in the presence of proteoglycan synthesis inhibitors. J Cell Sci. 1995;108(Pt 3):1307–15.

113. Barkho BZ, Song H, Aimone JB, Smrt RD, Kuwabara T, Nakashima K, Gage FH, Zhao X. Identification of astrocyte-expressed factors that modulate neural stem/progenitor cell differentiation. Stem Cells Dev. 2006;15:407–21.
114. Siebert JR, Osterhout DJ. The inhibitory effects of chondroitin sulfate proteoglycans on oligodendrocytes. J Neurochem. 2011;119:176–88.
115. Siebert JR, Stelzner DJ, Osterhout DJ. Chondroitinase treatment following spinal contusion injury increases migration of oligodendrocyte progenitor cells. Exp Neurol. 2011;231:19–29.
116. Hammond TR, Gadea A, Dupree J, Kerninon C, Nait-Oumesmar B, Aguirre A, Gallo V. Astrocyte-derived endothelin-1 inhibits remyelination through notch activation. Neuron. 2014;81:588–602.
117. Wang Y, Cheng X, He Q, Zheng Y, Kim DH, Whittemore SR, Cao QL. Astrocytes from the contused spinal cord inhibit oligodendrocyte differentiation of adult oligodendrocyte precursor cells by increasing the expression of bone morphogenetic proteins. J Neurosci. 2011;31:6053–8.
118. Tator CH, Fehlings MG. Review of the secondary injury theory of acute spinal cord trauma with emphasis on vascular mechanisms. J Neurosurg. 1991;75:15–26.
119. Garcia TA, Jonak CR, Binder DK. The role of aquaporins in spinal cord injury. Cells. 2023;12
120. Bozzo A, Marcoux J, Radhakrishna M, Pelletier J, Goulet B. The role of magnetic resonance imaging in the management of acute spinal cord injury. J Neurotrauma. 2011;28:1401–11.
121. Miyanji F, Furlan JC, Aarabi B, Arnold PM, Fehlings MG. Acute cervical traumatic spinal cord injury: MR imaging findings correlated with neurologic outcome – prospective study with 100 consecutive patients. Radiology. 2007;243:820–7.
122. Shepard MJ, Bracken MB. Magnetic resonance imaging and neurological recovery in acute spinal cord injury: observations from the National Acute Spinal Cord Injury Study 3. Spinal Cord. 1999;37:833–7.
123. Oshio K, Binder DK, Yang B, Schecter S, Verkman AS, Manley GT. Expression of aquaporin water channels in mouse spinal cord. Neuroscience. 2004;127:685–93.
124. Basso DM, Fisher LC, Anderson AJ, Jakeman LB, McTigue DM, Popovich PG. Basso Mouse Scale for locomotion detects differences in recovery after spinal cord injury in five common mouse strains. J Neurotrauma. 2006;23:635–59.
125. Saadoun S, Bell BA, Verkman AS, Papadopoulos MC. Greatly improved neurological outcome after spinal cord compression injury in AQP4-deficient mice. Brain. 2008;131:1087–98.
126. Kimura A, Hsu M, Seldin M, Verkman AS, Scharfman HE, Binder DK. Protective role of aquaporin-4 water channels after contusion spinal cord injury. Ann Neurol. 2010;67:794–801.
127. Koukoulithras I, Alkhazi A, Gkampenis A, Stamouli A, Plexousakis M, Drousia G, Xanthi E, Roussos C, Kolokotsios S. A systematic review of the interventions for Management of Pain in patients after spinal cord injury. Cureus. 2023;15:e42657.
128. Miranpuri GS, Bali P, Nguyen J, Kim JJ, Modgil S, Mehra P, Buttar S, Brown G, Yutuc N, Singh H, Wood A, Singh J, Anand A. Role of microglia and astrocytes in spinal cord injury induced neuropathic pain. Ann Neurosci. 2021;28:219–28.
129. Hu Y, Li W, Lu L, Cai J, Xian X, Zhang M, Li Q, Li L. An anti-nociceptive role for ceftriaxone in chronic neuropathic pain in rats. Pain. 2010;148:284–301.
130. Maeda S, Kawamoto A, Yatani Y, Shirakawa H, Nakagawa T, Kaneko S. Gene transfer of GLT-1, a glial glutamate transporter, into the spinal cord by recombinant adenovirus attenuates inflammatory and neuropathic pain in rats. Mol Pain. 2008;4:65.
131. Tawfik VL, Regan MR, Haenggeli C, Lacroix-Fralish ML, Nutile-McMenemy N, Perez N, Rothstein JD, DeLeo JA. Propentofylline-induced astrocyte modulation leads to alterations in glial glutamate promoter activation following spinal nerve transection. Neuroscience. 2008;152:1086–92.
132. Liu X, Liu H, Xu S, Tang Z, Xia W, Cheng Z, Li W, Jin Y. Spinal translocator protein alleviates chronic neuropathic pain behavior and modulates spinal astrocyte-neuronal function in rats with L5 spinal nerve ligation model. Pain. 2016;157:103–16.
133. Griggs RB, Donahue RR, Morgenweck J, Grace PM, Sutton A, Watkins LR, Taylor BK. Pioglitazone rapidly reduces neuropathic pain through astrocyte and nongenomic PPARgamma mechanisms. Pain. 2015;156:469–82.

134. Zhang FF, Morioka N, Kitamura T, Fujii S, Miyauchi K, Nakamura Y, Hisaoka-Nakashima K, Nakata Y. Lycopene ameliorates neuropathic pain by upregulating spinal astrocytic connexin 43 expression. Life Sci. 2016;155:116–22.
135. Lee-Kubli CA, Ingves M, Henry KW, Shiao R, Collyer E, Tuszynski MH, Campana WM. Analysis of the behavioral, cellular and molecular characteristics of pain in severe rodent spinal cord injury. Exp Neurol. 2016;278:91–104.
136. Kim H, Leng K, Park J, Sorets AG, Kim S, Shostak A, Embalabala RJ, Mlouk K, Katdare KA, Rose IVL, Sturgeon SM, Neal EH, Ao Y, Wang S, Sofroniew MV, Brunger JM, McMahon DG, Schrag MS, Kampmann M, Lippmann ES. Reactive astrocytes transduce inflammation in a blood-brain barrier model through a TNF-STAT3 signaling axis and secretion of alpha 1-antichymotrypsin. Nat Commun. 2022;13:6581.
137. McKeon RJ, Jurynec MJ, Buck CR. The chondroitin sulfate proteoglycans neurocan and phosphacan are expressed by reactive astrocytes in the chronic CNS glial scar. J Neurosci. 1999;19:10778–88.
138. Wang R, Zhou R, Chen Z, Gao S, Zhou F. The glial cells respond to spinal cord injury. Front Neurol. 2022;13:844497.
139. Yu SS, Li ZY, Xu XZ, Yao F, Luo Y, Liu YC, Cheng L, Zheng MG, Jing JH. M1-type microglia can induce astrocytes to deposit chondroitin sulfate proteoglycan after spinal cord injury. Neural Regen Res. 2022;17:1072–9.
140. Ziff OJ, Clarke BE, Taha DM, Crerar H, Luscombe NM, Patani R. Meta-analysis of human and mouse ALS astrocytes reveals multi-omic signatures of inflammatory reactive states. Genome Res. 2022;32:71–84.
141. Pitter KL, Tamagno I, Feng X, Ghosal K, Amankulor N, Holland EC, Hambardzumyan D. The SHH/Gli pathway is reactivated in reactive glia and drives proliferation in response to neurodegeneration-induced lesions. Glia. 2014;62:1595–607.
142. Liang J, Takeuchi H, Doi Y, Kawanokuchi J, Sonobe Y, Jin S, Yawata I, Li H, Yasuoka S, Mizuno T, Suzumura A. Excitatory amino acid transporter expression by astrocytes is neuroprotective against microglial excitotoxicity. Brain Res. 2008;1210:11–9.
143. Loov C, Hillered L, Ebendal T, Erlandsson A. Engulfing astrocytes protect neurons from contact-induced apoptosis following injury. PLoS One. 2012;7:e33090.
144. Morizawa YM, Hirayama Y, Ohno N, Shibata S, Shigetomi E, Sui Y, Nabekura J, Sato K, Okajima F, Takebayashi H, Okano H, Koizumi S. Reactive astrocytes function as phagocytes after brain ischemia via ABCA1-mediated pathway. Nat Commun. 2017;8:28.
145. Chen J, Poskanzer KE, Freeman MR, Monk KR. Live-imaging of astrocyte morphogenesis and function in zebrafish neural circuits. Nat Neurosci. 2020;23:1297–306.
146. Dervan AG, Roberts BL. Reaction of spinal cord central canal cells to cord transection and their contribution to cord regeneration. J Comp Neurol. 2003;458:293–306.
147. Wehner D, Tsarouchas TM, Michael A, Haase C, Weidinger G, Reimer MM, Becker T, Becker CG. Wnt signaling controls pro-regenerative Collagen XII in functional spinal cord regeneration in zebrafish. Nat Commun. 2017;8:126.
148. Zhou L, McAdow AR, Yamada H, Burris B, Klatt Shaw D, Oonk K, Poss KD, Mokalled MH. Progenitor-derived glia are required for spinal cord regeneration in zebrafish. Development. 2023;150:dev201162.
149. Filous AR, Miller JH, Coulson-Thomas YM, Horn KP, Alilain WJ, Silver J. Immature astrocytes promote CNS axonal regeneration when combined with chondroitinase ABC. Dev Neurobiol. 2010;70:826–41.
150. White RE, Yin FQ, Jakeman LB. TGF-alpha increases astrocyte invasion and promotes axonal growth into the lesion following spinal cord injury in mice. Exp Neurol. 2008;214:10–24.
151. Zukor K, Belin S, Wang C, Keelan N, Wang X, He Z. Short hairpin RNA against PTEN enhances regenerative growth of corticospinal tract axons after spinal cord injury. J Neurosci. 2013;33:15350–61.

Astrocyte Regulation of Neuronal Function and Survival in Stroke Pathophysiology

Bridget R. Boyle, Andrea P. Berghella, and Elena Blanco-Suarez

Abstract The interactions between astrocytes and neurons in the context of stroke play crucial roles in the disease's progression and eventual outcomes. After a stroke, astrocytes undergo significant changes in their morphology, molecular profile, and function, together termed reactive astrogliosis. Many of these changes modulate how astrocytes relate to neurons, inducing mechanisms both beneficial and detrimental to stroke recovery. For example, excessive glutamate release and astrocytic malfunction contribute to excitotoxicity in stroke, eventually causing neuronal death. Astrocytes also provide essential metabolic support and neurotrophic signals to neurons after stroke, ensuring homeostatic stability and promoting neuronal survival. Furthermore, several astrocyte-secreted molecules regulate synaptic plasticity in response to stroke, allowing for the rewiring of neural circuits to compensate for damaged areas. In this chapter, we highlight the current understanding of the interactions between astrocytes and neurons in response to stroke, explaining the varied mechanisms contributing to injury progression and the potential implications for future therapeutic interventions.

Bridget R. Boyle and Andrea P. Berghella contributed equally.

B. R. Boyle · A. P. Berghella
Department of Neuroscience, Vickie & Jack Farber Institute for Neuroscience, Sidney Kimmel Medical College, Thomas Jefferson University, Philadelphia, PA, USA

Jefferson College of Life Sciences, Thomas Jefferson University, Philadelphia, PA, USA

E. Blanco-Suarez (✉)
Department of Neuroscience, Vickie & Jack Farber Institute for Neuroscience, Sidney Kimmel Medical College, Thomas Jefferson University, Philadelphia, PA, USA

Jefferson College of Life Sciences, Thomas Jefferson University, Philadelphia, PA, USA

Department of Neurological Surgery, Vickie & Jack Farber Institute for Neuroscience, Sidney Kimmel Medical College, Thomas Jefferson University, Philadelphia, PA, USA
e-mail: elena.blancosuarez@jefferson.edu

E. Blanco-Suarez, I. Farhy-Tselnicker (eds.), *Astrocyte-Neuron Interactions in Health and Disease*, Advances in Neurobiology 39,
https://doi.org/10.1007/978-3-031-64839-7_10

Keywords Astrocyte · Stroke · Reactive astrogliosis · Excitotoxicity · Synaptic plasticity

Abbreviations

ABCA1	ATP-binding cassette transporter A1
ACM	Astrocyte-conditioned medium
AMPA	α-amino-3-hydroxy-5-methyl-4-isoxazolepropionic acid
ATP	Adenosine triphosphate
BCCA	Bilateral occlusion of the common carotid arteries
BDNF	Brain-derived neurotrophic factor
CCA	Common carotid artery
Chrdl1	Chordin-like 1
CNTF	Ciliary neurotrophic factor
CP-AMPAR	Ca^{2+}-permeable AMPAR
CSPGs	Chondroitin sulfate proteoglycans
DCPIB	4-[(2-butyl-6,7-dichloro-2-cyclopentyl-2,3-dihydro-1-oxo-1*H*--inden-5-yl)oxy] butanoic acid
EPO	Erythropoietin
ER	Endoplasmic reticulum
FGF2	Fibroblast growth factor 2
GABA	Gamma-aminobutyric acid
GATs	GABA transporters
GDNF	Glial cell line-derived neurotrophic factor
GFAP	Glial fibrillary acidic protein
GLAST	Glutamate-aspartate transporter
GLT-1	Glutamate transporter-1
GluTs	Glutamate transporters
GULP1	Engulfment adaptor PTB domain 1
HMGB1	High-mobility group box 1
IGF-1	Insulin-like growth factor-1
IL-15	Interleukin 15
IL-17A	interleukin-17A
IP3Rs	Inositol 1,4,5-triphosphate receptors
Lrp4	Lipoprotein receptor-related protein 4
MCAO	Middle cerebral artery occlusion
MEGF10	Multiple EGF-like-domains 10
MERKT	Proto-oncogene tyrosine-protein kinase MER
NASPM	1-naphthylacetyl spermine
NCX	Na^+-calcium Ca^{2+} exchanger
NF-κB	Nuclear factor kappa-light-chain-enhancer
NGF	Nerve growth factor
NKCC1	Sodium-potassium-chloride co-transporter

NMDA	N-methyl d-aspartate
Nrf2	Nuclear factor erythroid 2-related factor 2
NSPC	Neural stem/progenitor cell
ODNs	Oligodeoxynucleotides
OGD	Oxygen and glucose deprivation
pMCAO	Permanent middle cerebral arterial occlusion
PTPσ	Protein tyrosine phosphatase sigma
RGC	Retinal ganglion cell
ROS	Reactive oxygen species
SDF-1α	Stromal cell-derived factor 1α
SPARC	Secreted protein acidic and cysteine rich
SVZ	Subventricular zone
TGF-β	Transforming growth factor-beta
tMCAO	Transient middle cerebral arterial occlusion
TSP1, TSP2	Thrombospondin 1 and thrombospondin 2
VRAC	Volume-regulated anion channels
2-VO	Global ischemia by two-vessel occlusion
4-VO	Global ischemia by four-vessel occlusion

1 Introduction

The interactions between astrocytes and neurons change dramatically and take on unique significance in the context of stroke. During a stroke, blood flow to a region of the brain is acutely disrupted, eventually leading to tissue necrosis, functional deficits, and potentially death. Stroke is generally classified into two broad subtypes: hemorrhagic stroke, defined by extravasation of blood in the brain from a ruptured blood vessel, and ischemic stroke, due to the blockage of an artery, typically by a thrombus. Regardless of etiology, post-stroke sequelae follow a series of region- and time-dependent responses [1, 2]. In the area most severely impacted by ischemia, called the core of the injury or infarct, major cell death occurs, and it is considered irrecoverable. The tissue surrounding the core, called the peri-infarct (or penumbra in clinical settings), can be saved and undergoes complex changes that imbue it with a great capacity for neural repair [1, 3]. Time-dependent responses occur throughout several post-stroke stages that have been identified in both human and murine subjects [2, 4–6]. In the hyperacute phase immediately following a stroke (0–6 h post-injury in rodents; 0–24 h post-injury in humans), there is initiation of cell death pathways, edema, and disruption of the blood-brain barrier [7–10]. Subsequently, the acute phase (0–2 days post-injury in rodents; 1–7 days post-injury in humans) is characterized by further neuronal loss and commencement of inflammatory mechanisms and scar formation [11, 12]. The subacute phase (2–30 days post-injury in rodents; 7 days-6 months post-injury in humans) follows, featuring elevated levels of endogenous plasticity that enable a certain level of spontaneous repair. This phase plays a pivotal role in determining the extent of functional recovery witnessed in the last phase, referred to as the chronic phase (>30 days post-injury in rodents; >6 months post-injury in humans) when the injury and deficits stabilize [2, 4, 6, 13].

The involvement of astrocytes and their impact on neuronal recovery are common threads in the intricate cascade of events that follow the onset and progression of stroke. Astrocytic responses to stroke are multifaceted: Early after injury, a series of morphological and molecular changes take place, collectively known as reactive astrogliosis. Several astrocytic functions are altered, including blood-brain barrier maintenance, regulation of cerebral blood flow, glutamate homeostasis, metabolic support, modulation of synaptic plasticity, and neurogenesis. Astrocyte responses can be both beneficial and detrimental to neuronal restoration, depending on the mechanism at play, type of stroke, location, severity, and time post-stroke [14].

In this chapter, we explore the integral roles that astrocytes play in stroke, highlighting their functional interactions with neurons in this context. A better understanding of this crucial interplay will give insight into potential avenues to harness astrocytic function in the development of new neuroprotective and neurorestorative therapies.

2 Models of Stroke: The Study of Astrocyte-Neuron Interactions In Vitro and In Vivo

Throughout decades of research, different models of stroke, both in vitro and in vivo, have been developed and widely used. Every model has its advantages and disadvantages and researchers must carefully assess which model is best depending on their goals (see Table 1). For example, middle cerebral artery occlusion (MCAO) is considered the most clinically relevant model to reproduce human stroke in the rodent brain. However, other techniques such as photothrombosis or endothelin-1 occlusion generate well-defined focal strokes, and they may represent more convenient models for certain preclinical studies due to their higher reproducibility compared to MCAO.

3 Reactive Astrogliosis: Contributions to the Scar Border and Neuroinflammation

Following a stroke, astrocytes undergo a series of morphological, transcriptional, and functional changes termed reactive astrogliosis [42, 43]. This complex process is non-uniform, depending on several factors including the type of insult, time since the onset of the injury, and distance from the lesion [44].

Table 1 Models of stroke: the study of astrocyte-neuron interactions in vitro and in vivo

Stroke model		Technique	Description	Advantages	Disadvantages	References
Ischemic	In vitro	Oxygen and glucose deprivation (OGD)	Primary astrocytes alone, or co-cultured with neurons, are incubated with glucose-free medium in a deoxygenated environment. Cells may be reperfused with glucose-containing medium and oxygenated atmosphere	Can be used to study molecular pathways of cell death after ischemic conditions	Absence of blood vessels and blood flow Lack of peripheral cell infiltration	[15–17]
	In vivo	Permanent middle cerebral arterial occlusion (pMCAO)	The middle cerebral artery (MCA) is directly occluded following craniotomy using filament insertion, suture ligation, photothrombotic, or microvascular clip application A filament is inserted into the MCA and remained there to block blood flow	Exhibits large peri-infarct area One-day surgery	High mortality rate Variable infarct size depends on rodent strain	[18–20]
		Transient middle cerebral arterial occlusion (tMCAO)	A filament is inserted into the MCA to block blood flow for a set amount of time before reperfusion	Allows regulated reperfusion Exhibits large peri-infarct area	High mortality rate Infarct size is variable and depends on rodent strain Second surgery to remove filament	[18, 21–23]
		Photothrombosis (PT)	Localized stroke created by systemic injection of photoactivatable dye, rose Bengal, and stereotaxic location on skull exposed to light	Produce region and size-specific stroke Less invasive Highly reproducible Low mortality rate Can be adapted to target different areas	Small infarct No reperfusion injury Minimal peri-infarct region around injury	[24–26]
		Global ischemia by two-vessel occlusion (2-VO)	Reversible or permanent ischemia produced by bilateral common carotid artery (CCA) occlusion	One-step surgery Produces large infarct ischemia Low mortality Controllable reperfusion Feasible for chronic survival studies	Potential development of seizures High variability High mortality rate	[27–29]

Table 1 (continued)

Stroke model		Technique	Description	Advantages	Disadvantages	References
		Global ischemia by four-vessel occlusion (4-VO)	Two-part procedure A reversible CCA occlusion, done sequentially with permanent interruption of the vertebral arteries via electrocauterization	High reproducibility Low rate of seizure Partially reversible	Two-day required surgery Rodent strain-dependent mortality rate Permanently occluded vertebral arteries Very invasive	[30, 31]
Hemorrhagic	In vitro	Hemoglobin, heme, or hemin addition	Astrocyte and neuron co-culture treated with different concentrations of heme, hemin, or hemoglobin	Analyze secondary injury such as hematoma expansion and iron levels/ dynamics Easy and reproducible	Limited injury due to lack of severe hemotoxicity No vascular damage represented	[32–34]
	In vivo	Intracerebral blood injection	Stereotaxic intracerebral injection of autologous blood	Replicates hemotoxicity, immuno-reactivity, and astrocytic response Controlled size and injury location	Does not replicate rupture of blood vessels	[35, 36]
		Collagenase injection	Stereotaxic intracerebral injection of bacterial collagenase	Mimics spontaneous injury Triggers reactive astrocytes and gene expression changes Reproducible hematoma size and location	Slow and small bleeding that evolves over the first few hours	[37–39]
		Endovascular puncture of the intracranial arteries	A sharp suture is inserted into the anterior cerebral artery; the suture is then used to puncture the vessel and removed directly after	Replicates spontaneous aneurysm Controlled stroke location and severity Minimized surgical trauma	Variable stroke infarcts Causes more severe bleeding and mortality than seen in the clinic	[40, 41]

3.1 Formation of the Scar Border

Reactive astrocytes participate in the formation of the scar border that segregates healthy tissue from the injury core (Fig. 1) [12]. Contrasting research has made it difficult to define whether reactive astrogliosis and scar border formation are overall beneficial or detrimental processes for stroke recovery. For example, reactive astrocytes in the scar border secrete chondroitin sulfate proteoglycans (CSPGs), a family of extracellular matrix proteins known to be antagonistic to axonal growth [45]. Enzymatic degradation of CSPGs with chondroitinase ABC leads to enhanced neurite outgrowth through the scar border [46], as well as improved axonal sprouting, neuronal survival, and behavioral outcomes after stroke in rodents, even with delayed administration [47–51]. More targeted modulation, such as ablation of sox9, a transcription factor key for CSPG upregulation [52], or use of a peptide inhibitor against the CSPG receptor protein tyrosine phosphatase sigma (PTPσ) [53] show similarly beneficial results.

However, there is also evidence that attenuation of reactive astrogliosis stalls recovery from stroke. For example, double knockout of the glial fibrillary acidic protein (GFAP) and vimentin, two intermediate filament proteins and markers of reactive astrogliosis, significantly decreases glutamate transporter-1 (GLT-1)-mediated glutamate transport and increases injury volume, both detrimental phenotypes after middle cerebral artery transection [54]. The same double KO mice show

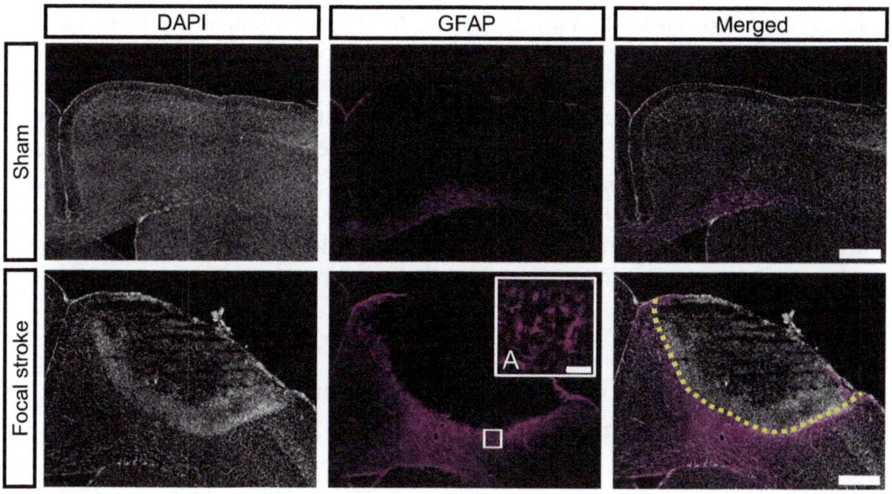

Fig. 1 Reactive astrocytes form a scar border surrounding the core of the ischemic injury. Immunofluorescence images showing DAPI (white) and GFAP (magenta) staining in the ipsilateral motor cortex 7 days after sham (upper panels) or photothrombotic focal stroke (lower panels) in the mouse brain. The dashed yellow line indicates the boundary of the core of the injury in the photothrombotic focal stroke (scale bar 500 μm). Zoomed-in image (A) shows GFAP+ astrocytes in the scar border in response to focal stroke (scale bar 50 μm). (Images provided by the Blanco-Suarez laboratory (2024))

a reduction in corticospinal tract axonal length, aberrant restoration of global neuronal connectivity, maladaptive plasticity responses, and deficient sensorimotor functional recovery after photothrombotic stroke in the motor cortex [55, 56]. Genetic stimulation of reactive astrogliosis via leucine zipper-bearing kinase overexpression enhances corticospinal axon sprouting post-stroke, representing a potential avenue for neuronal repair [57].

3.2 *Neuroinflammation*

Reactive astrocytes also contribute to the neuroinflammatory response after stroke, secreting a host of inflammatory factors including cytokines, chemokines, and reactive oxygen/nitrogen species [58, 59]. Transcriptomic analyses of reactive astrocytes after stroke have found upregulation in both neuroprotective and neurotoxic genetic profiles [60, 61], suggesting again both beneficial and detrimental roles. One such factor is the cytokine IL-15, found to be significantly upregulated in astrocytes both in postmortem human stroke tissue and in mouse models of stroke [62]. Induction of IL-15 specifically in astrocytes causes an increase in infarct size and exacerbates sensorimotor deficits after MCAO [62]. Knockout of IL-15 or the use of IL-15 neutralizing antibodies has opposite effects, resulting in smaller stroke volumes and improved motor activity [63]. In concordance with a detrimental role of astrocyte-mediated neuroinflammatory mechanisms in stroke, inactivation of astrocytic NF-κB, an upstream regulator of the neuroinflammatory response after stroke [64], promotes survival of retinal neurons following ischemic injury [65]. This evidence indicates that restriction of neuroinflammatory triggers from astrocytes may limit stroke damage and neuronal cell death, translating into better functional recovery. However, astrocytes can also function in an anti-inflammatory manner, underscoring their multifaceted role. For example, astrocyte-mediated TGF-β signaling contributes to stroke recovery, as blockage of this pathway in astrocytes enhances the infiltration of immune cells into the peri-infarct region, leading to worse motor outcomes and late infarct expansion [66].

4 Excitotoxicity: Astrocytic Influence on Neuronal Death and Neurotransmission After Stroke

Excitotoxicity is a mechanism triggered by excessive glutamate release from neurons and glial cells, usually in response to injurious signals, such as stroke, triggering cascades that ultimately lead to neuronal death [8, 9]. In healthy conditions, astrocytes regulate extracellular glutamate levels in the synaptic cleft via glutamate transporters (GluTs) [67], but this function is impaired under many pathological conditions. Following ischemic stroke, a significant decline in ATP production leads to the dysfunction of the Na^+/K^+ ion pumps responsible for maintaining the neuronal resting potential [68, 69], leading to neuronal depolarization, and excessive release of glutamate. This is further intensified by impaired glutamate reuptake by astrocytes due to energy failure [70, 71]. The excess glutamate activates postsynaptic glutamate receptors, facilitating excessive Ca^{2+} influx into neurons through Ca^{2+}-permeable receptors and increasing intracellular Ca^{2+} concentration to pathological levels, ultimately leading to neuronal death [72–75].

This section examines research on different mechanisms where astrocytes play a crucial role in the excitotoxic context, thus representing feasible targets to prevent neuronal cell death from stroke-induced excitotoxicity.

4.1 Astrocyte Glutamate Transporters in Stroke

Astrocyte glutamate transporters, such as glutamate transporter-1 (GLT-1) and glutamate-aspartate transporter (GLAST), are necessary to clear glutamate and to maintain the electrochemical gradient in the synaptic cleft after stroke and to prevent excitotoxicity [71, 76]. Early studies demonstrated that that absence of astrocytic GLAST facilitated excitotoxicity and led to seizures in mouse models of stroke [77]. In addition, stroke affects astrocyte function, prompting the release of glutamate through the reversal of astrocyte glutamate transporters [78]. In a genetic mouse model lacking GLT-1, neurons in the CA1 region of the hippocampus showed delayed cell death and higher glutamate concentrations early after transient ischemic stroke via bilateral occlusion of the common carotid arteries (BCCA) [79]. A separate study eliminating GLT-1 in rats using antisense oligodeoxynucleotides (ODNs) after MCAO showed similar results of significantly exacerbated neuronal cell death along with increased neurological impairments [80]. The findings from these studies suggest that the loss of astrocytic GLT-1 worsens excitotoxic neuronal death after stroke, indicating the crucial role that GLT-1 plays in neuroprotection. In response to ischemic insults, astrocytes from the CA1 area of the hippocampus increase reactive oxygen species (ROS) production [81], which contributes to loss of activity and immune reactivity of GLT-1 at hyperacute and acute time points after reperfusion, leading to neuronal cell death [82]. Extensive research has been conducted to counteract the detrimental effects of decreased GLT-1 levels seen in the early phase of these studies. A novel study using carnosine, a naturally occurring dipeptide in the CNS, was able to stabilize astrocytic GLT-1 expression and reduce glutamate excitotoxicity in vitro and in vivo ischemic stroke models by suppressing mitochondrial ROS generation [83]. Another pharmacological study on rats using ceftriaxone before the ischemic injury was able to upregulate GLT-1, resulting in decreased excitotoxicity and improved functional recovery [84]. Overexpression of GLT-1 in vivo prior to the ischemic insult reduced infarct size, decreased neuronal death, and consequently improved functional recovery [85]. All this evidence points to the important role that astrocytic GLT-1 plays in limiting excitotoxicity, one of the earliest contributing factors to neuronal cell death in response to stroke.

4.2 Ca^{2+} Influx During Excitotoxicity

Stroke-induced excitotoxicity facilitates the hyper-activation of various neuronal glutamate receptors, including N-methyl d-aspartate (NMDA) and α-amino-3-hydroxy-5-methyl-4-isoxazolepropionic acid (AMPA) receptors [72, 86]. NMDARs are known as the primary source of Ca^{2+} influx during excitotoxicity, leading to the investigation of various NMDAR antagonists [87]. However, pharmacological inhibition of NMDAR- Ca^{2+}-mediated excitotoxicity has shown limited potential due to adverse side effects [8, 9]. Due to this, researchers shifted their focus to AMPARs. AMPARs are heterotetramers composed of GluA1-4 subunits, with GluA2 being the most prevalent in the adult mammalian brain [88]. GluA2-containing AMPARs allow permeability only to K^+ and Na^+, whereas GluA2-lacking receptors are also permeable to Ca^{2+} [89, 90]. Ca^{2+}-permeable AMPARs (CP-AMPARs) have been shown to increase at the neuronal surface in response to ischemic-like conditions, while GluA2-containing AMPARS decrease in the most vulnerable neuronal populations [91, 92]. CP-AMPARs are more susceptible to contribute to excitotoxic damage following stroke [75, 93], as they facilitate excessive overload of intracellular Ca^{2+} in neurons triggering downstream Ca^{2+}-dependent neuronal death mechanisms [94]. The use of the CP-AMPAR antagonist, 1-naphthylacetyl spermine (NASPM), provides neuroprotection from stroke in hippocampal neurons [75, 95, 96], demonstrating that targeting CP-AMPARs helps prevent neuronal cell death. Recent research has shown astrocytes as AMPAR regulators via secretion of different proteins such as glypican 4 and 6 (Gpc4 and Gpc6), hevin, secreted protein acidic and cysteine rich (SPARC), or chordin-like 1 (Chrdl1) [97–99]. Thus, astrocytes have the potential of regulating synaptic AMPARs distribution and composition in response to stroke via secreted proteins, a topic that will be discussed later in the chapter.

Astrocytes, although more resilient to stroke compared to neurons [100], are not immune to elevated intracellular Ca^{2+} levels, which impair their normal function, including neurotransmitter regulation and glutamate homeostasis [94]. This increase in astrocytic Ca^{2+} in response to stroke leads to changes in gliotransmission that influence neuronal activity [101]. Research has been aimed at uncovering the mechanism of Ca^{2+} signaling in astrocytes in response to stroke. Stroke-induced Ca^{2+} increase not only comes from extracellular sources but is mainly contributed by release of Ca^{2+} from internal stores, influx through voltage-dependent Ca^{2+} channels, and/or reversal function of Na^+-Ca^{2+} exchanger (NCX). The release of Ca^{2+} from internal storage in astrocytes is primarily from the endoplasmic reticulum (ER) and mediated by the voltage-gated Ca^{2+} channel inositol 1,4,5-triphosphate receptors (IP3Rs) [102]. Specifically, IP3R2, which is found primarily in astrocytes, has shown to majorly contribute toward Ca^{2+} overload and neuronal death after MCAO and using a KO mouse model significantly attenuated intracellular Ca^{2+} levels and cell death [103]. Additional research showed that using an antagonist for the NCX and Na-K-Cl co-transporter (NKCC1) dampened intracellular Ca^{2+} levels after stroke [104, 105]. Although astrocytes are non-excitable cells, they can be stimulated via Ca^{2+} signaling [94]. Ca^{2+} signaling increases in the infarct and peri-infarct areas after in vivo acute stroke models [106, 107], and using a Ca^{2+} chelator, like BAPTA, results in decreased infarct volume and cell death [106]. In conclusion, research indicates that during ischemic conditions, there is an increase in astrocytic Ca^{2+} levels, which may exacerbate excitotoxicity, ultimately impacting brain function and recovery post-stroke. Further understanding and targeting astrocytic Ca^{2+} signaling could offer potential therapeutic avenues for mitigating the detrimental effects of stroke.

4.3 GABA Transporters (GATs)

Astrocytes also have the capacity to help maintain inhibitory signaling [108]. The main inhibitory neurotransmitter in the brain is gamma-aminobutyric acid (GABA), which is released from cortical motor neurons during the initial phase of stroke due to the lack of ATP [109]. Reactive astrocytes also release GABA into the synaptic cleft after stroke, increasing this neurotransmitter concentration well above its physiological levels [110]. GABA typically acts on synaptic and extrasynaptic $GABA_A$ and $GABA_B$ receptors to inhibit and modulate neurotransmission, while GABA transporters (GATs) on astrocytes work to regulate GABA levels in the synaptic cleft [109]. In stroke, the excessive release of GABA acting on their receptors results in phasic and eventually tonic inhibition, therefore changing from a transient inhibition to a persistent desensitizing GABAergic conductance, respectively [111]. This change in conductance affects neuronal excitability in response to stroke. Researchers are interested in understanding the role of phasic GABA-mediated inhibition during the early stages of stroke, considering the involvement of extrasynaptic $GABA_A$ receptors and tonic inhibition in regulating brain excitability and their recent connections to stroke recovery [112–114]. Previous work in in vitro and in vivo ischemic stroke animal models has shown that enhancing synaptic GABAergic signaling immediately after ischemia promotes neuroprotection [115]. Unfortunately, the successful use of GABA receptor antagonists to promote neuroprotection failed to translate into clinical studies [116].

It is suggested that in the initial phase of stroke the excess release of GABA from astrocytes is meant as an endogenous protective mechanism to dampen excitotoxicity by reducing Na^+ and Ca^{2+} entry into neurons and therefore decreasing depolarization [117], although research has shown that this release is not sufficient to provide neuroprotection [114]. GATs mediate other potential pathways that regulate persistent GABAergic inhibition. Four GATs have been identified (GAT 1–3 and BGT1) [118]. GAT1, primarily located on presynaptic neurons and astrocytes, is not affected by stroke injuries [114, 119], but GAT1 in vivo inhibition in gerbil and rat models of stroke showed increased motor recovery [114]. GAT3, found mainly in astrocytes [120], decreases in response to stroke in the peri-infarct area and contributes toward the GABA accumulation and increased GABAergic inhibition [114]. Furthermore, research showed that this decrease in GAT3 protein persists in the chronic phase, up to 42 days post-stroke [121]. In vivo research administering a GAT3 agonist, L-isoserine, after motor cortical photothrombosis showed to increase GAT3 expression and to improve motor outcomes [121]. This improvement is likely to occur through increased GABA uptake to lessen tonic inhibition and enhance cortical re-connections.

4.4 Volume-Regulated Anion Channels (VRACs)

Volume-regulated anion channels (VRACs) are ubiquitously expressed chloride/anion channels that are permeable to glutamate [122]. The role of VRAC in normal physiology is to regulate cell volume [123], but their activation in response to pathological cell swelling contributes to ischemic brain damage. After stroke, astrocytes are the primary cell type to exhibit an increase in volume via VRAC, leading them to release excessive glutamate and contributing to detrimental effects [123]. This swelling is typically counterbalanced by chloride/anions released by neurons and astrocytes [124].

Chemical transmitters, such as glutamate and aspartate released by astrocytes, are excitotoxic in stroke, and this once-protective mechanism turns into a damaging one that can increase cell death [123]. This evidence renders pharmacological research to target VRAC as a promising avenue. Several studies have reported that in an ischemic in vivo rodent model, tamoxifen, a nonspecific VRAC inhibitor, given prior to stroke resulted in neuroprotection by decreasing excitotoxic levels of glutamate, decreased infarct size, and decreased cell death in early phases after stroke [125–128]. However, given the nonspecificity of tamoxifen, it causes off-target effects, which makes it unsuitable for clinical applications [123]. More selective VRAC inhibitors such as 4-[(2-butyl-6,7-dichloro-2-cyclopentyl-2,3-dihydro-1-oxo-1*H*-inden-5-yl)oxy]butanoic acid (DCPIB) administered 15 min prior to stroke significantly reduce infarct volume and damage 24 h after stroke in vivo [129]. LRRC8A, also known as SWELL1, one of the obligatory subunits of VRACs [130], is essential in the anion channel function [130, 131]. This subunit plays an essential role in exacerbating stroke pathology, as it allows VRACs to function at low glucose and ATP levels while cell swelling occurs. Elimination of SWEL1 in astrocytes reduces the size of the stroke and allows functional improvement, as demonstrated in studies performed in vitro and in vivo [132, 133]. Overall, these studies show that targeting astrocytic VRAC has a crucial impact on neuronal death after stroke and that the inhibition of VRAC can significantly improve outcomes.

5 Metabolic Dysregulation After Stroke

Stroke induces a massive dysregulation of metabolic processes, acutely diminishing oxygen and glucose supply to the brain which impairs energy metabolism and increases oxidative stress [134]. Neurons are particularly vulnerable to these insults, as they rely on oxidative mechanisms to generate ATP, while astrocytes are mostly glycolytic and thus more resilient in this context [21, 135]. Seeing that astrocytes are known to provide metabolic support to neurons in health [136], they may be particularly needed in moments of energy imbalance, such as in stroke. In this section, we review various components that are dysregulated in response to stroke that, otherwise, are normally controlled by astrocytes to ensure proper metabolic support to neurons.

5.1 Antioxidant Production

Astrocytes provide metabolic support to neurons by supplying them with antioxidant molecules, specifically glutathione [137, 138]. Astrocytic glutathione is essential in preventing nitric oxide toxicity in co-cultured cortical neurons [139] and thus may be beneficial post-stroke. Indeed, glutathione levels increase in neurons in the peri-infarct area after MCAO, accompanied by a slight decrease in glutathione in peri-infarct astrocytes, potentially suggesting an enhancement of this support mechanism under hypoxic conditions [140]. Nuclear factor erythroid 2-related factor 2 (Nrf2) is a master regulator of endogenous antioxidant defenses and is upregulated in peri-infarct astrocytes and neurons in both humans and mice [141]. By reducing the degradation of Nrf2 in astrocytes with celastrol, researchers found decreases in oxidative stress and astrocyte activation after ischemic-like conditions in vitro (i.e., OGD), which rescued neurons from axonal damage and apoptosis [142]. DJ-1 is a multifunctional protein that regulates Nrf2 and its downstream pathways [143], and it has been associated with glutathione synthesis during oxidative stress [144]. DJ-1 is upregulated in astrocytes after stroke, and its inhibition leads to decreased total glutathione and Nrf2 levels, increased infarct volume, and worse neurological deficits in rats post-MCAO [22]. Astrocyte-specific DJ-1 knockdown in an in vitro stroke model similarly leads to decreased glutathione levels, Nrf2 expression, and viability of co-cultured neurons, suggesting astrocytic DJ-1 as a key regulator of glutathione production and downstream protective neuronal antioxidant mechanisms following stroke. In fact, DJ-1 overexpression yields neuroprotective results [22], pointing to the therapeutic potential of this pathway.

5.2 Lactate Supply

One auxiliary source of substrates for ATP production is the astrocyte-neuron lactate shuttle [145], which hypothesizes that glucose is metabolized to lactic acid in astrocytes and then supplied to neurons. The release of lactate from astrocytes and its uptake into neurons is mediated via monocarboxylate transporters [146]. While the importance of this energy source has been debated at rest [147, 148], astrocytic lactate may be an important mechanism that promotes neuroprotection in the context of ischemia (for a review, see [149, 150]). Indeed, experiments using fluorocitrate, an astrocyte-selective metabolic inhibitor, in hippocampal slices exposed to hypoxia found a significant reduction in lactate production alongside deficits in recovery of synaptic function [151]. This suggests that astrocytes are the main source of lactate in hypoxic conditions. Furthermore, as supplementation with lactate reverses the deficits, astrocyte-derived lactate may be necessary for restoration of neuronal function [151]. In line with this evidence, subsequent studies found that blocking lactate trafficking with a monocarboxylate transporter inhibitor, prior to transient global cerebral ischemia in rats, leads to a significantly greater degree of delayed neuronal damage [152]. Interestingly, astrocytes isolated from stroke-prone spontaneously hypertensive rats exhibit impaired lactate production and decreased monocarboxylate transporter 1 expression in response to hypoxia/reoxygenation, which could at least in part explain their vulnerability to stroke [153]. More recent work has confirmed intravenous or intraventricular lactate administration following MCAO to be neuroprotective, reducing neurological deficits and the injury volume [154]. Astrocyte-derived lactate may thus represent an essential energy source for neurons in the context of stroke, a potentially targetable mechanism for future therapeutic strategies [150].

5.3 Ketone Bodies and ATP Production

Ketone bodies are another energy source neurons utilize under extreme circumstances. Ketone body production has been shown to increase in astrocytes in response to hypoxia in vitro [155], and diet-induced ketotic rats show a 70% reduction in infarct volume following MCAO [156]. Recent work has shown that serum levels of β-hydroxybutyrate, an essential ketone body, in post-stroke patients are positively correlated with improved outcomes, based on the modified Rankin Scale, and β-hydroxybutyrate administration several days after photothrombosis promotes functional recovery in rodent models [157]. There is also evidence that astrocytes directly release ATP [158], which can in turn modulate synaptic transmission [159]. In the context of stroke, research has found that increasing ATP release from astrocytes via low-density lipoprotein receptor-related protein 4 (Lrp4) knockout reduces neuronal death, infarct volume, and neurological functional impairments [160].

5.4 Mitochondrial Transfer

A relatively newly proposed form of astrocyte-neuron crosstalk is intercellular mitochondrial transfer. It was first discovered that healthy retinal ganglion cells shed mitochondria in their axons, which are subsequently taken up and degraded by adjacent astrocytes, challenging the previous dogma that cells must degrade their own mitochondria [161]. Under pathophysiological conditions such as stroke, this mitochondrial release from neurons may be increased, with research finding elevated levels of defective mitochondria in the cerebrospinal fluid of post-MCAO mice [162]. In vitro studies challenging neurons with an acidic environment (as seen in acute stroke) showed that neurons increase their release of impaired mitochondria, which are again taken up by astrocytes [162]. In turn, engulfment of neuronal mitochondria leads astrocytes to increase mitochondrial biogenesis [162]. The significance of astrocytic clearance of damaged neuronal mitochondria post-stroke remains to be investigated, but could potentially constitute a protective pathway under high energy demands where neurons rely on astrocytes for mitochondrial clearance and energy support.

Astrocytes can also secrete functional mitochondria that can be taken up into the somas and axons of co-cultured neurons [163]. Interestingly, in a mouse model of ischemic stroke, astrocytic mitochondria are found within neurons 24 hours after the injury, and blocking this astrocyte-to-neuron mitochondrial transfer worsens neurological outcomes [163], suggesting that astrocytic mitochondria supply to neurons may be a protective bioenergetic mechanism post-stroke. These findings remain debated, however, with researchers questioning whether astrocyte-derived functional mitochondria are incorporated intracellularly in neurons or simply become cell-associated [164]. Nevertheless, more recent work has shown primary astrocytic mitochondrial transplantation into the striatum after MCAO to reduce infarct volume in rats [165], supporting a protective role.

6 Astrocytic Production of Neurotrophic Factors and Estrogen

In response to ischemia, reactive astrocytes produce a host of neurotrophic factors that may be neuroprotective; these include nerve growth factor (NGF) [166], erythropoietin (EPO) [167], brain-derived neurotrophic factor (BDNF) [168, 169], glial cell line-derived neurotrophic factor (GDNF) [170–172], and ciliary neurotrophic factor (CNTF) [173].

6.1 Brain-Derived Neurotrophic Factor (BDNF)

BDNF has garnered much research interest due to its important roles in neuronal survival, growth, and plasticity [174], all of which may confer therapeutic benefits in the context of stroke. While BDNF is mostly expressed in neurons under physiological conditions, a significant upregulation is seen in reactive astrocytes after stroke, both in in vitro and in vivo models [175]. Interestingly, this upregulation is paralleled by an upregulation in galectin-1 in astrocytes, which was shown to enhance the expression and secretion of astrocytic BDNF in a dose-dependent manner [175]. Rats infused with galectin-1 after ischemic stroke show reduced neuronal apoptosis in the peri-infarct region and improved functional recovery [175], suggesting a neuroprotective role for astrocytic BDNF in stroke. BDNF upregulation in astrocytes has also been shown to be necessary for adiponectin-induced neurogenesis after stroke, leading to reduced neurological deficits [176].

6.2 Glial Cell Line-Derived Neurotrophic Factor (GDNF)

GDNF shows potential benefits after stroke (for a review, see [177]). GDNF expression in the healthy brain is mostly restricted to neurons [178], but a marked upregulation is seen in astrocytes after stroke [170–172]. Several studies using a multitude of delivery methods have established GDNF administration to be neuroprotective in stroke [179–181], including as a promoter of neurogenesis [182]. However, only recently has the specific role of astrocyte-derived GDNF been elucidated in this context. GDNF conditional knockout in astrocytes in a photothrombotic stroke model leads to a significant increase in oxidative stress, larger infarct volume, hippocampal damage, and neurodegeneration, alongside decreased neurogenesis in the peri-infarct region and exacerbated motor behavioral deficits, indicating that astrocytic GDNF is crucial in preventing damage and promoting recovery after stroke [172].

6.3 Ciliary Neurotrophic Factor (CNTF)

CNTF is a potent neural cytokine with very low endogenous expression in the central nervous system; however, a dramatic upregulation is seen in astrocytes following injury [173, 183, 184]. The astrocytic increase in CNTF expression following stroke is likely due to a disrupted neuron-astrocyte contact-mediated regulatory interaction [183]. First, it was found that neurons reduce CNTF expression in co-cultured astrocytes; this change is prevented by light trypsinization or separation by a semipermeable membrane, but is not affected by soluble factors [183]. This suggests a contact-dependent mechanism of tonic neuronal inhibition of astrocytic CNTF. Indeed, blocking integrin in C6 cells (a glial cell line) is enough to induce CNTF expression, and selective neuronal loss in vivo using quinolinic acid causes increases in CNTF comparable to MCAO [183]. Experiments exposing mice to either transient hypoxia or proinflammatory cytokines seem to have little direct effect on CNTF expression, suggesting that it is neuronal damage in stroke that leads to increases in CNTF [183]. This CNTF upregulation was found to be neuroprotective after MCAO, as CNTF knockout mice have larger injury volumes [183]. This work suggests that contact-mediated integrin signaling between neurons and astrocytes represses CNTF expression under physiological conditions and that disruption in this interaction during stroke leads to a substantial upregulation and neuroprotective response from astrocytes [183]. Data from the same group found CNTF to be necessary for stroke-induced neurogenesis in the subventricular zone (SVZ) [185], which is regulated by astrocytic FAK-JNK signaling [186]. Although more research is warranted, the current evidence points to CNTF as a potential target to develop alternative therapies.

6.4 Estrogen

It is well known that sex differences exist in the prevalence, incidence, and severity of stroke [187]. Stroke is more common among men but is more severe in women, and its incidence increases significantly after menopause [187], drawing attention to a potential mediating role of sex hormones, including estrogen. It has also been documented that astrocytic responses to stroke differ by sex and are impacted by estrogen and estrus cycle [188]. The role of astrocytic aromatase, the enzyme necessary for 17β-estradiol synthesis (a major form of estrogen), has been examined in response to global cerebral ischemia in mice [189] after it was observed that both aromatase expression and 17 β-estradiol synthesis are elevated in astrocytes after stroke [190]. By eliminating aromatase expression (and thus 17β-estradiol production) in astrocytes in the context of ischemia, reactive astrogliosis is attenuated and neuronal loss and cognitive deficits are increased, indicating that astrocytic 17β-estradiol provides some level of protection in response to stroke [189]. This was confirmed by exogenous 17β-estradiol administration, which restores astrocyte activation and prevents neuronal damage and microglial activation in response to stroke [189]. Interestingly, a different study from the same group showed a similar phenotype in forebrain neuron-specific aromatase KO mice [191], including impaired reactive astrogliosis and astrocytic aromatization after stroke, suggesting that astrocytes rely on neuron-derived 17β-estradiol for proper initiation of injury responses. For example, the blunted astrocytic response was associated with a downregulation of astrocyte-derived GFAP, BDNF, insulin-like growth factor-1 (IGF-1), and GLT-1 [191], the roles of which have been previously discussed in this chapter. Looking to elucidate this astrocytic activation dependence on neuronal 17β-estradiol, researchers found that neuronal fibroblast growth factor 2 (FGF2), which suppresses astrocyte activation in a paracrine manner [192], is upregulated in the forebrain neuron-specific aromatase KO mice after ischemia [191]. In line with this finding, FGF2 blockade reverses the diminished astrocytic reactivity and aromatase expression previously seen in these mice [191]. Interestingly, exogenous 17β-estradiol is able to suppress FGF2 signaling, restoring reactive astrogliosis and ameliorating cognitive deficits post-stroke [191], suggesting that neuronal 17β-estradiol is an important dis-inhibitory signal for neuroprotective astrocyte activation and estrogen production. Looking to better understand the mechanisms of astrocytic estrogen signaling, researchers found mouse hippocampal astrocytes to primarily express the estrogen receptor β subtype and to increase their GFAP expression only upon administration of 17β-estradiol or a selective estrogen receptor β agonist, but not an estrogen receptor α agonist [193]. Interestingly, activation of estrogen receptor β, but not α, reduces neuronal apoptosis in co-cultured astrocytes and neurons subjected to OGD; similar protective effects are seen when giving a selective estrogen receptor β agonist in an in vivo model of global cerebral ischemia [193]. The above results suggest a neuroprotective role for estrogen post-stroke via astrocyte-neuron interactions. However, the use of estrogen replacement therapies for human stroke remains controversial, with data from clinical trials suggesting an increase in ischemic stroke risk among healthy postmenopausal women [194].

7 Astrocyte Contributions Toward Post-Stroke Synaptic Plasticity and Neuronal Growth

After ischemic stroke, synapses and dendritic spines are damaged and lost, leading to functional deficits later on [195–197]. Although synapses in the infarct area are lost, mechanisms in the peri-infarct area work to repair neuronal connections [197]. Astrocytes modulate synaptic connectivity, formation, maturation, and elimination after stroke [58]. Astrocytes are crucial in neuronal repair [198] and by secreting factors that aid in re-mapping neuronal synapses [199].

7.1 Astrocyte Regulation of Synaptic Phagocytosis/Pruning in Response to Stroke

Astrocytic maintenance of synapses is crucial for synaptic homeostasis in physiology and maintaining proper neural networks [58]. This function has been identified in brain areas of high plasticity, such as the hippocampus [200], and at crucial time points during development [201]. After stroke, there is a need to clear debris in order to allow neural circuits to remodel [202], which is typically done by microglia [203]. After stroke, astrocytes become primarily present in the peri-infarct area, where synaptic remodeling occurs, and become phagocytic at a delayed time point after stroke [204, 205]. ATP-binding cassette transporter A1 (ABCA1), multiple EGF-like-domains 10 (MEGF10), MERTK, and engulfment adaptor PTB domain 1 (GULP1) pathways are involved in astrocyte-mediated phagocytic pathways in response to stroke [200, 204, 205]. However, this engulfment of synapses might hinder functional recovery. A study removing crucial engulfment identifiers, MEGF10 and MERTK, blocked the phagocytosis of synapses [205]. Separate work in vivo in mice showed that disrupting ABCA1 also decreased phagocytosis of dead cells and small debris, comparable to what microglia normally do, and therefore suggesting astrocytes as important players in clearance of damaged tissue after stroke [204]. Overall, these findings show that astrocytic phagocytosis is crucial in post-stroke remodeling.

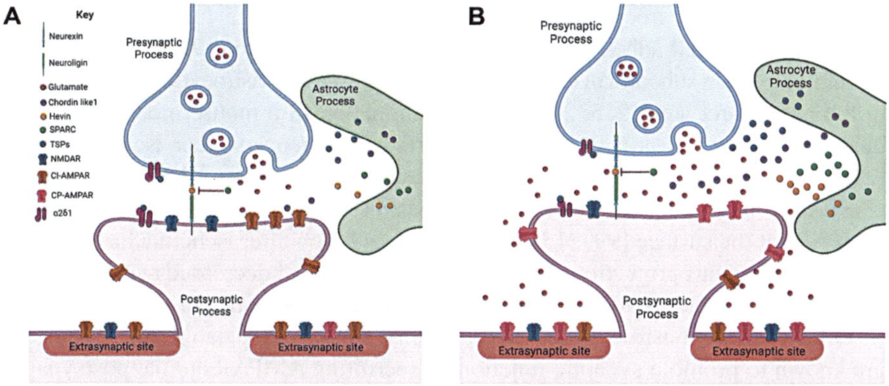

Fig. 2 Role of astrocyte-secreted proteins in stroke. (**a**) Normal physiology of an excitatory tripartite synapse, where homeostatic glutamate levels are released. Astrocyte-secreted factor hevin binds to connect presynaptic neurexin and postsynaptic neuroligin adhesion molecules to recruit NMDARs to the synaptic density, while SPARC works to inhibit hevin-induced synaptic formation. TSPs bind to presynaptic and postsynaptic calcium channel α2d1 and work to promote structurally normal but silent synapse formation. Chrdl1 works to induce synapse maturation, by recruiting GluA2-containing Ca^{2+}-impermeable-AMPARs (CI-AMPARs). (**b**) After stroke, excessive glutamate is released from the presynaptic terminal and overexcites CP-AMPARs in the postsynaptic density and extrasynaptic regions at the postsynaptic site. Multiple secreted factors (TSPs, hevin, SPARC) are upregulated in response to stroke in the peri-infarct region, contributing to neuronal protection, while others, like Chrdl1, are also upregulated limiting recovery. This highlights the dual role that astrocytes may play in response to injury and disease. (Generated using BioRender)

7.2 Astrocyte-Secreted Factors

Initial evidence showed that retinal ganglion cell (RGC) cultures require astrocyte-conditioned medium (ACM) in order to form functional synapses, which sparked an interest in the role of astrocyte-secreted factors [206]. Accumulating work has now shown that these secreted factors (lipids, proteins, and small molecules) are important in developmental and adult physiological synapse formation and maturation and mediate the synapse's recovery/repair after stroke (Fig. 2). For example, astrocyte-secreted protein Chrdl1 induces synaptic maturation in development by recruiting GluA2-containing AMPARs to the postsynaptic surface and limits experience-dependent plasticity in upper cortical layers [98]. Chrdl1 expression increases in response to ischemic injury [207, 208], and in a mouse model of stroke, elimination of Chrdl1 prevents the typical dendritic spine loss, cell death, and neuronal GluA2 reduction, resulting in improved motor recovery [207, 208].

Other astrocyte-secreted proteins such as thrombospondin 1 and 2 (TSP1, TSP2) are involved in synaptogenesis during development, by generating silent synapses and axon remodeling [209–211]. After stroke TSPs are upregulated [212], which has been shown to enhance the recovery of synaptic function after a stroke by promoting synaptic connections [16, 209]. This indicates that TSPs play an important role in synaptic and functional recovery after stroke.

Hevin, another astrocyte-secreted protein, works to induce functional excitatory synapses via the cell adhesion proteins neurexin and neuroligin [99, 213, 214]. After ischemic stroke in subcortical or cortical areas, expression of astrocytic Hevin increases in the peri-infarct area [215, 216] and its elimination in a mouse model resulted in increased synaptic loss and dampened neurological recovery after ischemic stroke [215], indicating that hevin is necessary to alleviate stroke-induced detrimental effects.

SPARC inhibits postsynaptic cell adhesion protein integrin β-3 stabilizing AMPARs at the surface [99, 217]. SPARC upregulation after ischemia has shown to play a role in neuroprotection, and its ablation resulted in decreased synapse formation, aberrant surface AMPAR clustering, and impaired synaptic plasticity [217, 218].

Other identified astrocyte-secreted proteins, such as glypican 4 and glypican 6, are known to promote synaptic function by recruiting AMPARs to the postsynaptic site [97, 219]. As mentioned earlier in this chapter, AMPARs are key players in neuronal vulnerability to stroke. Thus, synaptic-regulating astrocyte proteins and particularly those that are involved in AMPAR clustering are interesting candidates that hold the potential to prevent post-stroke synaptic aberrations, potentially representing new therapeutic approaches.

8 Astrocyte-Mediated Neurogenesis After Stroke

After a stroke, the adult brain produces new neurons in areas where neurogenesis typically does not occur, in an attempt to compensate for lost cells [220, 221]. Much work has looked to better understand the mechanisms underlying stroke-induced neurogenesis, as the general consensus is that augmentation or supplementation of this process could provide therapeutic benefits [220–222]. In this context, astrocytes could play a crucial role, owing to their ability to stimulate neurogenesis [223, 224], aid in the migration of neuronal precursors [225, 226], adopt stem cell-like properties [227], and even transdifferentiate into neurons [228].

After stroke, astrocytes support the survival, proliferation, and differentiation of neural stem/progenitor cells (NSPCs) [229, 230]. In vitro, conditioned media from astrocytes subjected to OGD increase NSPC proliferation, an effect that is mediated by astrocytic release of the damage-associated molecular pattern molecule high-mobility group box 1 (HMGB1) [229]. The secretome from astrocytes exposed to lipopolysaccharide (LPS), an inflammation-inducing factor, promotes neuronal differentiation of NSPCs [230]. Interleukin-17A (IL-17A) was identified as one of the main astrocyte-secreted proteins that mediate such effects, as NSPC differentiation is abolished when IL-17A is blocked [230]. In vivo, IL-17A was determined to be upregulated after MCAO, stemming specifically from astrocytes. In this context, IL-17A enhances subventricular zone NSPC survival, neuroblast migration, and neuronal differentiation, leading to elevated striatal neurogenesis [230]. Furthermore, IL-17A improves synaptogenesis in the peri-infarct region and ameliorates behavioral deficits post-stroke [230], suggesting that astrocytic IL-17A may stimulate recovery after stroke by promoting NSPC survival and differentiation into functional neurons. Astrocytic support of NSPCs has been further corroborated in transplant experiments, with astrocyte co-transplantation alongside NSPCs improving NSPC survival, proliferation, and differentiation in post-stroke rats [231].

Astrocytes are also involved in mediating the migration of NSPCs after stroke [232, 233]. Astrocytes local to an infarcted brain region upregulate the inflammatory chemoattractant stromal cell-derived factor 1α (SDF-1α), which induces NSPC proliferation and migration toward the core of the injury [232]. In an in vitro model, astrocytes were found to generate Ca^{2+} waves in response to injury that is propagated to co-cultured NSPCs [233]. These astrocytic Ca^{2+} waves induce an increase in the self-renewal and migratory behavior of NSPCs, via an upregulation of Notch signaling [233]. Gene expression analysis of the SVZ following MCAO suggests that this mechanism is also at play in vivo in stroke [233].

Efforts have also been made to characterize the stem cell potential of reactive astrocytes [227] and even their ability to transdifferentiate into neurons [234] following stroke. Reactive astrocytes in the peri-infarct region express proteins typically associated with NSPCs, including nestin, RC2, and Sox2 [235, 236]. Interestingly, reactive astrocytes isolated from peri-infarct tissue can produce neural spheres with multipotent and self-renewing properties, suggestive of astrocyte-derived NSPCs [235]. Generation of these NSPCs is dependent on Notch 1 signaling [235], with Sonic hedgehog signaling also being implicated [237]. Astrocyte-derived NSPCs can successfully differentiate into neurons, astrocytes, and oligodendrocytes in culture, but fail to produce neurons when transplanted in vivo [235], limiting their translational potential. Lineage tracing found native reactive astrocytes in the peri-infarct area to be unipotent and not multipotent [235], suggesting that only upon isolation from the stroke environment can reactive astrocytes de-differentiate into multipotent NSPCs. However, more recent research has been able to confirm a population of resident astrocytes that transdifferentiate into neurons in response to stroke [228, 238]. By inducing the expression of a reporter allele in native astrocytes and then following their genetic fate after MCAO, researchers found that some astrocytes in the striatum undergo a series of molecular changes, with intermediates expressing proneuronal transcription factors (Ascl1, Pax6), markers of neural stem/progenitor cells (nestin, Sox2), and signatures of immature neurons (doublecortin, Tuj-1). These ultimately give rise to cells that express markers of mature neurons, including NeuN, between 2 and 7 weeks after injury [228, 238]. A reduction in Notch 1 signaling is necessary to induce this neurogenic program in astrocytes, and abolishing Notch 1 signaling in astrocytes is sufficient to induce this transdifferentiation even in the absence of stroke [228]. In fact, further blockade of Notch signaling specifically in astrocytes enhances this neurogenic response to stroke up to 3.5-fold [239], suggesting Notch inhibition as a potential therapeutic strategy. Amazingly, this astrocytic transdifferentiation accounts for approximately one-third to two-thirds of the new neurons generated after stroke [228]. These astrocyte-derived neurons primarily develop into GABAergic interneurons or cholinergic neurons that establish functional synapses and are capable of firing action potentials and generating postsynaptic currents post-stroke [228, 238], supporting their likely relevance in restoring functional neural networks after injury.

The above work highlights the many roles astrocytes play in neurogenesis after stroke, not only as critical mediators of NSPC survival, proliferation, migration, and differentiation, but also as a potential source of new neurons themselves.

9 Conclusion

Stroke is a leading cause of death and disability worldwide, with current treatments limited to acute interventions and hampered by significant risk of complications [240–242]. To date, "neurocentric" approaches to develop novel therapeutics have not translated into effective clinical therapies, underscoring the need for new targets in stroke treatment [243, 244]. To this end, astrocytes have begun to attract interest as potential facilitators of stroke recovery, owing to their dramatic response to injury and their involvement in mechanisms critical to neuroprotection and neurorestoration [243, 244]. In this chapter, we have focused on the varied and intricate interactions that astrocytes establish with neurons in the context of stroke, highlighting mechanisms of reactive astrogliosis, excitotoxicity, metabolic and neurotrophic support, synaptic plasticity, and neurogenesis as determinants of stroke outcome. With continued investigation and understanding, we hope that astrocyte-neuron interactions may be leveraged to generate new stroke therapies and improve patient outcomes in the future.

Acknowledgements This work was supported by American Heart Association grants to EB-S (937695 and 23IPA1055207) and to BRB (24PRE1198593). The authors thank Dr. Eileen Collyer for critical reading of the manuscript. Figure 2 was created using BioRender.

References

1. Baron JC. The core/penumbra model: implications for acute stroke treatment and patient selection in 2021. Eur J Neurol. 2021;28(8):2794–803. https://doi.org/10.1111/ene.14916.
2. Bernhardt J, Hayward KS, Kwakkel G, Ward NS, Wolf SL, Borschmann K, et al. Agreed definitions and a shared vision for new standards in stroke recovery research: the stroke recovery and rehabilitation roundtable taskforce. Neurorehabil Neural Repair. 2017;31(9):793–9. https://doi.org/10.1177/1545968317732668.
3. Carmichael ST. Themes and strategies for studying the biology of stroke recovery in the poststroke epoch. Stroke. 2008;39(4):1380–8. https://doi.org/10.1161/STROKEAHA.107.499962.
4. Dobkin BH, Carmichael ST. The specific requirements of neural repair trials for stroke. Neurorehabil Neural Repair. 2016;30(5):470–8. https://doi.org/10.1177/1545968315604400.
5. Jorgensen HS, Nakayama H, Raaschou HO, Vive-Larsen J, Stoier M, Olsen TS. Outcome and time course of recovery in stroke. Part II: time course of recovery. The Copenhagen Stroke Study. Arch Phys Med Rehabil. 1995;76(5):406–12. https://doi.org/10.1016/s0003-9993(95)80568-0.
6. Krakauer JW, Carmichael ST, Corbett D, Wittenberg GF. Getting neurorehabilitation right: what can be learned from animal models? Neurorehabil Neural Repair. 2012;26(8):923–31. https://doi.org/10.1177/1545968312440745.

7. Bernardo-Castro S, Sousa JA, Bras A, Cecilia C, Rodrigues B, Almendra L, et al. Pathophysiology of blood-brain barrier permeability throughout the different stages of ischemic stroke and its implication on hemorrhagic transformation and recovery. Front Neurol. 2020;11:594672. https://doi.org/10.3389/fneur.2020.594672.
8. Choi, D. W. (2020). Excitotoxicity: still hammering the ischemic brain in 2020 [10.3389/fnins.2020.579953]. Front Neurosci, 14, 1104.
9. Choi DW. Excitotoxicity: still hammering the ischemic brain in 2020. Front Neurosci. 2020;14:579953. https://doi.org/10.3389/fnins.2020.579953.
10. Gerriets T, Walberer M, Ritschel N, Tschernatsch M, Mueller C, Bachmann G, et al. Edema formation in the hyperacute phase of ischemic stroke. Laboratory investigation. J Neurosurg. 2009;111(5):1036–42. https://doi.org/10.3171/2009.3.JNS081040.
11. Sekerdag E, Solaroglu I, Gursoy-Ozdemir Y. Cell death mechanisms in stroke and novel molecular and cellular treatment options. Curr Neuropharmacol. 2018;16(9):1396–415. https://doi.org/10.2174/1570159X16666180302115544.
12. Wang H, Song G, Chuang H, Chiu C, Abdelmaksoud A, Ye Y, Zhao L. Portrait of glial scar in neurological diseases. Int J Immunopathol Pharmacol. 2018;31:2058738418801406. https://doi.org/10.1177/2058738418801406.
13. Zeiler SR, Krakauer JW. The interaction between training and plasticity in the poststroke brain. Curr Opin Neurol. 2013;26(6):609–16. https://doi.org/10.1097/WCO.0000000000000025.
14. Sofroniew MV. Molecular dissection of reactive astrogliosis and glial scar formation. Trends Neurosci. 2009;32(12):638–47. https://doi.org/10.1016/j.tins.2009.08.002.
15. Tasca CI, Dal-Cim T, Cimarosti H. In vitro oxygen-glucose deprivation to study ischemic cell death. Methods Mol Biol. 2015;1254:197–210. https://doi.org/10.1007/978-1-4939-2152-2_15.
16. Wang B, Guo W, Huang Y. Thrombospondins and synaptogenesis. Neural Regen Res. 2012;7(22):1737–43. https://doi.org/10.3969/j.issn.1673-5374.2012.22.009.
17. Wang R, Zhang X, Zhang J, Fan Y, Shen Y, Hu W, Chen Z. Oxygen-glucose deprivation induced glial scar-like change in astrocytes. PLoS One. 2012;7(5):e37574. https://doi.org/10.1371/journal.pone.0037574.
18. Engel O, Kolodziej S, Dirnagl U, Prinz V. Modeling stroke in mice – middle cerebral artery occlusion with the filament model. J Vis Exp. 2011;47 https://doi.org/10.3791/2423.
19. Maeda K, Hata R, Hossmann K-A. Regional metabolic disturbances and cerebrovascular anatomy after permanent middle cerebral artery occlusion in C57black/6 and SV129 mice. Neurobiol Dis. 1999;6(2):101–8.
20. McBride DW, Zhang JH. Precision stroke animal models: the permanent MCAO model should be the primary model, not transient MCAO. Transl Stroke Res. 2017; https://doi.org/10.1007/s12975-017-0554-2.
21. Gurer G, Gursoy-Ozdemir Y, Erdemli E, Can A, Dalkara T. Astrocytes are more resistant to focal cerebral ischemia than neurons and die by a delayed necrosis. Brain Pathol. 2009;19(4):630–41. https://doi.org/10.1111/j.1750-3639.2008.00226.x.
22. Peng L, Zhao Y, Li Y, Zhou Y, Li L, Lei S, et al. Effect of DJ-1 on the neuroprotection of astrocytes subjected to cerebral ischemia/reperfusion injury. J Mol Med (Berl). 2019;97(2):189–99. https://doi.org/10.1007/s00109-018-1719-5.
23. Stadler J, Schurr H, Doyle D, Garmo L, Srinageshwar B, Spencer MR, et al. Temporal profile of reactive astrocytes after ischemic stroke in rats. Neuroglia. 2022;3(3):99–111.
24. Blanco-Suarez E. Photothrombotic model to create an infarct in the hippocampus. Methods Mol Biol. 2023;2616:29–38. https://doi.org/10.1007/978-1-0716-2926-0_4.
25. Labat-gest V, Tomasi S. Photothrombotic ischemia: a minimally invasive and reproducible photochemical cortical lesion model for mouse stroke studies. J Vis Exp. 2013;76 https://doi.org/10.3791/50370.
26. Watson BD, Dietrich WD, Busto R, Wachtel MS, Ginsberg MD. Induction of reproducible brain infarction by photochemically initiated thrombosis. Ann Neurol. 1985;17(5):497–504. https://doi.org/10.1002/ana.410170513.

27. Chen CY, Chen RJ, Lee GA. Two-vessel occlusion mouse model of cerebral ischemia-reperfusion. J Vis Exp. 2019;145 https://doi.org/10.3791/59078.
28. McBean DE, Winters V, Wilson AD, Oswald CB, Alps BJ, Armstrong JM. Neuroprotective efficacy of lifarizine (RS-87476) in a simplified rat survival model of 2 vessel occlusion. Br J Pharmacol. 1995;116(8):3093–8.
29. Smith M-L, Auer R, Siesjö B. The density and distribution of ischemic brain injury in the rat following 2–10 min of forebrain ischemia. Acta Neuropathol. 1984;64(4):319–32.
30. Yamaguchi M, Calvert JW, Kusaka G, Zhang JH. One-stage anterior approach for four-vessel occlusion in rat. Stroke. 2005;36(10):2212–4. https://doi.org/10.1161/01.STR.0000182238.08510.c5.
31. Yasuda Y, Shimoda T, Uno K, Tateishi N, Furuya S, Tsuchihashi Y, et al. Temporal and sequential changes of glial cells and cytokine expression during neuronal degeneration after transient global ischemia in rats. J Neuroinflammation. 2011;8(1):70. https://doi.org/10.1186/1742-2094-8-70.
32. Liu C, Guo Y, Deng S, Zhou S, Wu S, Chen T, et al. Hemorrhagic stroke-induced subtype of inflammatory reactive astrocytes disrupts blood-brain barrier. J Cereb Blood Flow Metab. 2024:271678X241235008. https://doi.org/10.1177/0271678X241235008.
33. Sun Z, Li Q, Li X, Shi Y, Nan C, Jin Q, et al. Casein kinase 2 attenuates brain injury induced by intracerebral hemorrhage via regulation of NR2B phosphorylation. Front Cell Neurosci. 2022;16:911973. https://doi.org/10.3389/fncel.2022.911973.
34. Syed B, Nirwane A, Yao Y. In vitro models of intracerebral hemorrhage. Brain Hemorrhages. 2022;3(3):105–7. https://doi.org/10.1016/j.hest.2022.06.002.
35. Koeppen AH, Dickson AC, McEvoy JA. The cellular reactions to experimental intracerebral hemorrhage. J Neurol Sci. 1995;134(Suppl):102–12. https://doi.org/10.1016/0022-510x(95)00215-n.
36. Ropper AH, Zervas NT. Cerebral blood flow after experimental basal ganglia hemorrhage. Ann Neurol. 1982;11(3):266–71. https://doi.org/10.1002/ana.410110306.
37. Rosenberg GA, Mun-Bryce S, Wesley M, Kornfeld M. Collagenase-induced intracerebral hemorrhage in rats. Stroke. 1990;21(5):801–7. https://doi.org/10.1161/01.str.21.5.801.
38. Sukumari-Ramesh S, Alleyne CH Jr, Dhandapani KM. Astrocyte-specific expression of survivin after intracerebral hemorrhage in mice: a possible role in reactive gliosis? J Neurotrauma. 2012;29(18):2798–804. https://doi.org/10.1089/neu.2011.2243.
39. Yokota H, Yoshikawa M, Hirabayashi H, Nakase H, Uranishi R, Nishimura F, et al. Expression of ciliary neurotrophic factor (CNTF), CNTF receptor alpha (CNTFR-alpha) following experimental intracerebral hemorrhage in rats. Neurosci Lett. 2005;377(3):170–5. https://doi.org/10.1016/j.neulet.2004.11.093.
40. Kooijman E, Nijboer CH, van Velthoven CT, Kavelaars A, Kesecioglu J, Heijnen CJ. The rodent endovascular puncture model of subarachnoid hemorrhage: mechanisms of brain damage and therapeutic strategies. J Neuroinflammation. 2014;11:2. https://doi.org/10.1186/1742-2094-11-2.
41. Leonardo CC, Robbins S, Dore S. Translating basic science research to clinical application: models and strategies for intracerebral hemorrhage. Front Neurol. 2012;3:85. https://doi.org/10.3389/fneur.2012.00085.
42. Burda JE, Sofroniew MV. Reactive gliosis and the multicellular response to CNS damage and disease. Neuron. 2014;81(2):229–48. https://doi.org/10.1016/j.neuron.2013.12.034.
43. Sims NR, Yew WP. Reactive astrogliosis in stroke: contributions of astrocytes to recovery of neurological function. Neurochem Int. 2017;107:88–103. https://doi.org/10.1016/j.neuint.2016.12.016.
44. Anderson MA, Ao Y, Sofroniew MV. Heterogeneity of reactive astrocytes. Neurosci Lett. 2014;565:23–9. https://doi.org/10.1016/j.neulet.2013.12.030.
45. Silver J, Miller JH. Regeneration beyond the glial scar. Nat Rev Neurosci. 2004;5(2):146–56. https://doi.org/10.1038/nrn1326.

46. Tuinstra HM, Ducommun MM, Briley WE, Shea LD. Gene delivery to overcome astrocyte inhibition of axonal growth: an in vitro model of the glial scar. Biotechnol Bioeng. 2013;110(3):947–57. https://doi.org/10.1002/bit.24750.
47. Chen XR, Liao SJ, Ye LX, Gong Q, Ding Q, Zeng JS, Yu J. Neuroprotective effect of chondroitinase ABC on primary and secondary brain injury after stroke in hypertensive rats. Brain Res. 2014;1543:324–33. https://doi.org/10.1016/j.brainres.2013.12.002.
48. Gherardini L, Gennaro M, Pizzorusso T. Perilesional treatment with chondroitinase ABC and motor training promote functional recovery after stroke in rats. Cereb Cortex. 2015;25(1):202–12. https://doi.org/10.1093/cercor/bht217.
49. Hill JJ, Jin K, Mao XO, Xie L, Greenberg DA. Intracerebral chondroitinase ABC and heparan sulfate proteoglycan glypican improve outcome from chronic stroke in rats. Proc Natl Acad Sci USA. 2012;109(23):9155–60. https://doi.org/10.1073/pnas.1205697109.
50. Soleman S, Yip PK, Duricki DA, Moon LD. Delayed treatment with chondroitinase ABC promotes sensorimotor recovery and plasticity after stroke in aged rats. Brain. 2012;135(Pt 4):1210–23. https://doi.org/10.1093/brain/aws027.
51. Wiersma AM, Fouad K, Winship IR. Enhancing spinal plasticity amplifies the benefits of rehabilitative training and improves recovery from stroke. J Neurosci. 2017;37(45):10983–97. https://doi.org/10.1523/JNEUROSCI.0770-17.2017.
52. Xu X, Bass B, McKillop WM, Mailloux J, Liu T, Geremia NM, et al. Sox9 knockout mice have improved recovery following stroke. Exp Neurol. 2018;303:59–71. https://doi.org/10.1016/j.expneurol.2018.02.001.
53. Luo F, Wang J, Zhang Z, You Z, Bedolla A, Okwubido-Williams F, et al. Inhibition of CSPG receptor PTPsigma promotes migration of newly born neuroblasts, axonal sprouting, and recovery from stroke. Cell Rep. 2022;40(4):111137. https://doi.org/10.1016/j.celrep.2022.111137.
54. Li L, Lundkvist A, Andersson D, Wilhelmsson U, Nagai N, Pardo AC, et al. Protective role of reactive astrocytes in brain ischemia. J Cereb Blood Flow Metab. 2008;28(3):468–81. https://doi.org/10.1038/sj.jcbfm.9600546.
55. Aswendt M, Wilhelmsson U, Wieters F, Stokowska A, Schmitt FJ, Pallast N, et al. Reactive astrocytes prevent maladaptive plasticity after ischemic stroke. Prog Neurobiol. 2022;209:102199. https://doi.org/10.1016/j.pneurobio.2021.102199.
56. Liu Z, Li Y, Cui Y, Roberts C, Lu M, Wilhelmsson U, et al. Beneficial effects of gfap/vimentin reactive astrocytes for axonal remodeling and motor behavioral recovery in mice after stroke. Glia. 2014;62(12):2022–33. https://doi.org/10.1002/glia.22723.
57. Chen M, Ingle L, Plautz EJ, Kong X, Tang R, Ghosh N, et al. LZK-dependent stimulation of astrocyte reactivity promotes corticospinal axon sprouting. Front Cell Neurosci. 2022;16:969261. https://doi.org/10.3389/fncel.2022.969261.
58. Li L, Zhou J, Han L, Wu X, Shi Y, Cui W, et al. The specific role of reactive astrocytes in stroke. Front Cell Neurosci. 2022;16:850866. https://doi.org/10.3389/fncel.2022.850866.
59. Linnerbauer M, Wheeler MA, Quintana FJ. Astrocyte crosstalk in CNS inflammation. Neuron. 2020;108(4):608–22. https://doi.org/10.1016/j.neuron.2020.08.012.
60. Rakers C, Schleif M, Blank N, Matuskova H, Ulas T, Handler K, et al. Stroke target identification guided by astrocyte transcriptome analysis. Glia. 2019;67(4):619–33. https://doi.org/10.1002/glia.23544.
61. Zamanian JL, Xu L, Foo LC, Nouri N, Zhou L, Giffard RG, Barres BA. Genomic analysis of reactive astrogliosis. J Neurosci. 2012;32(18):6391–410. https://doi.org/10.1523/JNEUROSCI.6221-11.2012.
62. Li M, Li Z, Yao Y, Jin WN, Wood K, Liu Q, et al. Astrocyte-derived interleukin-15 exacerbates ischemic brain injury via propagation of cellular immunity. Proc Natl Acad Sci USA. 2017;114(3):E396–405. https://doi.org/10.1073/pnas.1612930114.
63. Lee GA, Lin TN, Chen CY, Mau SY, Huang WZ, Kao YC, et al. Interleukin 15 blockade protects the brain from cerebral ischemia-reperfusion injury. Brain Behav Immun. 2018;73:562–70. https://doi.org/10.1016/j.bbi.2018.06.021.

64. Herrmann O, Baumann B, de Lorenzi R, Muhammad S, Zhang W, Kleesiek J, et al. IKK mediates ischemia-induced neuronal death. Nat Med. 2005;11(12):1322–9. https://doi.org/10.1038/nm1323.
65. Dvoriantchikova G, Barakat D, Brambilla R, Agudelo C, Hernandez E, Bethea JR, et al. Inactivation of astroglial NF-kappa B promotes survival of retinal neurons following ischemic injury. Eur J Neurosci. 2009;30(2):175–85. https://doi.org/10.1111/j.1460-9568.2009.06814.x.
66. Cekanaviciute E, Fathali N, Doyle KP, Williams AM, Han J, Buckwalter MS. Astrocytic transforming growth factor-beta signaling reduces subacute neuroinflammation after stroke in mice. Glia. 2014;62(8):1227–40. https://doi.org/10.1002/glia.22675.
67. Cuellar-Santoyo AO, Ruiz-Rodriguez VM, Mares-Barbosa TB, Patron-Soberano A, Howe AG, Portales-Perez DP, et al. Revealing the contribution of astrocytes to glutamatergic neuronal transmission. Front Cell Neurosci. 2022;16:1037641. https://doi.org/10.3389/fncel.2022.1037641.
68. Kim MJ, Hur J, Ham IH, Yang HJ, Kim Y, Park S, Cho YW. Expression and activity of the na-k ATPase in ischemic injury of primary cultured astrocytes. Korean J Physiol Pharmacol. 2013;17(4):275–81. https://doi.org/10.4196/kjpp.2013.17.4.275.
69. Rossi DJ, Brady JD, Mohr C. Astrocyte metabolism and signaling during brain ischemia. Nat Neurosci. 2007;10(11):1377–86. https://doi.org/10.1038/nn2004.
70. Belov Kirdajova D, Kriska J, Tureckova J, Anderova M. Ischemia-triggered glutamate excitotoxicity from the perspective of glial cells. Front Cell Neurosci. 2020;14:51. https://doi.org/10.3389/fncel.2020.00051.
71. Phillis JW, Ren J, O'Regan MH. Transporter reversal as a mechanism of glutamate release from the ischemic rat cerebral cortex: studies with DL-threo-beta-benzyloxyaspartate. Brain Res. 2000;880(1–2):224. https://doi.org/10.1016/s0006-8993(00)02755-4.
72. Arundine M, Tymianski M. Molecular mechanisms of calcium-dependent neurodegeneration in excitotoxicity. Cell Calcium. 2003;34(4–5):325–37. https://doi.org/10.1016/s0143-4160(03)00141-6.
73. D'Aes T, Marlier Q, Verteneuil S, Quatresooz P, Vandenbosch R, Malgrange B. Re-evaluating the relevance of the oxygen-glucose deprivation model in ischemic stroke: the example of Cdk inhibition. Int J Mol Sci. 2023;24(8) https://doi.org/10.3390/ijms24087009.
74. Dong XX, Wang Y, Qin ZH. Molecular mechanisms of excitotoxicity and their relevance to pathogenesis of neurodegenerative diseases. Acta Pharmacol Sin. 2009;30(4):379–87. https://doi.org/10.1038/aps.2009.24.
75. Guo C, Ma YY. Calcium permeable-AMPA receptors and excitotoxicity in neurological disorders. Front Neural Circuits. 2021;15:711564. https://doi.org/10.3389/fncir.2021.711564.
76. Pajarillo E, Rizor A, Lee J, Aschner M, Lee E. The role of astrocytic glutamate transporters GLT-1 and GLAST in neurological disorders: potential targets for neurotherapeutics. Neuropharmacology. 2019;161:107559. https://doi.org/10.1016/j.neuropharm.2019.03.002.
77. Rothstein JD, Dykes-Hoberg M, Pardo CA, Bristol LA, Jin L, Kuncl RW, et al. Knockout of glutamate transporters reveals a major role for astroglial transport in excitotoxicity and clearance of glutamate. Neuron. 1996;16(3):675–86. https://doi.org/10.1016/s0896-6273(00)80086-0.
78. Kostandy BB. The role of glutamate in neuronal ischemic injury: the role of spark in fire. Neurol Sci. 2012;33(2):223–37. https://doi.org/10.1007/s10072-011-0828-5.
79. Mitani A, Tanaka K. Functional changes of glial glutamate transporter GLT-1 during ischemia: an in vivo study in the hippocampal CA1 of normal mice and mutant mice lacking GLT-1. J Neurosci. 2003;23(18):7176–82. https://doi.org/10.1523/JNEUROSCI.23-18-07176.2003.
80. Rao VL, Dogan A, Todd KG, Bowen KK, Kim BT, Rothstein JD, Dempsey RJ. Antisense knockdown of the glial glutamate transporter GLT-1, but not the neuronal glutamate transporter EAAC1, exacerbates transient focal cerebral ischemia-induced neuronal damage in rat brain. J Neurosci. 2001;21(6):1876–83. https://doi.org/10.1523/JNEUROSCI.21-06-01876.2001.

81. Ouyang YB, Voloboueva LA, Xu LJ, Giffard RG. Selective dysfunction of hippocampal CA1 astrocytes contributes to delayed neuronal damage after transient forebrain ischemia. J Neurosci. 2007;27(16):4253–60. https://doi.org/10.1523/JNEUROSCI.0211-07.2007.
82. Chen JC, Hsu-Chou H, Lu JL, Chiang YC, Huang HM, Wang HL, et al. Down-regulation of the glial glutamate transporter GLT-1 in rat hippocampus and striatum and its modulation by a group III metabotropic glutamate receptor antagonist following transient global forebrain ischemia. Neuropharmacology. 2005;49(5):703–14. https://doi.org/10.1016/j.neuropharm.2005.05.002.
83. Shen Y, He P, Fan YY, Zhang JX, Yan HJ, Hu WW, et al. Carnosine protects against permanent cerebral ischemia in histidine decarboxylase knockout mice by reducing glutamate excitotoxicity. Free Radic Biol Med. 2010;48(5):727–35. https://doi.org/10.1016/j.freeradbiomed.2009.12.021.
84. Chu K, Lee ST, Sinn DI, Ko SY, Kim EH, Kim JM, et al. Pharmacological induction of ischemic tolerance by glutamate transporter-1 (EAAT2) upregulation. Stroke. 2007;38(1):177–82. https://doi.org/10.1161/01.STR.0000252091.36912.65.
85. Harvey BK, Airavaara M, Hinzman J, Wires EM, Chiocco MJ, Howard DB, et al. Targeted over-expression of glutamate transporter 1 (GLT-1) reduces ischemic brain injury in a rat model of stroke. PLoS One. 2011;6(8):e22135. https://doi.org/10.1371/journal.pone.0022135.
86. Medvedeva YV, Lin B, Shuttleworth CW, Weiss JH. Intracellular Zn2+ accumulation contributes to synaptic failure, mitochondrial depolarization, and cell death in an acute slice oxygen-glucose deprivation model of ischemia. J Neurosci. 2009;29(4):1105–14. https://doi.org/10.1523/JNEUROSCI.4604-08.2009.
87. Choi DW. Glutamate neurotoxicity and diseases of the nervous system. Neuron. 1988;1(8):623–34. https://doi.org/10.1016/0896-6273(88)90162-6.
88. Gaidin SG, Kosenkov AM. Calcium-permeable AMPA receptors: an underestimated pharmacological target for the therapy of brain pathologies. Neural Regen Res. 2023;18(12):2669–70. https://doi.org/10.4103/1673-5374.373714.
89. Henley JM, Wilkinson KA. Synaptic AMPA receptor composition in development, plasticity and disease. Nat Rev Neurosci. 2016;17(6):337–50. https://doi.org/10.1038/nrn.2016.37.
90. Nishizawa Y. Glutamate release and neuronal damage in ischemia. Life Sci. 2001;69(4):369–81. https://doi.org/10.1016/s0024-3205(01)01142-0.
91. Blanco-Suarez E, Hanley JG. Distinct subunit-specific alpha-amino-3-hydroxy-5-methyl-4-isoxazolepropionic acid (AMPA) receptor trafficking mechanisms in cultured cortical and hippocampal neurons in response to oxygen and glucose deprivation. J Biol Chem. 2014;289(8):4644–51. https://doi.org/10.1074/jbc.M113.533182.
92. Koszegi Z, Fiuza M, Hanley JG. Endocytosis and lysosomal degradation of GluA2/3 AMPARs in response to oxygen/glucose deprivation in hippocampal but not cortical neurons. Sci Rep. 2017;7(1):12318. https://doi.org/10.1038/s41598-017-12534-w.
93. Liu B, Liao M, Mielke JG, Ning K, Chen Y, Li L, et al. Ischemic insults direct glutamate receptor subunit 2-lacking AMPA receptors to synaptic sites. J Neurosci. 2006;26(20):5309–19. https://doi.org/10.1523/JNEUROSCI.0567-06.2006.
94. Ding S. Ca(2+) signaling in astrocytes and its role in ischemic stroke. Adv Neurobiol. 2014;11:189–211. https://doi.org/10.1007/978-3-319-08894-5_10.
95. Dixon RM, Mellor JR, Hanley JG. PICK1-mediated glutamate receptor subunit 2 (GluR2) trafficking contributes to cell death in oxygen/glucose-deprived hippocampal neurons. J Biol Chem. 2009;284(21):14230–5. https://doi.org/10.1074/jbc.M901203200.
96. Maiorov SA, Zinchenko VP, Gaidin SG, Kosenkov AM. Potential mechanism of GABA secretion in response to the activation of GluK1-containing kainate receptors. Neurosci Res. 2021;171:27–33. https://doi.org/10.1016/j.neures.2021.03.009.
97. Allen NJ, Bennett ML, Foo LC, Wang GX, Chakraborty C, Smith SJ, Barres BA. Astrocyte glypicans 4 and 6 promote formation of excitatory synapses via GluA1 AMPA receptors. Nature. 2012;486(7403):410–4. https://doi.org/10.1038/nature11059.

98. Blanco-Suarez E, Liu TF, Kopelevich A, Allen NJ. Astrocyte-secreted chordin-like 1 drives synapse maturation and limits plasticity by increasing synaptic GluA2 AMPA receptors. Neuron. 2018;100(5):1116–1132.e1113. https://doi.org/10.1016/j.neuron.2018.09.043.
99. Kucukdereli H, Allen NJ, Lee AT, Feng A, Ozlu MI, Conatser LM, et al. Control of excitatory CNS synaptogenesis by astrocyte-secreted proteins Hevin and SPARC. Proc Natl Acad Sci USA. 2011;108(32):E440–9. https://doi.org/10.1073/pnas.1104977108.
100. Lo EH, Dalkara T, Moskowitz MA. Mechanisms, challenges and opportunities in stroke. Nat Rev Neurosci. 2003;4(5):399–415. https://doi.org/10.1038/nrn1106.
101. Verkhratsky A, Parpura V, Pekna M, Pekny M, Sofroniew M. Glia in the pathogenesis of neurodegenerative diseases. Biochem Soc Trans. 2014;42(5):1291–301. https://doi.org/10.1042/bst20140107.
102. Parpura V, Grubisic V, Verkhratsky A. Ca(2+) sources for the exocytotic release of glutamate from astrocytes. Biochim Biophys Acta. 2011;1813(5):984–91. https://doi.org/10.1016/j.bbamcr.2010.11.006.
103. Rakers C, Petzold GC. Astrocytic calcium release mediates peri-infarct depolarizations in a rodent stroke model. J Clin Invest. 2017;127(2):511–6. https://doi.org/10.1172/JCI89354.
104. Bondarenko A, Svichar N, Chesler M. Role of Na+-H+ and Na+-Ca2+ exchange in hypoxia-related acute astrocyte death. Glia. 2005;49(1):143–52. https://doi.org/10.1002/glia.20107.
105. Liu Y, Kintner DB, Begum G, Algharabli J, Cengiz P, Shull GE, et al. Endoplasmic reticulum Ca2+ signaling and mitochondrial Cyt c release in astrocytes following oxygen and glucose deprivation. J Neurochem. 2010;114(5):1436–46. https://doi.org/10.1111/j.1471-4159.2010.06862.x.
106. Ding S, Wang T, Cui W, Haydon PG. Photothrombosis ischemia stimulates a sustained astrocytic Ca2+ signaling in vivo. Glia. 2009;57(7):767–76. https://doi.org/10.1002/glia.20804.
107. Winship IR, Murphy TH. In vivo calcium imaging reveals functional rewiring of single somatosensory neurons after stroke. J Neurosci. 2008;28(26):6592–606. https://doi.org/10.1523/JNEUROSCI.0622-08.2008.
108. Liu J, Feng X, Wang Y, Xia X, Zheng JC. Astrocytes: GABAceptive and GABAergic cells in the brain. Front Cell Neurosci. 2022;16:892497. https://doi.org/10.3389/fncel.2022.892497.
109. Lie MEK, Al-Khawaja A, Damgaard M, Haugaard AS, Schousboe A, Clarkson AN, Wellendorph P. Glial GABA transporters as modulators of inhibitory signalling in epilepsy and stroke. In: Ortega A, Schousboe A, editors. Glial amino acid transporters. Springer International Publishing; 2017. p. 137–67. https://doi.org/10.1007/978-3-319-55769-4_7.
110. Nam MH, Cho J, Kwon DH, Park JY, Woo J, Lee JM, et al. Excessive astrocytic GABA causes cortical hypometabolism and impedes functional recovery after subcortical stroke. Cell Rep. 2020;32(3):107975. https://doi.org/10.1016/j.celrep.2020.107975.
111. Farrant M, Nusser Z. Variations on an inhibitory theme: phasic and tonic activation of GABA(A) receptors. Nat Rev Neurosci. 2005;6(3):215–29. https://doi.org/10.1038/nrn1625.
112. Brickley SG, Mody I. Extrasynaptic GABA(A) receptors: their function in the CNS and implications for disease. Neuron. 2012;73(1):23–34. https://doi.org/10.1016/j.neuron.2011.12.012.
113. Clarkson AN. Perisynaptic GABA receptors the overzealous protector. Adv Pharmacol Sci. 2012;2012:708428. https://doi.org/10.1155/2012/708428.
114. Clarkson AN, Huang BS, Macisaac SE, Mody I, Carmichael ST. Reducing excessive GABA-mediated tonic inhibition promotes functional recovery after stroke. Nature. 2010;468(7321):305–9. https://doi.org/10.1038/nature09511.
115. Schwartz-Bloom RD, Sah R. Gamma-aminobutyric acid(A) neurotransmission and cerebral ischemia. J Neurochem. 2001;77(2):353–71. https://doi.org/10.1046/j.1471-4159.2001.00274.x.
116. Ginsberg MD. Neuroprotection for ischemic stroke: past, present and future. Neuropharmacology. 2008;55(3):363–89. https://doi.org/10.1016/j.neuropharm.2007.12.007.
117. Allen NJ, Karadottir R, Attwell D. Reversal or reduction of glutamate and GABA transport in CNS pathology and therapy. Pflugers Arch. 2004;449(2):132–42. https://doi.org/10.1007/s00424-004-1318-x.

118. Kristensen AS, Andersen J, Jorgensen TN, Sorensen L, Eriksen J, Loland CJ, et al. SLC6 neurotransmitter transporters: structure, function, and regulation. Pharmacol Rev. 2011;63(3):585–640. https://doi.org/10.1124/pr.108.000869.
119. Orfila JE, Grewal H, Dietz RM, Strnad F, Shimizu T, Moreno M, et al. Delayed inhibition of tonic inhibition enhances functional recovery following experimental ischemic stroke. J Cereb Blood Flow Metab. 2019;39(6):1005–14. https://doi.org/10.1177/0271678X17750761.
120. Conti F, Minelli A, Melone M. GABA transporters in the mammalian cerebral cortex: localization, development and pathological implications. Brain Res Brain Res Rev. 2004;45(3):196–212. https://doi.org/10.1016/j.brainresrev.2004.03.003.
121. Lie ME, Gowing EK, Johansen NB, Dalby NO, Thiesen L, Wellendorph P, Clarkson AN. GAT3 selective substrate l-isoserine upregulates GAT3 expression and increases functional recovery after a focal ischemic stroke in mice. J Cereb Blood Flow Metab. 2019;39(1):74–88. https://doi.org/10.1177/0271678X17744123.
122. Osei-Owusu J, Yang J, Vitery MDC, Qiu Z. Molecular biology and physiology of volume-regulated Anion Channel (VRAC). Curr Top Membr. 2018;81:177–203. https://doi.org/10.1016/bs.ctm.2018.07.005.
123. Mongin AA. Volume-regulated anion channel—a frenemy within the brain. Pflugers Arch – Eur J Physiol. 2016;468(3):421–41. https://doi.org/10.1007/s00424-015-1765-6.
124. Mongin AA. Disruption of ionic and cell volume homeostasis in cerebral ischemia: the perfect storm. Pathophysiology. 2007;14(3):183–93. https://doi.org/10.1016/j.pathophys.2007.09.009.
125. Feustel PJ, Jin Y, Kimelberg HK. Volume-regulated anion channels are the predominant contributors to release of excitatory amino acids in the ischemic cortical penumbra. Stroke. 2004;35(5):1164–8. https://doi.org/10.1161/01.STR.0000124127.57946.a1.
126. Kimelberg HK. Astrocytic swelling in cerebral ischemia as a possible cause of injury and target for therapy. Glia. 2005;50(4):389–97. https://doi.org/10.1002/glia.20174.
127. Kimelberg HK, Feustel PJ, Jin Y, Paquette J, Boulos A, Keller RW Jr, Tranmer BI. Acute treatment with tamoxifen reduces ischemic damage following middle cerebral artery occlusion. Neuroreport. 2000;11(12):2675–9. https://doi.org/10.1097/00001756-200008210-00014.
128. Kimelberg HK, Jin Y, Charniga C, Feustel PJ. Neuroprotective activity of tamoxifen in permanent focal ischemia. J Neurosurg. 2003;99(1):138–42. https://doi.org/10.3171/jns.2003.99.1.0138.
129. Zhang Y, Zhang H, Feustel PJ, Kimelberg HK. DCPIB, a specific inhibitor of volume regulated anion channels (VRACs), reduces infarct size in MCAo and the release of glutamate in the ischemic cortical penumbra. Exp Neurol. 2008;210(2):514–20. https://doi.org/10.1016/j.expneurol.2007.11.027.
130. Voss FK, Ullrich F, Münch J, Lazarow K, Lutter D, Mah N, et al. Identification of LRRC8 heteromers as an essential component of the volume-regulated anion channel VRAC. Science. 2014;344(6184):634–8. https://doi.org/10.1126/science.1252826.
131. Qiu Z, Dubin AE, Mathur J, Tu B, Reddy K, Miraglia LJ, et al. SWELL1, a plasma membrane protein, is an essential component of volume-regulated anion channel. Cell. 2014;157(2):447–58.
132. Wilson CS, Bach MD, Ashkavand Z, Norman KR, Martino N, Adam AP, Mongin AA. Metabolic constraints of swelling-activated glutamate release in astrocytes and their implication for ischemic tissue damage. J Neurochem. 2019;151(2):255–72. https://doi.org/10.1111/jnc.14711.
133. Yang J, Vitery MDC, Chen J, Osei-Owusu J, Chu J, Qiu Z. Glutamate-releasing SWELL1 channel in astrocytes modulates synaptic transmission and promotes brain damage in stroke. Neuron. 2019;102(4):813–27. e816. https://doi.org/10.1016/j.neuron.2019.03.029.
134. Sifat AE, Nozohouri S, Archie SR, Chowdhury EA, Abbruscato TJ. Brain energy metabolism in ischemic stroke: effects of smoking and diabetes. Int J Mol Sci. 2022;23(15) https://doi.org/10.3390/ijms23158512.

135. Magistretti PJ, Allaman I. A cellular perspective on brain energy metabolism and functional imaging. Neuron. 2015;86(4):883–901. https://doi.org/10.1016/j.neuron.2015.03.035.
136. Bonvento G, Bolanos JP. Astrocyte-neuron metabolic cooperation shapes brain activity. Cell Metab. 2021;33(8):1546–64. https://doi.org/10.1016/j.cmet.2021.07.006.
137. Dringen R, Pfeiffer B, Hamprecht B. Synthesis of the antioxidant glutathione in neurons: supply by astrocytes of CysGly as precursor for neuronal glutathione. J Neurosci. 1999;19(2):562–9. https://doi.org/10.1523/JNEUROSCI.19-02-00562.1999.
138. Wang XF, Cynader MS. Astrocytes provide cysteine to neurons by releasing glutathione. J Neurochem. 2000;74(4):1434–42. https://doi.org/10.1046/j.1471-4159.2000.0741434.x.
139. Chen Y, Vartiainen NE, Ying W, Chan PH, Koistinaho J, Swanson RA. Astrocytes protect neurons from nitric oxide toxicity by a glutathione-dependent mechanism. J Neurochem. 2001;77(6):1601–10. https://doi.org/10.1046/j.1471-4159.2001.00374.x.
140. Bragin DE, Zhou B, Ramamoorthy P, Muller WS, Connor JA, Shi H. Differential changes of glutathione levels in astrocytes and neurons in ischemic brains by two-photon imaging. J Cereb Blood Flow Metab. 2010;30(4):734–8. https://doi.org/10.1038/jcbfm.2010.9.
141. Takagi T, Kitashoji A, Iwawaki T, Tsuruma K, Shimazawa M, Yoshimura S, et al. Temporal activation of Nrf2 in the penumbra and Nrf2 activator-mediated neuroprotection in ischemia-reperfusion injury. Free Radic Biol Med. 2014;72:124–33. https://doi.org/10.1016/j.freeradbiomed.2014.04.009.
142. Hong Z, Cao J, Liu D, Liu M, Chen M, Zeng F, et al. Celastrol targeting Nedd4 reduces Nrf2-mediated oxidative stress in astrocytes after ischemic stroke. J Pharm Anal. 2023;13(2):156–69. https://doi.org/10.1016/j.jpha.2022.12.002.
143. Clements CM, McNally RS, Conti BJ, Mak TW, Ting JP. DJ-1, a cancer- and Parkinson's disease-associated protein, stabilizes the antioxidant transcriptional master regulator Nrf2. Proc Natl Acad Sci USA. 2006;103(41):15091–6. https://doi.org/10.1073/pnas.0607260103.
144. Zhou W, Freed CR. DJ-1 up-regulates glutathione synthesis during oxidative stress and inhibits A53T alpha-synuclein toxicity. J Biol Chem. 2005;280(52):43150–8. https://doi.org/10.1074/jbc.M507124200.
145. Pellerin L, Magistretti PJ. Glutamate uptake into astrocytes stimulates aerobic glycolysis: a mechanism coupling neuronal activity to glucose utilization. Proc Natl Acad Sci USA. 1994;91(22):10625–9. https://doi.org/10.1073/pnas.91.22.10625.
146. Pierre K, Pellerin L. Monocarboxylate transporters in the central nervous system: distribution, regulation and function. J Neurochem. 2005;94(1):1–14. https://doi.org/10.1111/j.1471-4159.2005.03168.x.
147. Dienel GA. Lack of appropriate stoichiometry: strong evidence against an energetically important astrocyte-neuron lactate shuttle in brain. J Neurosci Res. 2017;95(11):2103–25. https://doi.org/10.1002/jnr.24015.
148. Patel AB, Lai JC, Chowdhury GM, Hyder F, Rothman DL, Shulman RG, Behar KL. Direct evidence for activity-dependent glucose phosphorylation in neurons with implications for the astrocyte-to-neuron lactate shuttle. Proc Natl Acad Sci USA. 2014;111(14):5385–90. https://doi.org/10.1073/pnas.1403576111.
149. Brown AM, Ransom BR. Astrocyte glycogen as an emergency fuel under conditions of glucose deprivation or intense neural activity. Metab Brain Dis. 2015;30(1):233–9. https://doi.org/10.1007/s11011-014-9588-2.
150. Yamagata K. Lactate supply from astrocytes to neurons and its role in ischemic stroke-induced neurodegeneration. Neuroscience. 2022;481:219–31. https://doi.org/10.1016/j.neuroscience.2021.11.035.
151. Schurr A, Payne RS, Miller JJ, Rigor BM. Glia are the main source of lactate utilized by neurons for recovery of function posthypoxia. Brain Res. 1997;774(1–2):221–4. https://doi.org/10.1016/s0006-8993(97)81708-8.
152. Schurr A, Payne RS, Miller JJ, Tseng MT, Rigor BM. Blockade of lactate transport exacerbates delayed neuronal damage in a rat model of cerebral ischemia. Brain Res. 2001;895(1–2):268–72. https://doi.org/10.1016/s0006-8993(01)02082-0.

153. Yamagata K, Tagami M, Ikeda K, Noguchi T, Yamori Y, Nara Y. Reduced production of lactate during hypoxia and reoxygenation in astrocytes isolated from stroke-prone spontaneously hypertensive rats. Neurosci Lett. 2000;296(2–3):113–6. https://doi.org/10.1016/s0304-3940(00)01628-1.
154. Berthet C, Castillo X, Magistretti PJ, Hirt L. New evidence of neuroprotection by lactate after transient focal cerebral ischaemia: extended benefit after intracerebroventricular injection and efficacy of intravenous administration. Cerebrovasc Dis. 2012;34(5–6):329–35. https://doi.org/10.1159/000343657.
155. Takahashi S, Iizumi T, Mashima K, Abe T, Suzuki N. Roles and regulation of ketogenesis in cultured astroglia and neurons under hypoxia and hypoglycemia. ASN Neuro. 2014;6(5) https://doi.org/10.1177/1759091414550997.
156. Puchowicz MA, Zechel JL, Valerio J, Emancipator DS, Xu K, Pundik S, et al. Neuroprotection in diet-induced ketotic rat brain after focal ischemia. J Cereb Blood Flow Metab. 2008;28(12):1907–16. https://doi.org/10.1038/jcbfm.2008.79.
157. Lin YH, Yang D, Ni HY, Xu XM, Wu F, Lin L, et al. Ketone bodies promote stroke recovery via GAT-1-dependent cortical network remodeling. Cell Rep. 2023;42(4):112294. https://doi.org/10.1016/j.celrep.2023.112294.
158. Guthrie PB, Knappenberger J, Segal M, Bennett MV, Charles AC, Kater SB. ATP released from astrocytes mediates glial calcium waves. J Neurosci. 1999;19(2):520–8. https://doi.org/10.1523/JNEUROSCI.19-02-00520.1999.
159. Zhang JM, Wang HK, Ye CQ, Ge W, Chen Y, Jiang ZL, et al. ATP released by astrocytes mediates glutamatergic activity-dependent heterosynaptic suppression. Neuron. 2003;40(5):971–82. https://doi.org/10.1016/s0896-6273(03)00717-7.
160. Ye XC, Hu JX, Li L, Li Q, Tang FL, Lin S, et al. Astrocytic Lrp4 (low-density lipoprotein receptor-related protein 4) contributes to ischemia-induced brain injury by regulating ATP release and adenosine-A(2A)R (adenosine A2A receptor) signaling. Stroke. 2018;49(1):165–74. https://doi.org/10.1161/STROKEAHA.117.018115.
161. Davis CH, Kim KY, Bushong EA, Mills EA, Boassa D, Shih T, et al. Transcellular degradation of axonal mitochondria. Proc Natl Acad Sci USA. 2014;111(26):9633–8. https://doi.org/10.1073/pnas.1404651111.
162. Gao L, Liu F, Hou PP, Manaenko A, Xiao ZP, Wang F, et al. Neurons release injured mitochondria as “help-me” signaling after ischemic stroke. Front Aging Neurosci. 2022;14:785761. https://doi.org/10.3389/fnagi.2022.785761.
163. Hayakawa K, Esposito E, Wang X, Terasaki Y, Liu Y, Xing C, et al. Transfer of mitochondria from astrocytes to neurons after stroke. Nature. 2016;535(7613):551–5. https://doi.org/10.1038/nature18928.
164. Berridge MV, Schneider RT, McConnell MJ. Mitochondrial transfer from astrocytes to neurons following ischemic insult: guilt by association? Cell Metab. 2016;24(3):376–8. https://doi.org/10.1016/j.cmet.2016.08.023.
165. Lee EH, Kim M, Ko SH, Kim CH, Lee M, Park CH. Primary astrocytic mitochondrial transplantation ameliorates ischemic stroke. BMB Rep. 2023;56(2):90–5. https://doi.org/10.5483/BMBRep.2022-0108.
166. Lee TH, Kato H, Chen ST, Kogure K, Itoyama Y. Expression of nerve growth factor and trkA after transient focal cerebral ischemia in rats. Stroke. 1998;29(8):1687–96.; discussion 1697. https://doi.org/10.1161/01.str.29.8.1687.
167. Bernaudin M, Bellail A, Marti HH, Yvon A, Vivien D, Duchatelle I, et al. Neurons and astrocytes express EPO mRNA: oxygen-sensing mechanisms that involve the redox-state of the brain. Glia. 2000;30(3):271–8.
168. Bejot Y, Prigent-Tessier A, Cachia C, Giroud M, Mossiat C, Bertrand N, et al. Time-dependent contribution of non neuronal cells to BDNF production after ischemic stroke in rats. Neurochem Int. 2011;58(1):102–11. https://doi.org/10.1016/j.neuint.2010.10.019.
169. Sato Y, Chin Y, Kato T, Tanaka Y, Tozuka Y, Mase M, et al. White matter activated glial cells produce BDNF in a stroke model of monkeys. Neurosci Res. 2009;65(1):71–8. https://doi.org/10.1016/j.neures.2009.05.010.

170. Arvidsson A, Kokaia Z, Airaksinen MS, Saarma M, Lindvall O. Stroke induces widespread changes of gene expression for glial cell line-derived neurotrophic factor family receptors in the adult rat brain. Neuroscience. 2001;106(1):27–41. https://doi.org/10.1016/s0306-4522(01)00268-8.
171. Miyazaki H, Nagashima K, Okuma Y, Nomura Y. Expression of glial cell line-derived neurotrophic factor induced by transient forebrain ischemia in rats. Brain Res. 2001;922(2):165–72. https://doi.org/10.1016/s0006-8993(01)03013-x.
172. Zhang N, Zhang Z, He R, Li H, Ding S. GLAST-CreER(T2) mediated deletion of GDNF increases brain damage and exacerbates long-term stroke outcomes after focal ischemic stroke in mouse model. Glia. 2020;68(11):2395–414. https://doi.org/10.1002/glia.23848.
173. Park CK, Ju WK, Hofmann HD, Kirsch M, Ki Kang J, Chun MH, Lee MY. Differential regulation of ciliary neurotrophic factor and its receptor in the rat hippocampus following transient global ischemia. Brain Res. 2000;861(2):345–53. https://doi.org/10.1016/s0006-8993(00)02045-x.
174. Kowianski P, Lietzau G, Czuba E, Waskow M, Steliga A, Morys J. BDNF: a key factor with multipotent impact on brain signaling and synaptic plasticity. Cell Mol Neurobiol. 2018;38(3):579–93. https://doi.org/10.1007/s10571-017-0510-4.
175. Qu WS, Wang YH, Wang JP, Tang YX, Zhang Q, Tian DS, et al. Galectin-1 enhances astrocytic BDNF production and improves functional outcome in rats following ischemia. Neurochem Res. 2010;35(11):1716–24. https://doi.org/10.1007/s11064-010-0234-z.
176. Yu L, Wang J, Xia Y, Hou W, Yao X, Guo Y, et al. Adiponectin promotes neurogenesis after transient cerebral ischemia through STAT3 mediated BDNF upregulation in astrocytes. Neurochem Res. 2023;48(2):641–57. https://doi.org/10.1007/s11064-022-03790-y.
177. Zhang Z, Sun GY, Ding S. Glial cell line-derived neurotrophic factor and focal ischemic stroke. Neurochem Res. 2021;46(10):2638–50. https://doi.org/10.1007/s11064-021-03266-5.
178. Zhang Y, Chen K, Sloan SA, Bennett ML, Scholze AR, O'Keeffe S, et al. An RNA-sequencing transcriptome and splicing database of glia, neurons, and vascular cells of the cerebral cortex. J Neurosci. 2014;34(36):11929–47. https://doi.org/10.1523/JNEUROSCI.1860-14.2014.
179. Kitagawa H, Sasaki C, Sakai K, Mori A, Mitsumoto Y, Mori T, et al. Adenovirus-mediated gene transfer of glial cell line-derived neurotrophic factor prevents ischemic brain injury after transient middle cerebral artery occlusion in rats. J Cereb Blood Flow Metab. 1999;19(12):1336–44. https://doi.org/10.1097/00004647-199912000-00007.
180. Wang Y, Lin SZ, Chiou AL, Williams LR, Hoffer BJ. Glial cell line-derived neurotrophic factor protects against ischemia-induced injury in the cerebral cortex. J Neurosci. 1997;17(11):4341–8. https://doi.org/10.1523/JNEUROSCI.17-11-04341.1997.
181. Wong LF, Ralph GS, Walmsley LE, Bienemann AS, Parham S, Kingsman SM, et al. Lentiviral-mediated delivery of Bcl-2 or GDNF protects against excitotoxicity in the rat hippocampus. Mol Ther. 2005;11(1):89–95. https://doi.org/10.1016/j.ymthe.2004.08.026.
182. Kobayashi T, Ahlenius H, Thored P, Kobayashi R, Kokaia Z, Lindvall O. Intracerebral infusion of glial cell line-derived neurotrophic factor promotes striatal neurogenesis after stroke in adult rats. Stroke. 2006;37(9):2361–7. https://doi.org/10.1161/01.STR.0000236025.44089.e1.
183. Kang SS, Keasey MP, Cai J, Hagg T. Loss of neuron-astroglial interaction rapidly induces protective CNTF expression after stroke in mice. J Neurosci. 2012;32(27):9277–87. https://doi.org/10.1523/JNEUROSCI.1746-12.2012.
184. Stockli KA, Lottspeich F, Sendtner M, Masiakowski P, Carroll P, Gotz R, et al. Molecular cloning, expression and regional distribution of rat ciliary neurotrophic factor. Nature. 1989;342(6252):920–3. https://doi.org/10.1038/342920a0.
185. Kang SS, Keasey MP, Arnold SA, Reid R, Geralds J, Hagg T. Endogenous CNTF mediates stroke-induced adult CNS neurogenesis in mice. Neurobiol Dis. 2013;49:68–78. https://doi.org/10.1016/j.nbd.2012.08.020.
186. Jia C, Keasey MP, Lovins C, Hagg T. Inhibition of astrocyte FAK-JNK signaling promotes subventricular zone neurogenesis through CNTF. Glia. 2018;66(11):2456–69. https://doi.org/10.1002/glia.23498.

187. Appelros P, Stegmayr B, Terent A. Sex differences in stroke epidemiology: a systematic review. Stroke. 2009;40(4):1082–90. https://doi.org/10.1161/STROKEAHA.108.540781.
188. Cordeau P Jr, Lalancette-Hebert M, Weng YC, Kriz J. Live imaging of neuroinflammation reveals sex and estrogen effects on astrocyte response to ischemic injury. Stroke. 2008;39(3):935–42. https://doi.org/10.1161/STROKEAHA.107.501460.
189. Wang J, Sareddy GR, Lu Y, Pratap UP, Tang F, Greene KM, et al. Astrocyte-derived estrogen regulates reactive Astrogliosis and is neuroprotective following ischemic brain injury. J Neurosci. 2020;40(50):9751–71. https://doi.org/10.1523/JNEUROSCI.0888-20.2020.
190. Zhang QG, Wang R, Tang H, Dong Y, Chan A, Sareddy GR, et al. Brain-derived estrogen exerts anti-inflammatory and neuroprotective actions in the rat hippocampus. Mol Cell Endocrinol. 2014;389(1–2):84–91. https://doi.org/10.1016/j.mce.2013.12.019.
191. Lu Y, Sareddy GR, Wang J, Zhang Q, Tang FL, Pratap UP, et al. Neuron-derived estrogen is critical for astrocyte activation and neuroprotection of the ischemic brain. J Neurosci. 2020;40(38):7355–74. https://doi.org/10.1523/JNEUROSCI.0115-20.2020.
192. Zhang Y, Xu S, Liang KY, Li K, Zou ZP, Yang CL, et al. Neuronal mTORC1 is required for maintaining the nonreactive state of astrocytes. J Biol Chem. 2017;292(1):100–11. https://doi.org/10.1074/jbc.M116.744482.
193. Ma Y, Guo H, Zhang L, Tao L, Yin A, Liu Z, et al. Estrogen replacement therapy-induced neuroprotection against brain ischemia-reperfusion injury involves the activation of astrocytes via estrogen receptor beta. Sci Rep. 2016;6:21467. https://doi.org/10.1038/srep21467.
194. Wassertheil-Smoller S, Hendrix SL, Limacher M, Heiss G, Kooperberg C, Baird A, et al. Effect of estrogen plus progestin on stroke in postmenopausal women: the Women's Health Initiative: a randomized trial. JAMA. 2003;289(20):2673–84. https://doi.org/10.1001/jama.289.20.2673.
195. Brown CE, Wong C, Murphy TH. Rapid morphologic plasticity of peri-infarct dendritic spines after focal ischemic stroke. Stroke. 2008;39(4):1286–91.
196. Mostany R, Chowdhury TG, Johnston DG, Portonovo SA, Carmichael ST, Portera-Cailliau C. Local hemodynamics dictate long-term dendritic plasticity in peri-infarct cortex. J Neurosci. 2010;30(42):14116–26.
197. Zhao LR, Willing A. Enhancing endogenous capacity to repair a stroke-damaged brain: an evolving field for stroke research. Prog Neurobiol. 2018;163–164:5–26. https://doi.org/10.1016/j.pneurobio.2018.01.004.
198. Chiareli RA, Carvalho GA, Marques BL, Mota LS, Oliveira-Lima OC, Gomes RM, et al. The role of astrocytes in the neurorepair process. Front Cell Dev Biol. 2021;9:665795. https://doi.org/10.3389/fcell.2021.665795.
199. Patabendige A, Singh A, Jenkins S, Sen J, Chen R. Astrocyte activation in neurovascular damage and repair following ischaemic stroke. Int J Mol Sci. 2021;22(8) https://doi.org/10.3390/ijms22084280.
200. Lee J-H, Kim J-Y, Noh S, Lee H, Lee SY, Mun JY, et al. Astrocytes phagocytose adult hippocampal synapses for circuit homeostasis. Nature. 2021;590(7847):612–7.
201. Cahoy JD, Emery B, Kaushal A, Foo LC, Zamanian JL, Christopherson KS, et al. A transcriptome database for astrocytes, neurons, and oligodendrocytes: a new resource for understanding brain development and function. J Neurosci. 2008;28(1):264–78.
202. Lu Z, Elliott MR, Chen Y, Walsh JT, Klibanov AL, Ravichandran KS, Kipnis J. Phagocytic activity of neuronal progenitors regulates adult neurogenesis. Nat Cell Biol. 2011;13(9):1076–83.
203. Hu X, Li P, Guo Y, Wang H, Leak RK, Chen S, et al. Microglia/macrophage polarization dynamics reveal novel mechanism of injury expansion after focal cerebral ischemia. Stroke. 2012;43(11):3063–70. https://doi.org/10.1161/strokeaha.112.659656.
204. Morizawa YM, Hirayama Y, Ohno N, Shibata S, Shigetomi E, Sui Y, et al. Reactive astrocytes function as phagocytes after brain ischemia via ABCA1-mediated pathway. Nat Commun. 2017;8(1):28.
205. Shi X, Luo L, Wang J, Shen H, Li Y, Mamtilahun M, et al. Stroke subtype-dependent synapse elimination by reactive gliosis in mice. Nat Commun. 2021;12(1):6943.

206. Ullian EM, Sapperstein SK, Christopherson KS, Barres BA. Control of synapse number by glia. Science. 2001;291(5504):657–61. https://doi.org/10.1126/science.291.5504.657.
207. Blanco-Suarez E, Allen NJ. Astrocyte-secreted chordin-like 1 regulates spine density after ischemic injury. Sci Rep. 2022;12(1):4176. https://doi.org/10.1038/s41598-022-08031-4.
208. Collyer E, Boyle BR, Gomez-Galvez Y, Iacovitti L, Blanco-Suarez E. Absence of chordin-like 1 aids motor recovery in a mouse model of stroke. Exp Neurol. 2023;370:114548. https://doi.org/10.1016/j.expneurol.2023.114548.
209. Christopherson KS, Ullian EM, Stokes CCA, Mullowney CE, Hell JW, Agah A, et al. Thrombospondins are astrocyte-secreted proteins that promote CNS synaptogenesis. Cell. 2005;120(3):421–33. https://doi.org/10.1016/j.cell.2004.12.020.
210. Eroglu C. The role of astrocyte-secreted matricellular proteins in central nervous system development and function. J Cell Commun Signal. 2009;3(3):167–76. https://doi.org/10.1007/s12079-009-0078-y.
211. Risher WC, Kim N, Koh S, Choi JE, Mitev P, Spence EF, et al. Thrombospondin receptor alpha2delta-1 promotes synaptogenesis and spinogenesis via postsynaptic Rac1. J Cell Biol. 2018;217(10):3747–65. https://doi.org/10.1083/jcb.201802057.
212. Lin TN, Kim GM, Chen JJ, Cheung WM, He YY, Hsu CY. Differential regulation of thrombospondin-1 and thrombospondin-2 after focal cerebral ischemia/reperfusion. Stroke. 2003;34(1):177–86. https://doi.org/10.1161/01.str.0000047100.84604.ba.
213. Risher WC, Patel S, Kim IH, Uezu A, Bhagat S, Wilton DK, et al. Astrocytes refine cortical connectivity at dendritic spines. elife. 2014;3 https://doi.org/10.7554/eLife.04047.
214. Singh SK, Stogsdill JA, Pulimood NS, Dingsdale H, Kim YH, Pilaz L-J, et al. Astrocytes assemble thalamocortical synapses by bridging NRX1α and NL1 via hevin. Cell. 2016;164(1):183–96.
215. Kim J-H, Jung H-G, Kim A, Shim HS, Hyeon SJ, Lee Y-S, et al. Hevin–calcyon interaction promotes synaptic reorganization after brain injury. Cell Death Differ. 2021;28(9):2571–88.
216. Lively S, Moxon-Emre I, Schlichter LC. SC1/hevin and reactive gliosis after transient ischemic stroke in young and aged rats. J Neuropathol Exp Neurol. 2011;70(10):913–29.
217. Jones EV, Bernardinelli Y, Tse YC, Chierzi S, Wong TP, Murai KK. Astrocytes control glutamate receptor levels at developing synapses through SPARC–β-integrin interactions. J Neurosci. 2011;31(11):4154–65.
218. Jones EV, Bernardinelli Y, Zarruk JG, Chierzi S, Murai KK. SPARC and GluA1-containing AMPA receptors promote neuronal health following CNS injury. Front Cell Neurosci. 2018;12:22. https://doi.org/10.3389/fncel.2018.00022.
219. Farhy-Tselnicker I, van Casteren ACM, Lee A, Chang VT, Aricescu AR, Allen NJ. Astrocyte-secreted glypican 4 regulates release of neuronal pentraxin 1 from axons to induce functional synapse formation. Neuron. 2017;96(2):428–445.e413. https://doi.org/10.1016/j.neuron.2017.09.053.
220. Lindvall O, Kokaia Z. Neurogenesis following stroke affecting the adult brain. Cold Spring Harb Perspect Biol. 2015;7(11) https://doi.org/10.1101/cshperspect.a019034.
221. Rahman AA, Amruta N, Pinteaux E, Bix GJ. Neurogenesis after stroke: a therapeutic perspective. Transl Stroke Res. 2021;12(1):1–14. https://doi.org/10.1007/s12975-020-00841-w.
222. Arvidsson A, Collin T, Kirik D, Kokaia Z, Lindvall O. Neuronal replacement from endogenous precursors in the adult brain after stroke. Nat Med. 2002;8(9):963–70. https://doi.org/10.1038/nm747.
223. Lim DA, Alvarez-Buylla A. Interaction between astrocytes and adult subventricular zone precursors stimulates neurogenesis. Proc Natl Acad Sci USA. 1999;96(13):7526–31. https://doi.org/10.1073/pnas.96.13.7526.
224. Song H, Stevens CF, Gage FH. Astroglia induce neurogenesis from adult neural stem cells. Nature. 2002;417(6884):39–44. https://doi.org/10.1038/417039a.
225. Lois C, Garcia-Verdugo JM, Alvarez-Buylla A. Chain migration of neuronal precursors. Science. 1996;271(5251):978–81. https://doi.org/10.1126/science.271.5251.978.
226. Thomas LB, Gates MA, Steindler DA. Young neurons from the adult subependymal zone proliferate and migrate along an astrocyte, extracellular matrix-rich pathway. Glia. 1996;17(1):1–14. https://doi.org/10.1002/(SICI)1098-1136(199605)17:1<1::AID-GLIA1>3.0.CO;2-7.

227. Gotz M, Sirko S, Beckers J, Irmler M. Reactive astrocytes as neural stem or progenitor cells: in vivo lineage, in vitro potential, and genome-wide expression analysis. Glia. 2015;63(8):1452–68. https://doi.org/10.1002/glia.22850.
228. Magnusson JP, Goritz C, Tatarishvili J, Dias DO, Smith EM, Lindvall O, et al. A latent neurogenic program in astrocytes regulated by Notch signaling in the mouse. Science. 2014;346(6206):237–41. https://doi.org/10.1126/science.346.6206.237.
229. Li M, Sun L, Li Y, Xie C, Wan D, Luo Y. Oxygen glucose deprivation/reperfusion astrocytes promotes primary neural stem/progenitor cell proliferation by releasing high-mobility group box 1. Neurochem Res. 2014;39(8):1440–50. https://doi.org/10.1007/s11064-014-1333-z.
230. Lin Y, Zhang JC, Yao CY, Wu Y, Abdelgawad AF, Yao SL, Yuan SY. Critical role of astrocytic interleukin-17 A in post-stroke survival and neuronal differentiation of neural precursor cells in adult mice. Cell Death Dis. 2016;7(6):e2273. https://doi.org/10.1038/cddis.2015.284.
231. Luo L, Guo K, Fan W, Lu Y, Chen L, Wang Y, et al. Niche astrocytes promote the survival, proliferation and neuronal differentiation of co-transplanted neural stem cells following ischemic stroke in rats. Exp Ther Med. 2017;13(2):645–50. https://doi.org/10.3892/etm.2016.4016.
232. Imitola J, Raddassi K, Park KI, Mueller FJ, Nieto M, Teng YD, et al. Directed migration of neural stem cells to sites of CNS injury by the stromal cell-derived factor 1alpha/CXC chemokine receptor 4 pathway. Proc Natl Acad Sci USA. 2004;101(52):18117–22. https://doi.org/10.1073/pnas.0408258102.
233. Kraft A, Jubal ER, von Laer R, Doring C, Rocha A, Grebbin M, et al. Astrocytic calcium waves signal brain injury to neural stem and progenitor cells. Stem Cell Reports. 2017;8(3):701–14. https://doi.org/10.1016/j.stemcr.2017.01.009.
234. Magnusson JP, Frisen J. Stars from the darkest night: unlocking the neurogenic potential of astrocytes in different brain regions. Development. 2016;143(7):1075–86. https://doi.org/10.1242/dev.133975.
235. Shimada IS, LeComte MD, Granger JC, Quinlan NJ, Spees JL. Self-renewal and differentiation of reactive astrocyte-derived neural stem/progenitor cells isolated from the cortical peri-infarct area after stroke. J Neurosci. 2012;32(23):7926–40. https://doi.org/10.1523/JNEUROSCI.4303-11.2012.
236. Shimada IS, Peterson BM, Spees JL. Isolation of locally derived stem/progenitor cells from the peri-infarct area that do not migrate from the lateral ventricle after cortical stroke. Stroke. 2010;41(9):e552–60. https://doi.org/10.1161/STROKEAHA.110.589010.
237. Sirko S, Behrendt G, Johansson PA, Tripathi P, Costa M, Bek S, et al. Reactive glia in the injured brain acquire stem cell properties in response to sonic hedgehog. [corrected]. Cell Stem Cell. 2013;12(4):426–39. https://doi.org/10.1016/j.stem.2013.01.019.
238. Duan CL, Liu CW, Shen SW, Yu Z, Mo JL, Chen XH, Sun FY. Striatal astrocytes transdifferentiate into functional mature neurons following ischemic brain injury. Glia. 2015;63(9):1660–70. https://doi.org/10.1002/glia.22837.
239. Santopolo G, Magnusson JP, Lindvall O, Kokaia Z, Frisen J. Blocking notch-signaling increases neurogenesis in the striatum after stroke. Cells. 2020;9(7) https://doi.org/10.3390/cells9071732.
240. Collaborators GBDS. Global, regional, and national burden of stroke and its risk factors, 1990-2019: a systematic analysis for the Global Burden of Disease Study 2019. Lancet Neurol. 2021;20(10):795–820. https://doi.org/10.1016/S1474-4422(21)00252-0.
241. Phipps MS, Cronin CA. Management of acute ischemic stroke. BMJ. 2020;368:l6983. https://doi.org/10.1136/bmj.l6983.
242. Stinear, C. M., Lang, C. E., Zeiler, S., & Byblow, W. D. (2020). Advances and challenges in stroke rehabilitation. Lancet Neurol, 19(4), 348–360. https://doi.org/10.1016/S1474-4422(19)30415-6.
243. Barreto G, White RE, Ouyang Y, Xu L, Giffard RG. Astrocytes: targets for neuroprotection in stroke. Cent Nerv Syst Agents Med Chem. 2011;11(2):164–73. https://doi.org/10.2174/187152411796011303.
244. Liu Z, Chopp M. Astrocytes, therapeutic targets for neuroprotection and neurorestoration in ischemic stroke. Prog Neurobiol. 2016;144:103–20. https://doi.org/10.1016/j.pneurobio.2015.09.008.

The Interplay of Astrocytes and Neurons in Autism Spectrum Disorder

Amanda C. S. S. Cano, Debora Santos, and Patricia C. B. Beltrão-Braga

Abstract Autism spectrum disorder (ASD) comprises a complex neurodevelopmental condition characterized by an impairment in social interaction, involving communication deficits and specific patterns of behaviors, like repetitive behaviors. ASD is clinically diagnosed and usually takes time, typically occurring not before four years of age. Genetic mutations affecting synaptic transmission, such as neuroligin and neurexin, are associated with ASD and contribute to behavioral and cognitive deficits. Recent research highlights the role of astrocytes, the brain's most abundant glial cells, in ASD pathology. Aberrant Ca^{2+} signaling in astrocytes is linked to behavioral deficits and neuroinflammation. Notably, the cytokine IL-6 overexpression by astrocytes impacts synaptogenesis. Altered neurotransmitter levels, disruptions in the blood-brain barrier, and cytokine dysregulation further contribute to ASD complexity. Understanding these astrocyte-related mechanisms holds promise for identifying ASD subtypes and developing targeted therapies.

Keywords Autism spectrum disorder · Astrocytic dysfunctions · Astrocyte-neuron interplay · Neuroinflammation

A. C. S. S. Cano · D. Santos
Laboratory of Disease Modeling, Institute of Biomedical Sciences, University of São Paulo, São Paulo, Brazil
e-mail: amandacano@usp.br; deborawps@usp.br

P. C. B. Beltrão-Braga (✉)
Laboratory of Disease Modeling, Institute of Biomedical Sciences, University of São Paulo, São Paulo, Brazil

Institut Pasteur de São Paulo, São Paulo, Brazil
e-mail: patriciacbbbraga@usp.br

E. Blanco-Suarez, I. Farhy-Tselnicker (eds.), *Astrocyte-Neuron Interactions in Health and Disease*, Advances in Neurobiology 39,
https://doi.org/10.1007/978-3-031-64839-7_11

Abbreviations

AMPA	alpha-amino-3-hydroxy-5-methyl-4-isoxazole propionic acid
ASD	autism spectrum disorder
BA40	parietal cortex 40
BA9	Brodmann area 9
BBB	blood-brain barrier
BPA	bisphenol propane
CLDN	claudin
CNS	central nervous system
GABA	gamma-aminobutyric acid
GABBR	GABAB receptor
GAD	glutamic acid decarboxylase
GFAP	glial fibrillary acidic protein
GM-CSF	granulocyte-macrophage colony-stimulating factor
Grin	subunit of NMDA
HMGB1	high mobility group box 1
IFN	interferon
IL	interleukin
iPSC	induced pluripotent stem cell
MMP-9	matrix metalloprotease
MRS	magnetic resonance spectroscopy
NRXN	neurexin
NLGN	neuroligin
NMDA	N-methyl-d-aspartate
MIA	maternal immune activation
PPA	propionic acid
TGF	transforming growth factor
TJ	tight junctions
TLR4	toll-like receptor 4
TNF	tumor necrosis factor
TRIC	tricellulin
TSPO	translocator protein
VPA	valproic acid

1 Introduction

Autism spectrum disorder (ASD) was first noticed by a psychiatrist Grunya Ssucharewa, who diagnosed six children with autistic characteristics [1–3]. However, only after approximately 20 years the term "autism" came, suggested by Leo Kanner, who diagnosed 11 children with a specific pattern of behavior, presenting uncommon social deficits, saying, "*These children have come into the world with an innate inability to form the usual, biologically provided contact with people*" [4]. After that, individuals presenting deficits in communication and social interaction were recognized as having ASD, and since then, the research on ASD field has been investigating the causes and mechanisms implicated in this condition [5], acknowledged as caused by a composition between a susceptible genetic background and environmental disturbances during the fetus development [6, 7]. To date, ASD is characterized by the individual's impaired communication and difficulties in social interaction [8, 9]. Another important characteristic of ASD is the presence of repetitive and restrictive behaviors, as well as hyperactivity or hyporeactivity to sensory input [10–12].

For ASD diagnosis, the individual must present deficits in the three areas of communication and social interaction. The first area is linked to deficits in socio-emotional reciprocity, which occurs from the social approach and failure in conversation; the second is deficits in non-verbal communicative behaviors; and the third is deficits in developing and maintaining relationships. ASD presents a spectrum of behaviors classified into three levels according to daily activities and the necessity of support to realize them, being level 3 – requires very substantial support; level 2 – requires substantial support; and level 1 – needs support (Autism Spectrum Disorder, 2022). Considering that the diagnosis is clinical and usually requires a certain level of specialty, generally, ASD is not diagnosed before four years of age [13].

The prevalence of ASD varies around the world, being 1–2% [14]. Between 2011 and 2012, the prevalence of ASD was estimated at 2% in children aged 6–17 [15], which generates a cost of approximately US$1.4 to 2.4 million over a lifetime ([16]). It is important to mention that many parts of the world do not have an official definition of the prevalence of ASD.

Considering the clinical outcomes of ASD, many studies have focused on the patterns of functioning of the brain cells to unveil the pathways involved in the disorder. To do that, ASD research has been using models since biopsies of the CNS are not well accepted for research purposes. Therefore, postmortem tissue, animal models, and induced pluripotent stem cells (iPSCs) can be used to model ASD.

Postmortem tissue studies reveal morphological changes at the cellular level and in the organization of gray and white matter. Neurons are the effector cells in the brain, currently implicated in most neurological disorders, and also the target of studies for ASD elucidation. In the postmortem brains of two children with autistic behaviors, it was observed that the neurons had fewer branches compared to control neurons [17]. Neural pruning is an important process as it offers selection and maturation of synapses and neural circuits [18]. The peak of dendritic spine pruning occurs in early childhood, but a decline is observed in the following stages [19]. It was reported in frozen brain samples that the level of postsynaptic protein, PSD95, was higher in controls aged 2–10 than in controls aged 13 and 20 years. On the other hand, the decrease in these synaptic proteins was not observed in patients with ASD. In the postmortem ASD temporal lobe, increased dendritic spine density with reduced developmental spine pruning was observed in layer V pyramidal neurons, whose basal autophagy was low [18]. Autophagy in neurons protects against neurodegeneration and presents a programmed recycling of cellular components essential for synaptic integrity and neuronal functions [20]. It is worth noting that postmortem tissue samples are difficult to obtain and offer little insight into a disorder that arises throughout development [21, 22]. Besides, the tissue is already fixed and does not provide insights into what is happening within a live cell, so animal models have been used to study live cells in ASD.

Animal models for the study of ASD are wide-ranging, including nonhuman primates, domesticated animals (dogs, cattle, and cats), rodents (rats and mice), songbirds, fish, and invertebrates (*Drosophila melanogaster*, nematode, bees, *Aplysia*) [23, 24]. These models are divided into three groups: genetic, environmentally induced, and idiopathic animal models, aiming to mimic some of the causes of autism. Genetic animal models include knock-out, knock-in, and overexpressed animals. Environmentally induced animal models use chemicals (valproic acid (VPA), propionic acid (PPA), bisphenol propane (BPA), sevoflurane), infections (Borna disease virus), or other sources of stress, such as repeated cold temperature stress, maternal immune activation (MIA), and transplantation of gut microbiota from autism patients. Idiopathic animal models include BTBR T + Itpr3tfJ and BALB/cByJ mouse strains, which show symptoms similar to ASD [23]. Animal models are also often used to unravel proteins' roles and evaluate the alterations caused by genes mutated in ASD patients, for example, the synaptic scaffolding protein SHANK 1 and 3, a protein responsible for promoting the formation, maturation, and stability of dendritic spines, whose alterations cause autism-like behavior in knockout, overexpressed, and mutated mice, as well as in knockout and mutated zebrafish [25–29]. Although animal models for ASD studies have been widely used, it is important to consider that behavioral studies in animals are difficult to translate to human behavior, seen just as autistic-like behavior. Another important point to consider is that ASD is a complex multigenic and multifactorial disorder, challenging to mimic in animal models.

Considering the above, using live human cells from people with ASD could be a very powerful tool to advance ASD research. More recently, the discovery of iPSC [30] revealed a model with great potential for ASD modeling, keeping the patient's genetics and biochemistry. The iPSC can also be grown in 3D, giving rise to brain organoids, more complex models containing different cell types revealing intercellular interactions, and more faithfully representing the development and different aspects of the disorder.

So far, although the studies on autism were mainly focused on neurons, it seems that the disorder affects glial cells the most. More recent studies have shown that astrocytes may undergo metabolic changes that are directly implicated in the outcome of ASD. Astrocytes are the most abundant glial cells in the brain and play fundamental roles throughout brain functioning, as they regulate neuronal migration and dendritic development, control neurotransmitter concentration in the synaptic cleft, and are responsible for synaptogenesis and synaptic function [31]. This chapter will revise what we currently know about the interplay between neurons and astrocytes in the neuropathology of ASD.

2 Astrocytes and Their Role in Synaptogenesis in ASD

Astrocytes are glial cells that support synaptogenesis [32, 33]. It is already known that ASD individuals present alterations in synapses [34]. Considering both information, the role played by astrocytes for the synaptic alterations described for ASD is still unclear. Additionally, aberrant Ca2+ signaling has already been identified in ASD astrocytes, suggesting a possible mechanism for the characteristic behavioral and neural deficits of ASD [35, 36]. Interestingly, the transplantation of ASD astrocytes with aberrant Ca2+ signaling into postnatal mouse brains resulted in persistent digging behaviors and attenuated fear memory (Fig. 1) [35].

At the genetic level, several mutations playing a crucial role in the regulation of synaptic transmission have been identified as factors associated with ASD, including genes such as neuroligin (NLGN) and neurexin (NRXN) [37]. Neuroligin and neurexin are cell adhesion proteins that interact trans-synaptically [38] and are frequently mutated in ASD patients, leading to loss of functionality [39, 40]. NLGN4 knockout mice exhibited autism-like behavior, including deficits in social interactions and communication, as well as repetitive behaviors and interests [41–42]. NLGN2 knockout mice showed chronic changes in synaptic inhibition and cognitive deficits [43]. The mutation generated by the R451C substitution in neuroligin-3 expressed in mice showed impairment in social interactions but an improvement in spatial learning skills, as well as an increase inhibitory synaptic transmission, with

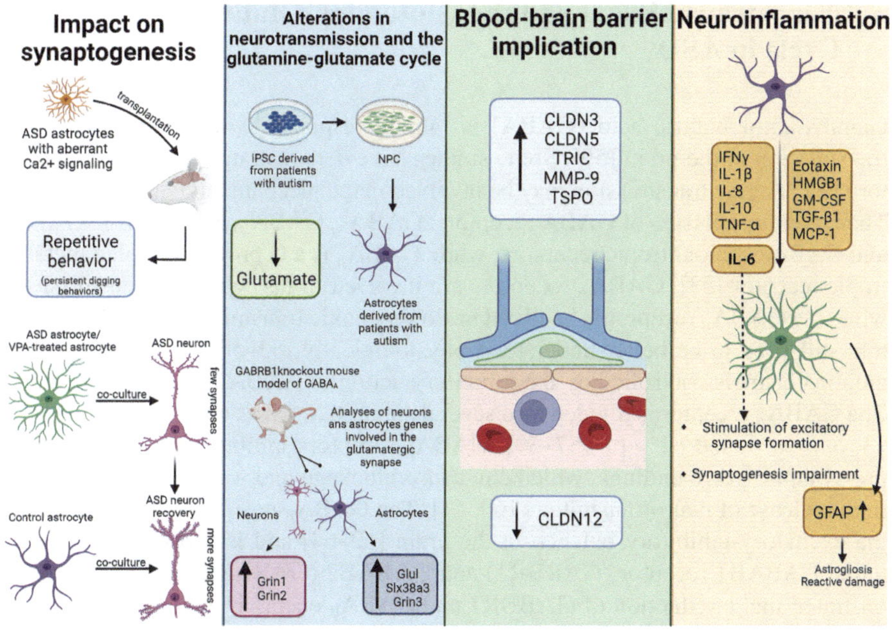

Fig. 1 Graphical abstract regarding the current knowledge about the interplay between neurons and astrocytes in ASD

no apparent effect on excitatory synapses [44]. Deletion of the gene responsible for NRXN1 expression in humans is associated with intellectual disability and speech delay [45]. NRXN1 inactivation directly impaired synaptic function in human neurons derived from human embryonic stem cells [46].

Even more clear influence in the interplay between neurons and astrocytes and their implications for ASD deficits in synaptogenesis was described using astrocytes derived from human iPSC from ASD children. In this work, astrocytes derived from ASD children decreased the number of synapses of neurons derived from neurotypical individuals (considered as controls) in co-culture conditions. On the other hand, neurons derived from ASD children co-cultured with astrocytes derived from controls presented an increase in the synapsis number, revealing that astrocytes were directly implicated in synaptogenesis [34] (Fig. 1).

Neuroinflammation is another factor described as impacting astrocyte function, directly affecting synaptogenesis. Overexpression of IL-6 in granule cells enhances excitatory synapse formation without affecting inhibitory synapses [47]. Conversely, VPA-treated astrocyte cultures reduced the number of inhibitory synapses but not excitatory synapses, specifically impairing inhibitory neuron synaptogenesis [48]. This hypothesis is further supported by the finding of an elevation in the ratio of excitatory neuron activity to inhibitory neurons in the medial prefrontal cortex of mice with autism-like behavior [49].

3 Neurotransmission and the Glutamine-Glutamate Cycle in ASD

Gamma-aminobutyric acid (GABA) is mainly responsible for inhibitory neurotransmission in the brain [50]. Some studies have demonstrated that altered inhibitory GABAergic transmission may be involved in the mechanism of autism [51, 52]. There are three classes of GABA receptors: $GABA_A$, $GABA_B$, and $GABA_C$. $GABA_A$ and $GABA_C$ are ionotropic receptors, while $GABA_B$ is a G protein-coupled metabotropic receptor [53]. $GABA_A$ receptors are involved in fast synaptic transmissions, while the $GABA_B$ receptor is involved in slow synaptic transmission. $GABA_C$ receptors still need to be better understood physiologically [53–56]. $GABA_A$ receptors are ion channels controlled by the inhibitory action of GABA in the brain [57, 58]. The $GABA_A$ receptor is divided into several subunits, such as $\alpha1 - \alpha6$, $\beta1 - \beta3$, $\gamma1 - \gamma3$, δ, ε, π, θ, and $\rho1 - \rho2$ [57–59]. $GABA_B$ receptors inhibit the Ca^{2+} channel in pre-synaptic nerve endings, which can cause blockage and a consequent reduction in the release of neurotransmitters [60, 61]. The $GABA_B$ receptor is responsible for the excitatory-inhibitory balance in the brain [62–64] and is formed by two subunits: GABAB1 receptor (GABBR1) and GABAB2 (GABBR2). It has been demonstrated that a reduction of GABBR1 and $GABA_B$ occurs in brain areas involved in ASD, such as the cerebellum, Brodmann area 9 (BA9) of the superior frontal section, and parietal cortex 40 (BA40). GABBR2, on the other hand, only presents changes in the cerebellum [62]. Furthermore, multiple seizures result in a long-term reduction of 15 days in GABBR1 expression and 30 days in GABBR2 expression [62, 65]. The expression level of the GABBR1 subunit of $GABA_A$ was evaluated in postmortem brains of patients with ASD, revealing lower levels of GABBR1 when compared to controls [50, 66].

In astrocytes derived from a knockout mouse model for GABRB1 of $GABA_A$, using single-cell RNA expression, the levels of the genes *Glul*, *Slx38a3,* and *Grin3* involved in the glutamatergic synapse were analyzed, revealing that these genes were upregulated. Additionally, in neurons using the same approach, the expression level of *Grin1* (subunit 1 of N-methyl-d-aspartate (NMDA)) and *Grin 2* (subunit 2 of NMDA) showed a trend of upregulation in hippocampal neurons (Fig. 1). This positive regulation in astrocytes and neurons suggests a contribution to the excitatory pathway of the glutamatergic synapse [52]. Magnetic resonance spectroscopy (MRS) showed an increase in glutamate in the anterior cingulate cortex (ACC) in adults with autism when compared to control individuals [67]. As well, the levels of the enzyme responsible for converting glutamate into GABA, glutamic acid decarboxylase (GAD), were reduced in its two isoforms (GAD65 and GAD67) in the parietal cortex (BA40) and cerebellum of autistic people [68]. Furthermore, the knockout mouse model for the GABRB1 subunit of $GABA_A$ showed an imbalance in the excitation/inhibition relationship, with an upregulation of N-methyl-d-aspartate (NMDA) currents and a reduction in $GABA_A$ currents in hippocampal neurons. The alpha-amino-3-hydroxy-5-methyl-4-isoxazolepropionic acid (AMPA) currents did not show significant differences [52]. These findings were consistent with previous mouse models with ASD [66, 69].

On the other hand, neurons derived from iPSC from individuals with ASD showed an imbalance in the glutamate-glutamine cycle, with a decrease in glutamate production by neurons derived from ASD children compared to neurons derived from controls. Additionally, the astrocytes derived from ASD children presented lower levels of glutamate in the culture supernatant (Fig. 1), revealing that both neurons and astrocytes presented an impairment in functioning, also corroborating the findings regarding the profile of synapses [34]. It is worth noting that this work was done using human cells derived from autistic children, which preserves the whole genetic background of the ASD person, which could be more in-depth with the broad genetic diversity of ASD instead of using a single-transgenic model. Considering the above, it is clear that there is a necessity for more research on the implication of neurotransmitter patterns of action in ASD.

4 The Blood-Brain Barrier Implication in ASD

The blood-brain barrier (BBB) is important for defending the central nervous system (CNS) by limiting access to substances that can determine neuronal activity. Therefore, dysfunctions in the BBB can contribute to neurological disorders such as ASD [70].

Claudins are proteins found in tight junctions (TJs) capable of maintaining the integrity of the BBB [71, 72]. Claudin 5 (CLDN5) is dominant in BBB endothelial cells, but claudins 1 (CLDN1), 3 (CLDN3), and 12 (CLDN12) are also expressed in those cells [71, 73–75]. CLDN3 and CLDN5 were upregulated in the cortex of individuals with ASD, while CLDN12 showed decreased expression [70]. The CLDN5 protein has a sealing function at TJ [76]. However, CLDN3 and CLDN12 need to be better understood. Some studies show that CLDN3 helps seal TJ [77], whereas CLDN12 cannot present homophilic interactions or homopolymeric chains in vitro, suggesting cis and trans interactions occurring with another TJ protein for CLDN12 to be incorporated [78]. Therefore, it was hypothesized that the increase in CLDN5 in individuals with ASD may be a way to compensate for the decrease in CLDN12 in a heterophilic interaction between the two claudins, CLDN5 and CLDN12 [70].

Gene expression levels of cerebral microvascular endothelial TJ components in the cortex and cerebellum of individuals with autism were also measured. Tricellulin (TRIC) was upregulated in the cortex of individuals with ASD. TRIC is associated with greater epithelial rigidity, in addition to a decrease in the permeability of macromolecules [79], which may suggest a compensatory mechanism for a ruptured BBB [70]. Furthermore, in the cortex of individuals with autism, matrix metalloproteinase (MMP-9) and 18 kDa translocator protein (TSPO) were significantly upregulated. MMP-9 induces rupture of the BBB, which can cause inflammatory diseases in the nervous system [80]. TSPO is a marker of brain inflammation, as the elevated level of TSPO in activated microglia and reactive astrocytes is associated with neuroinflammation and brain injury [81]. These observations agreed with the thesis that the brain of autistic individuals presents neuroinflammation (Fig. 1).

5 Neuroinflammation in ASD

The innate immune response of the CNS occurs through the activation of microglia and astrocytes, producing cytokines, chemokines, and other immune mediators. Astrocytes have ambiguous effects, capable of secreting pro-inflammatory or neuroprotective factors to regulate neuroinflammation [31].

In the context of ASD, excessive neuroinflammation has been reported, accompanied by the production of pro-inflammatory cytokines and an increase in reactive astrocytes [31]. A meta-analysis evaluated seventeen studies investigating cytokine concentrations in the blood of ASD patients and neurotypical individuals. Among the nineteen cytokines assessed, five showed a significant increase in ASD patients, namely, interleukin (IL)-1β, IL-6, IL-8, interferon (IFN)-γ, eotaxin, and transforming growth factor (TGF)-β1 [82]. It has been reported that children with ASD have an increased serum concentration of high mobility group box 1 protein (HMGB1), a protein with multiple functions that may serve as an indicator of early inflammation or a late mediator for the immune system [83]. Immunological assays in human brain tissue have shown an increase in pro-inflammatory cytokines such as TGF-β1,

MCP-1, IL-1β, IL-6, IL-8, IL-10, tumor necrosis factor (TNF)-α, granulocyte-macrophage colony-stimulating factor (GM-CSF), and IFN-γ (Fig. 1) [47, 84, 85].

Gene expression analysis in postmortem brains revealed increased levels of IL-1β and IFN-γ, while the expression of IL-6, IL-8, and IL-10 cytokines remained unchanged [70, 86]. In the hippocampus of mice with autism-like behaviors subjected to maternal separation stress, increased expression of HMGB1, TNF-α, IL-1β, and Toll-like receptor 4 (TLR4) was reported [87]. These conflicting results suggest that evaluating cytokine production by gene expression may not be the most appropriate method for this investigation or that other assays should complement this method.

Generalized neuroinflammation is described in autism. However, studies evaluated the brain in general, which makes it impossible to determine whether cytokine production was carried out by astrocytes or microglia. Therefore, investigation in monolayer culture of autistic astrocytes and microglia derived from iPSC originated from ASD patients revealed a strong tool for elucidating this topic. Astrocytes derived from iPSC from ASD children showed a significant increase in IL-6 levels, detected by ELISA and flow cytometry, causing a biological effect directly implicated in synaptogenesis impairment. Additionally, blocking IL-6 led to an improvement in synaptic transmission [34]. Overexpression of IL-6 in cerebellar granule cells of C57BL/6 J mouse pups impaired granule cell adhesion and migration. However, this overexpression stimulated excitatory synapse formation without affecting inhibitory synapses (Fig. 1) [47]. The effect caused by chronic IL-6 production by autistic astrocytes remains ambiguous, but resulting in excessive neuroinflammation may be responsible for the pathogenesis of ASD.

Glial fibrillary acidic protein (GFAP), a type III intermediate filament expressed mainly in astrocytes, is associated with astrogliosis and reactive damage, which may lead to an immune response and increased cytokine release [88]. Data on GFAP gene expression in ASD patients demonstrate an increase in different brain areas [31]. Rats treated with PPA or VPA and exhibiting behavior similar to autism also showed increased GFAP expression (Fig. 1) [1, 89].

Different degrees of ASD may be associated with different cytokine production profiles in different patients, which is believed by some that the profile could be used to identify ASD subtypes, creating endophenotypes that can share similar characteristics and may potentially respond to the same therapies [90].

6 Closing Remarks

Throughout this chapter, we highlighted that ASD causes several changes in the functioning and interaction between astrocytes and neurons. These individuals' synapses, the glutamate-glutamine cycle, the neurotransmitters, the cytokines, and the BBB are altered (Fig. 1), as explained throughout this chapter. ASD is a highly complex neurodevelopmental disorder, with its etiology considered to be multifactorial. A better comprehension of the brain of an ASD individual, combined with an early diagnosis, could be beneficial for interventions, contributing to an improvement in the behavior and benefiting cognitive development. Therefore, further investigations and advances in science on the pathophysiology of ASD involving CNS cells and in particular the interactions between astrocytes and neurons are essential for developing promising neurobiological, social, and educational strategies.

Acknowledgments PCBBB was supported by grants from FAPESP (2011/20683-0, 2013/08844-3, 2016/02978-6, 2017/27131-9, and 2018/16748-8), the NGO "The Tooth Fairy Project." ACSSC and DS are supported by CAPES (Finance Code 001). We thank the Institute of Biomedical Sciences from the University of São Paulo and Institut Pasteur de São Paulo for research infrastructure support. Figure 1 was made using BioRender.

References

1. Deckmann I, Santos-Terra J, Fontes-Dutra M, Körbes-Rockenbach M, Bauer-Negrini G, Schwingel GB, Riesgo R, Bambini-Junior V, Gottfried C. Resveratrol prevents brain edema, blood-brain barrier permeability, and altered aquaporin profile in autism animal model. Int J Dev Neurosci. 2021;81(7):579–604. https://doi.org/10.1002/jdn.10137. Epub 2021 Jul 10. PMID: 34196408.
2. Ssucharewa GE. Die schizoiden Psychopathien im Kindesalter. (part 1 of 2). Eur Neurol. 1926;60(3–4):235–47. https://doi.org/10.1159/000190478.
3. Zeldovich L. How history forgot the woman who defined autism. Spectrum News, v.7, 2018. Available at: https://www.spectrumnews.org/features/deep-dive/history-forgot-woman-defined-autism/. Accessed at: July, 15, 2024.
4. Kanner L. Autistic disturbances of affective contact. Acta Paedopsychiatr. 1968;35(4):100–136. PMID: 4880460.
5. Lamanna J, Meldolesi J. Autism spectrum disorder: brain areas involved, neurobiological mechanisms, diagnoses and therapies. Int J Mol Sci. 2024;25:2423. https://doi.org/10.3390/ijms25042423.
6. Ford TJL, Jeon BT, Lee H, Kim WY. Dendritic spine and synapse pathology in chromatin modifier-associated autism spectrum disorders and intellectual disability. Front Mol Neurosci. 2023;15:1048713. https://doi.org/10.3389/fnmol.2022.1048713. PMID: 36743289; PMCID: PMC9892461.
7. Hirota T, King BH. Autism spectrum disorder: a review. JAMA. 2023;329(2):157–68. https://doi.org/10.1001/jama.2022.23661. PMID: 36625807.
8. Lai MC, Lombardo MV, Baron-Cohen S. Autismo. Lanceta. 2014;383:896–910. https://doi.org/10.1016/S0140-6736(13)61539-1.
9. Ou J, Dong H, Dai S, Hou Y, Wang Y, Lu X, Xun G, Xia K, Zhao J, Shen Y. Development and validation of a risk score model for predicting autism based on pre- and perinatal factors. Front Psych. 2024;15:1291356. https://doi.org/10.3389/fpsyt.2024.1291356. PMID: 38435974; PMCID: PMC10904522.
10. American Psychiatric Association, DSM-5 Task Force. Diagnostic and statistical manual of mental disorders: DSM-5™. 5th ed. American Psychiatric Publishing, Inc; 2013. https://doi.org/10.1176/appi.books.9780890425596.
11. CDC, Centers for Disease Control and Prevention. Autism Spectrum Disorder (ASD). 2024. Accessed 6 Mar 2024. https://www.cdc.gov/ncbddd/autism/hcp-dsm.html.

12. Narvekar N, Carter Leno V, Pasco G, Begum Ali J, Johnson MH, Charman T, Jones EJH, STAARS Team. The roles of sensory hyperreactivity and hyporeactivity in understanding infant fearfulness and emerging autistic traits. J Child Psychol Psychiatry. 2024. https://doi.org/10.1111/jcpp.13941. Epub ahead of print. PMID: 38172076.
13. Sicherman N, Charite J, Eyal G, et al. Clinical signs associated with earlier diagnosis of children with autism Spectrum disorder. BMC Pediatr. 2021;21:96. https://doi.org/10.1186/s12887-021-02551-0.
14. World Health Organization (WHO). 2020. Pan American Health Organization. https://www.paho.org/pt/topicos/transtorno-do-espectro-autista.
15. Blumberg SJ, Bramlett MD, Kogan MD, Schieve LA, Jones JR, Lu MC. Changes in prevalence of parent-reported autism spectrum disorder in school-aged U.S. children: 2007 to 2011–2012. Natl Health Stat Report. 2013;65:1–11, 1 p following 11. PMID: 24988818.
16. Buescher AV, Cidav Z, Knapp M, Mandell DS. Costs of autism spectrum disorders in the United Kingdom and the United States. JAMA Pediatr. 2014;168(8):721–8. https://doi.org/10.1001/jamapediatrics.2014.210. PMID: 24911948.
17. Raymond GV, Bauman ML, Kemper TL. Hippocampus in autism: a Golgi analysis. Acta Neuropathol. 1996;91(1):117–9. https://doi.org/10.1007/s004010050401. PMID: 8773156.
18. Tang G, Gudsnuk K, Kuo SH, Cotrina ML, Rosoklija G, Sosunov A, Sonders MS, Kanter E, Castagna C, Yamamoto A, Yue Z, Arancio O, Peterson BS, Champagne F, Dwork AJ, Goldman J, Sulzer D. Loss of mTOR-dependent macroautophagy causes autistic-like synaptic pruning deficits. Neuron. 2014;83(5):1131–43. https://doi.org/10.1016/j.neuron.2014.07.040. Epub 2014 Aug 21. Erratum in: Neuron. 2014 Sep 17;83(6):1482. PMID: 25155956; PMCID: PMC4159743.
19. Penzes P, Cahill ME, Jones KA, VanLeeuwen JE, Woolfrey KM. Dendritic spine pathology in neuropsychiatric disorders. Nat Neurosci. 2011;14(3):285–93. https://doi.org/10.1038/nn.2741. PMID: 21346746; PMCID: PMC3530413.
20. Linda K, Lewerissa EI, Verboven AHA, Gabriele M, Frega M, Klein Gunnewiek TM, Devilee L, Ulferts E, Hommersom M, Oudakker A, Schoenmaker C, van Bokhoven H, Schubert D, Testa G, Koolen DA, de Vries BBA, Nadif Kasri N. Imbalanced autophagy causes synaptic deficits in a human model for neurodevelopmental disorders. Autophagy. 2022;18(2):423–42. https://doi.org/10.1080/15548627.2021.1936777. Epub 2021 Jul 21. PMID: 34286667; PMCID: PMC8942553.
21. Ardhanareeswaran K, Coppola G, Vaccarino F. The use of stem cells to study autism spectrum disorder. Yale J Biol Med. 2015;88(1):5–16. PMID: 25745370; PMCID: PMC4345539.
22. Varghese M, Keshav N, Jacot-Descombes S, Warda T, Wicinski B, Dickstein DL, Harony-Nicolas H, De Rubeis S, Drapeau E, Buxbaum JD, Hof PR. Autism spectrum disorder: neuropathology and animal models. Acta Neuropathol. 2017;134(4):537–66. https://doi.org/10.1007/s00401-017-1736-4. Epub 2017 Jun 5. PMID: 28584888; PMCID: PMC5693718.
23. Li Z, Zhu YX, Gu LJ, Cheng Y. Understanding autism spectrum disorders with animal models: applications, insights, and perspectives. Zool Res. 2021;42(6):800–24. https://doi.org/10.24272/j.issn.2095-8137.2021.251. PMID: 34755500; PMCID: PMC8645879.
24. Siemann JK, Veenstra-VanderWeele J, Wallace MT. Approaches to understanding multisensory dysfunction in autism spectrum disorder. Autism Res. 2020;13(9):1430–49. https://doi.org/10.1002/aur.2375. Epub 2020 Sep 1. PMID: 32869933; PMCID: PMC7721996.
25. Han K, Holder JL Jr, Schaaf CP, Lu H, Chen H, Kang H, Tang J, Wu Z, Hao S, Cheung SW, Yu P, Sun H, Breman AM, Patel A, Lu HC, Zoghbi HY. SHANK3 overexpression causes manic-like behaviour with unique pharmacogenetic properties. Nature 2013;503(7474):72–7. https://doi.org/10.1038/nature12630. Epub 2013 Oct 23. PMID: 24153177; PMCID: PMC3923348.
26. Kozol RA, Cukier HN, Zou B, Mayo V, De Rubeis S, Cai G, Griswold AJ, Whitehead PL, Haines JL, Gilbert JR, Cuccaro ML, Martin ER, Baker JD, Buxbaum JD, Pericak-Vance MA, Dallman JE. Two knockdown models of the autism genes SYNGAP1 and SHANK3 in zebrafish produce similar behavioral phenotypes associated with embryonic disruptions of brain

morphogenesis. Hum Mol Genet. 2015;24(14):4006–23. https://doi.org/10.1093/hmg/ddv138. Epub 2015 Apr 16. PMID: 25882707; PMCID: PMC4476447

27. Liu CX, Li CY, Hu CC, Wang Y, Lin J, Jiang YH, Li Q, Xu X. CRISPR/Cas9-induced shank3b mutant zebrafish display autism-like behaviors. Mol Autism. 2018;9:23. https://doi.org/10.1186/s13229-018-0204-x. PMID: 29619162; PMCID: PMC5879542.
28. Matas E, Maisterrena A, Thabault M, Balado E, Francheteau M, Balbous A, Galvan L, Jaber M. Major motor and gait deficits with sexual dimorphism in a Shank3 mutant mouse model. Mol Autism. 2021;12(1):2. https://doi.org/10.1186/s13229-020-00412-8. PMID: 33468258; PMCID: PMC7814442.
29. Mossa A, Pagano J, Ponzoni L, Tozzi A, Vezzoli E, Sciaccaluga M, Costa C, Beretta S, Francolini M, Sala M, Calabresi P, Boeckers TM, Sala C, Verpelli C. Developmental impaired Akt signaling in the Shank1 and Shank3 double knock-out mice. Mol Psychiatry. 2021;26(6):1928–44. https://doi.org/10.1038/s41380-020-00979-x. Epub 2021 Jan 5. PMID: 33402706; PMCID: PMC8440179.
30. Takahashi K, Yamanaka S. Induction of pluripotent stem cells from mouse embryonic and adult fibroblast cultures by defined factors. Cell. 2006;126(4):663–76. https://doi.org/10.1016/j.cell.2006.07.024. Epub 2006 Aug 10. PMID: 16904174.
31. Vakilzadeh G, Martinez-Cerdeño V. Pathology and astrocytes in autism. Neuropsychiatr Dis Treat. 2023;19:841–50. https://doi.org/10.2147/NDT.S390053. PMID: 37077706; PMCID: PMC10106330.
32. Panatier A, Vallée J, Haber M, Murai KK, Lacaille JC, Robitaille R. Astrocytes are endogenous regulators of basal transmission at central synapses. Cell. 2011;146:785–98.
33. Wahis J, Hennes M, Arckens L, Holt MG. Star power: the emerging role of astrocytes as neuronal partners during cortical plasticity. Curr Opin Neurobiol. 2021;67:174–82. https://doi.org/10.1016/j.conb.2020.12.001. Epub 2020 Dec 25. PMID: 33360483; PMCID: PMC8202513.
34. Russo FB, Freitas BC, Pignatari GC, Fernandes IR, Sebat J, Muotri AR, Beltrão-Braga PCB. Modeling the interplay between neurons and astrocytes in autism using human induced pluripotent stem cells. Biol Psychiatry. 2018;83(7):569–78. https://doi.org/10.1016/j.biopsych.2017.09.021. Epub 2017 Oct 3. PMID: 29129319.
35. Allen M, Huang BS, Notaras MJ, Lodhi A, Barrio-Alonso E, Lituma PJ, Wolujewicz P, Witztum J, Longo F, Chen M, Greening DW, Klann E, Ross ME, Liston C, Colak D. Astrocytes derived from ASD individuals alter behavior and destabilize neuronal activity through aberrant Ca2+ signaling. Mol Psychiatry. 2022;27(5):2470–84. https://doi.org/10.1038/s41380-022-01486-x. Epub 2022 Apr 1. PMID: 35365802; PMCID: PMC9135629.
36. Wang Q, Kong Y, Wu DY, Liu JH, Jie W, You QL, Huang L, Hu J, Chu HD, Gao F, Hu NY, Luo ZC, Li XW, Li SJ, Wu ZF, Li YL, Yang JM, Gao TM. Impaired calcium signaling in astrocytes modulates autism spectrum disorder-like behaviors in mice. Nat Commun. 2021;12(1):3321. https://doi.org/10.1038/s41467-021-23843-0. PMID: 34059669; PMCID: PMC8166865.
37. Südhof TC. Neuroligins and neurexins link synaptic function to cognitive disease. Nature. 2008;455(7215):903–11. https://doi.org/10.1038/nature07456. PMID: 18923512; PMCID: PMC2673233.
38. Krueger DD, Tuffy LP, Papadopoulos T, Brose N. The role of neurexins and neuroligins in the formation, maturation, and function of vertebrate synapses. Curr Opin Neurobiol. 2012;22(3):412–22. https://doi.org/10.1016/j.conb.2012.02.012. Epub 2012 Mar 15. PMID: 22424845.
39. Tromp A, Mowry B, Giacomotto J. Neurexins in autism and schizophrenia-a review of patient mutations, mouse models and potential future directions. Mol Psychiatry. 2021;26(3):747–60. https://doi.org/10.1038/s41380-020-00944-8. Epub 2020 Nov 15. PMID: 33191396.
40. Zhang C, Milunsky JM, Newton S, Ko J, Zhao G, Maher TA, Tager-Flusberg H, Bolliger MF, Carter AS, Boucard AA, Powell CM, Südhof TC. A neuroligin-4 missense mutation associated with autism impairs neuroligin-4 folding and endoplasmic reticulum export. J Neurosci. 2009;29(35):10843–54. https://doi.org/10.1523/JNEUROSCI.1248-09.2009. PMID: 19726642; PMCID: PMC2777970.

41. El-Kordi A, Winkler D, Hammerschmidt K, Kästner A, Krueger D, Ronnenberg A, Ritter C, Jatho J, Radyushkin K, Bourgeron T, Fischer J, Brose N, Ehrenreich H. Development of an autism severity score for mice using Nlgn4 null mutants as a construct-valid model of heritable monogenic autism. Behav Brain Res. 2013;251:41–9. https://doi.org/10.1016/j.bbr.2012.11.016. Epub 2012 Nov 23. PMID: 23183221.
42. Jamain S, Radyushkin K, Hammerschmidt K, Granon S, Boretius S, Varoqueaux F, Ramanantsoa N, Gallego J, Ronnenberg A, Winter D, Frahm J, Fischer J, Bourgeron T, Ehrenreich H, Brose N. Reduced social interaction and ultrasonic communication in a mouse model of monogenic heritable autism. Proc Natl Acad Sci USA. 2008;105(5):1710–5. https://doi.org/10.1073/pnas.0711555105. Epub 2008 Jan 28. PMID: 18227507; PMCID: PMC2234209.
43. Liang J, Xu W, Hsu YT, Yee AX, Chen L, Südhof TC. Conditional neuroligin-2 knockout in adult medial prefrontal cortex links chronic changes in synaptic inhibition to cognitive impairments. Mol Psychiatry. 2015;20(7):850–9. https://doi.org/10.1038/mp.2015.31. Epub 2015 Mar 31. PMID: 25824299.
44. Tabuchi K, Blundell J, Etherton MR, Hammer RE, Liu X, Powell CM, Südhof TC. A neuroligin-3 mutation implicated in autism increases inhibitory synaptic transmission in mice. Science. 2007;318(5847):71–6. https://doi.org/10.1126/science.1146221. Epub 2007 Sep 6. PMID: 17823315; PMCID: PMC3235367.
45. Schaaf CP, Boone PM, Sampath S, Williams C, Bader PI, Mueller JM, Shchelochkov OA, Brown CW, Crawford HP, Phalen JA, Tartaglia NR, Evans P, Campbell WM, Tsai AC, Parsley L, Grayson SW, Scheuerle A, Luzzi CD, Thomas SK, Eng PA, Kang SH, Patel A, Stankiewicz P, Cheung SW. Phenotypic spectrum and genotype-phenotype correlations of NRXN1 exon deletions. Eur J Hum Genet. 2012;20(12):1240–7. https://doi.org/10.1038/ejhg.2012.95. Epub 2012 May 23. PMID: 22617343; PMCID: PMC3499754.
46. Pak C, Danko T, Zhang Y, Aoto J, Anderson G, Maxeiner S, Yi F, Wernig M, Südhof TC. Human neuropsychiatric disease modeling using conditional deletion reveals synaptic transmission defects caused by heterozygous mutations in NRXN1. Cell Stem Cell. 2015;17(3):316–28. https://doi.org/10.1016/j.stem.2015.07.017. Epub 2015 Aug 13. PMID: 26279266; PMCID: PMC4560990.
47. Wei H, Zou H, Sheikh AM, Malik M, Dobkin C, Brown WT, Li X. IL-6 is increased in the cerebellum of autistic brain and alters neural cell adhesion, migration and synaptic formation. J Neuroinflammation. 2011;8:52. https://doi.org/10.1186/1742-2094-8-52. PMID: 21595886; PMCID: PMC3114764.
48. Takeda K, Watanabe T, Oyabu K, Tsukamoto S, Oba Y, Nakano T, Kubota K, Katsurabayashi S, Iwasaki K. Valproic acid-exposed astrocytes impair inhibitory synapse formation and function. Sci Rep. 2021;11(1):23. https://doi.org/10.1038/s41598-020-79520-7. PMID: 33420078; PMCID: PMC7794250.
49. Yizhar O, Fenno LE, Prigge M, Schneider F, Davidson TJ, O'Shea DJ, Sohal VS, Goshen I, Finkelstein J, Paz JT, Stehfest K, Fudim R, Ramakrishnan C, Huguenard JR, Hegemann P, Deisseroth K. Neocortical excitation/inhibition balance in information processing and social dysfunction. Nature. 2011;477(7363):171–8. https://doi.org/10.1038/nature10360. PMID: 21796121; PMCID: PMC4155501.
50. Fatemi SH, Reutiman TJ, Folsom TD, Rooney RJ, Patel DH, Thuras PD. mRNA and protein levels for GABAAalpha4, alpha5, beta1, and GABBR1 receptors are altered in brains from subjects with autism. J Autism Dev Disord. 2010;40(6):743–50. https://doi.org/10.1007/s10803-009-0924-z.
51. Caroline ER, Ratai E-M, Kanwisher N. Reduced GABAergic action in the autistic brain. Curr Biol. 2016;26(1):80–5. https://doi.org/10.1016/j.cub.2015.11.019, ISSN 0960-9822.
52. Jing W, Gao Y, Xiao L, Lin Y, Huang L, Chen J, Liang G, Li W, Yi W, Lao J, Zhang B, Gao T-M, Zhong M, Yang X. Increased NMDARs in neurons and glutamine synthetase in astrocytes underlying autistic-like behaviors of Gabrb1−/− mice. iScience. 2023;26(8):107476. https://doi.org/10.1016/j.isci.2023.107476, ISSN 2589-0042.

53. Qian X, Zhao X, Yu L, Yin Y, Zhang XD, Wang L, Li JX, Zhu Q, Luo JL. Current status of GABA receptor subtypes in analgesia. Biomed Pharmacother. 2023;168:115800. https://doi.org/10.1016/j.biopha.2023.115800. Epub 2023 Nov 6. PMID: 37935070.
54. Koh W, Kwak H, Cheong E, Lee CJ. GABA tone regulation and its cognitive functions in the brain. Nat Rev Neurosci. 2023;24(9):523–39. https://doi.org/10.1038/s41583-023-00724-7. Epub 2023 Jul 26. PMID: 37495761.
55. Luscher B, Maguire JL, Rudolph U, Sibille E. GABAA receptors as targets for treating affective and cognitive symptoms of depression. Trends Pharmacol Sci. 2023;44(9):586–600. https://doi.org/10.1016/j.tips.2023.06.009. Epub 2023 Aug 3. PMID: 37543478; PMCID: PMC10511219.
56. Schuler V, Lüscher C, Blanchet C, Klix N, Sansig G, Klebs K, Schmutz M, Heid J, Gentry C, Urban L, Fox A, Spooren W, Jaton AL, Vigouret J, Pozza M, Kelly PH, Mosbacher J, Froestl W, Käslin E, Korn R, Bischoff S, Kaupmann K, van der Putten H, Bettler B. Epilepsy, hyperalgesia, impaired memory, and loss of pre- and postsynaptic GABA(B) responses in mice lacking GABA(B(1)). Neuron. 2001;31(1):47–58. https://doi.org/10.1016/s0896-6273(01)00345-2. PMID: 11498050.
57. Brandon NJ, Smart TG, Moss SJ. Regulation of GABAA receptors by protein phosphorylation. In: Martin D, Olsen R, editors. GABA in the nervous system: the view at fifty years. Philadelphia: Lippincott, Williams and Wilkins; 2000. p. 191–206.
58. Fatemi SH, Reutiman TJ, Folsom TD, Thuras PD. GABA(A) receptor downregulation in brains of subjects with autism. J Autism Dev Disord. 2009a;39:223–30.
59. Ma DQ, Whitehead PL, Menold MM, Martin ER, Ashley-Koch AE, Mei H, Ritchie MD, Delong GR, Abramson RK, Wright HH, Cuccaro ML, Hussman JP, Gilbert JR, Pericak-Vance MA. Identification of significant association and gene-gene interaction of GABA receptor subunit genes in autism. Am J Hum Genet. 2005;77:377–88.
60. Holz GG, Rane SG, Dunlap K. GTP-binding proteins mediate transmitter inhibition of voltage-dependent calcium channels. Nature. 1986;319:670–2.
61. Kuriyama K, Hirouchi M, Kimura H. Neurochemical and molecular pharmacological aspects of the GABA(B) receptor. Neurochem Res. 2000;25:1233–9.
62. Fatemi SH, Folsom TD, Reutiman TJ, Thuras PD. Expression of GABA(B) receptors is altered in brains of subjects with autism. Cerebellum. 2009b;8:64–9.
63. Leung LS, Canning KJ, Shen B. Hippocampal afterdischarges after GABA(B)-receptor blockade in the freely moving rat. Epilepsia. 2005;46:203–16.
64. Li J, Olinger AB, Dassow MS, Abel MS. Up-regulation of GABA(B) receptor mRNA and protein in the hippocampus of cocaine-and lidocaine-kindled rats. Neuroscience. 2003;118:451–62.
65. Han Y, Qin J, Bu DF, Chang XZ, Yang ZX. Successive alterations of hippocampal gamma-aminobutyric acid B receptor subunits in a rat model of febrile seizure. Life Sci. 2006;78:2944–52.
66. Qin L, Williams JB, Tan T, Liu T, Cao Q, Ma K, Yan Z. Deficiency of autism risk factor ASH1L in prefrontal cortex induces epigenetic aberrations and seizures. Nat Commun. 2021;12:6589.
67. Oya M, Matsuoka K, Kubota M, Fujino J, Tei S, Takahata K, Tagai K, Yamamoto Y, Shimada H, Seki C, Itahashi T, Aoki YY, Ohta H, Hashimoto RI, Sugihara G, Obata T, Zhang MR, Suhara T, Nakamura M, Kato N, Takado Y, Takahashi H, Higuchi M. Increased glutamate and glutamine levels and their relationship to astrocytes and dopaminergic transmissions in the brains of adults with autism. Sci Rep. 2023;13(1):11655. https://doi.org/10.1038/s41598-023-38306-3. PMID: 37468523; PMCID: PMC10356952.
68. Fatemi SH, Halt AR, Realmuto G, Earle J, Kist DA, Thuras P, Merz A. Purkinje cell size is reduced in cerebellum of patients with autism. Cell Mol Neurobiol. 2002;22:171–5.
69. Bey AL, Wang X, Yan H, Kim N, Passman RL, Yang Y, Cao X, Towers AJ, Hulbert SW, Duffney LJ, et al. Brain region-specific disruption of Shank3 in mice reveals a dissociation for cortical and striatal circuits in autism-related behaviors. Transl Psychiatry. 2018;8:94.

70. Fiorentino M, Sapone A, Senger S, Camhi SS, Kadzielski SM, Buie TM, Kelly DL, Cascella N, Fasano A. Blood-brain barrier and intestinal epithelial barrier alterations in autism spectrum disorders. Mol Autism. 2016;7:49. https://doi.org/10.1186/s13229-016-0110-z. PMID: 27957319; PMCID: PMC5129651.
71. Jia W, Lu R, Martin TA, Jiang WG. O papel da claudina-5 na barreira hematoencefálica (BHE) e nas metástases cerebrais (revisão). Relatórios de Medicina Molecular. 2014;9(3):779–85.
72. Oue H, Hatakeyama R, Ishida E, Yokoi M, Tsuga K. Experimental tooth loss affects spatial learning function and blood-brain barrier of mice. Oral Dis. 2023;29(7):2907–16. https://doi.org/10.1111/odi.14379. Epub 2022 Oct 5. PMID: 36114741.
73. Cooper I, Cohen-Kashi-Malina K, Teichberg VI. Claudin-5 expression in in vitro models of the blood-brain barrier. Methods Mol Biol. 2011;762:347–54.
74. Neuhaus W, Wirth M, Plattner VE, Germann B, Gabor F, Noe CR. Expression of Claudin-1, Claudin-3 and Claudin-5 in human blood-brain barrier mimicking cell line ECV304 is inducible by glioma-conditioned media. Neurosci Lett. 2008;446:59–64.
75. Ohtsuki S, Yamaguchi H, Katsukura Y, Asashima T, Terasaki T. mRNA expression levels of tight junction protein genes in mouse brain capillary endothelial cells highly purified by magnetic cell sorting. J Neurochem. 2008;104:147–54.
76. Nitta T, Hata M, Gotoh S, Seo Y, Sasaki H, Hashimoto N, Furuse M, Tsukita S. Size-selective loosening of the blood–brain barrier in claudin-5-deficient mice. J Cell Biol. 2003;161(3):653–60.
77. Milatz S, Krug SM, Rosenthal R, Gunzel D, Muller D, Schulzke JD, Amasheh S, Fromm M. Claudin-3 acts as a sealing component of the tight junction for ions of either charge and uncharged solutes. Biochim Biophys Acta. 2010;1798(11):2048–57.
78. Piontek J, Fritzsche S, Cording J, Richter S, Hartwig J, Walter M, Yu D, Turner JR, Gehring C, Rahn HP, et al. Elucidating the principles of the molecular organization of heteropolymeric tight junction strands. Cell Mol Life Sci. 2011;68(23):3903–18.
79. Krug SM, Amasheh S, Richter JF, Milatz S, Gunzel D, Westphal JK, Huber O, Schulzke JD, Fromm M. Tricellulin forms a barrier to macromolecules in tricellular tight junctions without affecting ion permeability. Mol Biol Cell. 2009;20(16):3713–24.
80. Waubant E. Biomarkers indicative of blood–brain barrier disruption in multiple sclerosis. Dis Markers. 2006;22(4):235–44.
81. Lavisse S, Guillermier M, Herard AS, Petit F, Delahaye M, Van Camp N, Ben Haim L, Lebon V, Remy P, Dolle F, et al. Reactive astrocytes overexpress TSPO and are detected by TSPO positron emission tomography imaging. J Neurosci. 2012;32(32):10809–18.
82. Masi A, Quintana DS, Glozier N, Lloyd AR, Hickie IB, Guastella AJ. Cytokine aberrations in autism spectrum disorder: a systematic review and meta-analysis. Mol Psychiatry. 2015;20(4):440–6. https://doi.org/10.1038/mp.2014.59. Epub 2014 Jun 17. PMID: 24934179.
83. Makris G, Chouliaras G, Apostolakou F, Papageorgiou C, Chrousos GP, Papassotiriou I, Pervanidou P. Increased serum concentrations of High Mobility Group Box 1 (HMGB1) protein in children with autism spectrum disorder. Children (Basel). 2021;8(6):478. https://doi.org/10.3390/children8060478. PMID: 34198762; PMCID: PMC8228126.
84. Li X, Chauhan A, Sheikh AM, Patil S, Chauhan V, Li XM, Ji L, Brown T, Malik M. Elevated immune response in the brain of autistic patients. J Neuroimmunol. 2009;207(1–2):111–6. https://doi.org/10.1016/j.jneuroim.2008.12.002. Epub 2009 Jan 20. PMID: 19157572; PMCID: PMC2770268.
85. Vargas DL, Nascimbene C, Krishnan C, Zimmerman AW, Pardo CA. Neuroglial activation and neuroinflammation in the brain of patients with autism. Ann Neurol. 2005;57(1):67–81. https://doi.org/10.1002/ana.20315. Erratum in: Ann Neurol 2005 Feb;57(2):304. PMID: 15546155.
86. Patel N, Crider A, Pandya CD, Ahmed AO, Pillai A. Altered mRNA levels of glucocorticoid receptor, mineralocorticoid receptor, and co-chaperones (FKBP5 and PTGES3) in the middle frontal gyrus of autism spectrum disorder subjects. Mol Neurobiol. 2016;53(4):2090–9. https://doi.org/10.1007/s12035-015-9178-2. Epub 2015 Apr 26. PMID: 25912394.

87. Reisi-Vanani V, Lorigooini Z, Bijad E, Amini-Khoei H. Maternal separation stress through triggering of the neuro-immune response in the hippocampus induces autistic-like behaviors in male mice. Int J Dev Neurosci. 2023. https://doi.org/10.1002/jdn.10310. Epub ahead of print. PMID: 38110192.
88. Pekny M, Wilhelmsson U, Pekna M. The dual role of astrocyte activation and reactive gliosis. Neurosci Lett. 2014;565:30–8. https://doi.org/10.1016/j.neulet.2013.12.071. Epub 2014 Jan 7. PMID: 24406153.
89. Choi J, Lee S, Won J, Jin Y, Hong Y, Hur TY, Kim JH, Lee SR, Hong Y. Pathophysiological and neurobehavioral characteristics of a propionic acid-mediated autism-like rat model. PLoS One. 2018;13(2):e0192925. https://doi.org/10.1371/journal.pone.0192925. PMID: 29447237; PMCID: PMC5814017.
90. Masi A, Glozier N, Dale R, Guastella AJ. The immune system, cytokines, and biomarkers in autism spectrum disorder. Neurosci Bull. 2017;33(2):194–204. https://doi.org/10.1007/s12264-017-0103-8. Epub 2017 Feb 25. PMID: 28238116; PMCID: PMC5360854.

Astrocyte-Neuron Interactions Contributing to Amyotrophic Lateral Sclerosis Progression

Brigid K. Jensen

Abstract Amyotrophic lateral sclerosis (ALS) is a complex disease impacting motor neurons of the brain, brainstem, and spinal cord. Disease etiology is quite heterogeneous with over 40 genes causing the disease and a vast ~90% of patients having no prior family history. Astrocytes are major contributors to ALS, particularly through involvement in accelerating disease progression. Through study of genetic forms of disease including SOD1, TDP43, FUS, C9orf72, VCP, TBK1, and more recently patient-derived cells from sporadic individuals, many biological mechanisms have been identified to cause intrinsic or glial-mediated neurotoxicity to motor neurons. Overall, many of the normally supportive and beneficial roles that astrocytes contribute to neuronal health and survival instead switch to become deleterious and neurotoxic. While the exact pathways may differ based on disease-origin, altered astrocyte-neuron communication is a common feature of ALS. Within this chapter, distinct genetic forms are examined in detail, along with what is known from sporadic patient-derived cells. Overall, this chapter highlights the interplay between astrocytes and neurons in this complex disease and describes the key features underlying: astrocyte-mediated motor neuron toxicity, excitotoxicity, oxidative/nitrosative stress, protein dyshomeostasis, metabolic imbalance, inflammation, trophic factor withdrawal, blood-brain/blood-spinal cord barrier involvement, disease spreading, and the extracellular matrix/cell adhesion/TGF-β signaling pathways.

Keywords Amyotrophic lateral sclerosis (ALS) · Excitotoxicity · Metabolic imbalance · Inflammation · Trophic factors

B. K. Jensen (✉)
Neuroscience, Thomas Jefferson University, Philadelphia, PA, USA
e-mail: brigid.jensen@jefferson.edu

E. Blanco-Suarez, I. Farhy-Tselnicker (eds.), *Astrocyte-Neuron Interactions in Health and Disease*, Advances in Neurobiology 39,
https://doi.org/10.1007/978-3-031-64839-7_12

Abbreviations

ALS	Amyotrophic lateral sclerosis
AMPA	α-amino-3-hydroxy-5-methyl-4-isoxazolepropionic acid
AQP4	Aquaporin 4
ATP	Adenosine triphosphate
ATPase	Adenosine 5′-triphosphatase
BBB	Blood-brain barrier
BSCB	Blood-spinal cord barrier
C9orf72	Chromosome 9 open reading frame 72
CNS	Central nervous system
CSF	Cerebrospinal fluid
Cx30/Cx43	Connexin 30 or 43
DNA	Deoxyribonucleic acid
DPRs	Dipeptide repeat proteins
EAAT1/EAAT2/EAAT3	Excitatory amino acid transporter 1, 2, or 3
ECM	Extracellular matrix
ER	Endoplasmic reticulum
ET-1	Endothelin-1
ETA/ETB	Endothelin receptor A or B
EV	Extracellular vesicle
FUS	Fused in sarcoma
GDNF	Glial cell line-derived neurotrophic factor
GluA1/GluA2	Glutamate ionotropic receptor AMPA-type subunit 1 or 2
IBMPFD	Inclusion body myopathy with early-onset Paget disease and frontotemporal dementia
IFNγ	Interferon gamma
IL (ex IL-1β)	Interleukin
iNOS	Inducible nitric oxide synthase
IPSC	Induced pluripotent stem cell
IκB-α	Nuclear factor of kappa light polypeptide gene enhancer in B-cell inhibitor alpha
Kir4.1	Inwardly rectifying potassium channel 4.1
LC3	Microtubule-associated proteins 1A/1B light chain 3B
LIGHT/TNFSF14	Tumor necrosis factor superfamily member 14
LT-βR	Lymphotoxin-β-receptor
MAPK	Mitogen-activated protein kinases
MCT-1	Monocarboxylate transporter 1
mGluR1/mGluR5	Metabotropic glutamate receptor 1 or 5
miR	MicroRNA
MRI	Magnetic resonance imaging
MSP	Multisystem proteinopathy
mTNFα	Membrane-bound TNFα

mTOR	Mammalian target of rapamycin
Na/K ATPase	Sodium/potassium ATPase
Na_v	Voltage-sensitive sodium
NF-κB	Nuclear factor kappa B
NGF	Nerve growth factor
NLRP3	NLR family pyrin domain containing 3
NMJ	Neuromuscular junction
NPC	Neural progenitor cell
Nrf2	Nuclear factor erythroid-2-related factor 2
OPTN	Optineurin
PET	Positron emission tomography
PGE2	Prostaglandin E2
polyP	Inorganic polyphosphate
PTP1B	Protein tyrosine phosphatase 1B
RNA	Ribonucleic acid
RNAi	RNA interference
ROS/RNS	Reactive oxygen and nitrogen species
sALS	Sporadic ALS
SEMA3A	Semaphorin 3A
SOD1	Superoxide dismutase 1
TBK1	Tank-binding kinase 1
TDP43	Transactive response DNA-binding protein 43
TGF-β	Transforming growth factor β
TGFβ-RII	Transforming growth factor β receptor II
TNFR1/TNFR2	TNFα receptor 1 or 2
TNFα	Tumor necrosis factor alpha
TRAIL	TNF-related apoptosis-inducing ligand
UBQLN2	Ubiquilin-2
VCP	Valosin-containing protein
VEGF	Vascular endothelial growth factor
WNT	Wingless-related integration site

1 Introduction to ALS and Astrocytes in ALS

Amyotrophic lateral sclerosis (ALS) is a progressive, adult-onset neurodegenerative disease characterized by loss of motor neurons in the motor cortex, brainstem, and spinal cord, resulting in rapid decline in motor strength, muscle wasting, paralysis, and ultimately death [1]. Patients typically present with focal musculoskeletal symptoms, which spread to surrounding areas [2]. The site of onset is variable, ranging from distal limbs to bulbar ALS which first impacts muscles utilized for speech and swallowing [1]. Most patients quickly lose ambulatory ability and succumb to respiratory complications related to loss of diaphragm function within 3–5 years [1]. Disease etiology is complex, as 90% of cases are sporadic (sALS) and 10% are familial [3]. Over 40 ALS-causative genes have been described including superoxide dismutase 1 (SOD1), transactive response DNA-binding protein 43 (TDP43), fused in sarcoma (FUS), and chromosome 9 open reading frame 72 (C9orf72) [3, 4]. Implicated pathways include oxidative stress, RNA metabolism defects, inflammation, excitotoxicity, altered axonal transport, protein aggregation and dyshomeostasis, autophagy impairment, and mitochondrial dysfunction [5, 6]. Therapeutics are aimed at mediating these effects. Riluzole targets glutamate excitotoxicity and voltage-gated sodium channels. Radicava acts as a free radical scavenger for oxidative stress, and Relyvrio reduces protein misfolding [7–11]. The first disease-modifying therapeutic was recently approved, Tofersen (Qalsody), using antisense-oligonucleotides to downregulate SOD1 protein [12]. There is now considerable hope for patients with genetic ALS, as clinical trials are in-progress using similar strategies to target other genes.

The effects of mutant ALS proteins within motor neurons have been elegantly studied, implicating many pathways in the disease process. While ALS ultimately targets motor neurons, non-neuronal cells such as astrocytes are also dysfunctional and are crucial drivers of disease progression, contributing through mechanisms termed non-cell-autonomous toxicity [13–15]. When motor cortex and spinal cord are examined in patients, hallmarks of disease include loss of motor neurons, hypertrophy of astrocytes, and proteinaceous inclusions deposited in both neurons and glial cell populations [16–19]. Reactive astrogliosis indicative of neuroinflammatory processes is found in areas of inclusion pathology and neuronal loss [20–23]. Imaging using positron emission tomography (PET) and magnetic resonance imaging (MRI) methods has demonstrated reactive gliosis throughout the disease [24, 25]. To study ALS, in vitro and in vivo models have been established mimicking mutation-driven and sporadic etiologies. Many of these models also routinely display reactive gliosis and astrocytic inclusion pathology [26–31]. Astrocytes from rodent models and patient-derived induced pluripotent stem cells (IPSCs) have shown that many of the normally beneficial astrocytic roles encouraging neuronal health and survival are hijacked, instead evoking neurotoxicity [32–49]. While pathways involved differ based on disease-origin, altered astrocyte-neuron communication arises from both genetic and sALS, and the consensus is that astrocytes switch from a protective and supportive role, to one that is harmful. In this chapter, types of ALS will be examined in detail, highlighting the interplay between astrocytes and neurons in this complex disease.

2 SOD1-ALS

In 1993, mutations in SOD1 were identified as the first genetic cause of ALS [50]. SOD1 is a free radical scavenging enzyme that converts superoxide to hydrogen peroxide [50, 51]. Researchers quickly noted astrocytic pathology in SOD1-ALS tissue. The astrocyte-specific transporter for glutamate, excitatory amino acid transporter 2 (EAAT2) (glutamate transporter 1 (GLT-1)), was significantly lost in motor cortex and spinal cord [52–54]. EAAT2 maintains extracellular glutamate concentrations following neuronal firing, and reduction in levels or activity of these transporters can cause the excitotoxic state observed in ALS [55–57]. Rodent models using SOD1 variants have been extensively used, faithfully recapitulating the disease, resulting in rapid, robust, progressive phenotypes including weight loss, hindlimb weakness, extension splay reflex reduction, locomotor deficits, paralysis, and premature death [58, 59]. Extensive loss of motor neurons and innervated neuromuscular junctions (NMJs) underlies these effects [59]. Neuron-only expression of SOD1 variants was insufficient to produce this full phenotype and disease course, suggesting that while some toxicity occurs within motor neurons, glial cell populations are contributors to disease [60–63]. A majority of what we know regarding the interplay between astrocytes and neurons in ALS comes from these models.

2.1 Astrocytic Involvement in SOD1-ALS: A Focus on Excitotoxicity

In the SOD1^{G85R} model, astrocytic SOD1 inclusions and reactive astrogliosis preceded motor neuron pathology, and EAAT2 reduction coincided with disease progression [26]. This was confirmed in SOD1^{G93A} rats, where gliosis coincided with EAAT2 reduction and preceded neuronal death [64]. EAAT2 overexpression in SOD1^{G93A} animals delayed but did not prevent motor deficits and neuron loss, and blockade of EAAT2 cleavage/degradation slowed disease progression and extended lifespan [65, 66]. Mutant SOD1 has also been selectively removed from astrocytes. In SOD1^{G93A}, this slowed disease progression, preserved muscle function, and increased lifespan, and in SOD1^{G85R}, disease onset was also delayed [67–70]. Transplanting healthy astrocytes into SOD1^{G93A} mouse spinal cord resulted in reduced neuron loss, slower disease progression, and extended survival [71]. Introducing healthy human neural progenitors at disease onset also improved motor function and lifespan [72]. In contrast, SOD1^{G93A} astrocytes or neural precursors from SOD1 patients caused neuronal degeneration and motor deficits [73, 74]. Other factors also exacerbate the excitotoxic environment. In SOD1^{G93A} mice, neuronal glutamate release is elevated, likely through metabotropic mGluR1 and mGluR5 signaling [75–78]. mGluR1/5 are seen on neurons throughout spinal cord, and while these are normally at very low abundance in astrocytes, in ALS patient tissue and animal models, high expression is seen [79, 80]. A large proportion of AMPA glutamate receptors on recipient neurons lack the GluA2 subunit and are calcium-permeable, causing elevated calcium influx upon glutamate binding [81]. These events culminate in the failure of astrocytes to retrieve adequate glutamate from the synaptic cleft and cause excessive calcium influx, resulting in excitotoxic cascades and neuronal death [75, 76, 81]. Astrocytes also become susceptible to glutamate toxicity, as astrocyte degeneration in proximity to motor neurons occurs, resulting from mGluR5 signaling [82]. Reduction in astrocytic mGluR5 levels or activity attenuates motor neuron toxicity in vitro and in vivo [78, 83], showing that activation of this pathway leads to reactivity and release of toxic factors [78, 83].

2.2 Mechanisms of Excitotoxic, Oxidative, and Nitrosative Damage in SOD1-ALS

Astrocytes recovered from SOD1-ALS patients evoked toxicity when cultured with motor neurons [47, 84]. Many studies have since described astrocyte-mediated soluble factors which drive neuronal toxicity from cells derived from SOD1 transgenic mice and human IPSCs [33, 34, 41, 48, 85]. Neurotoxicity was specific, as mutant astrocytes only evoked motor neuron toxicity but did not affect other neuronal populations, and other cell populations did not impact motor neuron survival [34]. When exposed to SOD1^{G93A} astrocyte medium, motor neurons demonstrated persistent elevated sodium currents, repetitive action potential firing, increased frequency of calcium transients, and ultimately death [86]. Treating neurons with blockers of voltage-sensitive sodium channels (Na_v) returned calcium transients to baseline and inhibited both hyperexcitability and toxicity [86]. Inorganic polyphosphate (polyP) is another contributor to aberrant neuronal activity. PolyP is elevated in spinal cord and cerebrospinal fluid (CSF) from genetic (SOD1, TDP43, and C9orf72) and sALS patients [36]. Cultured astrocytes from mouse models of SOD1, TDP43, and C9orf72 and human TDP43-ALS IPSC astrocytes also all displayed elevated polyP, which invoked motor neuron toxicity via increased calcium transients, and aberrant spontaneous action potentials [36, 42]. For neurons, polyP shifts the voltage sensitivity of the Na_v channel, inhibits the voltage-gated potassium channel (K_v), and activates the voltage-gated calcium channel (Ca_V), with the net result of greatly enhanced neuronal excitability and network activity [87]. Neurons can also release polyP, allowing for signal amplification and impacts on neighboring glia [87]. In glia, polyP acts through purinergic P2Y1 receptor signaling, resulting in calcium influx, representing another cause for elevated calcium levels in ALS astrocytes [87]. Enzymatic polyP degradation or application of neutralizing antibodies were both effective at restoring neuronal calcium transients and preventing motor neuron toxicity [36]. Given these findings, astrocyte-driven neuronal hyperexcitability and excitotoxicity are important elements of SOD-1 pathogenesis.

Astrocytes form an interconnected network through gap junction proteins on their membranes, with each cell contributing one half of the complex, a hemichannel, composed of connexin proteins [88]. In the central nervous system (CNS) and astrocytes in particular, connexin 43 (Cx43) is predominant, allowing for the propagation of calcium waves and exchange of metabolic intermediates including glutamate, lactate, aspartate, and ATP (adenosine triphosphate) [89, 90]. In SOD1^{G93A} mice, elevated Cx43, Cx30, and aquaporin 4 (AQP4) were observed [91]. Cx43 expression progressively increases over the disease course in SOD1^{G93A} mice, and this increase is also seen in astrocytes from SOD1^{G93A} mice and in human IPSC astrocytes from genetic (SOD1 and C9orf72) and sALS patients [92, 93]. sALS patient tissue showed Cx43 increase in CSF, motor cortex, and across spinal cord, which correlated with speed of disease progression [92, 93]. SOD1^{G93A} astrocytes demonstrated elevated hemichannel activity, greater gap junction coupling, and increased intracellular calcium [92]. In IPSC astrocytes, baseline hemichannel-based release of ATP was increased [93]. Astrocyte-specific removal of Cx43 in SOD1^{G93A} mice improved survival and slowed disease progression [93]. In motor neuron coculture using astrocytes from SOD1^{G93A} mice or IPSCs, blocking Cx43 activity using GAP19 or tonabersat rectified aberrant neuronal firing and provided neuroprotection [92, 93]. Treating SOD1^{G93A} mice presymptomatically or after disease onset with tonabersat conveyed neuroprotection and preserved motor function [93]. ATP released by hemichannels may play a role in SOD1-ALS, particularly as ATP receptors are elevated in ALS tissue and in SOD1^{G93A} mice [94, 95]. ATP stimulation of SOD1^{G93A} astrocytes resulted in elevated calcium release from intracellular stores, which then leads to increase of astrocyte-released ATP, acting in a feedforward loop [96]. This pathway is initiated by the astrocytic purinergic P2X7 receptor, which is constitutively active in SOD1^{G93A} astrocytes, allowing for transport of glutamate and aspartate out of astrocytes, adding another mechanism for elevated extracellular glutamate [97]. P2X7 receptor antagonism proved beneficial in SOD1^{G93A} mice, delaying disease onset and maintaining motor behaviors, while ultimately not impacting motor neuron survival or lifespan [98]. Removing the receptor entirely accelerated disease onset and progression [99]. These outcomes prompt careful development of future antagonists for potential therapeutic strategies.

Damage from oxidative and nitrosative stress is pervasive in ALS, detectable in blood, urine, CSF, and tissue [100, 101]. Astrocytes normally protect neurons from such damage through glutathione, scavengers of reactive oxygen and nitrogen species (ROS/RNS), and antioxidant enzymes [102]. Given the role of SOD1 in detoxifying superoxide created through cellular processes including the mitochondrial electron transport chain, SOD1 mutations result in dysfunctional mitochondria, overabundance of ROS/RNS, and altered metabolic function [103, 104]. Motor neuron ROS/RNS and toxicity from SOD1^{G93A} or SOD1^{G86R} astrocytes were prevented by antioxidant pre-incubation [105, 106]. Blocking Na_v channels through compounds including riluzole protected from nitrosative stress and toxicity as well, showing that these effects are related to the excitotoxic setting [41]. In SOD1^{G93A}, reactive astrocytes upregulate nerve growth factor (NGF) and nitric oxide through inducible nitric oxide synthase (iNOS), which can activate apoptosis in motor neurons [107–111]. iNOS and NGF are elevated in SOD1^{G93A} astrocytes, and pretreatment with iNOS inhibitors prevents neurotoxicity [105, 111, 112]. sALS patient CSF also has elevated NGF, so this pathway may be involved in a broader patient population [111].

2.3 *Metabolic Imbalance in SOD1-ALS*

Gene expression evaluation of SOD1 mouse motor neurons revealed dysregulation of calcium sensing/signaling, mitochondrial respiration, and regenerative/injury and compliment activation pathways, while analysis of astrocytes showed many alterations in metabolism, particularly carbohydrate catabolism [111, 113–115]. Impairments of astrocytic glucose metabolism to provide neurons with glycolysis intermediates or ATP were seen, with phosphoglycerate kinase, isocitrate dehydrogenase, and ATP synthase complex F1 all downregulated [111]. These astrocytic metabolic shifts were dependent on coculture with motor neurons, supporting the idea that astrocytes are responsible for providing metabolic intermediates to promote neuronal health and homeostasis [111, 115]. Dysregulation of lactate was also indicated and was confirmed in spinal cords and astrocyte culture [111, 115]. Lactate can be shuttled from astrocytes to neurons during periods of enhanced neuronal activity or metabolic need [116]. Lactate release is dependent on glutamate transporters and the sodium/potassium ATPase (Na/K ATPase) [117], which were both decreased in SOD1 mouse motor neuron gene expression profiling [111]. Secreted lactate and the lactate solute carrier (Slc16a4) were both reduced in SOD1^{G93A} astrocytes and in astrocytes from SOD1-ALS patients [111]. Lactate supplementation to SOD1^{G93A} astrocyte-motor neuron coculture reduced toxicity, demonstrating that this deficiency impacts neuron health and survival [111]. In symptomatic SOD1^{G93A} mice and in SOD1-ALS and sALS tissue, astrocytes upregulate an α-adducin/Na/K ATPase complex, suggesting a compensatory mechanism during disease [118]. Knockdown of either component of this complex, or pharmacologically targeting Na/K ATPase activity through digoxin, decreased motor neuron degeneration [118]. Using an alternative approach to modulate metabolism, directly stimulating the mitochondrial pyruvate dehydrogenase complex using dichloroacetate, also ameliorated toxicity [119, 120]. When administered chronically in vivo, grip strength was improved, decreased astrocyte reactivity and preserved motor neurons were observed, and life span was increased [119]. Together, these results suggest that modulating glial metabolism may be beneficial in SOD1-ALS.

2.4 *Inflammation in SOD1-ALS*

Inflammation in neurodegenerative diseases was historically viewed as a consequence of dying neurons [121]. This attitude has now shifted to consider glial-derived molecules as contributors to disease. The regulator of inflammatory pathways, transcription factor nuclear factor kappa B (NF-κB), has been well described in ALS [33, 122]. The primary cytokine in blood and CSF in ALS is tumor necrosis factor alpha, TNFα [123, 124]. In SOD^{G93A} glial culture, TNFα and interferon gamma (IFNγ) were both highly elevated [112]. Prostaglandin D_2 (PGD_2) and glial maturation factor beta (GMFβ) were also increased, which promote an inflammatory state including secretion of TNFα [33, 125]. Upregulation of TNFα and its receptors TNFR1 and TNFR2 occur prior to symptom onset and motor neuron death in SOD^{G93A} models, suggesting that this pathway may be involved in disease development and progression [126–130]. In SOD^{G93A} culture, there was upregulation of membrane-bound, but not secreted TNFα, as well as elevated TNFR2 but not TNFR1 in both astrocytes and motor neurons [131]. While typically considered pro-survival, TNFR2 signaling is stimuli and situation-specific [132, 133]. TNFR2 deletion protected motor neurons from toxicity in culture, and in animals deletion partially preserved motor neurons and NMJs, but motor symptoms and survival were not improved [131]. This deletion decreased astrogliosis but not microgliosis in lumbar spinal cord, hinting that this TNFα effect may be astrocyte-specific [131]. Mechanistically, TNFR2 receptor activation in motor neurons may be linked to p38/MAPK activation, followed by subsequent cell death; however, a direct link between astrocyte-tethered mTNFα and stepwise activation of p38/MAPK has yet to be established [134]. A tight correlation was also found between TNFα, TNFR1, and glial cell line-derived neurotrophic factor (GDNF) during disease progression in SOD^{G93A} mice [127]. TNFR1 signaling in human and mouse astrocytes promoted GDNF synthesis and release, which was verified in SOD^{G93A} mice [127]. SOD^{G93A} mice with TNFR1 knockout prevented GDNF upregulation, but also accelerated disease progression and motor neuron loss [127]. Thus, the TNFα-TNFR1-GDNF pathway initiated at disease onset can be viewed as pro-survival, yet is insufficient. Astrocytic TNFα also contributes to elevated extracellular glutamate. TNFR1 signaling leads to prostaglandin E2 (PGE2) production, which elevates intracellular calcium and causes glutamate exocytosis [135]. TNFα also reduces EAAT2 levels, causing higher levels of retained extracellular glutamate [136, 137]. Neuronal TNFα signaling upregulates glutaminase, resulting in increased intracellular and secreted glutamate [138]. These elevated glutamate levels, in concert with mobilization of calcium-permeable AMPA receptors in response to TNFα, are a perfect storm for eliciting excitotoxic damage [138, 139]. The NLRP3 inflammasome is also upregulated in spinal cord of $SOD1^{G93A}$ rodents and is elevated in sALS tissue [140, 141]. Inflammasome assembly can be evoked by varied cellular events and results in complexes comprised of different core elements [142]. The NLRP3 inflammasome requires prerequisite NF-kB pathway activation, to promote NLRP3 expression [143]. Inflammasome components, NLRP3, adaptor protein apoptosis-associated speck-like protein (ASC), and pro-caspase 1, were detectable in presymptomatic $SOD1^{G93A}$ mice and were increased as disease progressed, specifically in spinal cord astrocytes [140]. Consequences of NLRP3 activation including caspase-1 and upregulation of neurotoxic IL-1β and IL-18 were also observed [140–142, 144].

IFNγ is also elevated in ALS patient CSF and serum [123, 145]. $SOD1^{G93A}$ astrocytes secrete IFNγ and cause selective motor neuron cell death through LIGHT (TNFSF14) and lymphotoxin-β-receptor (LT-βR) signaling [146]. $SOD1^{G93A}$ animals elevate IFNγ from disease onset onward, which correlates to LIGHT/LT-βR expression [146]. Astrocytes can also respond to interferon production by neurons undergoing damage or stress. In $SOD1^{G93A}$ mice, interferon-stimulated genes are found specifically in astrocytes surrounding motor neurons, even before symptom onset [147]. Reduction or ablation of INFα receptor 1, prevented this signaling and increased the lifespan of SOD1 mice [147].

2.5 Trophic Factor Considerations in SOD1-ALS

Astrocyte-derived vascular endothelial growth factor (VEGF) promotes angiogenesis and vasculogenesis and acts as a neuroprotective factor [148]. Reduction of VEGF in spinal cord resulted in a phenotype strikingly similar to ALS, with adult-onset progressive motor neuron degeneration [149]. When VEGF knockout animals were crossed with SOD1^{G93A} mice, disease progression and death were accelerated due to exacerbated motor neuron death, confirming the normally protective role of this astrocyte-derived molecule [150]. In contrast, transgenic mice with SOD1^{G93A} and overexpressed VEGF had delayed motor impairment and degeneration, and prolonged survival [151]. VEGF levels must be tightly regulated; while low levels of VEGF cause motor neuron degeneration as a result of reduced vascular perfusion/ischemic injury [149], increased VEGF expression has been linked to blood-brain barrier breakdown and subsequent neuronal damage [152]. Some groups have attempted to restore trophic support lost in disease. Transplantation of neural progenitor cells (NPCs) engineered to express GDNF has shown promise in the SOD1^{G93A} rat model, differentiating into astrocytes within the spinal cord and preserving motor neurons [153]. Similarly, human mesenchymal stem cells modified to secrete GDNF transplanted into skeletal muscle of the rat SOD1^{G93A} model increased the number of NMJs, attenuated motor neuron loss, significantly delayed disease progression, and increased lifespan [154]. A similar approach was taken using human astrocytes, which were injected into SOD1^{G93A} rodents [155]. Injected animals experienced delayed disease onset, improved rotarod motor performance, and increased survival [155]. Such results show promise for use of this strategy to restore a physiological compliment of trophic factors.

2.6 Additional Pathways to Disease in SOD1-ALS

Breakdown of the blood-brain barrier (BBB) and blood-spinal cord barrier (BSCB) are early events leading to neuronal damage in SOD1-ALS [156, 157]. First described as a regulator of vasoconstriction, endothelin-1 (ET-1) is secreted by activated astrocytes and regulates gliosis, immune activation, glutamate efflux, and EAAT2 repression [158, 159]. ET-1 is elevated in sALS patients and spinal astrocytes of SOD1^{G93A} mice [160, 161]. Endothelin receptors A and B (ETA/ETB) are found on neurons and glia, but neuronal receptors decreased as disease progressed [161]. Astrocyte-derived ET-1 caused motor neuron toxicity through either receptor, via upregulation of iNOS and nitric oxide [160–162]. Similarly, while associated with vasculogenesis, angiogenin functions as a signaling molecule released by motor neurons under oxidative stress to elicit a neuroprotective astrocytic response [163, 164]. It binds to syndecan-4 receptors and activates protein kinase Cα, which in turn activates transcription factor nuclear factor erythroid-2 related factor 2 (Nrf2) and the oxidative stress response [164]. In SOD1^{G93A}, astrocytic Nrf2 overexpression prevented motor neuron toxicity in vitro and delayed disease onset and extended survival in animals [165]. Loss-of-function ALS mutations have been identified in angiogenin, which further implicates this mechanism in ALS pathogenesis [166]. Impairments of the glymphatic system have also begun to emerge in ALS. This system is responsible for the CSF, running parallel to CNS vasculature and allowing for exchange of extracellular waste and metabolites [167, 168]. Pivotal to this function are astrocytic endfeet, which exclusively contain the water transporter, AQP4 [169]. Elevated AQP4 and decreased potassium channel Kir4.1 were found throughout brainstem and cortex of SOD1^{G93A} rats, with electrophysiological correlates confirmed in cultured astrocytes [170]. This would lead to impairment of astrocytic endfeet to maintain water and potassium homeostasis, impacting the function of the BBB/BSCB and having significant impacts on the neuronal environment. AQP4 upregulation was seen at disease onset in SOD1^{G93A} mice [91] and can be caused by the astrocyte reactivity seen across ALS cases [171]. Thus, dysregulation of CSF flow and metabolite exchange may impact neuronal health and function in sALS.

Similar to other proteins which can seed aggregates in recipient cells, SOD1 can initiate this pathological cascade [172]. This can occur through extracellular vesicles (EVs) released into the extracellular space or by aggregates released by dying cells that are incorporated by micropinocytosis [172]. SOD1 astrocytes display decreased normal secretory pathway activity in concert with increased EV release, suggesting compensation to limit intracellular aggregates [173]. EVs from brain and spinal cord of SOD1^{G93A} mice and SOD1-ALS spinal cord contain abundant misfolded and aggregated SOD1 and confer selective motor neuron toxicity [173, 174]. Based on cell markers, these EVs are derived from neurons and astrocytes, but not microglia [174]. As an increase in disease-causative proteins in EVs has been observed in other ALS models and in patients, this may be a pan-ALS phenotype [175]. Exosomes also carry microRNA (miRNA, miRs), which interact with recipient cell mRNAs to target them for degradation, thus conveying translational repression [144, 175]. Glial-derived miR-155 was enriched in EVs and was upregulated at end-stage disease in SOD1^{G93A} mice and in patient spinal cords [176–178]. Neuronal toxicity from SOD1^{G93A} EVs was prevented by treatment with a miR-155 antagomir [178]. miR-218 decreased in parallel with motor neuron loss in SOD1^{G93A} rats and was decreased in ALS patients [179]. Disease-resilient motor neurons maintained miR-218, while damaged motor neurons released miR-218 to impact neighboring astrocytes and become elevated in CSF [179]. miR-218 dramatically impact astrocytes, including EAAT2 downregulation [180]. Administration of a miR-218-directed antisense oligonucleotide (ASO) prevented EAAT2 loss, restored normal levels of CX43, and reduced gliosis [180].

Single-cell transcriptomics of SOD1^{G93A} motor neurons and astrocytes grown in a multicellular setting revealed that genes involved focal adhesion, extracellular matrix (ECM), and cytokine signaling were impacted in both cell populations [181]. Comparing mutant and healthy astrocytes found 124 additional dysregulated genes [181]. Cell-to-cell signaling dysfunction resulting from ECM interactions and transforming growth factor β (TGF-β) signaling were also identified [181]. TGF-β is a cytokine involved in diverse functions, particularly in cell proliferation and immune homeostasis with signaling occurring via ECM receptors or by impacting intracellular MAPK signaling pathways [182]. This pathway was altered in SOD1^{G93A} mice prior to symptom onset, with motor neurons showing elevated TGFβ-RII receptor in areas of astrogliosis and increased TGF-β in spinal cord [181, 183]. Astrocytic TGF-β overexpression in SOD1^{G93A} animals accelerated disease progression, and animals had more severe loss of motor function, showing that TGF-β was a negative regulator of neuroprotective responses [183]. Astrocytic SOD1 deletion prevented TGF-β upregulation, showing that this was a consequence of mutant SOD1 in astrocytes and not of surrounding neurodegeneration [183]. In transcriptome analysis of SOD1^{E100G} IPSC motor neurons, the downstream mediator of TGF-β SMAD2 was activated [184]. TGF-β was also implicated in motor neuron degeneration in other genetic ALS lines (C9orf72, VCP, and FUS) and in sALS and is elevated in sALS spinal cord [183, 184]. TGF-β release from reactive astrocytes has been tied to mTOR activation and autophagy disruption in motor neurons independent of genetic status, hinting that this may be a phenomenon in any disease with astrocyte reactivity [85].

Since the discovery of SOD1-ALS, much insight has been gleaned regarding the complex interplay between motor neurons and astrocytes in disease. Many mechanistic pathways and therapeutic targets have been identified and are in various stages of clinical testing.

3 ALS Related to DNA-/RNA-Binding Proteins: TDP43 and FUS

In 2006, the inclusions present in almost all ALS cases were found to be composed of aggregated TDP43 protein [17]. While it was initially unclear whether this was just a result of the disease process, TDP43 mutations can cause ALS [185–187]. Investigation of related protein FUS also led to identification of FUS-ALS mutations, and FUS inclusions were found in many patients lacking TDP43 pathology [188–191]. For both proteins, normal nucleocytoplasmic shuttling shifts to nuclear clearance and cytoplasmic aggregation in ALS [192–195]. These proteins engage in DNA, RNA, and protein interactions and are involved in transcription, translation, RNA processing, and trafficking [196–198]. Disease pathomechanisms have been extensively studied and recently reviewed [199, 200]. While TDP43/FUS mutations account for a small percentage of fALS, TDP43 or FUS aggregates are observed in 98% of cases, suggesting that their dysregulation is a fundamental feature of pathogenesis [201, 202]. Understanding neuroglial impacts of TDP43/FUS dysfunction is critical, particularly for patients with sporadic disease etiology.

3.1 TDP43-ALS

Rapidly following TDP43 identification, inclusions were found in astrocytes from patient samples, prompting exploration of possible astrocytic dysfunction and contribution to motor neuron damage [203–206]. RNA interference (RNAi) elimination of TDP43 resulted in age-dependent degeneration of spinal motor neurons, progressive weakness, paralysis, and ultimate death, suggesting that TDP43 loss of function may be sufficient to induce ALS pathology [207]. TDP43 was mainly reduced in astrocytes in this model, showing that TDP43 function is critical to maintaining CNS homeostatic balance [207]. TDP43 reduction in neurons resulted in a milder phenotype, signaling glial involvement in disease pathogenesis [208, 209]. With astrocytic TDP43 knockdown, animals displayed a reactive astrocyte phenotype and developed motor deficits, but did not have overt neuron loss or NMJ denervation [210]. Comparable in vitro studies also did not evoke motor neuron toxicity [211]. TDP43 knockdown in rat astrocytes resulted in robust gene alterations, particularly in activation of genes involved in reactivity and innate immune signaling [212]. In mice, TDP43 overexpression ($TDP43^{WT}$) is profoundly cytotoxic, causing progressive weakness, motor dysfunction, inflammation, NMJ denervation, and paralysis [213–217]. In rats, $TDP43^{WT}$ lead to increased glutamate and lactate in motor cortex, skewed the balance of excitatory/inhibitor neurotransmitters, and decreased the function of the citric acid cycle [218]. Both TDP43 and the NF-κB p65 subunit were elevated in spinal cords of $TDP43^{WT}$ mice and in sALS, and $TDP43^{WT}$ neurons were more susceptible to glia-based toxicity [219]. TDP43 was found to activate NF-κB, with these proteins colocalizing in neurons and glia [219]. Neuronal $TDP43^{WT}$ resulted in decreased synapsin 1 and altered the astrocytic glutamate-glutamine cycle, causing elevated glutamate and lactate, and reduced glutamine and aspartate [220]. In vitro astrocytic $TDP43^{WT}$ showed elevated TNFα, IL-1β, and IL-6 and reduced mitochondrial capacity [43]. These astrocytes caused neurotoxicity, which could be attenuated by neutralizing antibodies for these cytokines or by inhibiting inflammatory mediator protein tyrosine phosphatase 1B (PTP1B) [43]. Glial $TDP43^{WT}$ in *Drosophila* caused profound NMJ defects, locomotor deficits, and reduced lifespan, which were all rescued by downregulation of the fly PTP1B homolog [43].

TDP43-ALS models have yielded conflicting results when evaluating whether astrocytes are neurotoxic. The most widely used model, $TDP43^{A315T}$, results in progressive age-dependent motor deficits and reduced lifespan [214]. Neuron-specific $TDP43^{A315T}$ mice also developed progressive motor deficits [221]. Accumulation of diffuse cytoplasmic TDP43 and reactive gliosis was observed, which correlated to muscle wasting and loss of strength [221]. Short-term transgene suppression in symptomatic animals cleared all mutant protein, reduced astrogliosis, and resulted in retained motor performance [221]. Primary astrocytes from $TDP43^{A315T}$ mice showed cytoplasmic TDP43 aggregation and impaired glutamate uptake [27]. Conditioned media from $TDP43^{A315T}$ astrocytes was toxic to motor neurons through mechanisms involving nitrosative stress and voltage-gated sodium channel activity

[41]. In contrast, a second group found no toxic effects of TDP43^{A315T} astrocytes on motor neurons in direct coculture and observed no negative consequences when engrafting the glial precursor cell population from TDP43^{A315T} animals into spinal cord of wildtype rats [211]. Further study is needed to determine if neurons suppress toxic impacts of astrocytic TDP43^{A315}. The impact of EVs has also been evaluated using samples derived from patients and TDP43^{A315T} mice [222]. Exosomes contained both full length and the cleaved C-terminal fragment of TDP43 and evoked cytoplasmic TDP43 redistribution when applied to cells [222]. While preventing exosome secretion may be viewed as beneficial for preventing spread of pathology, inhibiting exosome secretion in vitro and in vivo led to exacerbated TDP43 aggregation and NMJ denervation [222]. These findings suggest that exosomal clearance is a protective measure and that strategies to prevent disease propagation may need to be directed toward lowering protein. When the C-terminal fragment was expressed in neurons, protein dyshomeostasis resulted from proteasomal halting [223]. Primary astrocytes containing C-fragment inclusions showed reductions in intracellular calcium and cyclic-AMP and increased lactate and activation of aerobic glycolysis, which combined with lowered levels of the MCT-1 lactate transporter, resulting in an alteration of astrocytic metabolism and neuronal metabolite support [224].

TDP43^{M337V} knock-in mice develop age-dependent progressive motor dysfunction, NMJ loss, and have shortened lifespans [225]. TDP43 cytoplasmic aggregates are not seen, yet mis-splicing events occur prior to disease onset, and motor neurons displayed cytoplasmic TDP43 mislocalization and aberrant stress granule dynamics [225]. Motor neuron analysis revealed that upon oxidative stress, TDP43 activates proteins in stress granule, endoplasmic reticulum, and secretory transport pathways, but these were reduced with TDP43^{M337V} [226]. Primary rodent and IPSC TDP43^{M337V} motor neurons also showed reduced EV secretion [226]. Astrocyte-restricted TDP43^{M337V} in rats led to astrocyte activation, loss of EAAT1/EAAT2, release of neurotoxic lipocalin-2 (Lcn2), progressive loss of motor neurons, denervation of NMJs, paralysis and death [44]. Lipocalin-2 is associated with inflammatory astrocytes and enhances reactive gliosis through autocrine signaling [227]. This molecule was also elevated in TDP43 knockdown animals [207], suggesting that perturbed TDP43 levels or function in either positive or negative directions can influence astrocytic reactivity. Surprisingly, TDP43^{M337V} IPSC astrocytes, which displayed cytoplasmic TDP43, did not impact survival of motor neurons; however, bearing this mutation did increase astrocytic risk of death [228]. Further work is needed to clarify how these astrocytes switch to a neurotoxic fate in vivo. Cell intrinsic toxicity was also evaluated in TDP43^{Q311K} animals, which display age-dependent motor dysfunction, loss of spinal motor neurons, and NMJs [229]. When TDP43^{Q311K} was removed from motor neurons, symptom onset was delayed, but disease progression and motor neuron and NMJ degeneration were still observed [230]. TDP43^{Q311K} astrocytes have also shown an inability to appropriately induce antioxidant responses both in vitro and in vivo [231].

3.2 FUS-ALS

Key facets of disease also result from astrocytic effects on neurons in FUS-ALS. Postnatal FUS knockout did not impact motor neuron survival, indicating that loss of function alone is not sufficient for degeneration and that a toxic gain of function is required for disease pathology [232]. In determining FUS-regulated genes, rodent spinal motor neurons and glia were subjected to FUS RNAi [233]. In addition to identifying specific motor neuron genes which may be impacted, >2000 differentially regulated astrocyte genes were discovered, which were enriched in cellular homeostasis and inflammation pathways [233]. Reactive astrocyte biology following FUS protein reduction was confirmed using a similar approach in frontal cortex of marmosets [234]. Reduction of FUS resulted in an increase in astrocyte number and reactivity, without gross neuronal loss [234]. In contrast, FUS overexpression (FUS^{WT}) in mouse or IPSC astrocytes impacted reactivity toward inflammatory stimuli such as IL-1β, which caused secretion of TNFα and 1L-6 and activated iNOS and PGE2 production [235]. As in SOD1-ALS, these astrocytic factors caused neuronal toxicity [235]. Similarly, as with TDP43, FUS is a coactivator of NF-kB p65, suggesting that cellular stress-mediated inflammatory signaling is a key function of these two proteins in glial cell populations [236]. These findings have revealed a "goldilocks" effect where either too much or too little FUS can both evoke a proinflammatory state in astrocytes.

Two mouse models, FUS^{P525L} and FUS^{R521C}, recapitulated juvenile and adult-onset FUS-ALS, respectively, exhibiting NMJ deficits preceding motor neuron cell death [232]. Motor neuron-restricted expression of these variants caused motor weakness and neurodegeneration; however, phenotype severity and disease progression were enhanced when mutant FUS was expressed globally [232]. These results indicate that mutant FUS in motor neurons is toxic, but glia exacerbate pathogenesis. In a model expressing truncation mutant $FUS^{\Delta NLS}$ which drives FUS to the cytoplasm, animals had gait deficits and age-dependent motor neuron loss [237]. Motor neuron deletion of this variant delayed but did not prevent symptoms [237]. FUS^{R521G} astrocytes had mislocalized cytosolic FUS and were toxic to motor neurons, with excitotoxicity resulting from altered GluA1/2 AMPA receptor subunit levels and elevated calcium influx [45]. Cytokine analysis revealed specific upregulation of TNFα, which could be targeted either by blocking NF-κB pathway activation, using SN50, or with a TNFα neutralizing antibody, which both prevented AMPAR changes and blocked toxicity [45]. NF-κB-based upregulation of TNFα was also seen in IPSC astrocytes bearing FUS^{H517Q} [238]. Astrocyte-mediated toxicity was also demonstrated using acute astrocyte-restricted expression of FUS^{R521G} in mouse spinal cord, which resulted in progressive motor neuron cell death and behavioral deficits [46]. In this in vivo model, TNFα neutralizing antibodies effectively prevented neuronal death and motor dysfunction [46]. A neuron-specific FUS^{R521G} model has shown FUS cytoplasmic localization, and juvenile cognitive impairments followed by age-dependent motor impairments [239]. Glial activation was pronounced in brain and spinal cord, and modulation of reactivity through NF-κB inhibition with Withaferin A greatly improved cognitive and motor performance [239]. Thus, even when mutant FUS is not present in glial cells, crosstalk from motor neurons promotes NF-κB activation and causes some of the pathogenic features of disease. The effects of IPSC FUS^{P525L} and FUS^{R521H} astrocytes on the formation of functional NMJs have been evaluated [49]. These astrocytes were reactive, displayed FUS cytosolic mislocalization, had abundant gene dysregulation, and secreted elevated IL-1β, TNFα, 1L-6, and IL-8 [49]. When cultured with motor neurons and myotubes, mutant FUS astrocytes impaired NMJ formation, while isogenic controls corrected this deficit, even when motor neurons carried the FUS mutations [49]. Transcriptional analysis showed convergence on astrocytic WNT/β--catenin pathway targets, which also became dysregulated in motor neurons following coculture [49].

Further research is still necessary to illuminate the roles of TDP43 and FUS in astrocyte biology and the contribution of disease mutations to astrocyte-mediated neuronal dysfunction and toxicity. Importantly, as inclusions of TDP43 or FUS also occur in sporadic ALS patients, it is important to determine whether loss of astrocytic TDP43/FUS activity is a factor driving astrocytic reactivity and neurotoxicity.

4 C9orf72-ALS

The most common genetic cause of ALS is a hexanucleotide repeat expansion in C9orf72, which accounts for ~40% of fALS and 5–20% of sALS [240–243]. Three nonexclusive disease mechanisms result from haploinsufficiency of the C9orf72 gene product and gain-of-function toxicities caused by both accumulated RNA transcripts and ectopic dipeptide products (DPRs) produced from this RNA [244]. Despite many attempts to capture the full pathology of C9-ALS in a mouse model, none have succeeded [245]. Intranuclear C9-RNA foci have been detected in glia, in patient tissue, and in IPSC astrocytes [40, 246]. While cytoplasmic accumulations of C9-expanded RNA and DPRs are not observed in glial cells in patient tissue or gain-of-function models [246–250], possibly due to enhanced clearance or cell proliferation, DPRs produced in neurons can be transferred to astrocytes, oligodendrocytes, and microglia [251]. TDP43 inclusions are also observed within glial cells in C9-ALS, with reactive gliosis and inflammatory cascades reported in patient tissue, as well as C9orf72 knockout and gain-of-function models [31, 252–256].

C9-ALS IPSC astrocytes confer motor neuron toxicity, possibly through autophagy impairment, due to motor neuron LC3-II protein reduction [32, 257, 258]. Expression analysis revealed antioxidant protein reduction including SOD1 and glutathione synthetase, along with increased oxidative stress and senescence within astrocytes [258]. Media from these astrocytes evoked motor neuron oxidative stress, illustrating that withdrawal of antioxidant factors has direct consequences on the homeostasis of neighboring neurons [258]. Using comparable methods, a separate study did not observe astrocyte-mediated toxicity, but did show progressive loss of action potential firing, along with a more depolarized resting membrane potential and reduced sodium and potassium currents [40]. Astrocyte secretome analysis identified molecules which may cause these changes, including receptor-integrating serine/threonine-protein kinase 1 (RIP1), mixed lineage kinase domain-like (MLKL), inflammatory cytokines, and ROS [40]. Dysregulated genes included downregulated sodium and potassium transporters, synapse assembly molecules, factors mediating cell adhesion and cell-to-cell signaling, and upregulated glutamate receptor signaling [40]. Motor neuron analysis showed dysregulated ECM and cell adhesion genes and elevated inflammation and TGF-β signaling genes [259]. C9-ALS IPSC astrocytes also release neurotoxic extracellular vesicles [38]. miRNAs involved in axonal maintenance were perturbed in these exosomes [38]. The most dysregulated was the downregulated miR-494-3p, a negative regulator of semaphorin 3A (SEMA3A) [38]. Lowering SEMA3 levels through a miRNA-mimic increased motor neuron survival, suggesting that this astrocyte-derived miRNA is essential for maintaining appropriate homeostasis [38]. miR-494-3p was also reduced in tissue from sALS patients, suggesting that this may be a pathway and molecule involved across ALS patient populations [38].

Disruption of metabolic balance has been suggested to result from loss of C9orf72 [260]. C9-ALS IPSC astrocytes or healthy astrocytes with reduced C9orf72 protein had altered glucose metabolism and buffering, resulting in increased cytoplasmic glutamate [261]. Glutamate cycling enzymes glutaminase and glutamine synthase were also reduced, and EAAT1/2 was downregulated [261]. These findings were supported by analysis of C9-ALS tissue also displayed reduced EAAT1/2 [261]. Thus, there is profound impairment of glutamate-handling pathways within astrocytes, likely impacting glutamate uptake from synapses. Following neuronal C9orf72 knockdown, neuron-specific glutamate transporter EAAT3 was also reduced, putting further strain on glutamate buffering capacities and potential for motor neuron excitotoxicity [261]. ET-1 and ETA/ETB receptors were also elevated, hinting at potential implications to vascular alterations [160, 261]. C9-ALS also results in other dysregulated metabolic intermediates, including ATP [39, 262, 263]. C9-ALS astrocytes and neurons have reduced adenosine deaminase, which creates inosine and directs this metabolite to ultimately produce energy through glycolysis [39]. Supplementing these astrocytes with inosine prevented motor neuron death, suggesting a pathway of astrocyte-mediated neuronal toxicity in C9-ALS related to neuronal metabolic support [39]. C9-ALS IPSC astrocytes also had defects in other cellular energy-based pathways such as glycogen breakdown and transport of mitochondrial metabolites [39, 262].

Since the discovery of C9orf72, many cell-autonomous effects have been described, and interactions between motor neurons and astrocytes are being revealed. Technology has recently allowed for the transcriptome and epigenome alterations in C9-ALS patients to be described at the single-cell level [264]. The novel dysregulated pathways identified in specific neuronal and glial populations will undoubtedly be pursued in the coming years.

5 Other Genetic Forms of ALS: Autophagy Pathways

Impairments in the autophagy-lysosome pathway have been heavily implicated in ALS, with recent additions to the ALS gene spectrum including optineurin (OPTN), ubiquilin-2 (UBQLN2), valosin-containing protein (VCP), and tank-binding kinase 1 (TBK1) [265–267]. In addition to dealing with intracellular homeostasis, astrocytes are responsible for the handling of biproducts from neighboring neurons experiencing cellular stress, such as protein aggregates or organelles released into the extracellular space or through transcellular uptake, putting further burden on astrocytic lysosomal and autophagy programs [268, 269]. Astrocytes with impaired lysosome and autophagy function can also independently contribute to neuronal degeneration, likely due to a loss of metabolic support [270]. Studies of OPTN remain limited, as animal models are lacking and patient IPSC lines are few. Neuronal expression of UBQLN2 variants through viral vectors caused aggregation in vitro and age-dependent motor phenotypes in vivo [271], but studies with ubiquitous expression or exploring astrocytic involvement have not been published.

TBK1 is a regulator of selective autophagy and is required for cargo recruitment and phosphorylation of adaptor proteins [272]. TBK1 is also involved in energy metabolism and in innate and adaptive immune signaling, as an inducer of type-1 interferons [272]. Creating a TBK1 animal model has been a challenge, as heterozygous deletion does not lead to motor neuron degeneration or autophagy defects [273]. Selective deletion of TBK1 from motor neurons also did not exhibit motor or neurological defects, but conditional deletion of TBK1 from all neurons yielded age-dependent cognitive and motor deficits [274, 275]. Thus far, studies of TBK1 in the context of SOD1-ALS have indicated that impacts on late-stage disease result from a failure of microglial activation from the interferon-inducible pathway [273, 276]. Additional studies exploring astrocytic contributions to TBK1-ALS related to metabolism and/or interferon signaling have not yet been performed.

VCP is an ATPase which extracts ubiquitinated proteins bound for degradation and is also required for autophagosome maturation [277]. As VCP is tied to both the proteasomal and autophagy protein degradation pathways, mutations greatly impact protein homeostasis [277]. A target of VCP-directed degradation is the negative regulator of NF-κB IκB-α, thus freeing NF-κB for activation of transcriptional responses [278, 279]. Given the extent of NF-κB-related neuroinflammation in ALS, exploring the impacts of VCP mutations on glial populations will be enlightening. Mutations in VCP also cause multisystem proteinopathy (MSP) which includes inclusion body myopathy, Paget's disease of bone, and frontotemporal dementia (previously known as IBMPFD) [277]. VCP^{R155H} and VCP^{A232E} mice develop pathology limited to brain, muscle, and bone and display clinical symptoms of MSP [280, 281]. ALS-related outcomes included TDP43 pathology, progressive muscle weakness, motor deficits, and reduced lifespan [281]. In vitro studies showed heightened responses to stimulators of NF-κB activity [281]. Cytoplasmic mislocalization of FUS was also demonstrated in spinal motor neurons of VCP^{A232E} animals, suggesting that effects of FUS dysregulation may need to be considered in

VCP-ALS [195]. For VCP^{R191Q} or VCP^{R115C} IPSC motor neurons, ER stress and cytoplasmic TDP43 were the mechanisms underlying mitochondrial dysfunction, oxidative stress, synaptic deficits, and toxicity [282]. While control IPSC astrocytes supported the health and longevity of control or mutant VCP motor neurons, mutant VCP astrocytes lost this effect [282].

6 Sporadic ALS (sALS)

Understanding sALS is challenging due to the unknown disease etiology, with studies limited to patient tissue, biofluids, and IPSCs. Nevertheless, common themes and pathway interactions have emerged, suggesting that many components of disease may instead be pan-ALS phenotypes. Astrocytes recovered from sALS patient spinal cord were toxic to motor neurons [47, 84]. sALS IPSC astrocytes also show astrocyte-derived toxicity to motor neurons via soluble released factors [32, 37]. When sALS astrocyte progenitors were transplanted into mouse spinal cords, animals displayed progressive grip strength and gait impairments, reduced motor neuron number, and increased NMJ denervation [35]. These findings conclude that even in the absence of disease-causative mutations, astrocytes from sALS individuals contribute to motor neuron toxicity. Intriguingly, in comparison to IPSC motor neurons, astrocytes demonstrated marked resilience to toxicity when challenged with TDP43 aggregate extracts from ALS patient spinal cord [283]. When cocultured, the presence of astrocytes was neuroprotective, though TDP43 aggregate propagation from motor neurons to astrocytes did occur, motor neurons displayed reduced TDP43 aggregation, cytoplasmic mislocalization, and toxicity [283].

sALS impacts inflammatory profiles, as patients had elevated TNFα and IFN-γ in CSF, and patient survival was negatively correlated with plasma levels of TNFα, IL-10, and TNF-related apoptosis-inducing ligand (TRAIL) [284]. Spinal cord mRNA also demonstrated elevated TNFα and TNFR1 [127]. Exposing rat astrocytes to ALS patient CSF resulted in increased TNFα and IL-6, and downregulated anti-inflammatory IL-10 [285]. When postmortem-isolated astrocytes from sALS patients were assessed for differentially expressed genes, inflammatory genes were highly activated [47]. Cultured media from sALS IPSC astrocytes also revealed elevated levels of secreted TNFα, IL-1β, and IL-6 [37]. A similar inflammatory profile could be elicited in control astrocytes through inhibition of the mTOR-autophagy pathway, with these sALS IPSC astrocytes also demonstrating endogenous impairments in autophagy function [37]. Stimulation of autophagy through application of rapamycin was effective in recovering autophagy function and reducing cytokine release from patient-derived cells [37]. Therefore, impairments in autophagy due to protein dyshomeostasis in all forms of ALS may be a key pathway for pan-ALS therapeutics, as boosting this function should lead to both decreased cellular protein burden and lowering of astrocytic inflammatory profiles.

Withdrawal of astrocytic trophic factors is also implicated in sALS. Rat astrocyte cultures treated with sALS-CSF recapitulated many features of genetic ALS astrocytes, namely dysregulation of glutamate, increase in ROS/RNS, and downregulation of GDNF and VEGF mRNA and protein [285]. Two clinical trials using supplementation with cells designed to secrete trophic factors have recently passed phase I/IIa safety studies [286, 287]. These reports advancing this idea to clinical stages show promise for future therapies.

7 Unifying Principles and Consideration of the Broader CNS Landscape

While each ALS genetic form and variant contributes specific disease mechanisms, common themes of excitotoxicity, inflammation, oxidative damage, withdrawal of trophic support, and protein dyshomeostasis result from dysfunctional astrocyte-neuron communication throughout all studied models (summarized in Table 1).

The development of "omics" technology, combined with IPSC-derived motor neurons and astrocytes, has also allowed for comprehensive analysis of the mRNA transcripts, cellular proteins, and secretory profiles associated with patient-derived cells. Meta-analysis using transcriptomic and proteomic data from SOD1, C9orf72, FUS, and VCP IPSC astrocytes identified over 1300 upregulated and 1500 downregulated genes in "pan-ALS" astrocytes [288]. Common pathways of upregulation of the immune response, inflammation, and TGF-β signaling were detected, as was dysregulation of the ECM, cellular stress pathways, glutamate uptake, trophic factors, and synaptic integrity [288]. Astrocyte-neuron interactions in ALS are dependent on the entire CNS environment, including microglia [289–291], blood-brain/blood-spinal cord barrier [292], and muscle cells [293]. Changes were also correlated with the features of a subset of reactive astrocytes, formed by exposure to microglial TNFα, IL-1α, and the immune complement pathway molecule C1q [288, 291]. This type of reactive astrocytes, generated by these factors, have been noted in sALS [291]. These three factors were simultaneously knocked out in $SOD1^{G93A}$ animals, with outcomes of striking rescue of motor performance, NMJ innervation, motor neuron numbers, and lifespan [290]. This finding again reiterates the importance of astrocytes in driving disease progression and provides insight into new therapeutic potential for ALS, through either targeting the microglia causing this reactive astrocyte switch or by directly targeting this specific subgroup of reactive astrocytes in disease.

Proposed therapeutic strategies for ALS include elimination of mutant protein effects by antisense oligonucleotide downregulation or aggregate resolution. However, it is unknown if this will attenuate the diverse astrocytic and motor neuron cellular changes described in this chapter, revert reactive and neurotoxic states, or repair aberrant astrocyte-neuron communication. As this approach requires targeting a specific protein, treatment strategies for sporadic ALS are also still needed. As therapeutic development evolves beyond simply targeting motor neurons, an

Table 1 Common mechanisms of astrocyte-neuron involvement in ALS

	SOD1	TDP43	FUS	C9orf72	Other genetic	Sporadic
Astrocyte-mediated motor neuron toxicity	✓	✓ Mutation dependent?	✓	✓	Probable (VCP)	✓
Excitotoxicity	✓	✓	✓	✓	Unknown	✓
Oxidative/nitrosative stress	✓	✓	Unknown	✓	✓ (VCP)	Unknown
Protein dyshomeostasis	✓	✓	✓	✓	✓ (VCP, TBK1, likely all autophagy-related variants)	✓
Metabolic imbalance	✓	✓	Unknown	✓	Possible (TBK1)	Unknown
Inflammation	✓	✓	✓	✓	Probable (VCP, TBK1)	✓
Trophic factor withdrawal	✓	Unknown	Possible (transcriptomics)	Possible (transcriptomics)	Probable (VCP)	✓
Blood-brain barrier involvement	✓	Unknown	Unknown	Possible (ET-1 signaling upregulated)	Unknown	Possible (ET-1 and AQP4 upregulation)
Disease spreading	✓	✓	✓	✓	Unknown	✓
Extracellular matrix/cell adhesion/TGF-β pathway	✓	Unknown	✓	✓	✓ (VCP)	✓

understanding of the ways in which the communication between astrocytes and neurons is disrupted is essential.

References

1. Younger DS, Brown RH Jr. Amyotrophic lateral sclerosis. Handb Clin Neurol. 2023;196:203–29.
2. Ravits J. Focality, stochasticity and neuroanatomic propagation in ALS pathogenesis. Exp Neurol. 2014;262 Pt B:121–6.
3. Wang H, Guan L, Deng M. Recent progress of the genetics of amyotrophic lateral sclerosis and challenges of gene therapy. Front Neurosci. 2023;17:1170996.
4. Gregory JM, Fagegaltier D, Phatnani H, et al. Genetics of amyotrophic lateral sclerosis. Curr Genet Med Rep. 2020;8:121–31.
5. Ovsepian SV, O'Leary VB, Martinez S. Selective vulnerability of motor neuron types and functional groups to degeneration in amyotrophic lateral sclerosis: review of the neurobiological mechanisms and functional correlates. Brain Struct Funct. 2024;229(1):1–14.
6. Zhou W, Xu R. Current insights in the molecular genetic pathogenesis of amyotrophic lateral sclerosis. Front Neurosci. 2023;17:1189470.
7. Sever B, et al. Comprehensive research on past and future therapeutic strategies devoted to treatment of amyotrophic lateral sclerosis. Int J Mol Sci. 2022;23(5):2400.
8. Cho H, Shukla S. Role of Edaravone as a treatment option for patients with amyotrophic lateral sclerosis. Pharmaceuticals (Basel). 2020;14(1):29.
9. Brooks BR, et al. Intravenous edaravone treatment in ALS and survival: an exploratory, retrospective, administrative claims analysis. EClinicalMedicine. 2022;52:101590.
10. Paganoni S, et al. Long-term survival of participants in the CENTAUR trial of sodium phenylbutyrate-taurursodiol in amyotrophic lateral sclerosis. Muscle Nerve. 2021;63(1):31–9.
11. Paganoni S, et al. Trial of sodium phenylbutyrate-taurursodiol for amyotrophic lateral sclerosis. N Engl J Med. 2020;383(10):919–30.
12. Miller TM, et al. Trial of antisense oligonucleotide tofersen for SOD1 ALS. N Engl J Med. 2022;387(12):1099–110.
13. Ilieva H, Polymenidou M, Cleveland DW. Non-cell autonomous toxicity in neurodegenerative disorders: ALS and beyond. J Cell Biol. 2009;187(6):761–72.
14. Lee J, et al. Astrocytes and microglia as non-cell autonomous players in the pathogenesis of ALS. Exp Neurobiol. 2016;25(5):233–40.
15. Philips T, Robberecht W. Neuroinflammation in amyotrophic lateral sclerosis: role of glial activation in motor neuron disease. Lancet Neurol. 2011;10(3):253–63.
16. Al-Chalabi A, et al. The genetics and neuropathology of amyotrophic lateral sclerosis. Acta Neuropathol. 2012;124(3):339–52.
17. Neumann M, et al. Ubiquitinated TDP-43 in frontotemporal lobar degeneration and amyotrophic lateral sclerosis. Science. 2006;314(5796):130–3.
18. Arai T, et al. TDP-43 is a component of ubiquitin-positive tau-negative inclusions in frontotemporal lobar degeneration and amyotrophic lateral sclerosis. Biochem Biophys Res Commun. 2006;351(3):602–11.
19. Philips T, et al. Oligodendrocyte dysfunction in the pathogenesis of amyotrophic lateral sclerosis. Brain. 2013;136(Pt 2):471–82.
20. Brettschneider J, et al. Pattern of ubiquilin pathology in ALS and FTLD indicates presence of C9ORF72 hexanucleotide expansion. Acta Neuropathol. 2012;123(6):825–39.
21. Nishihira Y, et al. Sporadic amyotrophic lateral sclerosis: two pathological patterns shown by analysis of distribution of TDP-43-immunoreactive neuronal and glial cytoplasmic inclusions. Acta Neuropathol. 2008;116(2):169–82.

22. Fujita K, et al. Increases in fragmented glial fibrillary acidic protein levels in the spinal cords of patients with amyotrophic lateral sclerosis. Neurochem Res. 1998;23(2):169–74.
23. Schiffer D, et al. Reactive astrogliosis of the spinal cord in amyotrophic lateral sclerosis. J Neurol Sci. 1996;139(Suppl):27–33.
24. Chio A, et al. Neuroimaging in amyotrophic lateral sclerosis: insights into structural and functional changes. Lancet Neurol. 2014;13(12):1228–40.
25. Johansson A, et al. Evidence for astrocytosis in ALS demonstrated by [11C](L)-deprenyl-D2 PET. J Neurol Sci. 2007;255(1–2):17–22.
26. Bruijn LI, et al. ALS-linked SOD1 mutant G85R mediates damage to astrocytes and promotes rapidly progressive disease with SOD1-containing inclusions. Neuron. 1997;18(2):327–38.
27. Barton SK, et al. Mutant TDP-43 expression triggers TDP-43 pathology and cell autonomous effects on primary astrocytes: implications for non-cell autonomous pathology in ALS. Neurochem Res. 2020;45(6):1451–9.
28. Nalbandian A, et al. The homozygote VCP(R(1)(5)(5)H/R(1)(5)(5)H) mouse model exhibits accelerated human VCP-associated disease pathology. PLoS One. 2012;7(9):e46308.
29. Nalbandian A, et al. A progressive translational mouse model of human valosin-containing protein disease: the VCP(R155H/+) mouse. Muscle Nerve. 2013;47(2):260–70.
30. Le NT, et al. Motor neuron disease, TDP-43 pathology, and memory deficits in mice expressing ALS-FTD-linked UBQLN2 mutations. Proc Natl Acad Sci USA. 2016;113(47):E7580–9.
31. Liu Y, et al. C9orf72 BAC mouse model with motor deficits and neurodegenerative features of ALS/FTD. Neuron. 2016;90(3):521–34.
32. Meyer K, et al. Direct conversion of patient fibroblasts demonstrates non-cell autonomous toxicity of astrocytes to motor neurons in familial and sporadic ALS. Proc Natl Acad Sci USA. 2014;111(2):829–32.
33. Di Giorgio FP, et al. Non-cell autonomous effect of glia on motor neurons in an embryonic stem cell-based ALS model. Nat Neurosci. 2007;10(5):608–14.
34. Nagai M, et al. Astrocytes expressing ALS-linked mutated SOD1 release factors selectively toxic to motor neurons. Nat Neurosci. 2007;10(5):615–22.
35. Qian K, et al. Sporadic ALS astrocytes induce neuronal degeneration in vivo. Stem Cell Rep. 2017;8(4):843–55.
36. Arredondo C, et al. Excessive release of inorganic polyphosphate by ALS/FTD astrocytes causes non-cell-autonomous toxicity to motoneurons. Neuron. 2022;110(10):1656–1670.e12.
37. BaofengFeng, et al. Autophagy-mediated inflammatory cytokine secretion in sporadic ALS patient iPSC-derived astrocytes. Oxidative Med Cell Longev. 2022;2022:6483582.
38. Varcianna A, et al. Micro-RNAs secreted through astrocyte-derived extracellular vesicles cause neuronal network degeneration in C9orf72 ALS. EBioMedicine. 2019;40:626–35.
39. Allen SP, et al. Astrocyte adenosine deaminase loss increases motor neuron toxicity in amyotrophic lateral sclerosis. Brain. 2019;142(3):586–605.
40. Zhao C, et al. Mutant C9orf72 human iPSC-derived astrocytes cause non-cell autonomous motor neuron pathophysiology. Glia. 2020;68(5):1046–64.
41. Rojas F, et al. Astrocytes expressing mutant SOD1 and TDP43 trigger motoneuron death that is mediated via sodium channels and nitroxidative stress. Front Cell Neurosci. 2014;8:24.
42. Rojas F, et al. Mature iPSC-derived astrocytes of an ALS/FTD patient carrying the TDP43(A90V) mutation display a mild reactive state and release polyP toxic to motoneurons. Front Cell Dev Biol. 2023;11:1226604.
43. Lee S, et al. The overexpression of TDP-43 in astrocytes causes neurodegeneration via a PTP1B-mediated inflammatory response. J Neuroinflammation. 2020;17(1):299.
44. Tong J, et al. Expression of ALS-linked TDP-43 mutant in astrocytes causes non-cell-autonomous motor neuron death in rats. EMBO J. 2013;32(13):1917–26.
45. Kia A, et al. Astrocytes expressing ALS-linked mutant FUS induce motor neuron death through release of tumor necrosis factor-alpha. Glia. 2018;66(5):1016–33.

46. Jensen BK, et al. Targeting TNFalpha produced by astrocytes expressing amyotrophic lateral sclerosis-linked mutant fused in sarcoma prevents neurodegeneration and motor dysfunction in mice. Glia. 2022;70(7):1426–49.
47. Haidet-Phillips AM, et al. Astrocytes from familial and sporadic ALS patients are toxic to motor neurons. Nat Biotechnol. 2011;29(9):824–8.
48. Ramirez-Jarquin UN, et al. Chronic infusion of SOD1(G93A) astrocyte-secreted factors induces spinal motoneuron degeneration and neuromuscular dysfunction in healthy rats. J Cell Physiol. 2017;232(10):2610–5.
49. Stoklund Dittlau K, et al. FUS-ALS hiPSC-derived astrocytes impair human motor units through both gain-of-toxicity and loss-of-support mechanisms. Mol Neurodegener. 2023;18(1):5.
50. Rosen DR. Mutations in Cu/Zn superoxide dismutase gene are associated with familial amyotrophic lateral sclerosis. Nature. 1993;364(6435):362.
51. Mondola P, et al. The Cu, Zn superoxide dismutase: not only a dismutase enzyme. Front Physiol. 2016;7:594.
52. Rothstein JD, et al. Selective loss of glial glutamate transporter GLT-1 in amyotrophic lateral sclerosis. Ann Neurol. 1995;38(1):73–84.
53. Bristol LA, Rothstein JD. Glutamate transporter gene expression in amyotrophic lateral sclerosis motor cortex. Ann Neurol. 1996;39(5):676–9.
54. Lin CL, et al. Aberrant RNA processing in a neurodegenerative disease: the cause for absent EAAT2, a glutamate transporter, in amyotrophic lateral sclerosis. Neuron. 1998;20(3):589–602.
55. Rothstein JD, et al. Knockout of glutamate transporters reveals a major role for astroglial transport in excitotoxicity and clearance of glutamate. Neuron. 1996;16(3):675–86.
56. Kim K, et al. Role of excitatory amino acid transporter-2 (EAAT2) and glutamate in neurodegeneration: opportunities for developing novel therapeutics. J Cell Physiol. 2011;226(10):2484–93.
57. Trotti D, et al. Amyotrophic lateral sclerosis-linked glutamate transporter mutant has impaired glutamate clearance capacity. J Biol Chem. 2001;276(1):576–82.
58. Peggion C, et al. SOD1 in ALS: taking stock in pathogenic mechanisms and the role of glial and muscle cells. Antioxidants (Basel). 2022;11(4):614.
59. Nardo G, et al. New insights on the mechanisms of disease course variability in ALS from mutant SOD1 mouse models. Brain Pathol. 2016;26(2):237–47.
60. Pramatarova A, et al. Neuron-specific expression of mutant superoxide dismutase 1 in transgenic mice does not lead to motor impairment. J Neurosci. 2001;21(10):3369–74.
61. Lino MM, Schneider C, Caroni P. Accumulation of SOD1 mutants in postnatal motoneurons does not cause motoneuron pathology or motoneuron disease. J Neurosci. 2002;22(12):4825–32.
62. Wang L, et al. Restricted expression of mutant SOD1 in spinal motor neurons and interneurons induces motor neuron pathology. Neurobiol Dis. 2008;29(3):400–8.
63. Clement AM, et al. Wild-type nonneuronal cells extend survival of SOD1 mutant motor neurons in ALS mice. Science. 2003;302(5642):113–7.
64. Howland DS, et al. Focal loss of the glutamate transporter EAAT2 in a transgenic rat model of SOD1 mutant-mediated amyotrophic lateral sclerosis (ALS). Proc Natl Acad Sci USA. 2002;99(3):1604–9.
65. Guo H, et al. Increased expression of the glial glutamate transporter EAAT2 modulates excitotoxicity and delays the onset but not the outcome of ALS in mice. Hum Mol Genet. 2003;12(19):2519–32.
66. Rosenblum LT, et al. Mutation of the caspase-3 cleavage site in the astroglial glutamate transporter EAAT2 delays disease progression and extends lifespan in the SOD1-G93A mouse model of ALS. Exp Neurol. 2017;292:145–53.
67. Yamanaka K, et al. Astrocytes as determinants of disease progression in inherited amyotrophic lateral sclerosis. Nat Neurosci. 2008;11(3):251–3.

68. Rochat C, et al. Astrocyte-targeting RNA interference against mutated superoxide dismutase 1 induces motoneuron plasticity and protects fast-fatigable motor units in a mouse model of amyotrophic lateral sclerosis. Glia. 2022;70(5):842–57.
69. Dirren E, et al. SOD1 silencing in motoneurons or glia rescues neuromuscular function in ALS mice. Ann Clin Transl Neurol. 2015;2(2):167–84.
70. Wang L, Gutmann DH, Roos RP. Astrocyte loss of mutant SOD1 delays ALS disease onset and progression in G85R transgenic mice. Hum Mol Genet. 2011;20(2):286–93.
71. Lepore AC, et al. Focal transplantation-based astrocyte replacement is neuroprotective in a model of motor neuron disease. Nat Neurosci. 2008;11(11):1294–301.
72. Kondo T, et al. Focal transplantation of human iPSC-derived glial-rich neural progenitors improves lifespan of ALS mice. Stem Cell Rep. 2014;3(2):242–9.
73. Papadeas ST, et al. Astrocytes carrying the superoxide dismutase 1 (SOD1G93A) mutation induce wild-type motor neuron degeneration in vivo. Proc Natl Acad Sci USA. 2011;108(43):17803–8.
74. Chen H, et al. Human-derived neural progenitors functionally replace astrocytes in adult mice. J Clin Invest. 2015;125(3):1033–42.
75. Milanese M, et al. Abnormal exocytotic release of glutamate in a mouse model of amyotrophic lateral sclerosis. J Neurochem. 2011;116(6):1028–42.
76. Bonifacino T, et al. Altered mechanisms underlying the abnormal glutamate release in amyotrophic lateral sclerosis at a pre-symptomatic stage of the disease. Neurobiol Dis. 2016;95:122–33.
77. Bonifacino T, et al. Enhanced function and overexpression of metabotropic glutamate receptors 1 and 5 in the spinal cord of the SOD1(G93A) mouse model of amyotrophic lateral sclerosis during disease progression. Int J Mol Sci. 2019;20(18):4552.
78. Bonifacino T, et al. In-vivo effects of knocking-down metabotropic glutamate receptor 5 in the SOD1(G93A) mouse model of amyotrophic lateral sclerosis. Neuropharmacology. 2017;123:433–45.
79. Aronica E, et al. Immunohistochemical localization of group I and II metabotropic glutamate receptors in control and amyotrophic lateral sclerosis human spinal cord: upregulation in reactive astrocytes. Neuroscience. 2001;105(2):509–20.
80. Vermeiren C, et al. Loss of metabotropic glutamate receptor-mediated regulation of glutamate transport in chemically activated astrocytes in a rat model of amyotrophic lateral sclerosis. J Neurochem. 2006;96(3):719–31.
81. Van Damme P, et al. GluR2 deficiency accelerates motor neuron degeneration in a mouse model of amyotrophic lateral sclerosis. J Neuropathol Exp Neurol. 2005;64(7):605–12.
82. Rossi D, et al. Focal degeneration of astrocytes in amyotrophic lateral sclerosis. Cell Death Differ. 2008;15(11):1691–700.
83. Torazza C, et al. Genetic downregulation of the metabotropic glutamate receptor type 5 dampens the reactive and neurotoxic phenotype of adult ALS astrocytes. Cells. 2023;12(15):1952.
84. Re DB, et al. Necroptosis drives motor neuron death in models of both sporadic and familial ALS. Neuron. 2014;81(5):1001–8.
85. Tripathi P, et al. Reactive astrocytes promote ALS-like degeneration and intracellular protein aggregation in human motor neurons by disrupting autophagy through TGF-beta1. Stem Cell Rep. 2017;9(2):667–80.
86. Fritz E, et al. Mutant SOD1-expressing astrocytes release toxic factors that trigger motoneuron death by inducing hyperexcitability. J Neurophysiol. 2013;109(11):2803–14.
87. Stotz SC, et al. Inorganic polyphosphate regulates neuronal excitability through modulation of voltage-gated channels. Mol Brain. 2014;7:42.
88. Bennett MV, et al. New roles for astrocytes: gap junction hemichannels have something to communicate. Trends Neurosci. 2003;26(11):610–7.
89. Orellana JA, Stehberg J. Hemichannels: new roles in astroglial function. Front Physiol. 2014;5:193.

90. Montero TD, Orellana JA. Hemichannels: new pathways for gliotransmitter release. Neuroscience. 2015;286:45–59.
91. Cui Y, et al. Extensive dysregulations of oligodendrocytic and astrocytic connexins are associated with disease progression in an amyotrophic lateral sclerosis mouse model. J Neuroinflammation. 2014;11:42.
92. Almad AA, et al. Connexin 43 in astrocytes contributes to motor neuron toxicity in amyotrophic lateral sclerosis. Glia. 2016;64(7):1154–69.
93. Almad AA, et al. Cx43 hemichannels contribute to astrocyte-mediated toxicity in sporadic and familial ALS. Proc Natl Acad Sci USA. 2022;119(13):e2107391119.
94. Casanovas A, et al. Strong P2X4 purinergic receptor-like immunoreactivity is selectively associated with degenerating neurons in transgenic rodent models of amyotrophic lateral sclerosis. J Comp Neurol. 2008;506(1):75–92.
95. Yiangou Y, et al. COX-2, CB2 and P2X7-immunoreactivities are increased in activated microglial cells/macrophages of multiple sclerosis and amyotrophic lateral sclerosis spinal cord. BMC Neurol. 2006;6:12.
96. Kawamata H, et al. Abnormal intracellular calcium signaling and SNARE-dependent exocytosis contributes to SOD1G93A astrocyte-mediated toxicity in amyotrophic lateral sclerosis. J Neurosci. 2014;34(6):2331–48.
97. Duan S, et al. P2X7 receptor-mediated release of excitatory amino acids from astrocytes. J Neurosci. 2003;23(4):1320–8.
98. Ruiz-Ruiz C, et al. Chronic administration of P2X7 receptor antagonist JNJ-47965567 delays disease onset and progression, and improves motor performance in ALS SOD1(G93A) female mice. Dis Model Mech. 2020;13(10):dmm045732.
99. Apolloni S, et al. Ablation of P2X7 receptor exacerbates gliosis and motoneuron death in the SOD1-G93A mouse model of amyotrophic lateral sclerosis. Hum Mol Genet. 2013;22(20):4102–16.
100. Kikuchi S, et al. Detection of N epsilon-(carboxymethyl)lysine (CML) and non-CML advanced glycation end-products in the anterior horn of amyotrophic lateral sclerosis spinal cord. Amyotroph Lateral Scler Other Motor Neuron Disord. 2002;3(2):63–8.
101. Wang Z, et al. Aberrations in oxidative stress markers in amyotrophic lateral sclerosis: a systematic review and meta-analysis. Oxidative Med Cell Longev. 2019;2019:1712323.
102. Chen Y, et al. The role of astrocytes in oxidative stress of central nervous system: a mixed blessing. Cell Prolif. 2020;53(3):e12781.
103. Vehvilainen P, Koistinaho J, Gundars G. Mechanisms of mutant SOD1 induced mitochondrial toxicity in amyotrophic lateral sclerosis. Front Cell Neurosci. 2014;8:126.
104. Damiano S, et al. Metabolism regulation and redox state: insight into the role of superoxide dismutase 1. Int J Mol Sci. 2020;21(18):6606.
105. Cassina P, et al. Mitochondrial dysfunction in SOD1G93A-bearing astrocytes promotes motor neuron degeneration: prevention by mitochondrial-targeted antioxidants. J Neurosci. 2008;28(16):4115–22.
106. Rojas F, et al. Reactive oxygen species trigger motoneuron death in non-cell-autonomous models of ALS through activation of c-Abl signaling. Front Cell Neurosci. 2015;9:203.
107. Pehar M, et al. Astrocytic production of nerve growth factor in motor neuron apoptosis: implications for amyotrophic lateral sclerosis. J Neurochem. 2004;89(2):464–73.
108. Lee J, Ryu H, Kowall NW. Differential regulation of neuronal and inducible nitric oxide synthase (NOS) in the spinal cord of mutant SOD1 (G93A) ALS mice. Biochem Biophys Res Commun. 2009;387(1):202–6.
109. Catania MV, et al. Increased expression of neuronal nitric oxide synthase spliced variants in reactive astrocytes of amyotrophic lateral sclerosis human spinal cord. J Neurosci. 2001;21(11):RC148.
110. Pehar M, et al. Mitochondrial superoxide production and nuclear factor erythroid 2-related factor 2 activation in p75 neurotrophin receptor-induced motor neuron apoptosis. J Neurosci. 2007;27(29):7777–85.

111. Ferraiuolo L, et al. Dysregulation of astrocyte-motoneuron cross-talk in mutant superoxide dismutase 1-related amyotrophic lateral sclerosis. Brain. 2011;134(Pt 9):2627–41.
112. Hensley K, et al. Primary glia expressing the G93A-SOD1 mutation present a neuroinflammatory phenotype and provide a cellular system for studies of glial inflammation. J Neuroinflammation. 2006;3:2.
113. Bandyopadhyay U, et al. RNA-Seq profiling of spinal cord motor neurons from a presymptomatic SOD1 ALS mouse. PLoS One. 2013;8(1):e53575.
114. Lobsiger CS, Boillee S, Cleveland DW. Toxicity from different SOD1 mutants dysregulates the complement system and the neuronal regenerative response in ALS motor neurons. Proc Natl Acad Sci USA. 2007;104(18):7319–26.
115. Madji Hounoum B, et al. Wildtype motoneurons, ALS-linked SOD1 mutation and glutamate profoundly modify astrocyte metabolism and lactate shuttling. Glia. 2017;65(4):592–605.
116. Wu A, Lee D, Xiong WC. Lactate metabolism, signaling, and function in brain development, synaptic plasticity, angiogenesis, and neurodegenerative diseases. Int J Mol Sci. 2023;24(17):13398.
117. Pellerin L, Magistretti PJ. Glutamate uptake into astrocytes stimulates aerobic glycolysis: a mechanism coupling neuronal activity to glucose utilization. Proc Natl Acad Sci USA. 1994;91(22):10625–9.
118. Gallardo G, et al. An alpha2-Na/K ATPase/alpha-adducin complex in astrocytes triggers non-cell autonomous neurodegeneration. Nat Neurosci. 2014;17(12):1710–9.
119. Miquel E, et al. Modulation of astrocytic mitochondrial function by dichloroacetate improves survival and motor performance in inherited amyotrophic lateral sclerosis. PLoS One. 2012;7(4):e34776.
120. Martinez-Palma L, et al. Mitochondrial modulation by dichloroacetate reduces toxicity of aberrant glial cells and gliosis in the SOD1G93A rat model of amyotrophic lateral sclerosis. Neurotherapeutics. 2019;16(1):203–15.
121. De Marchi F, et al. Emerging trends in the field of inflammation and proteinopathy in ALS/FTD spectrum disorder. Biomedicines. 2023;11(6):1599.
122. Kallstig E, McCabe BD, Schneider BL. The links between ALS and NF-kappaB. Int J Mol Sci. 2021;22(8):3875.
123. Tateishi T, et al. CSF chemokine alterations related to the clinical course of amyotrophic lateral sclerosis. J Neuroimmunol. 2010;222(1–2):76–81.
124. Poloni M, et al. Circulating levels of tumour necrosis factor-alpha and its soluble receptors are increased in the blood of patients with amyotrophic lateral sclerosis. Neurosci Lett. 2000;287(3):211–4.
125. Zaheer A, et al. A novel role of glia maturation factor: induction of granulocyte-macrophage colony-stimulating factor and pro-inflammatory cytokines. J Neurochem. 2007;101(2):364–76.
126. Yoshihara T, et al. Differential expression of inflammation- and apoptosis-related genes in spinal cords of a mutant SOD1 transgenic mouse model of familial amyotrophic lateral sclerosis. J Neurochem. 2002;80(1):158–67.
127. Brambilla L, et al. Disruption of the astrocytic TNFR1-GDNF axis accelerates motor neuron degeneration and disease progression in amyotrophic lateral sclerosis. Hum Mol Genet. 2016;25(14):3080–95.
128. Hensley K, et al. Temporal patterns of cytokine and apoptosis-related gene expression in spinal cords of the G93A-SOD1 mouse model of amyotrophic lateral sclerosis. J Neurochem. 2002;82(2):365–74.
129. Hensley K, et al. Message and protein-level elevation of tumor necrosis factor alpha (TNF alpha) and TNF alpha-modulating cytokines in spinal cords of the G93A-SOD1 mouse model for amyotrophic lateral sclerosis. Neurobiol Dis. 2003;14(1):74–80.
130. Veglianese P, et al. Activation of the p38MAPK cascade is associated with upregulation of TNF alpha receptors in the spinal motor neurons of mouse models of familial ALS. Mol Cell Neurosci. 2006;31(2):218–31.

131. Tortarolo M, et al. Lack of TNF-alpha receptor type 2 protects motor neurons in a cellular model of amyotrophic lateral sclerosis and in mutant SOD1 mice but does not affect disease progression. J Neurochem. 2015;135(1):109–24.
132. Perez C, et al. A nonsecretable cell surface mutant of tumor necrosis factor (TNF) kills by cell-to-cell contact. Cell. 1990;63(2):251–8.
133. Sipe KJ, et al. Expression of the 75 kDA TNF receptor and its role in contact-mediated neuronal cell death. Brain Res Mol Brain Res. 1998;62(2):111–21.
134. Dewil M, et al. Inhibition of p38 mitogen activated protein kinase activation and mutant SOD1(G93A)-induced motor neuron death. Neurobiol Dis. 2007;26(2):332–41.
135. Bezzi P, et al. Prostaglandins stimulate calcium-dependent glutamate release in astrocytes. Nature. 1998;391(6664):281–5.
136. Wang Z, et al. Reduced expression of glutamate transporter EAAT2 and impaired glutamate transport in human primary astrocytes exposed to HIV-1 or gp120. Virology. 2003;312(1):60–73.
137. Sitcheran R, et al. Positive and negative regulation of EAAT2 by NF-kappaB: a role for N-myc in TNFalpha-controlled repression. EMBO J. 2005;24(3):510–20.
138. Ye L, et al. IL-1beta and TNF-alpha induce neurotoxicity through glutamate production: a potential role for neuronal glutaminase. J Neurochem. 2013;125(6):897–908.
139. Leonoudakis D, Zhao P, Beattie EC. Rapid tumor necrosis factor alpha-induced exocytosis of glutamate receptor 2-lacking AMPA receptors to extrasynaptic plasma membrane potentiates excitotoxicity. J Neurosci. 2008;28(9):2119–30.
140. Johann S, et al. NLRP3 inflammasome is expressed by astrocytes in the SOD1 mouse model of ALS and in human sporadic ALS patients. Glia. 2015;63(12):2260–73.
141. Gugliandolo A, et al. NLRP3 inflammasome activation in a transgenic amyotrophic lateral sclerosis model. Inflammation. 2018;41(1):93–103.
142. Singh J, Habean ML, Panicker N. Inflammasome assembly in neurodegenerative diseases. Trends Neurosci. 2023;46(10):814–31.
143. O'Connor W Jr, et al. Cutting edge: CIAS1/cryopyrin/PYPAF1/NALP3/CATERPILLER 1.1 is an inducible inflammatory mediator with NF-kappa B suppressive properties. J Immunol. 2003;171(12):6329–33.
144. Chiarini A, et al. NLRP3 inflammasome's activation in acute and chronic brain diseases-an update on pathogenetic mechanisms and therapeutic perspectives with respect to other inflammasomes. Biomedicines. 2023;11(4):999.
145. Liu J, Gao L, Zang D. Elevated levels of IFN-gamma in CSF and serum of patients with amyotrophic lateral sclerosis. PLoS One. 2015;10(9):e0136937.
146. Aebischer J, et al. IFNgamma triggers a LIGHT-dependent selective death of motoneurons contributing to the non-cell-autonomous effects of mutant SOD1. Cell Death Differ. 2011;18(5):754–68.
147. Wang R, Yang B, Zhang D. Activation of interferon signaling pathways in spinal cord astrocytes from an ALS mouse model. Glia. 2011;59(6):946–58.
148. Calvo PM, et al. VEGF and neuronal survival. Neuroscientist. 2022;30:71–86. https://doi.org/10.1177/10738584221120803.
149. Oosthuyse B, et al. Deletion of the hypoxia-response element in the vascular endothelial growth factor promoter causes motor neuron degeneration. Nat Genet. 2001;28(2):131–8.
150. Lambrechts D, et al. VEGF is a modifier of amyotrophic lateral sclerosis in mice and humans and protects motoneurons against ischemic death. Nat Genet. 2003;34(4):383–94.
151. Wang Y, et al. Vascular endothelial growth factor overexpression delays neurodegeneration and prolongs survival in amyotrophic lateral sclerosis mice. J Neurosci. 2007;27(2):304–7.
152. Argaw AT, et al. Astrocyte-derived VEGF-A drives blood-brain barrier disruption in CNS inflammatory disease. J Clin Invest. 2012;122(7):2454–68.
153. Klein SM, et al. GDNF delivery using human neural progenitor cells in a rat model of ALS. Hum Gene Ther. 2005;16(4):509–21.

154. Suzuki M, et al. Direct muscle delivery of GDNF with human mesenchymal stem cells improves motor neuron survival and function in a rat model of familial ALS. Mol Ther. 2008;16(12):2002–10.
155. Izrael M, et al. Safety and efficacy of human embryonic stem cell-derived astrocytes following intrathecal transplantation in SOD1(G93A) and NSG animal models. Stem Cell Res Ther. 2018;9(1):152.
156. Zhong Z, et al. ALS-causing SOD1 mutants generate vascular changes prior to motor neuron degeneration. Nat Neurosci. 2008;11(4):420–2.
157. Winkler EA, et al. Blood-spinal cord barrier disruption contributes to early motor-neuron degeneration in ALS-model mice. Proc Natl Acad Sci USA. 2014;111(11):E1035–42.
158. Gadea A, Schinelli S, Gallo V. Endothelin-1 regulates astrocyte proliferation and reactive gliosis via a JNK/c-Jun signaling pathway. J Neurosci. 2008;28(10):2394–408.
159. Leonova J, et al. Endothelin-1 decreases glutamate uptake in primary cultured rat astrocytes. Am J Physiol Cell Physiol. 2001;281(5):C1495–503.
160. Ranno E, et al. Endothelin-1 is over-expressed in amyotrophic lateral sclerosis and induces motor neuron cell death. Neurobiol Dis. 2014;65:160–71.
161. Zhang Y, et al. Endothelin-1, over-expressed in SOD1(G93A) mice, aggravates injury of NSC34-hSOD1G93A cells through complicated molecular mechanism revealed by quantitative proteomics analysis. Front Cell Neurosci. 2022;16:1069617.
162. D'Antoni S, et al. Endothelin-1 induces degeneration of cultured motor neurons through a mechanism mediated by nitric oxide and PI3K/Akt pathway. Neurotox Res. 2017;32(1):58–70.
163. Sheng J, Xu Z. Three decades of research on angiogenin: a review and perspective. Acta Biochim Biophys Sin Shanghai. 2016;48(5):399–410.
164. Hoang TT, et al. Angiogenin activates the astrocytic Nrf2/antioxidant-response element pathway and thereby protects murine neurons from oxidative stress. J Biol Chem. 2019;294(41):15095–103.
165. Vargas MR, et al. Nrf2 activation in astrocytes protects against neurodegeneration in mouse models of familial amyotrophic lateral sclerosis. J Neurosci. 2008;28(50):13574–81.
166. Kishikawa H, Wu D, Hu GF. Targeting angiogenin in therapy of amyotropic lateral sclerosis. Expert Opin Ther Targets. 2008;12(10):1229–42.
167. Rangroo Thrane V, et al. Paravascular microcirculation facilitates rapid lipid transport and astrocyte signaling in the brain. Sci Rep. 2013;3:2582.
168. Iliff JJ, et al. Cerebral arterial pulsation drives paravascular CSF-interstitial fluid exchange in the murine brain. J Neurosci. 2013;33(46):18190–9.
169. Iliff JJ, et al. A paravascular pathway facilitates CSF flow through the brain parenchyma and the clearance of interstitial solutes, including amyloid beta. Sci Transl Med. 2012;4(147):147ra111.
170. Bataveljic D, et al. Changes in the astrocytic aquaporin-4 and inwardly rectifying potassium channel expression in the brain of the amyotrophic lateral sclerosis SOD1(G93A) rat model. Glia. 2012;60(12):1991–2003.
171. Papadopoulos MC, Verkman AS. Aquaporin water channels in the nervous system. Nat Rev Neurosci. 2013;14(4):265–77.
172. Grad LI, et al. Intercellular propagated misfolding of wild-type cu/Zn superoxide dismutase occurs via exosome-dependent and -independent mechanisms. Proc Natl Acad Sci USA. 2014;111(9):3620–5.
173. Basso M, et al. Mutant copper-zinc superoxide dismutase (SOD1) induces protein secretion pathway alterations and exosome release in astrocytes: implications for disease spreading and motor neuron pathology in amyotrophic lateral sclerosis. J Biol Chem. 2013;288(22):15699–711.
174. Silverman JM, et al. CNS-derived extracellular vesicles from superoxide dismutase 1 (SOD1) (G93A) ALS mice originate from astrocytes and neurons and carry misfolded SOD1. J Biol Chem. 2019;294(10):3744–59.
175. Arnold FJ, et al. Intercellular transmission of pathogenic proteins in ALS: exploring the pathogenic wave. Neurobiol Dis. 2023;184:106218.

176. Koval ED, et al. Method for widespread microRNA-155 inhibition prolongs survival in ALS-model mice. Hum Mol Genet. 2013;22(20):4127–35.
177. Butovsky O, et al. Targeting miR-155 restores abnormal microglia and attenuates disease in SOD1 mice. Ann Neurol. 2015;77(1):75–99.
178. Marton S, et al. SOD1(G93A) astrocyte-derived extracellular vesicles induce motor neuron death by a miRNA-155-5p-mediated mechanism. ASN Neuro. 2023;15:17590914231197527.
179. Hoye ML, et al. MicroRNA profiling reveals marker of motor neuron disease in ALS models. J Neurosci. 2017;37(22):5574–86.
180. Hoye ML, et al. Motor neuron-derived microRNAs cause astrocyte dysfunction in amyotrophic lateral sclerosis. Brain. 2018;141(9):2561–75.
181. Phatnani HP, et al. Intricate interplay between astrocytes and motor neurons in ALS. Proc Natl Acad Sci USA. 2013;110(8):E756–65.
182. Derynck R, Zhang YE. Smad-dependent and Smad-independent pathways in TGF-beta family signalling. Nature. 2003;425(6958):577–84.
183. Endo F, et al. Astrocyte-derived TGF-beta1 accelerates disease progression in ALS mice by interfering with the neuroprotective functions of microglia and T cells. Cell Rep. 2015;11(4):592–604.
184. Namboori SC, et al. Single-cell transcriptomics identifies master regulators of neurodegeneration in SOD1 ALS iPSC-derived motor neurons. Stem Cell Rep. 2021;16(12):3020–35.
185. Rutherford NJ, et al. Novel mutations in TARDBP (TDP-43) in patients with familial amyotrophic lateral sclerosis. PLoS Genet. 2008;4(9):e1000193.
186. Sreedharan J, et al. TDP-43 mutations in familial and sporadic amyotrophic lateral sclerosis. Science. 2008;319(5870):1668–72.
187. Kabashi E, et al. TARDBP mutations in individuals with sporadic and familial amyotrophic lateral sclerosis. Nat Genet. 2008;40(5):572–4.
188. Vance C, et al. Mutations in FUS, an RNA processing protein, cause familial amyotrophic lateral sclerosis type 6. Science. 2009;323(5918):1208–11.
189. Kwiatkowski TJ Jr, et al. Mutations in the FUS/TLS gene on chromosome 16 cause familial amyotrophic lateral sclerosis. Science. 2009;323(5918):1205–8.
190. Rademakers R, et al. Fus gene mutations in familial and sporadic amyotrophic lateral sclerosis. Muscle Nerve. 2010;42(2):170–6.
191. Suzuki N, et al. FUS/TLS-immunoreactive neuronal and glial cell inclusions increase with disease duration in familial amyotrophic lateral sclerosis with an R521C FUS/TLS mutation. J Neuropathol Exp Neurol. 2012;71(9):779–88.
192. Suk TR, Rousseaux MWC. The role of TDP-43 mislocalization in amyotrophic lateral sclerosis. Mol Neurodegener. 2020;15(1):45.
193. Fujimori K, et al. Modeling sporadic ALS in iPSC-derived motor neurons identifies a potential therapeutic agent. Nat Med. 2018;24(10):1579–89.
194. Ikenaka K, et al. Characteristic features of FUS inclusions in spinal motor neurons of sporadic amyotrophic lateral sclerosis. J Neuropathol Exp Neurol. 2020;79(4):370–7.
195. Tyzack GE, et al. Widespread FUS mislocalization is a molecular hallmark of amyotrophic lateral sclerosis. Brain. 2019;142(9):2572–80.
196. Deng H, Gao K, Jankovic J. The role of FUS gene variants in neurodegenerative diseases. Nat Rev Neurol. 2014;10(6):337–48.
197. Wood A, et al. Molecular mechanisms underlying TDP-43 pathology in cellular and animal models of ALS and FTLD. Int J Mol Sci. 2021;22(9):4705.
198. Weskamp K, Barmada SJ. TDP43 and RNA instability in amyotrophic lateral sclerosis. Brain Res. 2018;1693(Pt A):67–74.
199. Gao J, et al. Pathomechanisms of TDP-43 in neurodegeneration. J Neurochem. 2018;146:7–20.
200. Aksoy YA, et al. "STRESSED OUT": the role of FUS and TDP-43 in amyotrophic lateral sclerosis. Int J Biochem Cell Biol. 2020;126:105821.
201. Ling SC, Polymenidou M, Cleveland DW. Converging mechanisms in ALS and FTD: disrupted RNA and protein homeostasis. Neuron. 2013;79(3):416–38.

202. Mackenzie IR, Rademakers R, Neumann M. TDP-43 and FUS in amyotrophic lateral sclerosis and frontotemporal dementia. Lancet Neurol. 2010;9(10):995–1007.
203. Arai T, et al. Phosphorylated and cleaved TDP-43 in ALS, FTLD and other neurodegenerative disorders and in cellular models of TDP-43 proteinopathy. Neuropathology. 2010;30(2):170–81.
204. Zhang H, et al. TDP-43-immunoreactive neuronal and glial inclusions in the neostriatum in amyotrophic lateral sclerosis with and without dementia. Acta Neuropathol. 2008;115(1):115–22.
205. Mackenzie IR, et al. Pathological TDP-43 distinguishes sporadic amyotrophic lateral sclerosis from amyotrophic lateral sclerosis with SOD1 mutations. Ann Neurol. 2007;61(5):427–34.
206. Tan CF, et al. TDP-43 immunoreactivity in neuronal inclusions in familial amyotrophic lateral sclerosis with or without SOD1 gene mutation. Acta Neuropathol. 2007;113(5):535–42.
207. Yang C, et al. Partial loss of TDP-43 function causes phenotypes of amyotrophic lateral sclerosis. Proc Natl Acad Sci USA. 2014;111(12):E1121–9.
208. Wu LS, Cheng WC, Shen CK. Targeted depletion of TDP-43 expression in the spinal cord motor neurons leads to the development of amyotrophic lateral sclerosis-like phenotypes in mice. J Biol Chem. 2012;287(33):27335–44.
209. Iguchi Y, et al. Loss of TDP-43 causes age-dependent progressive motor neuron degeneration. Brain. 2013;136(Pt 5):1371–82.
210. Peng AYT, et al. Loss of TDP-43 in astrocytes leads to motor deficits by triggering A1-like reactive phenotype and triglial dysfunction. Proc Natl Acad Sci USA. 2020;117(46):29101–12.
211. Haidet-Phillips AM, et al. Altered astrocytic expression of TDP-43 does not influence motor neuron survival. Exp Neurol. 2013;250:250–9.
212. LaRocca TJ, et al. TDP-43 knockdown causes innate immune activation via protein kinase R in astrocytes. Neurobiol Dis. 2019;132:104514.
213. Wils H, et al. TDP-43 transgenic mice develop spastic paralysis and neuronal inclusions characteristic of ALS and frontotemporal lobar degeneration. Proc Natl Acad Sci USA. 2010;107(8):3858–63.
214. Wegorzewska I, et al. TDP-43 mutant transgenic mice develop features of ALS and frontotemporal lobar degeneration. Proc Natl Acad Sci USA. 2009;106(44):18809–14.
215. Stallings NR, et al. Progressive motor weakness in transgenic mice expressing human TDP-43. Neurobiol Dis. 2010;40(2):404–14.
216. Xu YF, et al. Wild-type human TDP-43 expression causes TDP-43 phosphorylation, mitochondrial aggregation, motor deficits, and early mortality in transgenic mice. J Neurosci. 2010;30(32):10851–9.
217. Yang C, et al. Low-level overexpression of wild type TDP-43 causes late-onset, progressive neurodegeneration and paralysis in mice. PLoS One. 2022;17(2):e0255710.
218. Hebron M, et al. Parkin reverses TDP-43-induced cell death and failure of amino acid homeostasis. J Neurochem. 2014;129(2):350–61.
219. Swarup V, et al. Deregulation of TDP-43 in amyotrophic lateral sclerosis triggers nuclear factor kappaB-mediated pathogenic pathways. J Exp Med. 2011;208(12):2429–47.
220. Heyburn L, et al. Tyrosine kinase inhibition reverses TDP-43 effects on synaptic protein expression, astrocytic function and amino acid dis-homeostasis. J Neurochem. 2016;139(4):610–23.
221. Ke YD, et al. Short-term suppression of A315T mutant human TDP-43 expression improves functional deficits in a novel inducible transgenic mouse model of FTLD-TDP and ALS. Acta Neuropathol. 2015;130(5):661–78.
222. Iguchi Y, et al. Exosome secretion is a key pathway for clearance of pathological TDP-43. Brain. 2016;139(Pt 12):3187–201.
223. Riemenschneider H, et al. Gel-like inclusions of C-terminal fragments of TDP-43 sequester stalled proteasomes in neurons. EMBO Rep. 2022;23(6):e53890.

224. Velebit J, et al. Astrocytes with TDP-43 inclusions exhibit reduced noradrenergic cAMP and ca(2+) signaling and dysregulated cell metabolism. Sci Rep. 2020;10(1):6003.
225. Gordon D, et al. Single-copy expression of an amyotrophic lateral sclerosis-linked TDP-43 mutation (M337V) in BAC transgenic mice leads to altered stress granule dynamics and progressive motor dysfunction. Neurobiol Dis. 2019;121:148–62.
226. Feneberg E, et al. An ALS-linked mutation in TDP-43 disrupts normal protein interactions in the motor neuron response to oxidative stress. Neurobiol Dis. 2020;144:105050.
227. Lee S, et al. Lipocalin-2 is an autocrine mediator of reactive astrocytosis. J Neurosci. 2009;29(1):234–49.
228. Serio A, et al. Astrocyte pathology and the absence of non-cell autonomy in an induced pluripotent stem cell model of TDP-43 proteinopathy. Proc Natl Acad Sci USA. 2013;110(12):4697–702.
229. Arnold ES, et al. ALS-linked TDP-43 mutations produce aberrant RNA splicing and adult-onset motor neuron disease without aggregation or loss of nuclear TDP-43. Proc Natl Acad Sci USA. 2013;110(8):E736–45.
230. Ditsworth D, et al. Mutant TDP-43 within motor neurons drives disease onset but not progression in amyotrophic lateral sclerosis. Acta Neuropathol. 2017;133(6):907–22.
231. Moujalled D, et al. TDP-43 mutations causing amyotrophic lateral sclerosis are associated with altered expression of RNA-binding protein hnRNP K and affect the Nrf2 antioxidant pathway. Hum Mol Genet. 2017;26(9):1732–46.
232. Sharma A, et al. ALS-associated mutant FUS induces selective motor neuron degeneration through toxic gain of function. Nat Commun. 2016;7:10465.
233. Fujioka Y, et al. FUS-regulated region- and cell-type-specific transcriptome is associated with cell selectivity in ALS/FTLD. Sci Rep. 2013;3:2388.
234. Endo K, et al. Silencing of FUS in the common marmoset (Callithrix jacchus) brain via stereotaxic injection of an adeno-associated virus encoding shRNA. Neurosci Res. 2018;130:56–64.
235. Ajmone-Cat MA, et al. Increased FUS levels in astrocytes leads to astrocyte and microglia activation and neuronal death. Sci Rep. 2019;9(1):4572.
236. Uranishi H, et al. Involvement of the pro-oncoprotein TLS (translocated in liposarcoma) in nuclear factor-kappa B p65-mediated transcription as a coactivator. J Biol Chem. 2001;276(16):13395–401.
237. Scekic-Zahirovic J, et al. Motor neuron intrinsic and extrinsic mechanisms contribute to the pathogenesis of FUS-associated amyotrophic lateral sclerosis. Acta Neuropathol. 2017;133(6):887–906.
238. Qosa H, et al. Astrocytes drive upregulation of the multidrug resistance transporter ABCB1 (P-Glycoprotein) in endothelial cells of the blood-brain barrier in mutant superoxide dismutase 1-linked amyotrophic lateral sclerosis. Glia. 2016;64(8):1298–313.
239. Pelaez MC, et al. Neuronal dysfunction caused by FUSR521G promotes ALS-associated phenotypes that are attenuated by NF-kappaB inhibition. Acta Neuropathol Commun. 2023;11(1):182.
240. DeJesus-Hernandez M, et al. Expanded GGGGCC hexanucleotide repeat in noncoding region of C9ORF72 causes chromosome 9p-linked FTD and ALS. Neuron. 2011;72(2):245–56.
241. Renton AE, et al. A hexanucleotide repeat expansion in C9ORF72 is the cause of chromosome 9p21-linked ALS-FTD. Neuron. 2011;72(2):257–68.
242. Gijselinck I, et al. A C9orf72 promoter repeat expansion in a Flanders-Belgian cohort with disorders of the frontotemporal lobar degeneration-amyotrophic lateral sclerosis spectrum: a gene identification study. Lancet Neurol. 2012;11(1):54–65.
243. Nguyen HP, Van Broeckhoven C, van der Zee J. ALS genes in the genomic era and their implications for FTD. Trends Genet. 2018;34(6):404–23.
244. Gendron TF, Petrucelli L. Disease mechanisms of C9ORF72 repeat expansions. Cold Spring Harb Perspect Med. 2018;8(4):a024224.
245. Batra R, Lee CW. Mouse models of C9orf72 hexanucleotide repeat expansion in amyotrophic lateral sclerosis/frontotemporal dementia. Front Cell Neurosci. 2017;11:196.

246. Lagier-Tourenne C, et al. Targeted degradation of sense and antisense C9orf72 RNA foci as therapy for ALS and frontotemporal degeneration. Proc Natl Acad Sci USA. 2013;110(47):E4530–9.
247. Gendron TF, et al. Antisense transcripts of the expanded C9ORF72 hexanucleotide repeat form nuclear RNA foci and undergo repeat-associated non-ATG translation in c9FTD/ALS. Acta Neuropathol. 2013;126(6):829–44.
248. DeJesus-Hernandez M, et al. In-depth clinico-pathological examination of RNA foci in a large cohort of C9ORF72 expansion carriers. Acta Neuropathol. 2017;134(2):255–69.
249. Ash PE, et al. Unconventional translation of C9ORF72 GGGGCC expansion generates insoluble polypeptides specific to c9FTD/ALS. Neuron. 2013;77(4):639–46.
250. Mackenzie IR, et al. Dipeptide repeat protein pathology in C9ORF72 mutation cases: clinico-pathological correlations. Acta Neuropathol. 2013;126(6):859–79.
251. Westergard T, et al. Cell-to-cell transmission of dipeptide repeat proteins linked to C9orf72-ALS/FTD. Cell Rep. 2016;17(3):645–52.
252. Cooper-Knock J, et al. Clinico-pathological features in amyotrophic lateral sclerosis with expansions in C9ORF72. Brain. 2012;135(Pt 3):751–64.
253. Schipper LJ, et al. Prevalence of brain and spinal cord inclusions, including dipeptide repeat proteins, in patients with the C9ORF72 hexanucleotide repeat expansion: a systematic neuropathological review. Neuropathol Appl Neurobiol. 2016;42(6):547–60.
254. O'Rourke JG, et al. C9orf72 is required for proper macrophage and microglial function in mice. Science. 2016;351(6279):1324–9.
255. Schludi MH, et al. Spinal poly-GA inclusions in a C9orf72 mouse model trigger motor deficits and inflammation without neuron loss. Acta Neuropathol. 2017;134(2):241–54.
256. Zhang YJ, et al. Poly(GR) impairs protein translation and stress granule dynamics in C9orf72-associated frontotemporal dementia and amyotrophic lateral sclerosis. Nat Med. 2018;24(8):1136–42.
257. Madill M, et al. Amyotrophic lateral sclerosis patient iPSC-derived astrocytes impair autophagy via non-cell autonomous mechanisms. Mol Brain. 2017;10(1):22.
258. Birger A, et al. Human iPSC-derived astrocytes from ALS patients with mutated C9ORF72 show increased oxidative stress and neurotoxicity. EBioMedicine. 2019;50:274–89.
259. Wong CO, Venkatachalam K. Motor neurons from ALS patients with mutations in C9ORF72 and SOD1 exhibit distinct transcriptional landscapes. Hum Mol Genet. 2019;28(16):2799–810.
260. Nelson AT, Trotti D. Altered bioenergetics and metabolic homeostasis in amyotrophic lateral sclerosis. Neurotherapeutics. 2022;19(4):1102–18.
261. Fomin V, et al. The C9ORF72 gene, implicated in amyotrophic lateral sclerosis and frontotemporal dementia, encodes a protein that functions in control of endothelin and glutamate signaling. Mol Cell Biol. 2018;38(22):e00155.
262. Allen SP, et al. C9orf72 expansion within astrocytes reduces metabolic flexibility in amyotrophic lateral sclerosis. Brain. 2019;142(12):3771–90.
263. Onesto E, et al. Gene-specific mitochondria dysfunctions in human TARDBP and C9ORF72 fibroblasts. Acta Neuropathol Commun. 2016;4(1):47.
264. Li J, et al. Divergent single cell transcriptome and epigenome alterations in ALS and FTD patients with C9orf72 mutation. Nat Commun. 2023;14(1):5714.
265. Amin A, et al. Amyotrophic lateral sclerosis and autophagy: dysfunction and therapeutic targeting. Cells. 2020;9(11):2413.
266. Renton AE, Chio A, Traynor BJ. State of play in amyotrophic lateral sclerosis genetics. Nat Neurosci. 2014;17(1):17–23.
267. Freischmidt A, et al. Haploinsufficiency of TBK1 causes familial ALS and fronto-temporal dementia. Nat Neurosci. 2015;18(5):631–6.
268. Melentijevic I, et al. C. Elegans neurons jettison protein aggregates and mitochondria under neurotoxic stress. Nature. 2017;542(7641):367–71.
269. Davis CH, et al. Transcellular degradation of axonal mitochondria. Proc Natl Acad Sci USA. 2014;111(26):9633–8.

270. Di Malta C, et al. Astrocyte dysfunction triggers neurodegeneration in a lysosomal storage disorder. Proc Natl Acad Sci USA. 2012;109(35):E2334–42.
271. Ceballos-Diaz C, et al. Viral expression of ALS-linked ubiquilin-2 mutants causes inclusion pathology and behavioral deficits in mice. Mol Neurodegener. 2015;10:25.
272. Oakes JA, Davies MC, Collins MO. TBK1: a new player in ALS linking autophagy and neuroinflammation. Mol Brain. 2017;10(1):5.
273. Brenner D, et al. Heterozygous Tbk1 loss has opposing effects in early and late stages of ALS in mice. J Exp Med. 2019;216(2):267–78.
274. Duan W, et al. Deletion of Tbk1 disrupts autophagy and reproduces behavioral and locomotor symptoms of FTD-ALS in mice. Aging (Albany NY). 2019;11(8):2457–76.
275. Gerbino V, et al. The loss of TBK1 kinase activity in motor neurons or in all cell types differentially impacts ALS disease progression in SOD1 mice. Neuron. 2020;106(5):789–805.e5.
276. Bruno C, et al. Haploinsufficiency of TANK-binding kinase 1 prepones age-associated neuroinflammatory changes without causing motor neuron degeneration in aged mice. Brain Commun. 2020;2(2):fcaa133.
277. Scarian E, et al. The role of VCP mutations in the Spectrum of amyotrophic lateral sclerosis-frontotemporal dementia. Front Neurol. 2022;13:841394.
278. Asai T, et al. VCP (p97) regulates NFkappaB signaling pathway, which is important for metastasis of osteosarcoma cell line. Jpn J Cancer Res. 2002;93(3):296–304.
279. Dai RM, et al. Involvement of valosin-containing protein, an ATPase co-purified with IkappaBalpha and 26 S proteasome, in ubiquitin-proteasome-mediated degradation of IkappaBalpha. J Biol Chem. 1998;273(6):3562–73.
280. Badadani M, et al. VCP associated inclusion body myopathy and paget disease of bone knock-in mouse model exhibits tissue pathology typical of human disease. PLoS One. 2010;5(10):e13183.
281. Custer SK, et al. Transgenic mice expressing mutant forms VCP/p97 recapitulate the full spectrum of IBMPFD including degeneration in muscle, brain and bone. Hum Mol Genet. 2010;19(9):1741–55.
282. Hall CE, et al. Progressive motor neuron pathology and the role of astrocytes in a human stem cell model of VCP-related ALS. Cell Rep. 2017;19(9):1739–49.
283. Smethurst P, et al. Distinct responses of neurons and astrocytes to TDP-43 proteinopathy in amyotrophic lateral sclerosis. Brain. 2020;143(2):430–40.
284. Olesen MN, et al. Inflammatory profiles relate to survival in subtypes of amyotrophic lateral sclerosis. Neurol Neuroimmunol Neuroinflamm. 2020;7(3):e697.
285. Mishra PS, et al. Astroglia acquires a toxic neuroinflammatory role in response to the cerebrospinal fluid from amyotrophic lateral sclerosis patients. J Neuroinflammation. 2016;13(1):212.
286. Baloh RH, et al. Transplantation of human neural progenitor cells secreting GDNF into the spinal cord of patients with ALS: a phase 1/2a trial. Nat Med. 2022;28(9):1813–22.
287. Gotkine M, et al. Safety and efficacy of first-in-man intrathecal injection of human astrocytes (AstroRx(R)) in ALS patients: phase I/IIa clinical trial results. J Transl Med. 2023;21(1):122.
288. Ziff OJ, et al. Meta-analysis of human and mouse ALS astrocytes reveals multi-omic signatures of inflammatory reactive states. Genome Res. 2022;32(1):71–84.
289. Geloso MC, et al. The dual role of microglia in ALS: mechanisms and therapeutic approaches. Front Aging Neurosci. 2017;9:242.
290. Guttenplan KA, et al. Knockout of reactive astrocyte activating factors slows disease progression in an ALS mouse model. Nat Commun. 2020;11(1):3753.
291. Liddelow SA, et al. Neurotoxic reactive astrocytes are induced by activated microglia. Nature. 2017;541(7638):481–7.
292. Ouali Alami N, et al. Multiplexed chemogenetics in astrocytes and motoneurons restore blood-spinal cord barrier in ALS. Life Sci Alliance. 2020;3(11):e201900571.
293. Shefner JM, et al. Skeletal muscle in amyotrophic lateral sclerosis. Brain. 2023;146(11):4425–36.

The Role of Astrocytes in Parkinson's Disease

Astrocytes in Parkinson's Disease

Roger Garcia, Sara Zarate, and Rahul Srinivasan

Abstract Parkinson's disease (PD) is a neurodegenerative disorder with a complex and multifactorial pathogenesis. This chapter delves into the critical role of astrocytes in PD. Once viewed as supporting cells in the central nervous system, astrocytes have emerged as key players in both maintaining neuronal health and contributing to neurodegeneration in PD. Their functions play a dual role in the progression of PD, ranging from protective functions like secretion of neurotrophic factors and clearance of α-synuclein to detrimental functions like promotion of neuroinflammation. This chapter is structured into three primary sections: the morphological and functional organization of astrocytes, astrocytic calcium signaling, and the role of astrocyte heterogeneity in PD. We provide a detailed exploration of astrocytic organelles, bidirectional astrocyte-neuron interactions, and the impact of astrocytic secretions such as antioxidant molecules and neurotrophic factors. Furthermore, we discuss the influence of astrocytes on non-neuronal cells, including interactions with microglia and the blood–brain barrier (BBB). By examining the multifaceted roles of astrocytes, in this chapter, we aim to bridge basic astrocyte biology with the clinical complexities of PD, offering insights into novel therapeutic strategies. The inclusion of astrocyte biology in our broader research approach will aid in the development of more effective treatment strategies for PD.

Keywords Parkinson's disease · Neurodegeneration · α-synuclein · Neuroinflammation · Blood–brain barrier · Calcium signaling · Heterogeneity

---

R. Garcia · S. Zarate
Department of Neuroscience & Experimental Therapeutics, Texas A&M University College of Medicine, Bryan, TX, USA

R. Srinivasan (✉)
Department of Neuroscience & Experimental Therapeutics, Texas A&M University College of Medicine, Bryan, TX, USA

Texas A&M Institute for Neuroscience (TAMIN), Texas A&M University, College Station, TX, USA
e-mail: rahul@tamu.edu

E. Blanco-Suarez, I. Farhy-Tselnicker (eds.), *Astrocyte-Neuron Interactions in Health and Disease*, Advances in Neurobiology 39,
https://doi.org/10.1007/978-3-031-64839-7_13

Abbreviations

6-OHDA	6-hydroxydopamine
α-syn	α-synuclein
A1	Adenosine A1 receptor
AQP4	Aquaporin 4
ARE	Antioxidant response element
BBB	Blood–brain barrier
bFGF	Basic fibroblast growth factor
C1q	Complement component 1q
Ca^{2+}	Calcium ions
CDNF	Cerebral dopamine neurotrophic factor
CNS	Central nervous system
Cx43	Connexin 43
CXCL1	Chemokine ligand 1
D1	Dopamine D1 receptor
D2	Dopamine D2 receptor
DLS	Dorsolateral striatum
EAAT	Excitatory amino acid transporter
FA	Fatty Acid
GDNF	Glial cell line-derived neurotrophic factor
GFAP	Glial fibrillary acidic protein
GSH	Glutathione
HLA	Human leukocyte antigen
IL-1α	Interleukin-1 alpha
IL-6	Interleukin-6
IL-8	Interleukin-8
iPSC	Induced pluripotent stem cell
Kir4.1	Inward rectifier potassium channel 4.1
LAMP1	Lysosomal associated membrane protein 1
LPS	Lipopolysaccharide
LRRK2	Leucine-rich repeat kinase 2
MANF	Mesencephalic astrocyte derived neurotrophic factor
MDS-UPDRS	Movement Disorder Society-Sponsored Unified PD Rating Scale
MHC	Major histocompatibility complex
MPTP	1-methyl-4-phenyl-1,2,3,6-tetrahydropyridine
NF-κB	Nuclear factor kappa B
NRF2	Nuclear factor erythroid 2-related factor 2
PD	Parkinson's disease
RAGE	Receptor for advanced glycation end-products
RIPK	Receptor interacting protein kinase
SCD1	Stearoyl-CoA 9-desaturase
SN	Substantia nigra
SNc	Substantia nigra pars compacta

SNr	Substantia nigra pars reticulata
TNF-α	Tumor necrosis factor-alpha
TNT	Tunneling nanotubules
VGCC	Voltage-gated calcium channel
VGKC	Voltage-gated potassium channel

1 Introduction

Parkinson's disease (PD) is a complex, progressive neurodegenerative disorder that primarily affects the motor system, leading to symptoms such as tremors, rigidity, bradykinesia, and postural instability. PD is the second most common neurodegenerative disorder worldwide, affecting 2–3% of the global population aged 65 and older [1]. Its etiology is multifaceted, involving genetic, environmental, and age-related factors.

Despite significant advancements in our understanding of PD, discovering a cure remains a formidable challenge. This difficulty is largely attributed to the disease's heterogeneous pathophysiology, much of which is still not fully understood, compounded by the challenge of developing models that accurately replicate the human condition. Moreover, recent treatments primarily aim to alleviate PD symptoms rather than tackle the root cause of the disease [2]. Therefore, the development of neuroprotective therapies capable of halting or reversing the progression of PD is a major focus of current research.

Traditional approaches to developing PD therapies have predominantly focused on neurons, overlooking the complex biology and interactions between glial cells and neurons within the central nervous system (CNS). This neuron-centric view has limited our understanding and treatment of PD. In order to improve diagnostics and therapeutics for PD, the complex etiology of this disease requires an inclusion of important roles for non-neuronal cells in the CNS. Consideration of the multitude of inter- and intracellular interactions within the brain could provide new insights and pathways for treatment. This includes acknowledging the significant impact of astrocytes, microglia, and other glial cells, which maintain crucial roles in neuronal health and disease progression. By expanding the focus beyond neurons to include these other vital cell types, novel therapeutic targets and strategies that more effectively address the underlying mechanisms of PD can be developed.

Astrocytes, the star-shaped glial cells in the brain, have recently come into focus for their role in PD. Traditionally viewed as supporting players in the CNS, astrocytes are now recognized as active participants in both the health and disease of neurons. In the context of PD, astrocytes contribute to maintaining neuronal health through various mechanisms, including neurotransmitter recycling, metabolic support, and regulation of blood–brain barrier (BBB) integrity. However, astrocytes represent a double-edged sword in PD pathogenesis. On one hand, they provide neuroprotective functions, such as the clearance of α-synuclein, a protein whose aggregates are a hallmark of PD. On the other hand, dysfunctional astrocytes can exacerbate disease progression through mechanisms like neuroinflammation and disrupted calcium signaling [3]. Taken together, this dual nature makes them an intriguing target for therapeutic interventions.

This chapter aims to explore the intricate role of astrocytes in PD. It is organized into three main sections. The first section delves into the morphological and functional organization of astrocytes, as it relates to PD. It examines the unique features of astrocytes and discusses how these structural characteristics influence their function in the healthy and diseased brain. Further, the role of astrocytic organelles, such as mitochondria and lysosomes, in PD pathogenesis is explored, providing insights into how the cellular structure impacts the disease. The second part of this chapter focuses on astrocytic calcium signaling and mechanisms by which dysfunctional signaling can promote neuron dysfunction and microglial reactivity and alter the BBB. The third and final section of this chapter focuses on the relevance of astrocyte heterogeneity in PD. Astrocytes are not a uniform cell population; rather, they exhibit significant heterogeneity in both form and function that can determine their relationship with dopaminergic neurons and other non-neuronal cells within the CNS. Additionally, this section discusses the broad spectrum of astrocyte reactivity and the role of each stage in PD.

By the end of this chapter, we hope that readers will gain a comprehensive understanding of the multifaceted roles of astrocytes in PD, as well as how these cells contribute to both the maintenance of neuronal health and the progression of neurodegenerative diseases. This chapter aims to bridge the gap between basic astrocyte biology and clinical PD, offering insights that could pave the way for novel therapeutic strategies. Through a detailed description of the dual roles of astrocytes, we hope to provide a perspective on these critical yet often overlooked cells in the brain, highlighting their potential as targets in the ongoing quest to understand and treat PD.

2 Astrocytes and PD: Understanding the Relationship Between Morphology, Function, and PD Pathogenesis

Astrocytes are morphologically distinct in their structure when compared to neurons in that they possess a cotton-ball-like appearance, with a small cell body which gives rise to primary and secondary processes that end in very fine ~50 nm wide branchlets and leaflets [4]. Each component of the astrocytic structure is studded with subcellular organelles that include mitochondria and lysosomes, capable of interacting with cytosolic proteins either normally present in the astrocyte or acquired by the astrocyte during pathological processes. It is therefore important to consider the interaction between these organelles and α-synuclein, which is the main pathological protein involved in neurodegeneration during PD. In addition, the elaborate processes of astrocytes are optimally structured for the secretion of proteins capable of functionally altering elements of the brain such as neurons, the

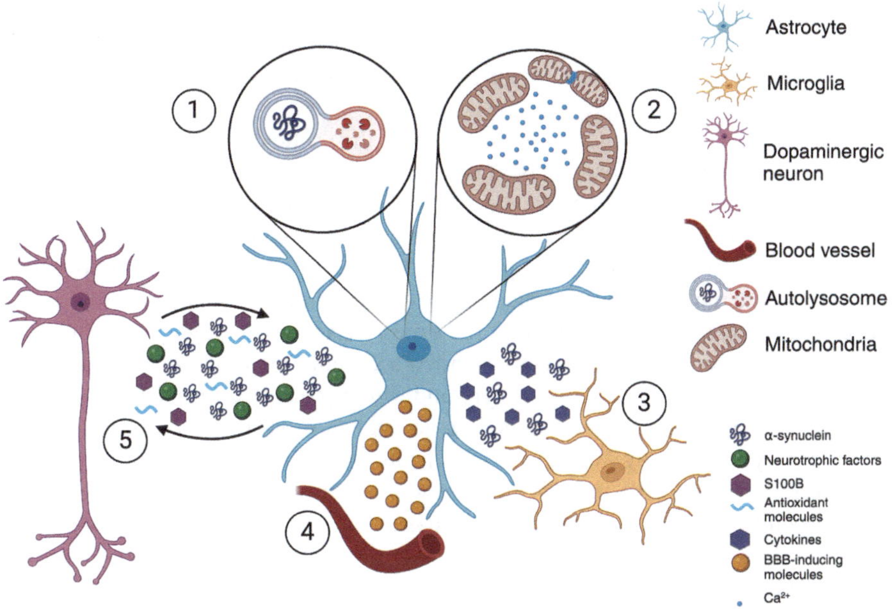

Fig. 1 Complexity of interactions between astrocytes and the central nervous system in Parkinson's disease. (1) Lysosomes are important for astrocytic degradation of α-synuclein, and dysfunction of these organelles leads to a diminished capacity to clear protein aggregates, which may further exacerbate neuronal degeneration in PD. (2) Dysfunctional astrocytic mitochondria are implicated in the pathogenesis of PD, with impaired mitochondrial fission, inefficient fatty acid metabolism, and dysregulated calcium fluxes observed in the astrocytic mitochondria of various PD models. (3) Crosstalk between astrocytes and microglia is mediated by secretion of pro-inflammatory and neurotrophic cytokines. This interaction promotes efficient degradation of α-syn by both cell types. (4) BBB disruption contributes to PD, and astrocytes are important contributors to the BBB through secretion of growth factors. (5) Astrocytes interact with dopaminergic neurons via secretion of various proteins including neurotrophic factors, antioxidants, and S100B. These proteins can have both protective and toxic effects on neurons in a manner that determines the extent of PD progression

blood–brain barrier, or other glial cells, as summarized in Fig. 1. In the sections below, we consider these important aspects of astrocyte biology in relation to PD.

2.1 *Astrocytic Organelles and Secreted Cytosolic Proteins in PD*

2.1.1 α-Synuclein and Cellular Proteostasis in Astrocytes

α-Synuclein (α-syn) is a small protein expressed in both neurons and astrocytes. Despite the gaps in knowledge regarding its physiological functions, it is established that the accumulation of α-syn aggregates in dopaminergic neurons is a hallmark of several neurodegenerative disorders, including PD [5]. Like many aspects of astrocyte biology, the involvement of astrocytes in α-syn processing and degradation has a dual nature, serving both neuroprotective and neurodestructive roles in response to these protein aggregates.

One neuroprotective function of astrocytes is the proficient degradation of α-syn, which is facilitated by lysosomal degradation. Loria et al. found that after astrocytic internalization of α-syn, the protein colocalizes with the lysosomal-associated membrane protein 1 (LAMP1), suggesting that lysosomal degradation of fibrils occurs [6]. Given this important degradative function of astrocytes, it is important to understand the mechanisms by which the protein accumulates within astrocytes. We have recently discussed the idea that the origin of astrocytic aggregation of α-syn can be categorized into three forms: de novo accumulation of α-syn, phagocytic or pinocytic uptake of extracellular α-syn, and direct transfer from neurons to astrocytes through processes such as tunneling nanotubules (TNTs) or exosomes [7]. De novo astrocytic α-syn aggregation results from endogenous expression of the protein. However, the levels that accumulate within astrocytes is typically below the level required for aggregation and transmission of aggregates [8]. Nonetheless, it has been shown that the exposure of astrocytes to inflammatory cytokines results in the upregulation of endogenous α-syn expression, which can contribute to PD progression [9]. Phagocytosis or pinocytosis is widely accepted as the most likely method of astrocytic α-syn accumulation. Lee et al. recently demonstrated that primary astrocytes cultured in media with α-syn produced by SH-SY5Y cells exhibited uptake of extracellular α-syn, which resulted in aggregate formation [10]. Furthermore, Tsunemi et al. demonstrated that astrocytes, when cocultured with induced pluripotent stem cell (iPSC)-derived dopaminergic neurons, rapidly internalize α-syn and exhibit higher lysosomal degradation rates compared to neurons [11]. Lastly, methods of direct transfer of α-syn from neurons to astrocytes have recently been reported. TNTs are thin channels composed of actin, which connect the cytosol of neurons and astrocytes and have been found to facilitate the bidirectional transfer of α-syn [12]. Interestingly, astrocytes may also form these nanotubule-based physical connections with microglia, which allows for α-syn transfer between these different glial cells [13]. In addition to TNTs, direct transmission of α-syn can also occur via exosomal transport. Venturini et al. found that exosomes can be released from astrocytic processes and are able to target neurons, suggesting another mechanism by which astrocytes can promote the spread of α-syn [14]. Furthermore, Emmanouilidou et al. demonstrated that exosomes containing α-syn can be generated by neurons, which supports the idea that α-syn can be transferred in a bidirectional manner via exosomes [15].

While astrocytic uptake and degradation of α-syn can be protective, accumulation of the protein can also initiate a neurotoxic response. The internalization of preformed fibrils can trigger pathologic activation of midbrain astrocytes, characterized by inflammatory transcriptional responses, including the necroptotic RIPK-dependent activation of NF-κB signaling pathways [16]. Additionally, astrocytes exposed to α-syn fibrils can become antigen-presenting cells and demonstrate upregulation of HLA genes encoding MHC class I and II proteins [17]. Taken together, these findings illustrate mechanisms by which α-syn accumulation within astrocytes can drive neuroinflammation, both by direct activation of inflammatory pathways and by promotion of a phenotype resembling antigen-presenting cells.

2.1.2 Lysosomes in Astrocytes and their Role in PD

Lysosomes are vesicular organelles whose primary function is the degradation of intracellular macromolecules for elimination and recycling. Lysosomal degradation of proteins and lipids maintains cellular homeostasis, and the abnormal accumulation of these macromolecules can contribute to cellular dysfunction [18]. In the context of PD, lysosomal dysfunction in astrocytes has emerged as an important contributor to disease pathogenesis. A notable consequence of this dysfunction is the impaired ability of astrocytes to degrade extracellular neuron-derived α-syn, which can promote neuroinflammation. This was evidenced in a study involving astrocytes generated from iPSCs derived from patients with Gaucher disease, a lysosomal storage disorder caused by a deficiency in the hydrolase β-glucocerebrosidase [19, 20]. The study revealed that uptake of α-syn in Gaucher disease iPSC-derived astrocytes culminated in an elevated secretion of pro-inflammatory cytokines such as IL-6, CXCL1, and IL-8. These findings highlight the critical role of lysosomal degradation in clearing neuron-secreted α-syn, a process vital in reducing neuroinflammation and mitigating the progression of PD.

Additionally, mutations in the major PD-related gene, *leucine-rich repeat kinase 2 (LRRK2)*, are linked to astrocytic lysosomal dysfunction. Observations by Henry et al. demonstrated that the LRRK2 protein regulates the size, number, and function of lysosomes in astrocytes [21]. Notably, the astrocytic expression of the LRRK2 G2019S mutation, the most prevalent pathological variant of LRRK2, resulted in enlarged lysosomes and diminished lysosomal degradative capacity. This phenotype was reversed through the genetic elimination of the kinase and autophosphorylation activity of LRRK2, underscoring the therapeutic potential of LRRK2 kinase inhibitors in rectifying lysosomal defects associated with PD.

2.1.3 The Role of Astrocytic Mitochondria in PD

Mitochondria are dynamic, multifunctional organelles, playing pivotal roles in energy generation, lipid and amino acid metabolism, calcium homeostasis, and the regulation of programmed cell death. Dysfunctional astrocytic mitochondria are implicated in the pathogenesis of PD, as evidenced in specific genetic knockdown models of the disease [22]. For example, the knockdown of *PARK7,* a gene linked to familial early onset PD, in cultured astrocytes was found to impair mitochondrial motility and fission, processes essential for ensuring the proliferation and distribution of functional mitochondria [23]. Interestingly, in brain tissue from sporadic PD patients, the protein encoded by *PARK7*, DJ-1, is highly upregulated in reactive astrocytes but not in neurons [24].

Mitochondria facilitate many aspects of cellular metabolism, and astrocytic fatty acid metabolism is an important mechanism for maintaining neuronal health. Fatty acids (FA) and lipid droplets have been associated with α-syn toxicity in cellular models [25]; therefore, dysfunction of FA metabolism in astrocytes may promote the toxicity of this protein in PD patients. Notably, a lipidomic analysis of α-syn-mediated neurotoxicity identified the enzyme stereoacyl-CoA desaturase-1 (SCD1) as a potential therapeutic target in PD [26]. SCD1 converts stearic acid to oleic acid, a FA molecule that accumulates in lipid droplets in neurons, and it was previously demonstrated that inhibition of SCD1 can reduce α-syn-mediated dopaminergic neuron degeneration [25]. Astrocytic mitochondria have been reported to substantially contribute to beta-oxidation of FA within lipid droplets that are transferred from neurons to astrocytes via apolipoprotein E-positive lipid particles [27]. Therefore, a potential mechanism of astrocytic mitochondrial involvement in PD is the attenuation of α-syn toxicity via the metabolism of toxic FA molecules, as was recently proposed by Bantle et al. [28].

Fragments from damaged mitochondria may also have a role in the exacerbation of PD. Joshi et al. found that mitochondrial fragments from activated microglia can induce astrocyte reactivity [29]. Furthermore, these reactive astrocytes can secrete their own mitochondrial fragments to neurons, which can result in neuronal damage. Given the role of these mitochondrial fragments in propagating neuroinflammation, it is likely that this mechanism contributes to the inflammatory environment found in PD pathology.

An important feature of mitochondria is the ability of the organelle to regulate intracellular calcium and contribute to astrocytic calcium signaling. In the calcium signaling section of this chapter, we will discuss the role of astrocytic mitochondria in these signaling mechanisms and their contribution to PD pathogenesis.

2.2 *Astrocyte-Secreted Proteins in PD*

2.2.1 The Role of Astrocytic Antioxidant Molecules in PD

Oxidative stress plays a pivotal role in the onset and progression of PD. Increased levels of oxidized macromolecules such as DNA and lipids have been found in brain tissues of PD patients [30, 31]. Furthermore, oxidative damage to dopaminergic neurons in the substantia nigra pars compacta (SNc) activates a cascade of downstream events that results in dopaminergic neuron death [32]. Astrocytes in the substantia nigra (SN) are involved in the generation of antioxidant molecules, which reduce the exposure of dopaminergic neurons to oxidative stress, and a reduction of astrocytic production of these molecules can contribute to dopaminergic neuron degeneration.

Glutathione (GSH) is an antioxidant molecule that helps protect the neuron against intracellular reactive oxygen species (ROS) [33]. GSH can be synthesized by both neurons and astrocytes using cysteine as a precursor; however, cysteine is rapidly oxidized to cystine in the extracellular environment. Because neurons do not express cystine transporter proteins [34], they are unable to take up this molecule and reduce it to cysteine for GSH synthesis. Astrocytes take up cystine via the cystine/glutamate exchange transporter, after which they reduce cystine to cysteine and utilize it to synthesize and secrete GSH [35]. Extracellular GSH is then used to reduce extracellular cystine to cysteine, which is taken up by neurons for intraneuronal GSH synthesis. Separately, astrocytic GSH is degraded in the extracellular space, resulting in the production of cysteinylglycine. Cysteinylglycine is also taken up by neurons and can also be utilized for GSH synthesis [36].

Proteins involved in the regulation of the oxidative stress response in astrocytes have important roles in the protection of dopaminergic neurons and PD. *PARK7* encodes the protein DJ-1, which is involved in the oxidative stress response, and point mutations in the *PARK7* gene have been identified as contributing genetic factors to autosomal recessive PD [37]. DJ-1 is primarily expressed in astrocytes and is responsible for activating the nuclear factor erythroid 2-related factor 2 (NRF2)-mediated antioxidant response and acting as a sensor for oxidative stress [24, 38]. Activation of NRF2 leads to increased expression of antioxidant proteins, including those involved in GSH synthesis, by binding to the antioxidant response element (ARE) [39]. Shih et al. demonstrated that overexpression of Nrf2 in glia alone resulted in augmented synthesis and release of glial GSH, which protected neurons against oxidative stress in a neuron-glial coculture experiment [40]. Together, these PD-associated proteins are imperative for astrocytic synthesis and secretion of GSH, which is needed for the protection of neurons against oxidative stress.

2.2.2 The Role of Astrocyte-Secreted Neurotrophic Factors in PD

Astrocytes produce several neurotrophic factors that act in a paracrine fashion by binding to dopaminergic neuron receptors and promoting survival. These factors include proteins such as basic fibroblast growth factor (bFGF), cerebral dopamine neurotrophic factor (CDNF), glial cell-line derived neurotrophic factor (GDNF), and mesencephalic astrocyte-derived neurotrophic factor (MANF). Basic FGF is secreted by midbrain astrocytes and promotes the survival of dopaminergic neurons in the SNc [41]. CDNF and MANF are homologous proteins that act as potent neuroprotective factors that enhance the repair of dopaminergic neurons [42, 43]. In addition to their actions as neurotrophic extracellular factors, these proteins regulate the ER stress-induced unfolded protein response in dopaminergic neurons [44]. GDNF is a neurotrophic factor that is mainly expressed in neurons rather than astrocytes in the healthy brain. However, GDNF synthesis is upregulated in astrocytes in the injured brain, which can provide neuroprotective effects depending on the level and duration of glial cell activation [45].

2.2.3 The Role of Astrocyte-Secreted S100B in PD

Astrocytes synthesize and release proteins such as GSH and neurotrophic factors, which protect dopaminergic neurons against damage from oxidative and unfolded protein-related stress [33]. However, astrocytes can also secrete proteins that promote PD progression. One example of this can be seen in the protein S100B, a molecule which is released by astrocytes and is involved in the pathogenesis of PD [46]. Interestingly, patients with PD were found to have increased levels of S100B in their cerebrospinal fluid and had significantly higher expression of S100B in the SNc. S100B can accelerate neurodegeneration by acting on downstream targets, such as activating receptors for advanced glycation end-products (RAGE)-mediated pro-inflammatory pathways in astrocytes and microglia [46]. Moreover, recent studies by Bancroft et al. have demonstrated that extracellular S100B can inhibit fast-inactivating A-type voltage-gated potassium channels (VGKC) in dopaminergic neurons, which can increase neuronal L-type voltage-gated calcium channel (VGCC) fluxes [47]. This altered dopaminergic neuron L-type VGCC function may cause a predisposition to PD development by increasing neuronal vulnerability to neurodegeneration [48, 49]. In sum, through activation of pro-inflammatory pathways and alteration of VGKC and VGCC activity, S100B is an important contributor to dopaminergic neuron loss in PD.

2.3 Astrocyte Interaction with Other Cell Types in PD

When one considers interactions between astrocytes and neurons in the context of PD, it is important to recognize the fact that astrocytic influences on other cell types in the brain have a direct impact on the nature and extent of neuron dysfunction in PD. Therefore, the following two subsections focus on mechanisms underlying the interaction of astrocytes with other cell types during PD.

2.3.1 Interactions Between Astrocytes and Microglia in PD

Crosstalk between astrocytes and microglia has a significant role in the pathogenesis of multiple neurodegenerative disorders, including PD. Liddelow et al. demonstrated that astrocyte reactivity can be induced by exposure to pro-inflammatory molecules like Il-1α, TNF-α, and C1q released by activated microglia [50]. Furthermore, Yun et al. showed that pharmacological blockade of microglia-mediated astrocyte reactivity provides neuroprotection in various PD models [51].

Additionally, the interaction between astrocytes and microglia is essential for the clearance of α-syn. Experiments have shown that α-syn aggregates are significantly reduced in cocultures of astrocytes and microglia compared to monocultures of either cell type alone [13]. This enhanced degradation of α-syn is facilitated by astrocytic transfer of α-syn to microglia through secretory mechanisms and through direct cell-to-cell contact between the two cell types, as discussed in the previous section on astrocytic α-syn. Furthermore, microglial phagocytosis can be stimulated by astrocytic cytokine production, although further studies are needed to characterize the specific molecules mediating this stimulation [52]. These findings underscore the importance of microglia-astrocyte communication in PD and highlight the therapeutic potential of modulating this interaction.

2.3.2 Interactions Between Astrocytes and Blood–Brain Barrier in PD

The blood–brain barrier (BBB) is a critical barrier that allows for selective permeability of specific molecules and restriction of others. As a result, maintaining a functional BBB is crucial for CNS function, and dysfunction of this barrier can initiate and exacerbate various neurodegenerative disorders, including PD [53]. Astrocytes are intricately involved in the maintenance of the BBB. For example, in a study utilizing endothelial cell cultures, it was found that when cocultured with astrocytes, endothelial cells are more likely to form close membrane appositions, produce structures resembling tight junctions, and form functional barriers with decreased permeability to large molecules [54]. This strengthening of the BBB is likely mediated by astrocytic secretion of angiogenic growth factors, which can induce phenotypic expression of BBB characteristics such as tight junctions in endothelial cells [55].

Changes in BBB permeability have been observed in PD animal models and postmortem tissue samples from PD patients [56, 57]. Due to their close interactions with neurons and endothelial cells, astrocytes are key mediators that can detect changes in BBB permeability that affect neurodegenerative processes in PD. An example of this was demonstrated in a study using 3D human BBB chips with patient-derived iPSCs, where astrocytes from patients with the PD-related *LRRK2* G2019S mutation exhibited pro-inflammatory characteristics and impaired support for functional capillary formation [58]. Functional interactions between astrocytes and components of the BBB will be further discussed in our section on astrocytic calcium signals, where we will describe signaling between astrocytic end feet and the BBB, as well as the role of this interaction in PD.

3 Astrocytic Calcium Signals in PD

In the previous sections, we considered subcompartments of astrocytes at the level of individual organelles and at the level of astrocytic interactions with other components of the CNS, such as the BBB. In this section, we focus on calcium signals, which have emerged as an important aspect of astrocyte biology in relation to PD. Recent studies have demonstrated that calcium signals in astrocytes are an important unifying characteristic that affects not only individual astrocytic organelles but also interactions of astrocytes with various components in the brain [59].

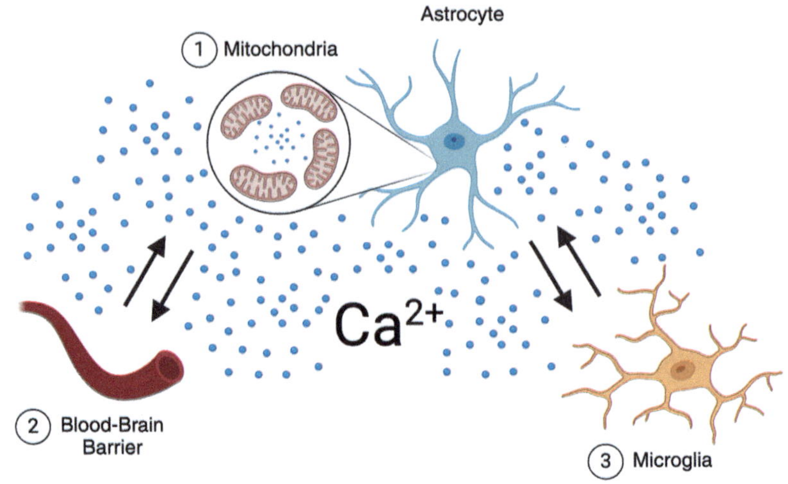

Fig. 2 Astrocytic calcium signaling in Parkinson's disease. (1) Calcium signaling within mitochondria contributes to PD in various model systems, affecting neuronal function in both the dorsolateral striatum (DLS) and the midbrain. (2) Astrocytic end feet interact with endothelial cells at the blood–brain barrier, where they modulate vasoconstriction and vasodilation via Ca^{2+} signaling and facilitate vascular repair after injury. Moreover, these spontaneous Ca^{2+} events undergo altered kinetics with aging, due to reduced expression of the buffer protein calreticulin. (3) Astrocytic Ca^{2+} signaling determines the localization of aquaporin 4 (AQP4) to the astrocyte plasma membrane. Dysfunctional Ca^{2+} signaling can disrupt astrocytic AQP4 localization and function, which can enhance microglial activation and worsens PD pathology by maintaining an environment of neuroinflammation

Astrocytic calcium signaling is an important mechanism for the maintenance of normal CNS function and can occur in response to various stimuli [60]. These signals result from the release of Ca^{2+} from subcellular compartments, including the endoplasmic reticulum and mitochondria, as well as extracellular sources [61–63]. Because these signals are crucial for brain function, dysfunction of such signaling can result in the acceleration of PD. Dysfunctional calcium signaling can initiate and exacerbate neurodegeneration in PD and explore mechanisms underlying cell-to-cell communication between astrocytes, neurons, and non-neuronal cells of the CNS. These concepts are summarized in Fig. 2.

3.1 The Role of Astrocytic Calcium Signaling in PD

Astrocytes play a complex and integral role in neurotransmission, and disruption of their bidirectional interaction with neurons has emerged as a key factor in the development of PD. Our current understanding of the role of astrocytes in neurotransmission involves both glutamatergic and dopaminergic signaling. Interestingly, dysfunction of this neurotransmission is closely related to aberrant calcium signaling [59].

Glutamate, the central nervous system's primary excitatory neurotransmitter, plays a crucial role in motor, cognitive, and autonomic functions. Dysfunctional regulation of glutamate can contribute to the development of PD, particularly through mechanisms such as glutamate excitotoxicity [64]. At the synapse, glutamate is released and then recycled through the glutamate–glutamine cycle. Following its neuronal release and the subsequent activation of postsynaptic action potentials, glutamate is rapidly cleared from the synaptic cleft by transporters located on both neurons and glial cells. Astrocytes convert glutamate into the inert amino acid, glutamine, via the enzyme glutamine synthetase. This astrocyte-derived glutamine is then released back into the extracellular space, where it is available for neuronal uptake and can be used to synthesize glutamate [65].

Glutamate is taken up from the synaptic cleft via excitatory amino acid transporters (EAATs), which are specialized glutamate transporters expressed in various cell types. EAAT1 is predominantly expressed in astrocytes while EAAT2 is present in both neurons and astrocytes [66]. An excessive accumulation of glutamate at synapses can be toxic, leading to neuronal apoptosis due to Ca^{2+} overload. This excitotoxicity is a characteristic of many neurodegenerative disorders, including PD, indicating that dysfunction or reduced expression of EAATs in the CNS, particularly in astrocytes, may have a crucial role in the pathogenesis and progression of the disease [67].

Mutations in *PARK7*, a PD-associated gene that expresses the DJ-1 protein, have been shown to result in the disruption of the astrocytic lipid raft assembly, which leads to decreased expression of EAAT2 and potentially disrupts astrocytic uptake of glutamate [68]. Additionally, studies using a rodent 6-hydroxydopamine (6-OHDA) model of PD have revealed that exposure to the neurotoxic agent results in a reduction of EAAT2 expression and a consequent decrease in glutamate uptake [69].

The role of astrocytes in dopaminergic signaling is less understood, especially in the context of PD. Utilizing optogenetic techniques and live mouse brain slice experiments, Corkrum et al. demonstrated that astrocytes in the nucleus accumbens are responsive to synaptic dopamine. This response is mediated through the activation of astrocytic D1 receptors, resulting in intracellular Ca^{2+} elevations and subsequent ATP/adenosine release. This ATP release, in turn, induces a depression of excitatory synaptic transmission through the activation of presynaptic neuronal A1 receptors [70]. In contrast, a recent study by Xin et al. revealed that in the ventral midbrain, astrocytic calcium activity is reduced by the activation of the astrocytic D2 receptor. Furthermore, inhibition of the astrocytic D2 receptor resulted in decreased calcium responses to glutamate [71].

Taken together, these studies provide compelling evidence of astrocyte involvement in dopaminergic neurotransmission through calcium signaling, which highlights new avenues for research into the cellular mechanisms underlying disrupted neurotransmission in PD.

3.2 The Role of Calcium Signaling in Astrocytic Mitochondria During PD

Dysfunction of calcium regulation within mitochondria has been demonstrated in various PD model systems. In the dorsolateral striatum (DLS), activation of astrocytic D1, D2, and glutamate receptors results in a dual response in mitochondrial Ca^{2+} influx frequencies, with an increase observed in some mitochondria and a simultaneous decrease observed in others within the same astrocyte [62]. In the midbrain of a Parkin-knockout mouse line, astrocytes exhibited a four-fold increase in the percentage of damaged mitochondria compared to neurons [72]. Moreover, in the same study, mouse models overexpressing doubly mutated human α-syn displayed midbrain astrocytic mitochondria with a diminished Ca^{2+} storage capacity. Interestingly, these mutations inhibited the ability of cultured midbrain astrocytes to promote neuronal differentiation, suggesting that disrupted neuron-astrocyte interactions potentially mediated by soluble molecules such as neurotrophic factors can be caused by mitochondrial Ca^{2+} dysfunction in PD. Together, these findings underscore the importance of astrocytic mitochondrial Ca^{2+} fluxes in mediating neuron-astrocyte interactions, as well as how dysregulation of these dynamics within mitochondria can exacerbate PD pathogenesis.

3.3 The Role of Astrocyte–Astrocyte Calcium Signaling in PD

A key feature of astrocytes is their ability to form extensive networks through astrocyte-to-astrocyte communication. This is facilitated by gap junctions, which enable the transcellular exchange of ions and molecules by sharing cytoplasmic content [73]. Astrocytes typically express high levels of gap junction proteins, such as connexins. These proteins form the structural basis of gap junctions by assembling into hexamer connexons or hemichannels, which align between two adjacent cell membranes to create gap junction channels [74].

Astrocyte reactivity has been observed to modify the expression and functionality of connexin proteins [75]. For instance, conditions like oxidative stress and inflammation have been shown to trigger the opening of connexin hemichannels [76, 77]. These changes in astrocytic connexins are significant in the context of PD pathogenesis, as evidenced by studies examining connexin expression and function in PD models. In a 1-methyl-4-phenyl-1,2,3,6-tetrahydropyridine (MPTP)-induced PD model, the expression of connexin 43 (Cx43), a crucial component of astrocytic gap junctions, was found to be increased in the striatum [78]. Additionally, a more recent study revealed that astrocytic Cx43 hemichannel permeability is heightened in the midbrain of this PD model and is associated with increased intracellular Ca^{2+} levels in astrocytes of live slices [79]. While excessive Cx43 opening seems to exacerbate disease severity, other gap junction components like connexin 30 may have protective roles. For example, deficiency in connexin 30 was linked to diminished protection by astrocytes and an exacerbation of MPTP-induced neurodegeneration [80].

Further research has explored the impact of α-syn on connexin function, revealing that its accumulation promotes the opening of Cx43 hemichannels in mouse cortical astrocytes. This leads to altered intracellular Ca^{2+} dynamics and reduced astrocyte survival [81]. Collectively, these findings suggest that astrocyte–astrocyte communication via gap junctions and Ca^{2+} signaling is crucial for maintaining homeostasis in the CNS, and dysregulation of these processes is involved in the pathophysiology of PD.

3.4 The Role of Astrocytic Calcium Signaling at the Blood–Brain Barrier in PD

Astrocytic end feet interact with endothelial cells at the BBB to ensheath capillaries. These end feet can modulate vasoconstriction and vasodilation via Ca^{2+} signaling and facilitate vascular repair after injury [82]. Interestingly, the kinetics and dynamics of spontaneous Ca^{2+} events in these end feet change with age, particularly in brain regions implicated in PD, like the DLS. In DLS astrocytic end feet of aging mice, spontaneous Ca^{2+} signals are larger, longer-lasting, and propagate faster compared to those in younger mice [83]. The expression of the Ca^{2+} buffer protein calreticulin is significantly reduced in aging DLS astrocytic end feet, which may underlie the abnormal calcium signaling observed. Importantly, these changes were associated with decreased expression of aquaporin 4 (AQP4) in DLS blood vessels, a water channel protein which is highly expressed in astrocytes and has been associated with PD. This suggests that aberrant astrocytic Ca^{2+} signaling may be contributing to increased BBB permeability in this brain region, which may promote the development of PD. These findings suggest a potential mechanism by which dysfunctional astrocyte-endothelial cell calcium signaling may contribute to age-related neurodegenerative disorders, including PD.

3.5 The Role of Calcium Signaling Between Astrocytes and Microglia in PD

Microglial activation is a key characteristic of PD, as reactivity has been documented in rodent models [84], nonhuman primates [85], and clinical PD patients [86] and can directly influence neurodegeneration in PD. Therefore, it is important to examine calcium signaling-mediated interactions between these astrocytes and microglia, as well as the role that dysfunctional signaling between these two types may have in PD pathogenesis. Recently, Sun et al. found that AQP4 knockout mice exhibit increased microglial reactivity after exposure to MPTP when compared to wild-type mice [87]. Notably, this study demonstrated that the observed increase in microglial reactivity occurs due to increased neuroinflammatory cytokine secretion compared to wild-type mice. AQP4 is highly expressed in astrocytes, and the protein must be localized to the plasma membrane to allow for the passage of water molecules within a cell. Multiple studies have shown that the protein depends on Ca^{2+} signals for precise localization to the plasma membrane [88, 89]. Taken together, these studies support the notion that dysfunctional astrocytic Ca^{2+} signaling can disrupt astrocytic AQP4 localization and function, which may enhance microglial activation and worsen PD pathology by promoting a neuroinflammatory environment.

4 The Role of Astrocyte Heterogeneity in PD

4.1 The Relationship Between Astrocyte Morphological Heterogeneity and PD

Astrocytes exhibit remarkable heterogeneity in both morphology and function, with variations observed within a single brain region and across different PD-relevant brain regions. For example, Chai et al. found notable differences in the morphology of astrocytes between the mouse striatum and hippocampus [90]. Astrocytes in the striatum were found to have larger territorial sizes and to contact nearly twice as many neuronal somas as those in the hippocampus. This morphological difference has direct implications for how striatal astrocytes would interact with medium spiny neurons and cholinergic interneurons, both of which play important roles in the pathogenesis of PD [91].

Morphological heterogeneity can also be found at different time points within specific PD-relevant regions. Bondi et al. recently found that in the substantia nigra of mice, there is subregion-specific heterogeneity in the age-related structural remodeling of astrocytes [92]. In the SNc, the morphological complexity of GFAP+ astrocytes was significantly increased when old mice (age 20–24 months) were compared to middle-aged (14–17 months) and young mice (4–6 months). In contrast, GFAP+ astrocytes in the substantia nigra pars reticulata (SNr) did not undergo structural remodeling, with no significant changes in morphological complexity between age groups. Together, these findings demonstrate heterogeneity in astrocyte structure that varies between PD-associated brain regions after aging, reflecting the complexity of the contribution of these cells to age-related PD development.

4.2 The Role of Functional Astrocyte Heterogeneity in PD

Beyond morphology, astrocytes also demonstrate significant heterogeneity in their functional roles and molecular signatures. While the heterogeneity of astrocytes in the healthy brain is well-documented, the role of varying stages of astrocyte reactivity in PD pathogenesis remains elusive. The astrocytic response to injury results in a broad spectrum of astrocyte reactivity, encompassing a wide range of transcriptomic profiles that suggest distinct astrocytic populations involved in various stages of neurodegenerative diseases [93]. Escartin et al. note that the previous classification of reactive astrocytes into distinct A1 and A2 subgroups has been challenged by data, suggesting that these transcriptomic signatures represent only two of many possible astrocyte transcriptomes that result from reactivity. Therefore, it is highly likely that the wide range of astrocyte heterogeneity that occurs in various disease states also occurs in PD and can influence the progression of the disease.

This heterogeneity in astrocytic response to injury has been further characterized by the measurement of proteins and cytokines expressed throughout the spectrum of astrocyte reactivity. Zamanian et al. found that in two different CNS injury models, reactive astrocytes expressed increased levels of various proteins in comparison to quiescent astrocytes [94]. These included intermediate filament proteins such as GFAP and vimentin, as well as inflammatory cytokines such as interleukin-6 and cytokine receptors such as IL-1 receptor 1. Moreover, preferential expression of reactive glial genes was observed between the two different injury models, with some being expressed more in ischemic-stroke-induced astrogliosis and others being expressed more in LPS-induced astrogliosis.

The significance of this heterogeneity in PD is underscored in multiple models. For instance, Lee et al. observed that in mice overexpressing α-syn, the accumulation of protein aggregates led to the release of pro-inflammatory cytokines and chemokines by astrocytes, implicating reactive astrocytes in the pathogenesis of PD [10]. Similarly, preventing the conversion of astrocytes to a reactive phenotype has been shown to confer neuroprotection in mouse models of PD. In an α-syn-preformed fibril mouse model, the administration of the GLP1R agonist NYL01 inhibited the microglia-mediated induction of astrocyte reactivity [51]. This intervention was associated with protection against loss of dopaminergic neurons and behavioral deficits. Additionally, NYL01 was found to alleviate behavioral deficits and extend the lifespan of a human A53T α-syn transgenic mouse model of PD [51]. Additionally, in a macaque model using the neurotoxin MPTP, reactive astrocytes were identified in the substantia nigra pars compacta in the earliest stages of PD development, underscoring their role in the disease's pathology [95].

Beyond reactivity, the functional heterogeneity of astrocytes in PD is demonstrated in region-specific astrocytic gene expression. Utilizing two extensive RNA expression datasets from postmortem human samples, Soreq et al. found a substantial decrease in the expression of astrocyte-specific genes in the SN with aging, which was not observed in other brain regions such as the hippocampus and cortex [96]. Given the significant transcriptional changes in this PD-relevant brain region that are associated with aging, a major risk factor for PD, it is possible that such changes influence vulnerability toward SNc neurodegeneration during PD pathogenesis. Furthermore, Chai et al. compared astrocyte populations in the mouse striatum and hippocampus, revealing other notable functional differences in PD-relevant brain regions [90]. Astrocytes in the two regions displayed distinct Ca^{2+} signaling dynamics, and transcriptomic and proteomic analyses further highlighted the unique molecular characteristics of each population. While both populations shared the expression of genes like GLT-1 and Kir4.1, hippocampal astrocytes exhibited significantly higher levels of GFAP, whereas striatal astrocytes showed elevated expression of the gene-encoding μ-crystallin. This study suggests that rather than utilizing GFAP as a universal astrocyte marker, it is more appropriate to utilize it as a region-specific marker of reactivity. Additionally, this study underscores the potential for identifying novel, region-specific biomarkers for astrocytes.

Similarly, region-specific astrocyte heterogeneity exists in the ability of these cells to perform specific PD-relevant functions, such as α-syn clearance [97]. Yang et al. found that when compared to cortical astrocytes, astrocytes from the ventral midbrain cocultured with α-syn-overexpressing dopaminergic neurons demonstrated an increased ability to reduce α-syn aggregates. These findings highlight the importance of region of origin when developing astrocyte-based PD therapeutics, such as astrocyte transplantation.

Collectively, these studies highlight the importance of astrocytic heterogeneity in PD. From these findings, it is evident that the varied morphology, molecular characteristics, and function of astrocytes have significant contributions to PD development, a concept which is summarized in Fig. 3. Thus, it is essential to understand the complexities of astrocyte biology and astrocytic interactions with the CNS when considering the development of novel therapeutic targets.

5 Future Directions

Astrocytes maintain a myriad of functions that are pivotal to the health and maintenance of the central nervous system. However, our understanding that their dysfunction can initiate and exacerbate the pathogenesis of PD is only beginning to emerge. Thus, investigating these cells not only as contributors to PD development but also as therapeutic targets offers a promising avenue in the quest to discover and develop disease-modifying therapies.

Our growing understanding of astrocytes in PD has already led to the exploration of therapeutic strategies focused on their biology. Proposed approaches have included utilizing astrocyte-secreted neurotrophic factors to protect neurons, as described previously in this chapter. Early efforts to utilize this mechanism in the clinic have been challenging, as clinical trials using direct brain infusion of neurotrophic factors for PD therapy have failed [98]. Therefore, researchers have recently aimed to enhance astrocytic production of neurotrophic factors via administration of modified viruses that express these molecules. For example, a recent randomized

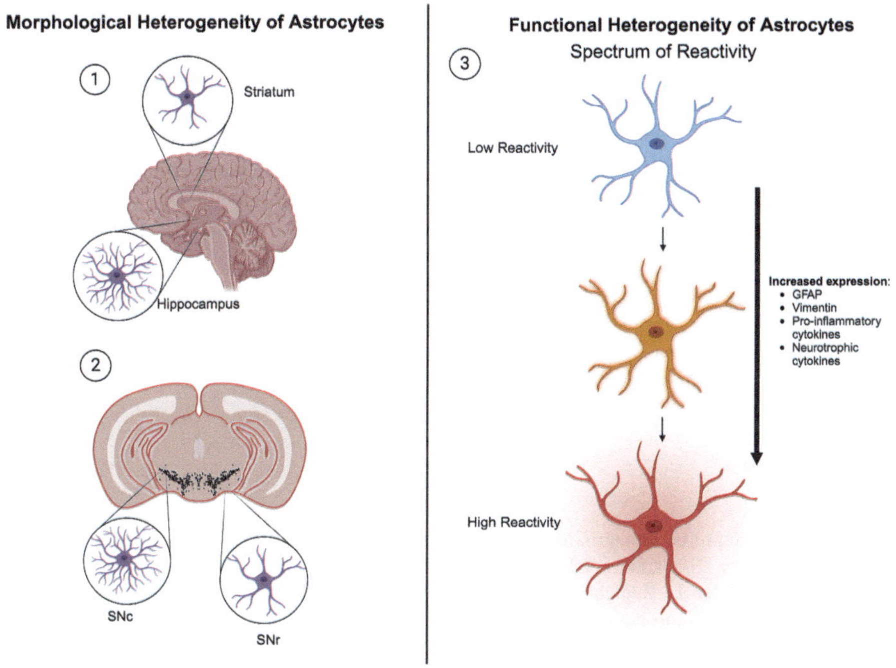

Fig. 3 Morphological and functional heterogeneity of astrocytes in relation to Parkinson's disease. (1) Astrocytes in the striatum have larger territorial sizes and contact nearly twice as many neuronal somas as those in the hippocampus. (2) In the mouse substantia nigra, there is subregion-specific heterogeneity in the age-related structural remodeling of astrocytes, with the morphological complexity of astrocytes increasing with age in the SNc but not the SNr. (3) The astrocytic response to injury is characterized by a broad spectrum of astrocyte reactivity, encompassing a wide range of molecular profiles that suggests distinct astrocytic populations involved in various stages of PD progression. As astrocytic reactivity increases, their molecular profile changes from secretion of neurotrophic cytokines to pro-inflammatory cytokines

controlled trial utilized adeno-associated virus-mediated gene delivery to promote astrocytic expression of the molecule neurturin, a structural and functional analog of GDNF. However, the study did not reveal any benefit to PD patients [99]. Similar challenges have been encountered when testing other potentially neuroprotective therapies that target astrocytes. While prior studies suggested that the cholesterol-lowering agent simvastatin might prevent astrocytes from adopting a neurotoxic phenotype [100, 101], a recent trial by Stevens et al. found that simvastatin did not effectively modify the disease, with patients showing worse performance on the MDS-UPDRS part 3 score compared to the control group [102].

Despite these initial setbacks, many astrocyte-focused therapeutic targets in PD remain largely unexplored in clinical trials. Furthermore, by continuing to leverage the interactions between astrocytes and dopaminergic neurons, new therapeutic strategies can be developed, such as upregulation of neuroprotective astrocyte-secreted factors like the antioxidant GSH and inhibition of detrimental proteins like S100B.

As research into novel PD therapies continues, the study of astrocyte biology in the context of this disease is essential. Continued investigation within this field of research has the potential to lead to innovative disease-modifying therapies; therefore, the goal of developing effecting PD therapeutics will depend on gaining a deeper understanding of the complexity of astrocytes and their interactions with neurons.

Acknowledgments This work was partially funded by National Institute of Neurological Disorders and Stroke, Grant/Award Number: R01NS115809. Figures were constructed using Biorender.com. Thank you to Crystal Chi for designing the coronal mouse midbrain section icon.

References

1. Poewe W, Seppi K, Tanner CM, Halliday GM, Brundin P, Volkmann J, et al. Parkinson disease. Nat Rev Dis Primers. 2017;3:17013.
2. Stocchi F, Vacca L, Radicati FG. How to optimize the treatment of early stage Parkinson's disease. Transl Neurodegener. 2015;4(1):4.
3. Booth HDE, Hirst WD, Wade-Martins R. The role of astrocyte dysfunction in Parkinson's disease pathogenesis. Trends Neurosci. 2017;40(6):358–70.
4. Baldwin KT, Murai KK, Khakh BS. Astrocyte Morphology. Trends Cell Biol. 2023;34(7):547–565.
5. Stefanis L. α-Synuclein in Parkinson's disease. Cold Spring Harb Perspect Med. 2012;2(2):a009399.
6. Loria F, Vargas JY, Bousset L, Syan S, Salles A, Melki R, Zurzolo C. α-Synuclein transfer between neurons and astrocytes indicates that astrocytes play a role in degradation rather than in spreading. Acta Neuropathol. 2017;134(5):789–808.
7. Ozoran H, Srinivasan R. Astrocytes and alpha-synuclein: friend or foe? J Parkinsons Dis. 2023;13:1289–301.
8. Mori F, Tanji K, Yoshimoto M, Takahashi H, Wakabayashi K. Demonstration of alpha-synuclein immunoreactivity in neuronal and glial cytoplasm in normal human brain tissue using proteinase K and formic acid pretreatment. Exp Neurol. 2002;176(1):98–104.
9. Tanji K, Imaizumi T, Yoshida H, Mori F, Yoshimoto M, Satoh K, Wakabayashi K. Expression of alpha-synuclein in a human glioma cell line and its up-regulation by interleukin-1beta. Neuroreport. 2001;12(9):1909–12.
10. Lee HJ, Suk JE, Patrick C, Bae EJ, Cho JH, Rho S, et al. Direct transfer of alpha-synuclein from neuron to astroglia causes inflammatory responses in synucleinopathies. J Biol Chem. 2010;285(12):9262–72.
11. Tsunemi T, Ishiguro Y, Yoroisaka A, Valdez C, Miyamoto K, Ishikawa K, et al. Astrocytes protect human dopaminergic neurons from α-synuclein accumulation and propagation. J Neurosci. 2020;40(45):8618–28.
12. Rostami J, Holmqvist S, Lindström V, Sigvardson J, Westermark GT, Ingelsson M, et al. Human astrocytes transfer aggregated alpha-synuclein via tunneling nanotubes. J Neurosci. 2017;37(49):11835–53.
13. Rostami J, Mothes T, Kolahdouzan M, Eriksson O, Moslem M, Bergström J, et al. Crosstalk between astrocytes and microglia results in increased degradation of α-synuclein and amyloid-β aggregates. J Neuroinflammation. 2021;18(1):124.

14. Venturini A, Passalacqua M, Pelassa S, Pastorino F, Tedesco M, Cortese K, et al. Exosomes from astrocyte processes: signaling to neurons. Front Pharmacol. 2019;10:1452.
15. Emmanouilidou E, Melachroinou K, Roumeliotis T, Garbis SD, Ntzouni M, Margaritis LH, et al. Cell-produced alpha-synuclein is secreted in a calcium-dependent manner by exosomes and impacts neuronal survival. J Neurosci. 2010;30(20):6838–51.
16. Chou TW, Chang NP, Krishnagiri M, Patel AP, Lindman M, Angel JP, et al. Fibrillar α-synuclein induces neurotoxic astrocyte activation via RIP kinase signaling and NF-κB. Cell Death Dis. 2021;12(8):756.
17. Russ K, Teku G, Bousset L, Redeker V, Piel S, Savchenko E, et al. TNF-α and α-synuclein fibrils differently regulate human astrocyte immune reactivity and impair mitochondrial respiration. Cell Rep. 2021;34(12):108895.
18. Navarro-Romero A, Montpeyó M, Martinez-Vicente M. The emerging role of the lysosome in Parkinson's disease. Cells. 2020;9(11):2399.
19. Platt FM, d'Azzo A, Davidson BL, Neufeld EF, Tifft CJ. Lysosomal storage diseases. Nat Rev Dis Primers. 2018;4(1):27.
20. Aflaki E, Stubblefield BK, McGlinchey RP, McMahon B, Ory DS, Sidransky E. A characterization of Gaucher iPS-derived astrocytes: potential implications for Parkinson's disease. Neurobiol Dis. 2020;134:104647.
21. Henry AG, Aghamohammadzadeh S, Samaroo H, Chen Y, Mou K, Needle E, Hirst WD. Pathogenic LRRK2 mutations, through increased kinase activity, produce enlarged lysosomes with reduced degradative capacity and increase ATP13A2 expression. Hum Mol Genet. 2015;24(21):6013–28.
22. Bose A, Beal MF. Mitochondrial dysfunction in Parkinson's disease. J Neurochem. 2016;139(Suppl 1):216–31.
23. Larsen NJ, Ambrosi G, Mullett SJ, Berman SB, Hinkle DA. DJ-1 knock-down impairs astrocyte mitochondrial function. Neuroscience. 2011;196:251–64.
24. Bandopadhyay R, Kingsbury AE, Cookson MR, Reid AR, Evans IM, Hope AD, et al. The expression of DJ-1 (PARK7) in normal human CNS and idiopathic Parkinson's disease. Brain. 2004;127(Pt 2):420–30.
25. Vincent BM, Tardiff DF, Piotrowski JS, Aron R, Lucas MC, Chung CY, et al. Inhibiting stearoyl-CoA desaturase ameliorates α-synuclein cytotoxicity. Cell Rep. 2018;25(10):2742–54.e31.
26. Fanning S, Haque A, Imberdis T, Baru V, Barrasa MI, Nuber S, et al. Lipidomic analysis of α-synuclein neurotoxicity identifies stearoyl CoA desaturase as a target for Parkinson treatment. Mol Cell. 2019;73(5):1001–14.e8.
27. Ioannou MS, Jackson J, Sheu SH, Chang CL, Weigel AV, Liu H, et al. Neuron-astrocyte metabolic coupling protects against activity-induced fatty acid toxicity. Cell. 2019;177(6):1522–35.e14.
28. Bantle CM, Hirst WD, Weihofen A, Shlevkov E. Mitochondrial dysfunction in astrocytes: a role in Parkinson's disease? Front Cell Dev Biol. 2020;8:608026.
29. Joshi AU, Minhas PS, Liddelow SA, Haileselassie B, Andreasson KI, Dorn GW 2nd, Mochly-Rosen D. Fragmented mitochondria released from microglia trigger A1 astrocytic response and propagate inflammatory neurodegeneration. Nat Neurosci. 2019;22(10):1635–48.
30. Bosco DA, Fowler DM, Zhang Q, Nieva J, Powers ET, Wentworth P, et al. Elevated levels of oxidized cholesterol metabolites in Lewy body disease brains accelerate α-synuclein fibrilization. Nat Chem Biol. 2006;2(5):249–53.
31. Nakabeppu Y, Tsuchimoto D, Yamaguchi H, Sakumi K. Oxidative damage in nucleic acids and Parkinson's disease. J Neurosci Res. 2007;85(5):919–34.
32. Dias V, Junn E, Mouradian MM. The role of oxidative stress in Parkinson's disease. J Parkinsons Dis. 2013;3(4):461–91.
33. Miyazaki I, Asanuma M. Neuron-astrocyte interactions in Parkinson's disease. Cells. 2020;9(12):2623.

34. Kranich O, Hamprecht B, Dringen R. Different preferences in the utilization of amino acids for glutathione synthesis in cultured neurons and astroglial cells derived from rat brain. Neurosci Lett. 1996;219(3):211–4.
35. Wang XF, Cynader MS. Astrocytes provide cysteine to neurons by releasing glutathione. J Neurochem. 2000;74(4):1434–42.
36. Dringen R, Gutterer JM, Gros C, Hirrlinger J. Aminopeptidase N mediates the utilization of the GSH precursor CysGly by cultured neurons. J Neurosci Res. 2001;66(5):1003–8.
37. Bonifati V, Rizzu P, Squitieri F, Krieger E, Vanacore N, van Swieten JC, et al. DJ-1(PARK7), a novel gene for autosomal recessive, early onset parkinsonism. Neurol Sci. 2003;24(3):159–60.
38. Neves M, Grãos M, Anjo SI, Manadas B. Modulation of signaling pathways by DJ-1: an updated overview. Redox Biol. 2022;51:102283.
39. Hayes JD, Chanas SA, Henderson CJ, McMahon M, Sun C, Moffat GJ, et al. The Nrf2 transcription factor contributes both to the basal expression of glutathione S-transferases in mouse liver and to their induction by the chemopreventive synthetic antioxidants, butylated hydroxyanisole and ethoxyquin. Biochem Soc Trans. 2000;28(2):33–41.
40. Shih AY, Johnson DA, Wong G, Kraft AD, Jiang L, Erb H, et al. Coordinate regulation of glutathione biosynthesis and release by Nrf2-expressing glia potently protects neurons from oxidative stress. J Neurosci. 2003;23(8):3394–406.
41. Engele J, Bohn MC. The neurotrophic effects of fibroblast growth factors on dopaminergic neurons in vitro are mediated by mesencephalic glia. J Neurosci. 1991;11(10):3070–8.
42. Lindholm P, Voutilainen MH, Laurén J, Peränen J, Leppänen VM, Andressoo JO, et al. Novel neurotrophic factor CDNF protects and rescues midbrain dopamine neurons in vivo. Nature. 2007;448(7149):73–7.
43. Voutilainen MH, Bäck S, Pörsti E, Toppinen L, Lindgren L, Lindholm P, et al. Mesencephalic astrocyte-derived neurotrophic factor is neurorestorative in rat model of Parkinson's disease. J Neurosci. 2009;29(30):9651–9.
44. Lindholm P, Saarma M. Cerebral dopamine neurotrophic factor protects and repairs dopamine neurons by novel mechanism. Mol Psychiatry. 2022;27(3):1310–21.
45. Duarte Azevedo M, Sander S, Tenenbaum L. GDNF, a neuron-derived factor upregulated in glial cells during disease. J Clin Med. 2020;9(2):456.
46. Sathe K, Maetzler W, Lang JD, Mounsey RB, Fleckenstein C, Martin HL, et al. S100B is increased in Parkinson's disease and ablation protects against MPTP-induced toxicity through the RAGE and TNF-α pathway. Brain. 2012;135(11):3336–47.
47. Bancroft EA, De La Mora M, Pandey G, Zarate SM, Srinivasan R. Extracellular S100B inhibits A-type voltage-gated potassium currents and increases L-type voltage-gated calcium channel activity in dopaminergic neurons. Glia. 2022;70(12):2330–47.
48. Verma A, Ravindranath V. CaV1.3 L-type calcium channels increase the vulnerability of substantia Nigra dopaminergic neurons in MPTP mouse model of Parkinson's disease. Front Aging Neurosci. 2020;11:382.
49. Wang R, Ma Z, Wang J, Xie J. L-type Cav1.2 calcium channel is involved in 6-hydroxydopamine-induced neurotoxicity in rats. Neurotox Res. 2012;21(3):266–70.
50. Liddelow SA, Guttenplan KA, Clarke LE, Bennett FC, Bohlen CJ, Schirmer L, et al. Neurotoxic reactive astrocytes are induced by activated microglia. Nature. 2017;541(7638):481–7.
51. Yun SP, Kam TI, Panicker N, Kim S, Oh Y, Park JS, et al. Block of A1 astrocyte conversion by microglia is neuroprotective in models of Parkinson's disease. Nat Med. 2018;24(7):931–8.
52. Lian H, Litvinchuk A, Chiang AC, Aithmitti N, Jankowsky JL, Zheng H. Astrocyte-microglia cross talk through complement activation modulates amyloid pathology in mouse models of Alzheimer's disease. J Neurosci. 2016;36(2):577–89.
53. Cabezas R, Avila M, Gonzalez J, El-Bachá RS, Báez E, García-Segura LM, et al. Astrocytic modulation of blood brain barrier: perspectives on Parkinson's disease. Front Cell Neurosci. 2014;8:211.

54. Hayashi Y, Nomura M, Yamagishi S, Harada S, Yamashita J, Yamamoto H. Induction of various blood-brain barrier properties in non-neural endothelial cells by close apposition to co-cultured astrocytes. Glia. 1997;19(1):13–26.
55. Abbott NJ, Rönnbäck L, Hansson E. Astrocyte–endothelial interactions at the blood–brain barrier. Nat Rev Neurosci. 2006;7(1):41–53.
56. Carvey PM, Zhao CH, Hendey B, Lum H, Trachtenberg J, Desai BS, et al. 6-Hydroxydopamine-induced alterations in blood-brain barrier permeability. Eur J Neurosci. 2005;22(5):1158–68.
57. Gray MT, Woulfe JM. Striatal blood-brain barrier permeability in Parkinson's disease. J Cereb Blood Flow Metab. 2015;35(5):747–50.
58. de Rus JA, Alpaugh M, Denis HL, Tancredi JL, Boutin M, Decaestecker J, et al. The contribution of inflammatory astrocytes to BBB impairments in a brain-chip model of Parkinson's disease. Nat Commun. 2023;14(1):3651.
59. Bancroft EA, Srinivasan R. Emerging roles for aberrant astrocytic calcium signals in Parkinson's disease. Front Physiol. 2022;12:812212.
60. Semyanov A, Henneberger C, Agarwal A. Making sense of astrocytic calcium signals – from acquisition to interpretation. Nat Rev Neurosci. 2020;21(10):551–64.
61. Okubo Y, Kanemaru K, Suzuki J, Kobayashi K, Hirose K, Iino M. Inositol 1,4,5-trisphosphate receptor type 2-independent Ca(2+) release from the endoplasmic reticulum in astrocytes. Glia. 2019;67(1):113–24.
62. Huntington TE, Srinivasan R. Astrocytic mitochondria in adult mouse brain slices show spontaneous calcium influx events with unique properties. Cell Calcium. 2021;96:102383.
63. Srinivasan R, Huang BS, Venugopal S, Johnston AD, Chai H, Zeng H, et al. Ca(2+) signaling in astrocytes from Ip3r2(-/-) mice in brain slices and during startle responses in vivo. Nat Neurosci. 2015;18(5):708–17.
64. Lewerenz J, Maher P. Chronic glutamate toxicity in neurodegenerative diseases-what is the evidence? Front Neurosci. 2015;9:469.
65. Sonnewald U, Schousboe A. Introduction to the glutamate-glutamine cycle. Adv Neurobiol. 2016;13:1–7.
66. Malik AR, Willnow TE. Excitatory amino acid transporters in physiology and disorders of the central nervous system. Int J Mol Sci. 2019;20(22):5671.
67. Iovino L, Tremblay ME, Civiero L. Glutamate-induced excitotoxicity in Parkinson's disease: the role of glial cells. J Pharmacol Sci. 2020;144(3):151–64.
68. Kim JM, Cha SH, Choi YR, Jou I, Joe EH, Park SM. DJ-1 deficiency impairs glutamate uptake into astrocytes via the regulation of flotillin-1 and caveolin-1 expression. Sci Rep. 2016;6:28823.
69. Chotibut T, Meadows S, Kasanga EA, McInnis T, Cantu MA, Bishop C, Salvatore MF. Ceftriaxone reduces L-dopa-induced dyskinesia severity in 6-hydroxydopamine parkinson's disease model. Mov Disord. 2017;32(11):1547–56.
70. Corkrum M, Covelo A, Lines J, Bellocchio L, Pisansky M, Loke K, et al. Dopamine-evoked synaptic regulation in the nucleus accumbens requires astrocyte activity. Neuron. 2020;105(6):1036–47.e5.
71. Xin W, Schuebel KE, Jair KW, Cimbro R, De Biase LM, Goldman D, Bonci A. Ventral midbrain astrocytes display unique physiological features and sensitivity to dopamine D2 receptor signaling. Neuropsychopharmacology. 2019;44(2):344–55.
72. Schmidt S, Linnartz B, Mendritzki S, Sczepan T, Lübbert M, Stichel CC, Lübbert H. Genetic mouse models for Parkinson's disease display severe pathology in glial cell mitochondria. Hum Mol Genet. 2011;20(6):1197–211.
73. Huang X, Su Y, Wang N, Li H, Li Z, Yin G, et al. Astroglial connexins in neurodegenerative diseases. Front Mol Neurosci. 2021;14:657514.
74. Nielsen MS, Axelsen LN, Sorgen PL, Verma V, Delmar M, Holstein-Rathlou NH. Gap junctions. Compr Physiol. 2012;2(3):1981–2035.
75. Giaume C, Naus CC, Sáez JC, Leybaert L. Glial connexins and pannexins in the healthy and diseased brain. Physiol Rev. 2021;101(1):93–145.

76. Retamal MA, Cortés CJ, Reuss L, Bennett MV, Sáez JC. S-nitrosylation and permeation through connexin 43 hemichannels in astrocytes: induction by oxidant stress and reversal by reducing agents. Proc Natl Acad Sci USA. 2006;103(12):4475–80.
77. Retamal MA, Froger N, Palacios-Prado N, Ezan P, Sáez PJ, Sáez JC, Giaume C. Cx43 hemichannels and gap junction channels in astrocytes are regulated oppositely by proinflammatory cytokines released from activated microglia. J Neurosci. 2007;27(50):13781–92.
78. Rufer M, Wirth S, Hofer A, Dermietzel R, Pastor A, Kettenmann H, Unsicker K. Regulation of connexin-43, GFAP, and FGF-2 is not accompanied by changes in astroglial coupling in MPTP-lesioned, FGF-2-treated Parkisonian mice. J Neurosci Res. 1996;46(5):606–17.
79. Maatouk L, Yi C, Carrillo-de Sauvage MA, Compagnion AC, Hunot S, Ezan P, et al. Glucocorticoid receptor in astrocytes regulates midbrain dopamine neurodegeneration through connexin hemichannel activity. Cell Death Differ. 2019;26(3):580–96.
80. Fujita A, Yamaguchi H, Yamasaki R, Cui Y, Matsuoka Y, Yamada KI, Kira JI. Connexin 30 deficiency attenuates A2 astrocyte responses and induces severe neurodegeneration in a 1-methyl-4-phenyl-1,2,3,6-tetrahydropyridine hydrochloride Parkinson's disease animal model. J Neuroinflammation. 2018;15(1):227.
81. Díaz EF, Labra VC, Alvear TF, Mellado LA, Inostroza CA, Oyarzún JE, et al. Connexin 43 hemichannels and pannexin-1 channels contribute to the α-synuclein-induced dysfunction and death of astrocytes. Glia. 2019;67(8):1598–619.
82. Takano T, Tian GF, Peng W, Lou N, Libionka W, Han X, Nedergaard M. Astrocyte-mediated control of cerebral blood flow. Nat Neurosci. 2006;9(2):260–7.
83. Zarate SM, Huntington TE, Bagher P, Srinivasan R. Aging reduces calreticulin expression and alters spontaneous calcium signals in astrocytic endfeet of the mouse dorsolateral striatum. NPJ Aging. 2023;9(1):5.
84. Hoenen C, Gustin A, Birck C, Kirchmeyer M, Beaume N, Felten P, et al. Alpha-synuclein proteins promote pro-inflammatory cascades in microglia: stronger effects of the A53T mutant. PLoS One. 2016;11(9):e0162717.
85. Barkholt P, Sanchez-Guajardo V, Kirik D, Romero-Ramos M. Long-term polarization of microglia upon α-synuclein overexpression in nonhuman primates. Neuroscience. 2012;208:85–96.
86. Gerhard A, Pavese N, Hotton G, Turkheimer F, Es M, Hammers A, et al. In vivo imaging of microglial activation with [11C](R)-PK11195 PET in idiopathic Parkinson's disease. Neurobiol Dis. 2006;21(2):404–12.
87. Sun H, Liang R, Yang B, Zhou Y, Liu M, Fang F, et al. Aquaporin-4 mediates communication between astrocyte and microglia: implications of neuroinflammation in experimental Parkinson's disease. Neuroscience. 2016;317:65–75.
88. Kitchen P, Salman MM, Halsey AM, Clarke-Bland C, MacDonald JA, Ishida H, et al. Targeting aquaporin-4 subcellular localization to treat central nervous system edema. Cell. 2020;181(4):784–99.e19.
89. Salman MM, Kitchen P, Woodroofe MN, Brown JE, Bill RM, Conner AC, Conner MT. Hypothermia increases aquaporin 4 (AQP4) plasma membrane abundance in human primary cortical astrocytes via a calcium/transient receptor potential vanilloid 4 (TRPV4)- and calmodulin-mediated mechanism. Eur J Neurosci. 2017;46(9):2542–7.
90. Chai H, Diaz-Castro B, Shigetomi E, Monte E, Octeau JC, Yu X, et al. Neural circuit-specialized astrocytes: transcriptomic, proteomic, morphological, and functional evidence. Neuron. 2017;95(3):531–549.e9.
91. Ztaou S, Amalric M. Contribution of cholinergic interneurons to striatal pathophysiology in Parkinson's disease. Neurochem Int. 2019;126:1–10.
92. Bondi H, Chiazza F, Masante I, Bortolotto V, Canonico PL, Grilli M. Heterogenous response to aging of astrocytes in murine substantia Nigra pars compacta and pars reticulata. Neurobiol Aging. 2023;123:23–34.
93. Escartin C, Galea E, Lakatos A, O'Callaghan JP, Petzold GC, Serrano-Pozo A, et al. Reactive astrocyte nomenclature, definitions, and future directions. Nat Neurosci. 2021;24(3):312–25.

94. Zamanian JL, Xu L, Foo LC, Nouri N, Zhou L, Giffard RG, Barres BA. Genomic analysis of reactive astrogliosis. J Neurosci. 2012;32(18):6391–410.
95. Barcia C, Ros CM, Annese V, Gómez A, Ros-Bernal F, Aguado-Yera D, et al. IFN-γ signaling, with the synergistic contribution of TNF-α, mediates cell specific microglial and astroglial activation in experimental models of Parkinson's disease. Cell Death Dis. 2011;2(4):e142.
96. Soreq L, Rose J, Soreq E, Hardy J, Trabzuni D, Cookson MR, et al. Major shifts in glial regional identity are a transcriptional hallmark of human brain aging. Cell Rep. 2017;18(2):557–70.
97. Yang Y, Song J-J, Choi YR, Kim S-H, Seok M-J, Wulansari N, et al. Therapeutic functions of astrocytes to treat α-synuclein pathology in Parkinson's disease. Proc Natl Acad Sci. 2022;119(29):e2110746119.
98. Gill SS, Patel NK, Hotton GR, O'Sullivan K, McCarter R, Bunnage M, et al. Direct brain infusion of glial cell line-derived neurotrophic factor in Parkinson disease. Nat Med. 2003;9(5):589–95.
99. Marks WJ Jr, Bartus RT, Siffert J, Davis CS, Lozano A, Boulis N, et al. Gene delivery of AAV2-neurturin for Parkinson's disease: a double-blind, randomised, controlled trial. Lancet Neurol. 2010;9(12):1164–72.
100. Tong H, Zhang X, Meng X, Lu L, Mai D, Qu S. Simvastatin inhibits activation of NADPH oxidase/p38 MAPK pathway and enhances expression of antioxidant protein in Parkinson disease models. Front Mol Neurosci. 2018;11:165.
101. Du R-W, Bu W-G. Simvastatin prevents neurodegeneration in the MPTP mouse model of Parkinson's disease via inhibition of A1 reactive astrocytes. Neuroimmunomodulation. 2021;28(2):82–9.
102. Stevens KN, Creanor S, Jeffery A, Whone A, Zajicek J, Foggo A, et al. Evaluation of simvastatin as a disease-modifying treatment for patients with Parkinson disease: a randomized clinical trial. JAMA Neurol. 2022;79(12):1232–41.

Astrocyte–Neuron Interactions in Alzheimer's Disease

Clara Muñoz-Castro and Alberto Serrano-Pozo

Abstract Besides its two defining misfolded proteinopathies—Aβ plaques and tau neurofibrillary tangles—Alzheimer's disease (AD) is an exemplar of a neurodegenerative disease with prominent reactive astrogliosis, defined as the set of morphological, molecular, and functional changes that astrocytes suffer as the result of a toxic exposure. Reactive astrocytes can be observed in the vicinity of plaques and tangles, and the relationship between astrocytes and these AD neuropathological lesions is bidirectional so that each AD neuropathological hallmark causes specific changes in astrocytes, and astrocytes modulate the severity of each neuropathological feature in a specific manner. Here, we will review both how astrocytes change as a result of their chronic exposure to AD neuropathology and how those astrocytic changes impact each AD neuropathological feature. We will emphasize the repercussions that AD-associated reactive astrogliosis has for the astrocyte–neuron interaction and highlight areas of uncertainty and priorities for future research.

Keywords Alzheimer's disease · Amyloid plaques · Astrocyte · Neurofibrillary tangles · Neuron · Reactive astrogliosis

Abbreviations

4R-tau	Four-repeat tau
AAV	Adeno-associated virus
Aβ	Amyloid-β

C. Muñoz-Castro
Instituto de Biomedicina de Sevilla, IBiS/Hospital Universitario Virgen del Rocío/CSIC/
Departamento de Bioquímica y Biología Molecular, Universidad de Sevilla, Seville, Spain
e-mail: cmunoz5@us.es

A. Serrano-Pozo (✉)
Massachusetts General Hospital Neurology Department, Boston, MA, USA

Harvard Medical School, Boston, MA, USA
e-mail: aserrano1@mgh.harvard.edu

E. Blanco-Suarez, I. Farhy-Tselnicker (eds.), *Astrocyte-Neuron Interactions in Health and Disease*, Advances in Neurobiology 39,
https://doi.org/10.1007/978-3-031-64839-7_14

AβPP	Aβ precursor protein
AD	Alzheimer's disease
ADORA1	Adenosine 1 receptor A1
APOE	Apolipoprotein E
APOJ	Apolipoprotein J
AQP4	Aquaporin-4
ARTAG	Aging-related tau astrogliopathy
A2AR	Adenosine 2A receptor
BAG3	Bcl-2-associated athanogene family molecular chaperone regulator 3
Best1	Bestrophin-1
CA	Constitutively active
CA1	Cornus ammonis 1 hippocampal subfield
CAA	Cerebral amyloid angiopathy
CLU	Clusterin
CMA	Chaperone-mediated autophagy
CSF	Cerebrospinal fluid
CTRL	Control
CXCL10	C-X-C motif chemokine 10
Cx43	Connexin-43
DG	Dentate gyrus granular cell layer
FGF1/2	Fibroblast growth factor 1/2
FITC	Fluorescein isothiocyanate
GAD67	Glutamate decarboxylase 67 kDa
GAT3/4	GABA transporter 3/4
GFA	Granular/fuzzy astrocytes
GFAP	Glial fibrillary acidic protein
GPC4	Glypican-4
HMOX1	Heme oxygenase 1
H2O2	Hydrogen peroxide
HSF1	Heat shock factor 1
IDE	Insulin-degrading enzyme
IL-3	Interleukin-3
IL3Rα	IL-3 receptor α
IL-6	Interleukin-6
ISF	Interstitial fluid
ITGB1	Integrin-β1 receptor
K+	Potassium
KI	Knock-in
KO	Knock-out
LAMP2A	Lysosome-associated membrane protein 2A
LCN2	Lipocalin-2
LDLR	Low-density lipoprotein receptor
LRP1/4	Lipoprotein receptor-related protein 1/4
LV	Lentivirus

MAO-B	Monoamine oxidase-B
MRI	Magnetic resonance imaging
mGluR5	Metabotropic glutamate receptor 5
MMP2/9	Matrix metalloprotease 2/9
NFE2L2 or NRF2	Nuclear factor erythroid 2-related factor 2
PAP	Peri-synaptic astrocyte process
PET	Positron emission tomography
pffs	Pre-formed fibrils
PRDX6	Peroxiredoxin-6
pTau	Hyperphosphorylated tau
P2Y1R	Purinergic 2Y1 receptor
ROS	Reactive oxygen species
SOD1/2	Superoxide dismutase 1/2
TFEB	Transcription factor EB
TNC	Tenascin-C
TRPA1	Transient receptor potential ankyrin 1
TSA	Thorn-shaped astrocytes
α2-NKA	α2-Na+/K+ adenosine triphosphatase

1 Introduction to Alzheimer's Disease

1.1 Definition of Alzheimer's Disease

Alzheimer's disease (AD) is the most common neurodegenerative disease and the most common cause of dementia, typically presenting with an amnestic syndrome followed by deficits in executive, visuospatial, and language domains [1]. Its neuropathological diagnosis requires the presence of two microscopic lesions—senile plaques and neurofibrillary tangles—in sufficient numbers and widespread throughout the cerebral cortex [2–4]. Although for many decades the ascertainment of these two lesions was only possible at postmortem autopsy examination, it is now possible to accurately diagnose AD in living individuals by demonstrating the accumulation of plaques and tangles in the brain via cerebrospinal fluid (CSF) and/or plasma biomarkers and positron emission tomography (PET) imaging [5]. The latter imaging technique has made it possible to confirm the early predictions of inherently cross-sectional postmortem neuropathological studies suggesting that plaques deposit throughout the cortex even before cognitive decline begins, whereas tangles accumulate following a stereotyped spatiotemporal pattern through interconnected areas (i.e., from medial temporal lobe to cortical association areas to cortical primary areas) and track better with both clinical presentation and severity of cognitive decline [6–8].

1.2 The Neuropathological Landscape of Alzheimer's Disease

1.2.1 Aβ Pathology

Senile plaques are extracellular deposits primarily made of the amyloid-β (Aβ) peptide, which is a by-product of the Aβ precursor protein (AβPP) released after its sequential cleavage by two enzymes: β-secretase and γ-secretase. AβPP is mainly expressed by neurons; therefore, Aβ is mainly secreted by neurons [1]. Aβ plaques have been traditionally classified as diffuse, which are fleecy deposits made of an ultrastructurally amorphous material, vs. compact or dense-core plaques, which are made of Aβ fibrils (i.e., enriched in β-pleated sheet conformation, hence their positive staining with dyes such as Congo red or Thioflavin-S). The latter correspond to the senile plaques described by Alois Alzheimer. Aβ can also deposit in the wall of small- and medium-sized arterioles and capillaries, resulting in cerebral amyloid angiopathy (CAA), which is an independent contributor to cognitive decline [9] and can lead to ischemic (cortical microinfarcts, enlarged perivascular spaces, white matter changes), hemorrhagic (i.e., lobar cerebral microbleeds, cortical superficial siderosis), and inflammatory (i.e., CAA-related inflammation) complications [10].

1.2.2 Tau Pathology

Neurofibrillary tangles are intraneuronal somato-dendritic deposits mainly composed of abnormally misfolded and hyperphosphorylated tau (heretofore phospho-tau or pTau), which is a microtubule binding protein thought to stabilize microtubules in the axon to facilitate axonal transport. Neuropil threads—another tau lesion—are axonal segments filled with aggregated pTau. They are thought to precede neurofibrillary tangle formation, which involves the translocation of pTau from the axon to the somato-dendritic compartment of the neurons [11].

1.2.3 Neurodegenerative Features

The neurodegeneration associated with AD comprises a massive loss of neurons, synapses (i.e., both presynaptic axonal boutons and postsynaptic dendritic spines), and dendritic arbors, resulting in macroscopically evident cortical thinning and ventricle enlargement (*ex vacuo* hydrocephalus). The neuronal death is not indiscriminate but rather predominantly targets excitatory projecting neurons in vulnerable regions such as layer II stellate neurons in the entorhinal cortex, pyramidal neurons in hippocampal CA1 and subiculum, and pyramidal neurons in layers II and V of multimodal association neocortical areas [4]. The observation of atrophy of these medial temporal lobe structures and cortical areas via magnetic resonance imaging (MRI) has diagnostic and prognostic value in the appropriate clinical setting

[12–14]. The synapse loss is not just a consequence of the neuron loss, but surviving neurons show fewer dendritic spines, suggesting that it precedes neuronal demise [15, 16]. This synapse loss is particularly prominent in the halo of dense-core amyloid plaques [17, 18], but tangle-bearing neurons also exhibit loss of dendritic spines [19, 20]. Additionally, another neurodegenerative feature is the dystrophic neurites that decorate Aβ plaques, particularly those compact; these are tortuous and/or bulbous axonal and, to a lesser extent, dendritic segments which can be depicted via silver staining and immunohistochemistry with various pTau, neurofilament, ubiquitin, and organelle-specific antibodies, because they accumulate ubiquitinated proteins and pTau aggregates, lysosomes (e.g., LAMP1), and mitochondria (e.g., VDAC1) [4]. It is noteworthy that these so-called neuritic plaques are the best exemplar of Aβ accumulation leading to pTau aggregation and that they impair neural network function [21, 22] and independently contribute to cognitive impairment [9]. Besides compact and diffuse, plaques can also be classified into neuritic vs. non-neuritic based on the presence or absence of dystrophic neurites. Although most compact plaques are neuritic and most diffuse plaques are non-neuritic, there are also compact non-neuritic and diffuse neuritic plaques [23, 24].

The relationship between astrocytes and these AD neuropathological changes is bidirectional, with each AD feature causing specific changes in astrocytes and astrocytes modulating the severity of each neuropathological hallmark in a specific manner. Here, we will review both how astrocytes change as a result of their chronic exposure to AD neuropathology and how those astrocytic changes impact AD neuropathology. We will emphasize the repercussions that those changes have for the astrocyte–neuron interaction and highlight areas of uncertainty.

2 Basic Features of Reactive Astrogliosis in Alzheimer's Disease

2.1 Alzheimer's Disease as Epitome of Reactive Astrogliosis

Alois Alzheimer's famous drawing illustrating a senile plaque stained with silver stain included a few astrocytes with prominent morphology surrounding it, so-called "reactive astrocytes" [25]. But it was not until the 1980s with the advent of immunohistochemistry and antibodies against the glial fibrillary acidic protein (GFAP)—a major constituent of the astrocyte intermediate filament cytoskeleton, which increases many-fold in response to injury—that neuropathologists described the full extent and severity of reactive astrogliosis in the AD brain [26, 27]. Indeed, based on GFAP immunohistochemistry, AD could be considered the epitome of a chronic neurodegenerative disease with prominent reactive astrogliosis, perhaps only comparable to amyotrophic lateral sclerosis and prion disease. Although in AD GFAP-immunoreactive astrocytes typically surround both dense-core Aβ plaques and neurofibrillary tangles [28] (Fig. 1), GFAP upregulation reflects only one

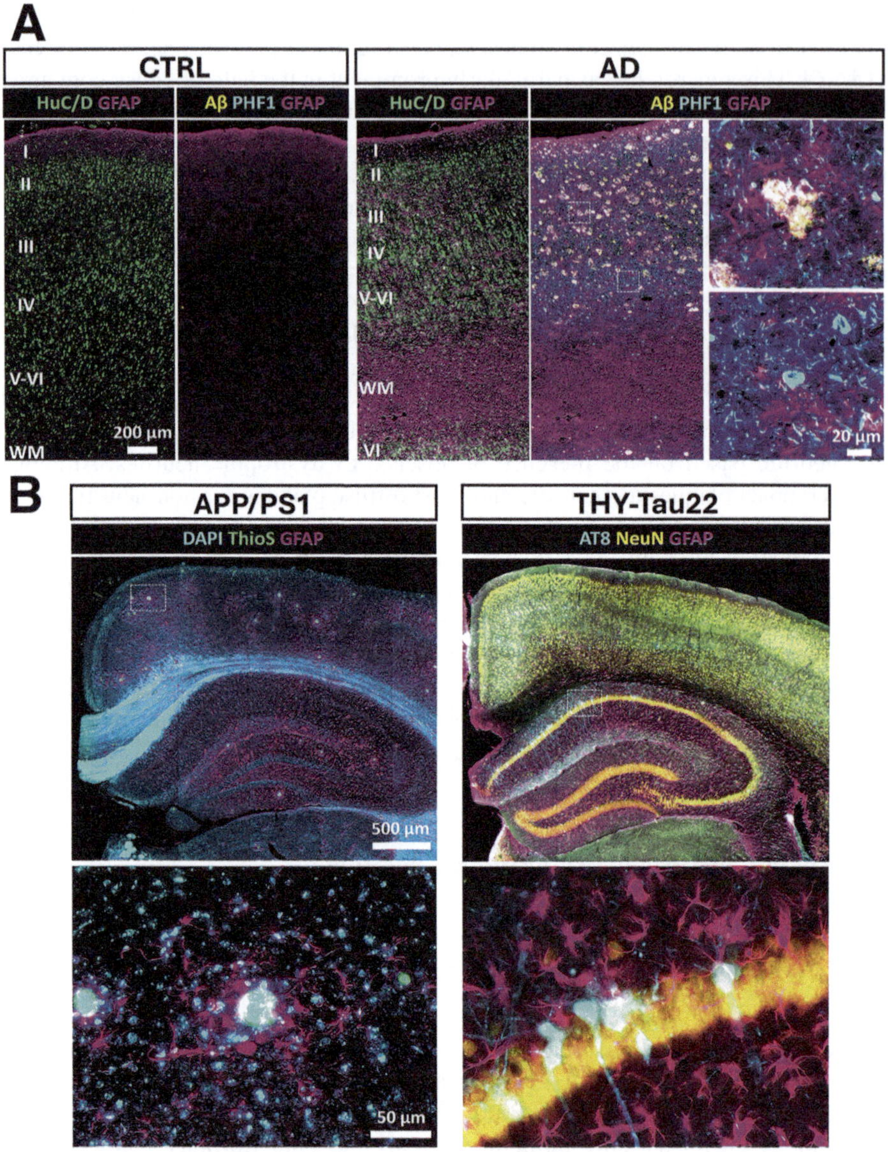

Fig. 1 Spatial relationships of reactive astrocytes and AD neuropathological changes. GFAP+ astrocytes in the vicinity of Aβ plaques (immunostained for Aβ or stained with Thioflavin-S [ThioS]) and neurofibrillary tangles (immunostained for PHF1 or AT8) in (**a**) the human AD cortex and (**b**) the cortex and hippocampus of AD transgenic mouse models of β-amyloidosis (*APP*swe/*PSEN1*dE9) and tauopathy (THY-Tau22). Neuronal nuclei are immunostained with HuC/D in (**a**) and NeuN in (**b**). Abbreviation: *CTRL* control

feature of reactive astrogliosis—the remodeling of the astrocyte intermediate filament cytoskeleton—and does not capture the complexity and heterogeneity of astrocyte responses [29–31]. Thus, the construct "reactive astrogliosis" was recently redefined to capture the full set of context- and time-dependent morphological, molecular, and functional changes that astrocytes exhibit in the setting of acute or chronic central nervous system insults [32]. Moreover, GFAP immunoreactivity is not always synonymous with reactive astrogliosis since subpial interlaminar astrocytes in cortical layer I and fibrous astrocytes in the white matter are GFAP-immunoreactive in the normal-aging human brain [26, 33].

2.2 *Alzheimer's Disease-Associated Reactive Astrogliosis Is Nonproliferative*

Unlike acute CNS injuries, where astrocytes proliferate and migrate toward the site of injury and form a glial scar to prevent the damage from extending to the surrounding intact neuropil [34], the proliferation of astrocytes in AD is very limited. This conclusion is supported by both 5′-bromo-deoxy-uridine (5'-BrdU) injection experiments in multiple mouse models of β-amyloidosis showing very rare proliferating astrocytes around plaques [35–37] and careful quantitative immunohistochemical studies rendering similar numbers of astrocytes in postmortem AD and control brains [38]. That is likely why eliminating proliferative astrocytes in a transgenic AD mouse model had little effect on Aβ plaque burden or size [39].

Unlike microglia, which are highly motile and recruited to plaques [40], migration of astrocytes toward plaques and tangles is thought to be very limited; in other words, it is the astrocytes located in areas adjacent to where these lesions form that become reactive [41] and any loss of the typical "tiling" spatial distribution of astrocytes in postmortem AD brains could be explained by the dramatic cortical atrophy resulting from neuronal loss. That is likely why, unlike microglia, the number of reactive astrocytes observed in the peri-plaque area does not correlate with plaque size [42]. Thus, even though GFAP-laden reactive astrocytes surround dense-core plaques, the concept of glial scar formed by newly proliferated astrocytes migrating toward injury does not strictly apply to AD [34]. Indeed, while the intermediate filament proteins Gfap and Vimentin are required for glial scar formation around acute CNS injuries [43], genetic deletion of Gfap and Vimentin resulted in conflicting outcomes with regard to Aβ plaque growth [44, 45].

2.3 *Alzheimer's Disease-Associated Reactive Astrogliosis Is Complex and Heterogeneous*

The advent of transcriptomics methods, including single nucleus and single-cell RNA sequencing (sn/scRNA-seq), has enabled a fuller characterization of the molecular changes that astrocytes suffer in AD, offering insights well beyond the spatial information provided by GFAP immunohistochemistry. Although transcriptomics-functional correlations are imperfect, transcriptomic studies in mouse models genetically engineered to develop Aβ plaques and neurofibrillary tangles have demonstrated distinct responses of astrocytes to Aβ and pTau [46, 47], including a mix of deleterious and neuroprotective signatures [47]. Similarly, human snRNA-seq studies have revealed the existence of diverse states of astrocytes with various potential functional correlates [48–55]. Spatial transcriptomic approaches will eventually confirm that these changes are occurring in the vicinity of plaques and tangles [56–59]. Interestingly, human snRNA-seq studies have uncovered a loss of astrocyte homeostatic functions, particularly at very advanced stages of the disease [53, 60], which mouse models may only partially recapitulate [61], implying that, in addition to neurons, reactive astrocytes chronically exposed to the toxic microenvironment of plaques and tangles ultimately suffer a dramatic dysfunction. Thus, the picture emerging from these studies is that reactive astrogliosis in AD is complex and heterogeneous, with distinct gene expression programs triggered by Aβ and pTau exposures, neuroprotective (adaptive) and neurotoxic (maladaptive) responses, and functional gains and losses, likely compounded by differences in intrinsic astrocyte properties across brain regions and by duration of the exposure (i.e., disease stage) [53].

3 What Does Alzheimer's Disease Neuropathology Do to Astrocytes?

How do astrocytes change in the presence of Aβ plaques, neurofibrillary tangles, and AD-associated neurodegeneration? Are there differences between Aβ and pTau-induced astrocyte responses? Neuropathological studies in human postmortem brains [31], experimental manipulations of astrocytes in AD transgenic mouse models and, more recently, sn/sc-RNA-seq studies in mouse models and postmortem human control and AD brains have helped answer some of these questions.

3.1 Effects of Aβ on Astrocytes

Reactive astrocytes in contact with dense-core, usually neuritic, Aβ plaques undergo a number of molecular and functional changes, whereas those close to diffuse Aβ deposits do not seem to be perturbed [23, 25]. The most prominent changes are described in the following sections.

3.1.1 Cytoskeleton Remodeling

One of the most widely known features of AD-associated reactive astrogliosis is the upregulation of GFAP, particularly in the vicinity of dense-core Aβ plaques. While *GFAPα*, encoding GFAP isoform 1 with 432 amino acids (i.e., full-length), is the most abundant GFAP splice isoform, AD reactive astrocytes also upregulate other GFAP splice isoforms as well as the intermediate filament proteins Vimentin, Nestin, and Synemin [62]. SnRNA-seq studies have also revealed upregulation of expression of genes encoding components of the microtubule (i.e., microtubule-associated proteins such as *MAPT*) and actin cytoskeletons [53]. It should be noted that a hypertrophic cytoskeleton is not synonymous with a hypertrophic astrocyte and that the astrocyte volume may not change that much despite the visual effect of GFAP immunoreactivity [32]. Moreover, astrocytes distant from plaques may suffer atrophy of their GFAP cytoskeleton [63].

3.1.2 Glutamate Transport Defects

The effect of Aβ plaques on astrocyte glutamate metabolism has been investigated in detail, although studies have rendered mixed results. Some studies have reported that AD reactive astrocytes downregulate the glutamate transporter EAAT2/GLT-1 and glutamine synthetase [48, 64]. This is consistent with the finding of an excess of extracellular glutamate around Aβ plaques [65] and could partly explain the observation of hyperactive neurons near them [66]. However, other studies have found no change in EAAT2/GLT-1 brain levels in AD vs. control donors [67] or have suggested a more complex picture in which a subset of astrocytes may exhibit a compensatory increase of EAAT2/GLT-1 and/or EAAT1/GLAST1 [30].

3.1.3 Calcium Dysregulation

Astrocytes from plaque-depositing transgenic mice exhibit calcium dyshomeostasis. This consists of higher resting levels of calcium independent of their proximity to plaques, more frequent calcium transients which are synchronously coordinated across long distances and not coupled with neuronal activity, as well as calcium waves originating from the plaques and extending up to 200 μm [68, 69]. Some of these alterations can be triggered by soluble Aβ oligomers in the absence of plaques [70]. This astrocyte calcium dyshomeostasis is thought to be mainly due to an increased calcium influx from the interstitial space mediated by G-protein-coupled receptors such as adenosine 2A receptor (A2AR) [71], purinergic 2Y1 receptor (P2Y1R) [72, 73], and the metabotropic glutamate receptor mGluR5 [74–76], as well as by the transient receptor potential ankyrin 1 (TRPA1) [77–79]. However, the calcium buffering capacity of the astrocyte endoplasmic reticulum and mitochondria intracellular stores is likely reduced as well and contributes to the increased cytosolic calcium too. Calcium waves propagate across astrocytes through connexon gap junctions made of six connexin hemichannels like connexin-43 (Cx43, encoded by *GJA1*), which are upregulated in plaque-associated reactive astrocytes; thus, deletion [80, 81] or pharmacological inhibition [82] of Cx43 is neuroprotective and ameliorates cognitive deficits. Indeed, astrocyte calcium dyshomeostasis impairs neural network function and cognition [71, 73, 83, 84]. Moreover, calcium excess in astrocytes can activate the Ca^{+2}-dependent phosphatase calcineurin and the calcineurin/NFAT signaling pathway and drive neuroinflammation and network hyperexcitability, which can be restored with calcineurin inhibitors or blocking the calcineurin-NFAT interaction [83–85].

3.1.4 Inflammation

The plaque microenvironment contains a milieu of pro-inflammatory cytokines. Besides microglia, plaque-associated reactive astrocytes are a major source of these cytokines. Specifically, reactive astrocytes can secrete complement factors C1q and C3, interleukin-6 (IL-6), and C-X-C motif chemokine 10 (CXCL10), among other cytokines [86]. Several transcription factors have been implicated in this inflammatory response and examined as potential therapeutic targets including the NF-κB [87–90], the JAK2/STAT3 [91–95], and the calcineurin/NFAT [83–85] pathways.

3.1.5 Oxidative Stress

The plaque microenvironment also contains high concentrations of reactive oxygen species (ROS) [96], and reactive astrocytes surrounding Aβ plaques contribute to this oxidative stress via several proposed mechanisms underscoring astrocyte mitochondrial dysfunction: (1) Depletion of glutathione and downregulation of glutathione transferase [97]; (2) activation of NADPH oxidase [98]; and (3) upregulation of monoamine oxidase-B (MAO-B), an enzyme that catabolizes monoamines giving rise to hydrogen peroxide (H_2O_2) as a by-product [99]. However, reactive astrocytes also exhibit upregulation of antioxidant defenses, including metallothioneins, heme oxygenase 1 (HMOX1), and peroxiredoxin-6 (PRDX6) [31, 53, 100]. The nuclear factor erythroid 2-related factor 2 (NFE2L2), also known as NRF2, is the main transcription factor orchestrating this antioxidant response [47, 88, 101]. Of note, NRF2 also induces an anti-inflammatory response by counteracting NF-κB [88].

3.1.6 Proteostasis Stress

Reactive astrocytes around Aβ plaques upregulate heat shock proteins such as HSP (also known as HSPB1) and αB-crystallin [53, 102–104]. These are small chaperones that prevent the misfolding of proteins and the subsequent formation of toxic oligomers, and their expression is mainly induced by the transcription factor heat shock factor 1 (HSF1) [105].

3.1.7 Loss of Perivascular Aquaporin-4

Aquaporin-4 (AQP4) has emerged as a key player in the removal of misfolded proteins Aβ and tau (among other waste solutes) from the brain through the so-called perivascular glymphatic pathway [106]. In the mouse brain, AQP4 expression is exquisitely polarized to the astrocytic vascular endfeet, and its deletion [107–113], inhibition [107, 114], or mislocalization [110, 115] is associated with increased Aβ and tau deposition. In the normal human brain, AQP4 is mainly expressed by periventricular white matter astrocytes, as well as perivascular and subpial cortical astrocytes. In the AD brain, however, cortical protoplasmic astrocytes surrounding Aβ plaques upregulate AQP4 [53], and its predominant perivascular localization is lost, which could result in impaired perivascular drainage of both Aβ and tau [116].

3.1.8 Extracellular Matrix Remodeling

Astrocytes are a major cellular source of the extracellular matrix, and AD reactive astrocytes upregulate and release several components of the extracellular matrix, especially in the plaque microenvironment, including CD44, matrix metalloproteases (MMPs), and tenascin-C (TNC) [31, 53]. Notably, the hyaluronic acid receptor CD44 is among the top most consistently upregulated proteins in AD vs. control brains in a compilation of bulk proteomic studies, but its role in AD is not well known [117].

3.2 *Effects of pTau on Astrocytes*

Much less is known about the effect of pTau (both soluble aggregates and neurofibrillary tangles) on astrocytes partly because the spatial association of GFAP-immunoreactive astrocytes with neurofibrillary tangles in the AD brain is less evident than that with Aβ plaques [28] and partly due to the difficulty of distinguishing between the effects of pTau vs. those of pTau-induced neurodegeneration on astrocyte phenotype.

3.2.1 Cytoskeleton Remodeling

Similar to Aβ, exposure of astrocytes to pTau aggregates leads to an upregulation of intermediate filaments GFAP and Vimentin as well as other cytoskeletal proteins, resulting in subsequent reactive morphological changes of astrocytes [46]. However, this phenotype is likely dependent on pTau levels and/or species and exacerbated by pTau-induced neuronal loss because mouse models of tauopathy with limited neuronal loss do not exhibit increased GFAP levels [118].

3.2.2 Glutamate Transport Defects

Besides Aβ plaques, pTau has also been implicated in the dysfunction of astrocyte glutamate metabolism. pTau propagation along neural networks following Braak stages has been associated with changes in expression levels of *SLC1A2*, encoding EAAT2/GLT-1 [53, 119]. It has been reported that EAAT2/GLT-1 expression by astrocytes is inversely correlated with Braak neurofibrillary tangle stages [64] and that preservation of EAAT2/GLT-1 expression by astrocytes in the entorhinal cortex could be a signature of resilience and help cope with AD neuropathology [120].

3.2.3 Proteostasis Stress

As with Aβ, accumulation of insoluble pTau also triggers a heat shock protein response in astrocytes, with increased expression of HSP27 in astrocytes near neurofibrillary tangles [121].

3.3 Effects of Neurodegeneration on Astrocytes

Neuronal demise results in the release of predominantly intraneuronal cytosolic molecules to the interstitial space, which can signal to nearby astrocytes and microglia and trigger their response. For example, ATP released from dying neurons can lead to reactive astrogliosis through activating purinergic receptors [71–73, 76]. Free K^+ in the interstitial fluid after neurons die can be buffered by astrocyte ATP-sensitive inward-rectifier K^+ channels such as Kir4.1, encoded by the gene *KCNJ10* [122, 123]. The accumulation of pTau aggregates in neurons can upregulate the adenosine 1 receptor A1 (encoded by *ADORA1*), which can promote neuronal release of lipocalin-2 (LCN2) and activate nearby astrocytes [124]. Ultimately, neuronal corpses are removed by microglia engulfment, and this neuronophagia is coordinated by astrocytes [125].

4 What Do Reactive Astrocytes Do to Alzheimer's Disease Neuropathology?

Do reactive astrocytes have any impact on Aβ and pTau pathologies or downstream neurodegeneration? If so, how? Are they neuroprotective or neurotoxic? Genetic (overexpression and knock down/out) or pharmacological (agonistic and inhibitory) manipulations of multiple molecular targets in astrocytes have helped answer these questions (Table 1); however, much more work is needed to fully elucidate the net effect of reactive astrogliosis at each disease stage.

Table 1 Effects of genetic and pharmacological manipulations of reactive astrocytes in AD transgenic mouse models based on the literature

Target	Mouse model/Manipulation	Astrocytes	Aβ/pTau	Neuro-degeneration	Other findings	Reference
Cytoskeleton rearrangement						
Gfap/Vimentin	APPswe/PS1dE9 × *Gfap*-KO APPswe/PS1dE9 × *Gfap*/*Vim*-KO	↓ plaque coverage, = morphology, ↑ neuronal support, lysosomal degradation, & inflammation genes	= Aβ	–	= microglial proliferation & activation	Kamphuis et al. 2015 [44]
	APPswe/PS1dE9 × *Gfap*/*Vim*- KO	↓ plaque coverage, ↓ process hypertrophy & plaque contact	↑Aβ	↑ dystrophic neurites	↑ plaque coverage by microglia	Kraft et al. 2013 [45]
Synaptic modulation						
Glt-1	APPswe/PS1-L166P + Ceftriaxone (upregulates *Slc1a2*, encoding Glt-1)	= Gfap	–	–	Restores glutamate dynamics	Hefendehl et al. 2016 [65]
Gat3/4	5 × FAD + SNAP-5114 (blocks astrocytic GABA release)	–	–	↑ LTP & memory	–	Wu et al. 2014 [196]
Inflammation						
NF-κB	APPswe/ind (line 102)-TTA × *Gfap-IκBα* $^{fl/-}$ (NF-κB activation)	↑ Gfap, ↑ C3	↑ Aβ, = Aβ production	–	↑ Iba1+ microglia	Lian et al. 2016 [89]
	APP23 ×*Gfap − TTA*× *IKK2-CA* (NF-κB activation)	↑ Gfap, ↑ C3, ↑ CXCL10, ↑ LCN2	↓ Aβ, ↓ AβPP, ↓ βCTF	–	↑ M2-microglia, ↑ Aβ microglia phagocytosis,	Yang et al. 2022 [90]
Calcineurin/NFAT	APPswe/PS1dE9 injected with AAV-*Gfap-VIVIT* (blocks calcineurin-NFAT interaction)	↓ Gfap	↓ Aβ, ↓ Bace1, ↑ neprilysin	↑ memory, synaptic function & LTP	↓ Iba1 microglia	Furman et al. 2012 [85]
	5 × FAD injected with AAV-*Gfap-VIVIT* (blocks calcineurin-NFAT interaction)	↓ nuclear Nfat4, ↑ Glt-1	↓ Aβ	↑ memory, ↓ neuronal hyperactivity, ↑ synaptic function	–	Sompol et al. 2017 [84]
Jak2	APPswe/PS1dE9 injected with AAV-*Gfap-Jak2*^{T875N} (constitutively active mutant)	–	↑ Aβ plaques, = Aβ production	–	–	Ceyzériat et al. 2018 [92]
	3 × Tg injected with AAV-*Gfap-Jak2*^{T875N} (constitutively active mutant)	↑ Gfap, ↑ Vim	= Aβ, = pTau	–	= Iba1 microglia per plaque	Guillemaud et al. 2020 [95]

(continued)

Table 1 (continued)

Target	Mouse model/Manipulation	Astrocytes	Aβ/pTau	Neuro-degeneration	Other findings	Reference
Socs3 (endogenous JAK inhibitor)	3 × Tg injected with LV-*Socs3*	↓ Stat3, ↓ Gfap, = S100β	–	–	–	Ben Haim et al. 2015 [91]
	APPswe/PS1dE9 injected with AAV-*Gfap-Socs3*	↓ Stat3, ↓ Gfap, ↓ Vim, ↓ reactive & inflammatory genes	↓ Aβ plaques, = Aβ production	↑ spatial learning & memory, ↑ LTP	↓ inflammatory genes, = Iba1 microglia per plaque	Ceyzériat et al. 2018 [92]
	3 × Tg injected with AAV-*Gfap-Socs3*	↓ Gfap, ↓ Vim (trend)	= Aβ, = pTau	= memory, ↓ anxiety	= Iba1 microglia per plaque	Guillemaud et al. 2020 [95]
Il-3	5 × FAD×*Aldh1l1-Il3*-KO	–	↑ Aβ	↓ memory	↓ microglial activation & microglia per plaque	McAlpine et al. 2021 [150]
Oxidative stress						
Nrf2	APPswe/PS1dE9 × *Nrf2*-KO	↑ Gfap	↑ Aβ, ↑ pTau	↓ spatial learning & memory	↑ Iba1, inflammation & oxidative stress	Ren et al. 2020 [101]
	5 × FAD×*Nrf2*-KO	↑ neurotoxic reactive markers	–	–	–	Nakano-Kobayashi et al. 2023 [88]
	5 × FAD + Algernon2 (potentiates Nrf2 pathway)	↓ neurotoxic reactive markers	–	↑ cognition	–	
	APPswe/PS1dE9 × *Gfap-Nrf2*	↑ proteostasis, carbohydrate metabolism, protein translation & synaptic genes	↓ Aβ, = APP processing	↓ neuro-degeneration & cognitive deficits	↑ autophagy	Jiwaji et al. 2022 [47]
	Tau P301S × *Gfap-Nrf2*	–	↓ pTau	↓ physical deterioration	–	
Prdx6	APPswe/PS1dE9 × *Prdx6*$^{+/-}$ (haplodeficiency)	↓ Gfap plaque coverage	↑ Aβ	↑ dystrophic neurites	↓ Iba1+ & Cd68+ microglia per plaque	Pankiewicz et al. 2020 [100]
	APPswe/PS1dE9 × *Prdx6*$^{129X1/SvJ}$ (overexpression)	↑ Gfap plaque coverage	↓ Aβ	↓ dystrophic neurites	↑ Iba1+ & Cd68+ microglia per plaque	

Maob	APPswe/PS1dE9 + Selegiline (Maob irreversible inhibitor)	–	–	↑ learning, memory & synaptic plasticity	–	Jo et al. 2014 [197]
Lipid metabolism						
Apoe	APPswe/PS1dE9 × *Gfap-Apoe*-KO	↓ Gfap	↓ Aβ	↑ spatial learning & memory	–	Zheng et al. 2017 [135]
	APPPS1–21 ×*APOE3* – KI× *Aldh1l1-APOE3*-KO	↓ APOE, = Gfap	↓ Aβ	↓ overall dystrophic neurites, ↑ dystrophic neurites per plaque	↓ microglial activation	Mahan et al. 2022 [133]
	APPPS1–21 ×*APOE4* – KI× *Aldh1l1-APOE4*-KO	↓ APOE, ↓ overall Gfap in females, but = Gfap around Aβ plaques	↓ Aβ	↓ overall dystrophic neurites, ↑ dystrophic neurites per plaque	↓ microglial activation	
	5×FAD × *APOE4* – KI× *Aldh1l1-APOE4*-KO	↓ ApoE ↓ overall Gfap, = Gfap plaque coverage, ↓ C3	↓ brain Aβ, ↑ CAA	↓ dystrophic neurites	↑ cerebrovascular function, = Iba1, ↓ DAM phenotype	Xiong et al. 2023 [134]
	APPswe/PS1dE9 × *Gfap-APOE3*-KI	↓ Gfap	=Aβ, = Aβ clearance	= dystrophic neurites, ↑ Psd95, = Synaptophysin	= Iba1	Liu et al. 2017 [132]
	APPswe/PS1dE9 × *Gfap-APOE4*-KI	↑ Gfap	↑Aβ, ↓ Aβ clearance	↑ dystrophic neurites = Psd95, = Synaptophysin	↑ Iba1	
	Tau P301L injected with AAV-*Gfap-APOE4*	–	↑ pTau	–	–	Jablonski et al. 2021 [131]

(continued)

Table 1 (continued)

Target	Mouse model/Manipulation	Astrocytes	Aβ/pTau	Neuro-degeneration	Other findings	Reference
Clusterin/Apoj	APPswe/PS1dE9 × *Clu*$^{+/-}$	↑ Gfap	↑ Aβ	↑ dystrophic neurites	↑Iba1 & Cst7	Wojtas et al. 2020 [137]
	APPswe/PS1dE9 injected with AAV-*Gfap-Clu*	↓ Gfap	↓ Aβ	↓ dystrophic neurites	↓ Iba1 & Cst7	
	5 × FAD injected with AAV-*Gfap-Clu*	↓ Gfap	↓ Aβ	↑ synaptic function, ↓ dystrophic neurites	↓ Iba1 & Cd68	Chen et al. 2021 [136]
Ldlr	5 × FAD×*Ldlr*-KO	↓ Gfap	↑ Aβ, = Aβ production	–	↑ Apoe, ↓ Iba1	Katsouri & Georgopoulos 2011 [141]
	5×FAD × *Apoe* – KO×*Ldlr*-KO	↓ Gfap	↑ Aβ, = Aβ production	–	↓ Iba1	
Lrp1	APPswe/PS1dE9 × *Gfap-Lrp1*-KO	↑ Gfap	↑ Aβ, = Aβ production & ↓ Aβ clearance	–	↑Iba1 & inflammation	Liu et al. 2017 [139]
Lrp4	5 × FAD×*Gfap-Lrp4*-KO	–	↑ Aβ, = Aβ production	↑ cognitive and synaptic deficits	–	Zhang et al. 2020 [142]
Tmem164	5 × FAD injected with AAV-*Gfap-Tmem164-P2A-mCherry-WPRE*	↑ Gfap, ↓ C3, ↑ lysosomal autophagy, ↓ *Elovl1* (fatty acid elongation)	↓ Aβ	↓ neuron loss, ↑ cognition	–	Zhang et al. 2023 [180]

Water homeostasis						
Aqp4	APP695/PS1dE9 × *Aqp4*-KO	↑ atrophy, ↓ Lrp1 in peri-plaque astrocytes	↑ Aβ, ↑ CAA	↑ cognitive impairment & synaptic loss	↓ Il-1β & BDNF, = Iba1, Il-6, & TNFα	Xu et al. 2015 [113]
	APP695/PS1dE9 × *Aqp4*^No_X^ (two stop codons inserted with CRISPR-Cas9)	↓ perivascular Aqp4, = total Aqp4, = Gfap	↑ ISF Aβ	–	–	Sapkota et al. 2022 [111]
	APP695/PS1dE9 × *Aqp4*-KO + voluntary exercise (vs. APP695/PS1dE9 + voluntary exercise)	↑ Gfap	↑ Aβ	↓ cognitive function & synaptic plasticity	↑ Iba1, ↓ glymphatic transport, ↓ BDNF	Liu et al. 2022 [109]
	5 × FAD×*Aqp4*-KO	↓ Gfap plaque coverage	↑ Aβ	↑ intraneuronal Aβ & dystrophic neurites	↓ Iba1 microglia plaque coverage	Smith et al. 2019 [112]
	5 × FAD×*Aqp4*-KO	–	↑ Aβ, ↑ perivascular Aβ	–	–	Pedersen et al. 2023 [110]
	5 × FAD×*Snta1*-KO	↓ perivascular Aqp4	↑ Aβ, ↑ perivascular Aβ	–	–	
	Tg2576 × *Snta1*-KO	↓ perivascular Aqp4	↑ Aβ	–	↓ CSF-ISF exchange	Simon et al. 2022 [115]
	Tg2576 injected with AAV-*Gfap-AQP4*-M1 or M23 (*AQP4* splice isoforms)	↑ perivascular Aqp4	= Aβ deposition	–	= CSF-ISF exchange	
	Aqp4-KO with intrastriatal injection of ^{125}I-Aβ_{1-40}	–	↓ Aβ clearance	–	–	Iliff et al. 2012 [106]
	C57BL/6 J injected with Aβ40 peptide + TGN-020 (Aqp4 inhibitor)	–	↑ CAA	–	= blood brain barrier	Rosu et al. 2020 [114]
	Tau P301S × *Aqp4*-KO	–	↑ pTau, ↑ CSF tau	↑ neuro-degeneration	= Lrp1	Ishida et al. 2022 [108]
	C57BL/6 with intrastriatal infusion of tau + TGN-020 (Aqp4 inhibitor)	–	↓ CSF tau	–	–	Harrison et al. 2020 [107]
	Aqp4-KO with intrastriatal infusion of tau	–	↓ CSF tau	–	–	

(continued)

Table 1 (continued)

Target	Mouse model/Manipulation	Astrocytes	Aβ/pTau	Neuro-degeneration	Other findings	Reference
Ion homeostasis						
Adora2a	APP-J20 × *Gfap-Adora2a*	–	–	↑ spatial memory	–	Orr et al. 2015 [71]
P2y1r	APPswe/PS1-L166P + MRS2179 (P2y1r inhibitor)	↓ hyperactivity	–	–	–	Delekate et al. 2014 [72]
	APPswe/PS1-L166P × *Ip3r2*-KO	–	–	↑ spatial memory & network hyper-excitability	–	Reichenbach et al. 2018 [73]
	APPswe/PS1-L166P + MRS2179 (P2y1r inhibitor)	↑ Gfap plaque coverage	= Aβ	↑ LTP, spatial memory & learning, ↑ synapses, ↓ dystrophic neurites	= microglia	
Connexins	APPswe/PS1-L166P + carbenoxolone (connexin inhibitor)	↓ hyperactivity	–	–	–	Delekate et al. 2014 [72]
	APPswe/PS1dE9 × *Gfap-Gja1*-KO	↓ astroglial reaction	= Aβ deposition	↑ memory & synaptic plasticity, ↓ Nmdar expression	= pro- & anti-inflammatory factors	Ren et al. 2018 [80]
	APPswe/PS1dE9 × *Gfap-Gja1*-KO	↓ Ca^{2+} influx, gliotransmitter & glutamate release	–	↓ neuronal damage and dystrophic neurites	–	Yi et al. 2016 [81]
	APPswe/PS1dE9 + Boldine (connexin inhibitor)	↓ hemichannel activity, = gap junctional communication	= Aβ	↓ neuronal damage	↓ microglia hemichannel activity, ↓ intracellular Ca^{2+} & ATP/glutamate release	Yi et al. 2017 [82]

Trpa1	APPswe/PS1dE9 × *Trpa1*-KO	↑ Gfap, ↓ Abca1, ↓ Lrp1, = Apoe	↓ Aβ, ↑ βCTF	↑ spatial learning & memory	↓ inflammatory cytokines, ↓ NF-κB & NFAT activity	Lee et al. 2016 [78]
	APP/PS1–21 + HC030031 (Trpa1 blocker)	↓ astrocyte hyperactivity	–	↓ neuronal hyperactivity	–	Bosson et al. 2017 [77]
	APP/PS1–21 + HC030031 (Trpa1 blocker)	↓ astrocyte hyperactivity	↑ plaque compaction	↓ neuronal hyperactivity & dysfunction, ↑ synaptic integrity and memory	–	Paumier et al. 2022 [79]
Atp1a2	Tau P301S + Digoxin (astrocyte α2-NKA inhibitor)	↓ Gfap, ↓ Lcn2	↓ pTau	↓ brain atrophy	↓ Iba1 & inflammatory cytokines	Mann et al. 2022 [178]
Proteostasis						
Tfeb	APPswe/PS1dE9 injected with AAV-*Gfap-Tfeb*	↑ Lysosomal biogenesis	↓ Aβ	–	–	Xiao et al. 2014 [144]
	rTg4510 injected with AAV-*Gfap-Tfeb*	= Gfap	= Tau pathology	–	= Iba1	Martini-Stoica et al. 2018 [161]
	Tau P301S injected with AAV-*Gfap-Tfeb*	↓ Gfap	↓ Tau pathology & spreading	–	↓ Iba1	
	Tau P301S injected with AAV-*Gfap-Tfeb* and Tau pffs	↓ Gfap ↑ pTau engulfment	↓ Tau spreading	–	↓ Iba1	
Pparα	5 × FAD + Gemfibrozil + Retinoic acid (Pparα/Tfeb activators)	Switches astrocyte to reactive state, ↑ Ldlr, ↑ lysosomal biogenesis & autophagy	↓ Aβ	↑ spatial learning and memory	–	Raha et al. 2021 [143]
	5 × FAD× *Gfap-Pparα*-KO + Gemfibrozil + Retinoic acid (Pparα/Tfeb activators)	= astrocyte reactive state	= Aβ	= spatial learning and memory	–	

(continued)

Table 1 (continued)

Target	Mouse model/Manipulation	Astrocytes	Aβ/pTau	Neuro-degeneration	Other findings	Reference
Bmal1	Tau P301S × *Aldh1l1-Bmal1*-KO	↑ Gfap	↓ pTau	↑ CA1 & DG thickness	–	Sheehan et al. 2023 [163]
Bag3	Tau P301S injected with AAV-*Gfap-shBag3* + FITC-Tau	= Gfap	↑ FITC-Tau	–	–	
Extracellular matrix						
Mmp	APPswe + GM6001 (Mmp inhibitor)	–	↑ ISF Aβ	–	–	Yin et al. 2006 [147]
Gpc4	WT injected with AAV-*EGFP-P2A*-Tau P301L + *Gpc4* shRNA	–	↓ Tau propagation	–	–	Saroja et al. 2022 [159]
	Tau P301S injected with *Gpc4* CRISPR/dCas9	↑ Gpc4	↑ Tau	–	–	
Others						
Bace1 (β-Secretase)	5 × FAD×*Gfap-Bace1*-KO	↑ Clu	↓ Aβ, = production	–	–	Zhou et al. 2023 [138]
Cell cycle	APP23 ×*Gfap-TK* + Gancyclovir (causes apoptosis of proliferative astrocytes)	↓ proliferating astrocytes	↑ non-aggregated Aβ, = Aβ plaque burden	↓ neuronal and synaptic density, and memory	↑ pro-inflammatory cytokines	Katsouri et al. 2020 [39]

AAV adeno-associated virus, *CA* constitutively active, *CA1 Cornus ammonis* 1 hippocampal subfield, *CAA* cerebral amyloid angiopathy, *CSF* cerebrospinal fluid, *DG* dentate gyrus granular cell layer, *FITC* fluorescein isothiocyanate, *ISF* interstitial fluid, *KI* knock-in, *KO* knock-out, *LV* lentivirus, pffs pre-formed fibrils

4.1 Effects of Reactive Astrocytes on Aβ

Reactive astrocytes surround and penetrate Aβ plaques with their processes [23, 27, 42] and wrap CAA-laden vessels [126]. Astrocytes containing granules of nonfibrillar Aβ can be seen in the AD brain, especially near diffuse Aβ deposits located in subpial and perivascular areas [127–129], as well as in the dyshoric changes associated with CAA-laden vessels [130], suggesting that they can take up extracellular Aβ. Indeed, astrocytes can modulate extracellular Aβ levels by secreting apolipoprotein E (APOE) [131–135] and clusterin (CLU, also known as APOJ) [136–138], which besides transporting cholesterol and other lipids, bind Aβ. Aβ-APOE complexes are internalized by astrocytes via receptor-mediated endocytosis involving APOE receptors such as lipoprotein receptor-related protein 1 (LRP1) [139], the low-density lipoprotein receptor (LDLR) [140, 141], and the lipoprotein receptor-related protein 4 (LRP4) [142]. Internalized Aβ is ultimately degraded within the astrocyte lysosomes, and enhancing astrocyte lysosomal biogenesis by upregulating the transcription factor EB (TFEB) facilitates Aβ uptake and clearance [143, 144]. It has been suggested that both aging [145] and the *APOEε4* allele [132, 146]—the allele associated with increased genetic risk of sporadic AD—impair astrocyte uptake and degradation of Aβ. Besides receptor-mediated endocytosis, astrocytes can control Aβ levels by promoting its proteolytic degradation in the interstitial space by Aβ-degrading enzymes such as MMP2 and MMP9 [147], neprilysin [148], and insulin-degrading enzyme (IDE) [149], a mechanism that is governed by APOE and the exercise-induced hormone Irisin [148, 149]. Indeed, the Aβ peptide found in diffuse deposits surrounding Aβ-containing subpial astrocytes is N-terminally truncated, suggesting its proteolytic cleavage by astrocytes [127]. Further, astrocytes may promote microglial clearance of Aβ by secreting interleukin-3 (IL-3), which binds to the microglial surface receptor IL-3 receptor α (IL3Rα, also known as CD123) [150]. Lastly, reactive astrocytes could impact Aβ levels in an indirect manner by causing neuronal dysfunction since Aβ generation and secretion to the extracellular space by neurons is thought to be coupled with neuronal activity [151].

4.2 Effects of Reactive Astrocytes on pTau

Reactive astrocytes also surround tangle-bearing neurons [28]. In particular, extracellular ghost tangles are infiltrated by astrocyte processes, and these GFAP-immunoreactive ghost tangles are considered the end-stage within the neurofibrillary tangle lifetime [86, 152–154]. Moreover, a subset of patients with AD exhibit pTau-immunoreactive astrocytes with thorn-shape or granular/fuzzy morphologies in subpial, perivascular, and white matter areas, which can also be found in otherwise normal brains with so-called aging-related tau astrogliopathy (ARTAG) [155]. How do these astrocyte tau inclusions form? Astrocytes express detectable amounts of *MAPT* mRNA (encoding the microtubule-associated protein tau), and reactive astrocytes appear to upregulate *MAPT* as AD advances [53]; therefore, the pTau-immunoreactive thorn-shaped astrocytes (TSA) and granular/fuzzy astrocytes (GFA) found in some AD brains could be explained by an increased endogenous tau synthesis. Moreover, in the brain of patients with progressive supranuclear palsy, tufted astrocytes—one of its pathological hallmarks, defined by pTau aggregation in the astrocyte cell body and proximal branches, giving a star-like tuft appearance—seem to express higher *MAPT* levels than tau-free astrocytes [156], which further supports this possibility.

Another explanation for these astrocytic pTau inclusions is the uptake of extracellular pTau by reactive astrocytes, which are capable of eliminating pTau through several proteostasis mechanisms. While heparan sulfate proteoglycans [157] and LRP1 [158] have been implicated in pTau uptake by neurons, their implication in pTau internalization by astrocytes has not been convincingly demonstrated. However, astrocyte-secreted glypican-4 (GPC4) has been involved in tau internalization and phosphorylation by neurons, promoting LRP1 trafficking to the neuronal plasma membrane through its interaction with APOE [159]. Astrocytes can take up fibrillar tau via the integrin-β1 receptor (ITGB1) [160], whereas the lysosomal biogenesis orchestrator TFEB has been implicated in both uptake and lysosomal degradation [161]. Reactive astrocytes can also degrade pTau by upregulating the Bcl-2-associated athanogene family molecular chaperone regulator 3 (BAG3), a major orchestrator of macroautophagy [162, 163]. Lastly, GFAP has been implicated in chaperone-mediated autophagy (CMA) through its interaction with the lysosome-associated membrane protein 2A (LAMP2A) at the lysosomal membrane [164, 165], but whether astrocytes degrade pTau via CMA remains to be confirmed. Interestingly, reactive astrocytes may also alleviate neuronal pTau burden and pTau-induced neurodegeneration in a non-cell autonomous manner by secreting free HSP27 (not as cargo in extracellular vesicles) to the extracellular space, which is then taken up by neurons in a paracrine fashion [104].

Intriguingly, the vast majority of astrocyte pTau-immunoreactive morphologies (i.e., TSA and GFA in ARTAG and AD, tufted astrocytes in progressive supranuclear palsy, astrocytic plaques in corticobasal degeneration, globular astroglial inclusions in globular glial tauopathy, and many ramified astrocytes in Pick's disease) are made of four-repeat (4R) tau [155], but the reasons explaining this observation—i.e., 4R/3R imbalance in *MAPT* pre-mRNA exon 10 splicing vs. more efficient uptake of 4R-tau and/or degradation of 3R-tau in astrocytes—remain to be elucidated.

Ultimately, if the astrocyte proteostasis mechanisms fail to clear this pTau and it accumulates within the astrocyte cytoplasm, this can be deleterious for astrocyte physiology (i.e., inducing calcium dyshomeostasis, mitochondrial oxidative stress, and electron transport chain deficits) and, secondarily, for neuronal synaptic function and memory formation [166, 167]. Moreover, failure of astrocytes to degrade internalized tau could result in the release of tau seeds and facilitation of neuron-to-neuron tau propagation [168].

An indirect mechanism by which reactive astrocytes may impact pTau levels is through causing neuronal dysfunction. Tau levels appear to modulate neuronal activity because tau deficiency prevents seizures [169] and several tau-overexpressing transgenic mice exhibit neuronal hyperexcitability [170–173]. Conversely, neuronal activity promotes tau release to the extracellular space and neuron-to-neuron propagation [174, 175], which is facilitated by the presence of synaptic contacts [176]. Thus, perturbation of the normal tuning of neuronal activity by reactive astrogliosis could conceivably contribute to pTau release and propagation. However, much more work is needed to determine the role of reactive astrocytes (protective vs. deleterious) in pTau propagation along AD-vulnerable neural networks.

4.3 *Effects of Reactive Astrocytes on Neurodegeneration*

While there is consensus that astrocytes are eminently neuroprotective in the healthy brain, whether the reactive astrogliosis observed in the AD brain is neuroprotective or a subset of reactive astrocytes becomes neurotoxic and accelerates neuronal loss remains highly controversial. One reason for this lack of consensus is the great complexity and heterogeneity of astrocyte response to AD neuropathological changes, which is compounded by the spatial (i.e., brain region) and temporal (i.e., disease stage) axes, as revealed by snRNA-seq studies [53]. Thus, a biphasic role of reactive astrogliosis is plausible, whereby astrocytes would initially mount an adaptive and neuroprotective response to AD neuropathology, which would become maladaptive later on as they suffer the chronic exposure to toxic stimuli such as Aβ plaques and tau neurofibrillary tangles. It is also plausible that astrocytes from different brain regions may respond differently and/or be differentially equipped to cope with AD neuropathology. Notwithstanding these uncertainties, it is clear that the astrocyte-neuron interaction is dramatically altered in AD as a result of reactive astrogliosis.

4.3.1 Effects on Neuronal Loss

Reactive astrocytes may contribute to neurodegeneration through a gain of toxic function, namely the secretion of various neurotoxic molecules. Among these neurotoxic mediators are ROS such as H_2O_2—a by-product of monoamine catabolism by the mitochondrial enzyme MAO-B [177]; pro-inflammatory cytokines (e.g., IL-6, CXCL10); the iron-binding cytokine LCN2, whose secretion can be promoted by tau-mediated activation of the astrocytic α2-Na+/K+ adenosine triphosphatase (α2-NKA) [178]; and very long-chain saturated lipids resulting from the activity of the enzyme ELOVL1, which is enhanced by the downregulation of its suppressor TMEM164 in AD [179, 180]. Reactive astrocytes may also contribute to neuronal death through a loss of their many neuroprotective homeostatic functions (e.g., trophic and metabolic support and glutamate neurotransmission) as shown by snRNA-seq and bulk RNA-seq studies from human postmortem brains [53, 60, 181]. However, reactive astrocytes may also offer extra neuroprotection via increasing the expression of trophic factors such as fibroblast growth factors 1 and 2 (FGF1 and FGF2) and antioxidant defenses such as heme oxygenase 1 (HMOX1), metallothioneins (e.g., MT1A, MT2A, MT3), superoxide dismutases 1 and 2 (SOD1 and SOD2), and peroxiredoxin-6 (PRDX6) [31].

4.3.2 Effects on Synapse Dysfunction/Loss

The role of reactive astrocytes in AD-related synaptic dysfunction and loss has just begun to emerge despite the long-known facts that astrocyte fine processes are an integral part of the tripartite excitatory synapse [182], that astrocytes have crucial developmental roles in synaptogenesis (e.g., via thrombospondin secretion) [183] and synaptic pruning (e.g., via MEGF10 and MERTK) [184], and that synaptic loss is the strongest correlate of the severity of cognitive decline in AD [4].

Reactive astrocytes can contribute to AD-related synaptic loss by directly eliminating synapses. Together with reactive microglia, astrocytes have been shown to engulf "tagged" synapses—both presynaptic boutons and postsynaptic dendritic spines—at increased rates in the AD brains [185, 186]. Aβ [18, 187] and tau oligomers [185, 186], the complement factors C1q and C3, and APOE [17, 187, 188] have been recognized among other tags marking synapses for accelerated pruning. It has been suggested that Aβ-laden synapses are less amenable to astrocyte uptake than pTau-containing synapses [189]. The MFG-E8, MEGF10, and MERTK receptors have been implicated in synapse engulfment by astrocytes *in vitro* [186, 189].

Reactive astrocytes can also contribute to synaptic dysfunction and neural network failure through several mechanisms:

1. Glutamate neuronal excitotoxicity: Reactive astrocytes surrounding Aβ plaques may downregulate the glutamate transporter EAAT2/GLT-1 and the enzyme glutamine synthetase, thereby failing to re-uptake glutamate from the synaptic cleft and detoxify it, and contributing to neuronal hyperexcitability via glutamate spill-over and activation of extra-synaptic NMDA receptors [65]; moreover, plaque-associated reactive astrocytes may also actively release glutamate to the synaptic cleft [190–193];
2. Impaired potassium (K^+) buffering due to downregulation of the astrocyte inward-rectifier K^+ channel Kir4.1 [123], although this remains controversial [122];
3. Increased D-serine synthesis and release: Reactive astrocytes upregulate serine racemase, a predominantly neuronal enzyme that catalyzes the conversion of L-serine into D-serine, a co-agonist of NMDA receptors that modulates glutamatergic neurotransmission [194, 195]; and.
4. GABA-mediated tonic inhibition of excitatory neurons: Reactive astrocytes can produce GABA from glutamate by upregulating glutamate decarboxylase 67 kDa (GAD67) [196] and also from the diamine substrate Putrescine via

MAO-B enzymatic activity [197, 198]; astrocytic GABA can then be released to the synaptic cleft via either GABA transporters 3 and 4 (GAT3/4) [196] or the bestrophin-1 (Best1) channel [197].

4.3.3 Effects on Plaque-Associated Dystrophic Neurites

Lastly, the contribution of reactive astrocytes to plaque-associated dystrophic neurites remains relatively obscure. Reactive astrocytes may engulf plaque-associated dystrophic neurites [199]. The cytoskeletal intermediate filament proteins GFAP and Vimentin might be implicated in this process because the combined deletion of both genes results in higher numbers of dystrophic neurites per plaque [45].

5 Conclusions and Areas of Uncertainty

Given their intimate relationship with neurons, astrocytes are uniquely positioned to modulate various key aspects of AD pathophysiology such as the accumulation of the neuronally derived, aggregation-prone, misfolded proteins Aβ and pTau, the negative impact that both lesions have on neurons (e.g., synaptic dysfunction and loss, oxidative stress, and neuroinflammation), and the highly selective regional vulnerability to network failure and neurofibrillary degeneration (represented by the Braak neurofibrillary tangle stages) that characterizes AD. However, the reactive astrogliosis associated with AD is a complex, multifaceted process, with both neuroprotective (adaptive) and deleterious (maladaptive) effects—the latter due to the gain of toxic functions and the loss of homeostatic functions—which are dynamic along the spatial and temporal progression axes of AD (Fig. 2).

Ongoing and future work will aim at shedding light on several remaining obscure areas. How does each aspect of AD-associated reactive astrogliosis impact neuronal activity at the network level? Besides secreted mediators acting via ligand-receptor binding, what other mechanisms of astrocyte-neuron communication (e.g., extracellular vesicles, tunneling nanotubes) are altered in AD? How are the interactions between astrocytes and other cell types (i.e., microglia, oligodendrocytes, and endothelial cells) affected by AD-associated reactive astrogliosis and, if altered, how do they perturb neuronal physiology? Addressing these questions will require the development of novel tools [200] but will likely pay dividends in the form of astrocyte-directed therapeutics aiming to revert the abnormal functioning of the astrocyte-neuron unit observed in AD [201].

Acknowledgments CM-C was supported by the V Plan Propio US-Acceso Universidad de Sevilla; AS-P was supported by the National Institute on Aging (NIA K08AG064039), The Karen Toffler Charitable Trust and the Harrison Gardner, Jr. Innovation Award. Figure 2 was created with Servier Medical Art, which is licensed under CC BY 4.0 (https://smart.servier.com).

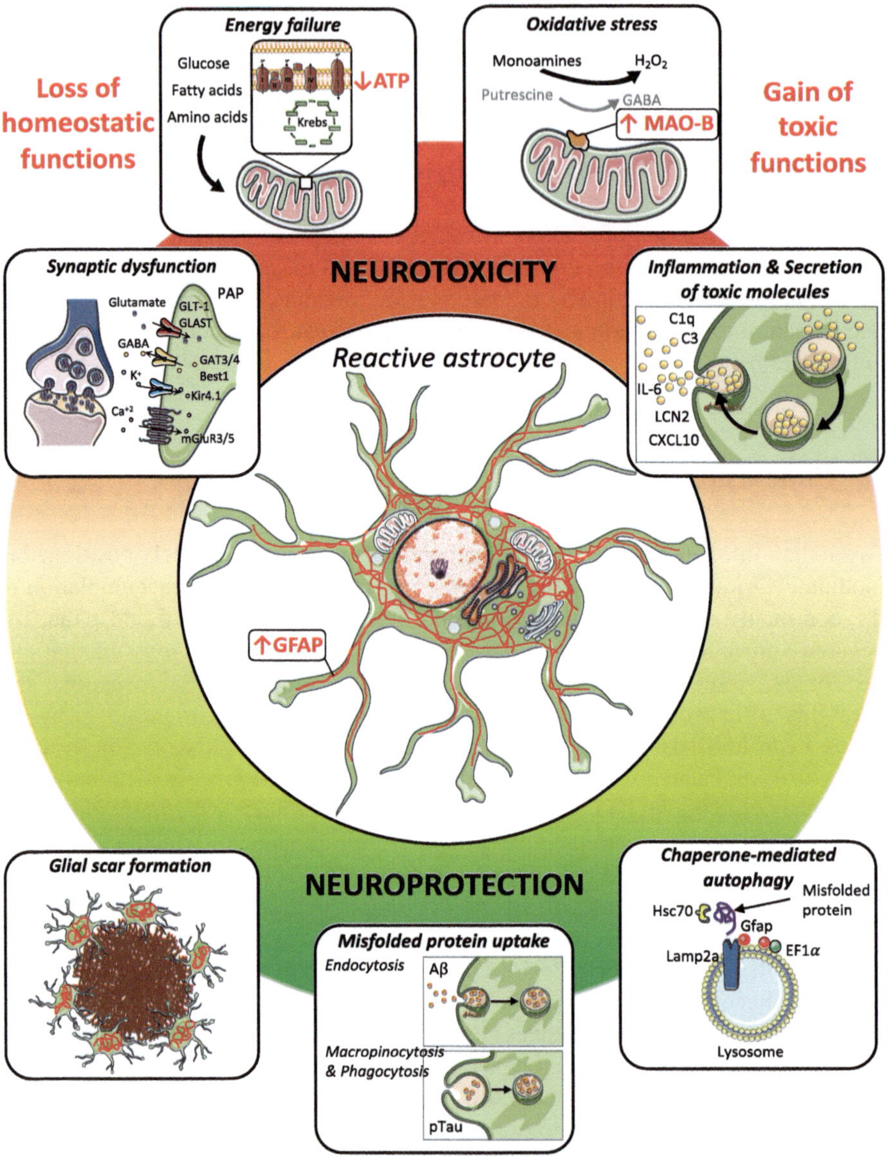

Fig. 2 Complexity of reactive astrogliosis in AD. Reactive astrocytes may initially have neuroprotective effects such as the barricading and removal of Aβ and pTau aggregates (adaptive response); however, as AD advances, they may turn neurotoxic by releasing pro-inflammatory cytokines, reactive oxygen species (ROS), and other toxic molecules, as well as by losing their normal homeostatic functions such as energy metabolism and synaptic activity modulation (maladaptive response). Abbreviation: *PAP* peri-synaptic astrocyte process

References

1. Knopman DS, Amieva H, Petersen RC, et al. Alzheimer disease. Nat Rev Dis Primers. 2021;7:33. https://doi.org/10.1038/s41572-021-00269-y.
2. Hyman BT, Phelps CH, Beach TG, et al. National Institute on Aging-Alzheimer's Association guidelines for the neuropathologic assessment of Alzheimer's disease. Alzheimers Dement. 2012;8:1–13. https://doi.org/10.1016/j.jalz.2011.10.007.
3. Montine TJ, Phelps CH, Beach TG, et al. National Institute on Aging-Alzheimer's Association guidelines for the neuropathologic assessment of Alzheimer's disease: a practical approach. Acta Neuropathol. 2012;123:1–11. https://doi.org/10.1007/s00401-011-0910-3.
4. Serrano-Pozo A, Frosch MP, Masliah E, Hyman BT. Neuropathological alterations in Alzheimer disease. Cold Spring Harb Perspect Med. 2011;1:a006189. https://doi.org/10.1101/cshperspect.a006189.
5. Therriault J, Schindler SE, Salvadó G, et al. Biomarker-based staging of Alzheimer disease: rationale and clinical applications. Nat Rev Neurol. 2024; https://doi.org/10.1038/s41582-024-00942-2.
6. Arnold SE, Hyman BT, Flory J, et al. The topographical and neuroanatomical distribution of neurofibrillary tangles and neuritic plaques in the cerebral cortex of patients with Alzheimer's disease. Cereb Cortex. 1991;1:103–16. https://doi.org/10.1093/cercor/1.1.103.
7. Braak H, Braak E. Neuropathological stageing of Alzheimer-related changes. Acta Neuropathol. 1991;82:239–59. https://doi.org/10.1007/BF00308809.
8. Braak H, Alafuzoff I, Arzberger T, et al. Staging of Alzheimer disease-associated neurofibrillary pathology using paraffin sections and immunocytochemistry. Acta Neuropathol. 2006;112:389–404. https://doi.org/10.1007/s00401-006-0127-z.
9. Serrano-Pozo A, Qian J, Monsell SE, et al. Examination of the clinicopathologic continuum of Alzheimer disease in the autopsy cohort of the National Alzheimer Coordinating Center. J Neuropathol Exp Neurol. 2013;72:1182–92. https://doi.org/10.1097/NEN.0000000000000016.
10. Charidimou A, Boulouis G, Gurol ME, et al. Emerging concepts in sporadic cerebral amyloid angiopathy. Brain. 2017;140:1829–50. https://doi.org/10.1093/brain/awx047.
11. Braak E, Braak H, Mandelkow EM. A sequence of cytoskeleton changes related to the formation of neurofibrillary tangles and neuropil threads. Acta Neuropathol. 1994;87:554–67. https://doi.org/10.1007/BF00293315.
12. Albert MS, DeKosky ST, Dickson D, et al. The diagnosis of mild cognitive impairment due to Alzheimer's disease: recommendations from the National Institute on Aging-Alzheimer's Association workgroups on diagnostic guidelines for Alzheimer's disease. Alzheimers Dement. 2011;7:270–9. https://doi.org/10.1016/j.jalz.2011.03.008.
13. Dubois B, Feldman HH, Jacova C, et al. Advancing research diagnostic criteria for Alzheimer's disease: the IWG-2 criteria. Lancet Neurol. 2014;13:614–29. https://doi.org/10.1016/S1474-4422(14)70090-0.
14. McKhann GM, Knopman DS, Chertkow H, et al. The diagnosis of dementia due to Alzheimer's disease: recommendations from the National Institute on Aging-Alzheimer's Association workgroups on diagnostic guidelines for Alzheimer's disease. Alzheimers Dement. 2011;7:263–9. https://doi.org/10.1016/j.jalz.2011.03.005.
15. Gertz HJ, Cervos-Navarro J, Ewald V. The septo-hippocampal pathway in patients suffering from senile dementia of Alzheimer's type. Evidence for neuronal plasticity? Neurosci Lett. 1987;76:228–32. https://doi.org/10.1016/0304-3940(87)90720-8.
16. Mehraein P, Yamada M, Tarnowska-Dziduszko E. Quantitative study on dendrites and dendritic spines in Alzheimer's disease and senile dementia. Adv Neurol. 1975;12:453–8.
17. Koffie RM, Hashimoto T, Tai H-C, et al. Apolipoprotein E4 effects in Alzheimer's disease are mediated by synaptotoxic oligomeric amyloid-β. Brain. 2012;135:2155–68. https://doi.org/10.1093/brain/aws127.

18. Koffie RM, Meyer-Luehmann M, Hashimoto T, et al. Oligomeric amyloid beta associates with postsynaptic densities and correlates with excitatory synapse loss near senile plaques. Proc Natl Acad Sci USA. 2009;106:4012–7. https://doi.org/10.1073/pnas.0811698106.
19. Kopeikina KJ, Polydoro M, Tai H-C, et al. Synaptic alterations in the rTg4510 mouse model of tauopathy. J Comp Neurol. 2013;521:1334–53. https://doi.org/10.1002/cne.23234.
20. Kopeikina KJ, Wegmann S, Pitstick R, et al. Tau causes synapse loss without disrupting calcium homeostasis in the rTg4510 model of tauopathy. PLoS One. 2013;8:e80834. https://doi.org/10.1371/journal.pone.0080834.
21. Stern EA, Bacskai BJ, Hickey GA, et al. Cortical synaptic integration in vivo is disrupted by amyloid-beta plaques. J Neurosci. 2004;24:4535–40. https://doi.org/10.1523/JNEUROSCI.0462-04.2004.
22. Yuan P, Zhang M, Tong L, et al. PLD3 affects axonal spheroids and network defects in Alzheimer's disease. Nature. 2022;612:328–37. https://doi.org/10.1038/s41586-022-05491-6.
23. Pike CJ, Cummings BJ, Cotman CW. Early association of reactive astrocytes with senile plaques in Alzheimer's disease. Exp Neurol. 1995;132:172–9. https://doi.org/10.1016/0014-4886(95)90022-5.
24. Thal DR, Rüb U, Schultz C, et al. Sequence of Abeta-protein deposition in the human medial temporal lobe. J Neuropathol Exp Neurol. 2000;59:733–48. https://doi.org/10.1093/jnen/59.8.733.
25. Beach TG. A history of Senile plaques: from Alzheimer to amyloid imaging. J Neuropathol Exp Neurol. 2022;81:387–413. https://doi.org/10.1093/jnen/nlac030.
26. Beach TG, Walker R, McGeer EG. Patterns of gliosis in Alzheimer's disease and aging cerebrum. Glia. 1989;2:420–36. https://doi.org/10.1002/glia.440020605.
27. Itagaki S, McGeer PL, Akiyama H, et al. Relationship of microglia and astrocytes to amyloid deposits of Alzheimer disease. J Neuroimmunol. 1989;24:173–82. https://doi.org/10.1016/0165-5728(89)90115-x.
28. Serrano-Pozo A, Mielke ML, Gómez-Isla T, et al. Reactive glia not only associates with plaques but also parallels tangles in Alzheimer's disease. Am J Pathol. 2011;179:1373–84. https://doi.org/10.1016/j.ajpath.2011.05.047.
29. Das S, Li Z, Noori A, et al. Meta-analysis of mouse transcriptomic studies supports a context-dependent astrocyte reaction in acute CNS injury versus neurodegeneration. J Neuroinflammation. 2020;17:227. https://doi.org/10.1186/s12974-020-01898-y.
30. Muñoz-Castro C, Noori A, Magdamo CG, et al. Cyclic multiplex fluorescent immunohistochemistry and machine learning reveal distinct states of astrocytes and microglia in normal aging and Alzheimer's disease. J Neuroinflammation. 2022;19:30. https://doi.org/10.1186/s12974-022-02383-4.
31. Viejo L, Noori A, Merrill E, et al. Systematic review of human post-mortem immunohistochemical studies and bioinformatics analyses unveil the complexity of astrocyte reaction in Alzheimer's disease. Neuropathol Appl Neurobiol. 2022;48:e12753. https://doi.org/10.1111/nan.12753.
32. Escartin C, Galea E, Lakatos A, et al. Reactive astrocyte nomenclature, definitions, and future directions. Nat Neurosci. 2021;24:312–25. https://doi.org/10.1038/s41593-020-00783-4.
33. Colombo JA, Reisin HD, Jones M, Bentham C. Development of interlaminar astroglial processes in the cerebral cortex of control and Down's syndrome human cases. Exp Neurol. 2005;193:207–17. https://doi.org/10.1016/j.expneurol.2004.11.024.
34. Burda JE, Sofroniew MV. Reactive gliosis and the multicellular response to CNS damage and disease. Neuron. 2014;81:229–48. https://doi.org/10.1016/j.neuron.2013.12.034.
35. Kamphuis W, Orre M, Kooijman L, et al. Differential cell proliferation in the cortex of the APPswePS1dE9 Alzheimer's disease mouse model. Glia. 2012;60:615–29. https://doi.org/10.1002/glia.22295.
36. Luccarini I, Grossi C, Traini C, et al. Aβ plaque-associated glial reaction as a determinant of apoptotic neuronal death and cortical gliogenesis: a study in APP mutant mice. Neurosci Lett. 2012;506:94–9. https://doi.org/10.1016/j.neulet.2011.10.056.

37. Sirko S, Behrendt G, Johansson PA, et al. Reactive glia in the injured brain acquire stem cell properties in response to sonic hedgehog. [corrected]. Cell Stem Cell. 2013;12:426–39. https://doi.org/10.1016/j.stem.2013.01.019.
38. Serrano-Pozo A, Gómez-Isla T, Growdon JH, et al. A phenotypic change but not proliferation underlies glial responses in Alzheimer disease. Am J Pathol. 2013;182:2332–44. https://doi.org/10.1016/j.ajpath.2013.02.031.
39. Katsouri L, Birch AM, Renziehausen AWJ, et al. Ablation of reactive astrocytes exacerbates disease pathology in a model of Alzheimer's disease. Glia. 2020;68:1017–30. https://doi.org/10.1002/glia.23759.
40. Hierro-Bujalance C, Bacskai BJ, Garcia-Alloza M. In vivo imaging of microglia with multiphoton microscopy. Front Aging Neurosci. 2018;10:218. https://doi.org/10.3389/fnagi.2018.00218.
41. Galea E, Morrison W, Hudry E, et al. Topological analyses in APP/PS1 mice reveal that astrocytes do not migrate to amyloid-β plaques. Proc Natl Acad Sci USA. 2015;112:15556–61. https://doi.org/10.1073/pnas.1516779112.
42. Serrano-Pozo A, Muzikansky A, Gómez-Isla T, et al. Differential relationships of reactive astrocytes and microglia to fibrillar amyloid deposits in Alzheimer disease. J Neuropathol Exp Neurol. 2013;72:462–71. https://doi.org/10.1097/NEN.0b013e3182933788.
43. Brenner M. Role of GFAP in CNS injuries. Neurosci Lett. 2014;565:7–13. https://doi.org/10.1016/j.neulet.2014.01.055.
44. Kamphuis W, Kooijman L, Orre M, et al. GFAP and vimentin deficiency alters gene expression in astrocytes and microglia in wild-type mice and changes the transcriptional response of reactive glia in mouse model for Alzheimer's disease. Glia. 2015;63:1036–56. https://doi.org/10.1002/glia.22800.
45. Kraft AW, Hu X, Yoon H, et al. Attenuating astrocyte activation accelerates plaque pathogenesis in APP/PS1 mice. FASEB J. 2013;27:187–98. https://doi.org/10.1096/fj.12-208660.
46. De Bastiani MA, Bellaver B, Brum WS, et al. Hippocampal GFAP-positive astrocyte responses to amyloid and tau pathologies. Brain Behav Immun. 2023;110:175–84. https://doi.org/10.1016/j.bbi.2023.03.001.
47. Jiwaji Z, Tiwari SS, Avilés-Reyes RX, et al. Reactive astrocytes acquire neuroprotective as well as deleterious signatures in response to Tau and Aß pathology. Nat Commun. 2022;13:135. https://doi.org/10.1038/s41467-021-27702-w.
48. Grubman A, Chew G, Ouyang JF, et al. A single-cell atlas of entorhinal cortex from individuals with Alzheimer's disease reveals cell-type-specific gene expression regulation. Nat Neurosci. 2019;22:2087–97. https://doi.org/10.1038/s41593-019-0539-4.
49. Lau S-F, Cao H, Fu AKY, Ip NY. Single-nucleus transcriptome analysis reveals dysregulation of angiogenic endothelial cells and neuroprotective glia in Alzheimer's disease. Proc Natl Acad Sci USA. 2020;117:25800–9. https://doi.org/10.1073/pnas.2008762117.
50. Leng K, Li E, Eser R, et al. Molecular characterization of selectively vulnerable neurons in Alzheimer's disease. Nat Neurosci. 2021;24:276–87. https://doi.org/10.1038/s41593-020-00764-7.
51. Mathys H, Davila-Velderrain J, Peng Z, et al. Single-cell transcriptomic analysis of Alzheimer's disease. Nature. 2019;570:332–7. https://doi.org/10.1038/s41586-019-1195-2.
52. Sadick JS, O'Dea MR, Hasel P, et al. Astrocytes and oligodendrocytes undergo subtype-specific transcriptional changes in Alzheimer's disease. Neuron. 2022;110:1788–1805.e10. https://doi.org/10.1016/j.neuron.2022.03.008.
53. Serrano-Pozo A, Li Z, Woodbury ME, et al. Astrocyte transcriptomic changes along the spatiotemporal progression of Alzheimer's disease. Nat Neurosci 2024 (in press). https://doi.org/10.1101/2022.12.03.518999
54. Smith AM, Davey K, Tsartsalis S, et al. Diverse human astrocyte and microglial transcriptional responses to Alzheimer's pathology. Acta Neuropathol. 2022;143:75–91. https://doi.org/10.1007/s00401-021-02372-6.

55. Zhou Y, Song WM, Andhey PS, et al. Human and mouse single-nucleus transcriptomics reveal TREM2-dependent and TREM2-independent cellular responses in Alzheimer's disease. Nat Med. 2020;26:131–42. https://doi.org/10.1038/s41591-019-0695-9.
56. Chen S, Chang Y, Li L, et al. Spatially resolved transcriptomics reveals genes associated with the vulnerability of middle temporal gyrus in Alzheimer's disease. Acta Neuropathol Commun. 2022;10:188. https://doi.org/10.1186/s40478-022-01494-6.
57. Chen W-T, Lu A, Craessaerts K, et al. Spatial transcriptomics and in situ sequencing to study Alzheimer's disease. Cell. 2020;182:976–991.e19. https://doi.org/10.1016/j.cell.2020.06.038.
58. Das S, Li Z, Wachter A, et al. Distinct transcriptomic responses to Aβ plaques, neurofibrillary tangles, and APOE in Alzheimer's disease. Alzheimers Dement. 2024;20:74–90. https://doi.org/10.1002/alz.13387.
59. Zeng H, Huang J, Zhou H, et al. Integrative in situ mapping of single-cell transcriptional states and tissue histopathology in a mouse model of Alzheimer's disease. Nat Neurosci. 2023;26:430–46. https://doi.org/10.1038/s41593-022-01251-x.
60. Dai DL, Li M, Lee EB. Human Alzheimer's disease reactive astrocytes exhibit a loss of homeostastic gene expression. Acta Neuropathol Commun. 2023;11:127. https://doi.org/10.1186/s40478-023-01624-8.
61. Habib N, McCabe C, Medina S, et al. Disease-associated astrocytes in Alzheimer's disease and aging. Nat Neurosci. 2020;23:701–6. https://doi.org/10.1038/s41593-020-0624-8.
62. Kamphuis W, Middeldorp J, Kooijman L, et al. Glial fibrillary acidic protein isoform expression in plaque related astrogliosis in Alzheimer's disease. Neurobiol Aging. 2014;35:492–510. https://doi.org/10.1016/j.neurobiolaging.2013.09.035.
63. Olabarria M, Noristani HN, Verkhratsky A, Rodríguez JJ. Concomitant astroglial atrophy and astrogliosis in a triple transgenic animal model of Alzheimer's disease. Glia. 2010;58:831–8. https://doi.org/10.1002/glia.20967.
64. Simpson JE, Ince PG, Lace G, et al. Astrocyte phenotype in relation to Alzheimer-type pathology in the ageing brain. Neurobiol Aging. 2010;31:578–90. https://doi.org/10.1016/j.neurobiolaging.2008.05.015.
65. Hefendehl JK, LeDue J, Ko RWY, et al. Mapping synaptic glutamate transporter dysfunction in vivo to regions surrounding Aβ plaques by iGluSnFR two-photon imaging. Nat Commun. 2016;7:13441. https://doi.org/10.1038/ncomms13441.
66. Busche MA, Eichhoff G, Adelsberger H, et al. Clusters of hyperactive neurons near amyloid plaques in a mouse model of Alzheimer's disease. Science. 2008;321:1686–9. https://doi.org/10.1126/science.1162844.
67. Garcia-Esparcia P, Diaz-Lucena D, Ainciburu M, et al. Glutamate transporter GLT1 expression in Alzheimer disease and dementia with Lewy bodies. Front Aging Neurosci. 2018;10:122. https://doi.org/10.3389/fnagi.2018.00122.
68. Huffels CFM, Osborn LM, Cappaert NLM, Hol EM. Calcium signaling in individual APP/PS1 mouse dentate gyrus astrocytes increases ex vivo with Aβ pathology and age without affecting astrocyte network activity. J Neurosci Res. 2022;100:1281–95. https://doi.org/10.1002/jnr.25042.
69. Kuchibhotla KV, Lattarulo CR, Hyman BT, Bacskai BJ. Synchronous hyperactivity and intercellular calcium waves in astrocytes in Alzheimer mice. Science. 2009;323:1211–5. https://doi.org/10.1126/science.1169096.
70. Kelly P, Sanchez-Mico MV, Hou SS, et al. Neuronally derived soluble Abeta evokes cell-wide astrocytic calcium dysregulation in absence of amyloid plaques in vivo. J Neurosci. 2023;43:4926–40. https://doi.org/10.1523/JNEUROSCI.1988-22.2023.
71. Orr AG, Hsiao EC, Wang MM, et al. Astrocytic adenosine receptor A2A and Gs-coupled signaling regulate memory. Nat Neurosci. 2015;18:423–34. https://doi.org/10.1038/nn.3930.
72. Delekate A, Füchtemeier M, Schumacher T, et al. Metabotropic P2Y1 receptor signalling mediates astrocytic hyperactivity in vivo in an Alzheimer's disease mouse model. Nat Commun. 2014;5:5422. https://doi.org/10.1038/ncomms6422.

73. Reichenbach N, Delekate A, Breithausen B, et al. P2Y1 receptor blockade normalizes network dysfunction and cognition in an Alzheimer's disease model. J Exp Med. 2018;215:1649–63. https://doi.org/10.1084/jem.20171487.
74. Iyer AM, van Scheppingen J, Milenkovic I, et al. Metabotropic glutamate receptor 5 in Down's syndrome hippocampus during development: increased expression in astrocytes. Curr Alzheimer Res. 2014;11:694–705. https://doi.org/10.2174/1567205011666140812115423.
75. Lim D, Iyer A, Ronco V, et al. Amyloid beta deregulates astroglial mGluR5-mediated calcium signaling via calcineurin and Nf-kB. Glia. 2013;61:1134–45. https://doi.org/10.1002/glia.22502.
76. Shrivastava AN, Kowalewski JM, Renner M, et al. β-amyloid and ATP-induced diffusional trapping of astrocyte and neuronal metabotropic glutamate type-5 receptors. Glia. 2013;61:1673–86. https://doi.org/10.1002/glia.22548.
77. Bosson A, Paumier A, Boisseau S, et al. TRPA1 channels promote astrocytic Ca2+ hyperactivity and synaptic dysfunction mediated by oligomeric forms of amyloid-β peptide. Mol Neurodegener. 2017;12:53. https://doi.org/10.1186/s13024-017-0194-8.
78. Lee K-I, Lee H-T, Lin H-C, et al. Role of transient receptor potential ankyrin 1 channels in Alzheimer's disease. J Neuroinflammation. 2016;13:92. https://doi.org/10.1186/s12974-016-0557-z.
79. Paumier A, Boisseau S, Jacquier-Sarlin M, et al. Astrocyte-neuron interplay is critical for Alzheimer's disease pathogenesis and is rescued by TRPA1 channel blockade. Brain. 2022;145:388–405. https://doi.org/10.1093/brain/awab281.
80. Ren R, Zhang L, Wang M. Specific deletion connexin43 in astrocyte ameliorates cognitive dysfunction in APP/PS1 mice. Life Sci. 2018;208:175–91. https://doi.org/10.1016/j.lfs.2018.07.033.
81. Yi C, Mei X, Ezan P, et al. Astroglial connexin43 contributes to neuronal suffering in a mouse model of Alzheimer's disease. Cell Death Differ. 2016;23:1691–701. https://doi.org/10.1038/cdd.2016.63.
82. Yi C, Ezan P, Fernández P, et al. Inhibition of glial hemichannels by boldine treatment reduces neuronal suffering in a murine model of Alzheimer's disease. Glia. 2017;65:1607–25. https://doi.org/10.1002/glia.23182.
83. Abdul HM, Sama MA, Furman JL, et al. Cognitive decline in Alzheimer's disease is associated with selective changes in calcineurin/NFAT signaling. J Neurosci. 2009;29:12957–69. https://doi.org/10.1523/JNEUROSCI.1064-09.2009.
84. Sompol P, Furman JL, Pleiss MM, et al. Calcineurin/NFAT signaling in activated astrocytes drives network hyperexcitability in Aβ-bearing mice. J Neurosci. 2017;37:6132–48. https://doi.org/10.1523/JNEUROSCI.0877-17.2017.
85. Furman JL, Sama DM, Gant JC, et al. Targeting astrocytes ameliorates neurologic changes in a mouse model of Alzheimer's disease. J Neurosci. 2012;32:16129–40. https://doi.org/10.1523/JNEUROSCI.2323-12.2012.
86. Perez-Nievas BG, Serrano-Pozo A. Deciphering the astrocyte reaction in Alzheimer's Disease. Front Aging Neurosci. 2018;10:114. https://doi.org/10.3389/fnagi.2018.00114.
87. Lian H, Yang L, Cole A, et al. NFκB-activated astroglial release of complement C3 compromises neuronal morphology and function associated with Alzheimer's disease. Neuron. 2015;85:101–15. https://doi.org/10.1016/j.neuron.2014.11.018.
88. Nakano-Kobayashi A, Canela A, Yoshihara T, Hagiwara M. Astrocyte-targeting therapy rescues cognitive impairment caused by neuroinflammation via the Nrf2 pathway. Proc Natl Acad Sci USA. 2023;120:e2303809120. https://doi.org/10.1073/pnas.2303809120.
89. Lian H, Litvinchuk A, Chiang AC-A, et al. Astrocyte-microglia cross talk through complement activation modulates amyloid pathology in mouse models of Alzheimer's disease. J Neurosci. 2016;36:577–89. https://doi.org/10.1523/JNEUROSCI.2117-15.2016.
90. Yang S, Magnutzki A, Alami NO, et al. IKK2/NF-κB activation in astrocytes reduces amyloid β deposition: a process associated with specific microglia polarization. Cells. 2021;10:2669. https://doi.org/10.3390/cells10102669.

91. Ben Haim L, Ceyzériat K, Carrillo-de Sauvage MA, et al. The JAK/STAT3 pathway is a common inducer of astrocyte reactivity in Alzheimer's and Huntington's diseases. J Neurosci. 2015;35:2817–29. https://doi.org/10.1523/JNEUROSCI.3516-14.2015.
92. Ceyzériat K, Ben Haim L, Denizot A, et al. Modulation of astrocyte reactivity improves functional deficits in mouse models of Alzheimer's disease. Acta Neuropathol Commun. 2018;6:104. https://doi.org/10.1186/s40478-018-0606-1.
93. Leng K, Rose IVL, Kim H, et al. CRISPRi screens in human iPSC-derived astrocytes elucidate regulators of distinct inflammatory reactive states. Nat Neurosci. 2022;25:1528–42. https://doi.org/10.1038/s41593-022-01180-9.
94. Reichenbach N, Delekate A, Plescher M, et al. Inhibition of Stat3-mediated astrogliosis ameliorates pathology in an Alzheimer's disease model. EMBO Mol Med. 2019;11:e9665. https://doi.org/10.15252/emmm.201809665.
95. Guillemaud O, Ceyzériat K, Saint-Georges T, et al. Complex roles for reactive astrocytes in the triple transgenic mouse model of Alzheimer disease. Neurobiol Aging. 2020;90:135–46. https://doi.org/10.1016/j.neurobiolaging.2020.02.010.
96. Garcia-Alloza M, Dodwell SA, Meyer-Luehmann M, et al. Plaque-derived oxidative stress mediates distorted neurite trajectories in the Alzheimer mouse model. J Neuropathol Exp Neurol. 2006;65:1082–9. https://doi.org/10.1097/01.jnen.0000240468.12543.af.
97. Abramov AY, Canevari L, Duchen MR. Changes in intracellular calcium and glutathione in astrocytes as the primary mechanism of amyloid neurotoxicity. J Neurosci. 2003;23:5088–95. https://doi.org/10.1523/JNEUROSCI.23-12-05088.2003.
98. Abramov AY, Canevari L, Duchen MR. Beta-amyloid peptides induce mitochondrial dysfunction and oxidative stress in astrocytes and death of neurons through activation of NADPH oxidase. J Neurosci. 2004;24:565–75. https://doi.org/10.1523/JNEUROSCI.4042-03.2004.
99. Jaisa-aad, M, Muñoz-Castro, C, Healey, MA, et al. Characterization of monoamine oxidase-B (MAO-B) as a biomarker of reactive astrogliosis in Alzheimer's disease and related dementias. Acta Neuropathol. 2024;147(1):66. https://doi.org/10.1007/s00401-024-02712-2
100. Pankiewicz JE, Diaz JR, Martá-Ariza M, et al. Peroxiredoxin 6 mediates protective function of astrocytes in Aβ proteostasis. Mol Neurodegener. 2020;15:50. https://doi.org/10.1186/s13024-020-00401-8.
101. Ren P, Chen J, Li B, et al. Nrf2 ablation promotes Alzheimer's disease-like pathology in APP/PS1 transgenic mice: the role of neuroinflammation and oxidative stress. Oxidative Med Cell Longev. 2020;2020:3050971. https://doi.org/10.1155/2020/3050971.
102. Renkawek K, Bosman GJ, de Jong WW. Expression of small heat-shock protein hsp 27 in reactive gliosis in Alzheimer disease and other types of dementia. Acta Neuropathol. 1994;87:511–9. https://doi.org/10.1007/BF00294178.
103. Renkawek K, Voorter CE, Bosman GJ, et al. Expression of alpha B-crystallin in Alzheimer's disease. Acta Neuropathol. 1994;87:155–60. https://doi.org/10.1007/BF00296185.
104. Yang F, Beltran-Lobo P, Sung K, et al. Reactive astrocytes secrete the chaperone HSPB1 to mediate neuroprotection. Sci Adv. 2024;10:eadk9884. https://doi.org/10.1126/sciadv.adk9884.
105. Neef DW, Jaeger AM, Thiele DJ. Heat shock transcription factor 1 as a therapeutic target in neurodegenerative diseases. Nat Rev Drug Discov. 2011;10:930–44. https://doi.org/10.1038/nrd3453.
106. Iliff JJ, Wang M, Liao Y, et al. A paravascular pathway facilitates CSF flow through the brain parenchyma and the clearance of interstitial solutes, including amyloid β. Sci Transl Med. 2012;4:147ra111. https://doi.org/10.1126/scitranslmed.3003748.
107. Harrison IF, Ismail O, Machhada A, et al. Impaired glymphatic function and clearance of tau in an Alzheimer's disease model. Brain. 2020;143:2576–93. https://doi.org/10.1093/brain/awaa179.
108. Ishida K, Yamada K, Nishiyama R, et al. Glymphatic system clears extracellular tau and protects from tau aggregation and neurodegeneration. J Exp Med. 2022;219:e20211275. https://doi.org/10.1084/jem.20211275.

109. Liu Y, Hu P-P, Zhai S, et al. Aquaporin 4 deficiency eliminates the beneficial effects of voluntary exercise in a mouse model of Alzheimer's disease. Neural Regen Res. 2022;17:2079–88. https://doi.org/10.4103/1673-5374.335169.
110. Pedersen TJ, Keil SA, Han W, et al. The effect of aquaporin-4 mis-localization on Aβ deposition in mice. Neurobiol Dis. 2023;181:106100. https://doi.org/10.1016/j.nbd.2023.106100.
111. Sapkota D, Florian C, Doherty BM, et al. Aqp4 stop codon readthrough facilitates amyloid-β clearance from the brain. Brain. 2022;145:2982–90. https://doi.org/10.1093/brain/awac199.
112. Smith AJ, Duan T, Verkman AS. Aquaporin-4 reduces neuropathology in a mouse model of Alzheimer's disease by remodeling peri-plaque astrocyte structure. Acta Neuropathol Commun. 2019;7:74. https://doi.org/10.1186/s40478-019-0728-0.
113. Xu Z, Xiao N, Chen Y, et al. Deletion of aquaporin-4 in APP/PS1 mice exacerbates brain Aβ accumulation and memory deficits. Mol Neurodegener. 2015;10:58. https://doi.org/10.1186/s13024-015-0056-1.
114. Rosu G-C, Catalin B, Balseanu TA, et al. Inhibition of Aquaporin 4 decreases amyloid Aβ40 drainage around cerebral vessels. Mol Neurobiol. 2020;57:4720–34. https://doi.org/10.1007/s12035-020-02044-8.
115. Simon M, Wang MX, Ismail O, et al. Loss of perivascular aquaporin-4 localization impairs glymphatic exchange and promotes amyloid β plaque formation in mice. Alzheimers Res Ther. 2022;14:59. https://doi.org/10.1186/s13195-022-00999-5.
116. Zeppenfeld DM, Simon M, Haswell JD, et al. Association of perivascular localization of aquaporin-4 with cognition and Alzheimer disease in aging brains. JAMA Neurol. 2017;74:91–9. https://doi.org/10.1001/jamaneurol.2016.4370.
117. Askenazi M, Kavanagh T, Pires G, et al. Compilation of reported protein changes in the brain in Alzheimer's disease. Nat Commun. 2023;14:4466. https://doi.org/10.1038/s41467-023-40208-x.
118. Benskey MJ, Panoushek S, Saito T, et al. Behavioral and neuropathological characterization over the adult lifespan of the human tau knock-in mouse. Front Aging Neurosci. 2023;15:1265151. https://doi.org/10.3389/fnagi.2023.1265151.
119. Montal V, Diez I, Kim C-M, et al. Network Tau spreading is vulnerable to the expression gradients of APOE and glutamatergic-related genes. Sci Transl Med. 2022;14:eabn7273. https://doi.org/10.1126/scitranslmed.abn7273.
120. Kobayashi E, Nakano M, Kubota K, et al. Activated forms of astrocytes with higher GLT-1 expression are associated with cognitive normal subjects with Alzheimer pathology in human brain. Sci Rep. 2018;8:1712. https://doi.org/10.1038/s41598-018-19442-7.
121. Filipcik P, Cente M, Zilka N, et al. Intraneuronal accumulation of misfolded tau protein induces overexpression of Hsp27 in activated astrocytes. Biochim Biophys Acta. 2015;1852:1219–29. https://doi.org/10.1016/j.bbadis.2015.03.003.
122. Huffels CFM, Osborn LM, Hulshof LA, et al. Amyloid-β plaques affect astrocyte Kir4.1 protein expression but not function in the dentate gyrus of APP/PS1 mice. Glia. 2022;70:748–67. https://doi.org/10.1002/glia.24137.
123. Ding F, Sun Q, Long C, et al. Dysregulation of extracellular potassium distinguishes healthy ageing from neurodegeneration. Brain. 2024;147:awae075. https://doi.org/10.1093/brain/awae075.
124. Zhou L-T, Liu D, Kang H-C, et al. Tau pathology epigenetically remodels the neuron-glial cross-talk in Alzheimer's disease. Sci Adv. 2023;9:eabq7105. https://doi.org/10.1126/sciadv.abq7105.
125. Damisah EC, Hill RA, Rai A, et al. Astrocytes and microglia play orchestrated roles and respect phagocytic territories during neuronal corpse removal in vivo. Sci Adv. 2020;6:eaba3239. https://doi.org/10.1126/sciadv.aba3239.
126. Mandybur TI. Cerebral amyloid angiopathy and astrocytic gliosis in Alzheimer's disease. Acta Neuropathol. 1989;78:329–31. https://doi.org/10.1007/BF00687764.

127. Thal DR, Schultz C, Dehghani F, et al. Amyloid beta-protein (Abeta)-containing astrocytes are located preferentially near N-terminal-truncated Abeta deposits in the human entorhinal cortex. Acta Neuropathol. 2000;100:608–17. https://doi.org/10.1007/s004010000242.
128. Thal DR, Härtig W, Schober R. Diffuse plaques in the molecular layer show intracellular A beta(8-17)-immunoreactive deposits in subpial astrocytes. Clin Neuropathol. 1999;18:226–31.
129. Utter S, Tamboli IY, Walter J, et al. Cerebral small vessel disease-induced apolipoprotein E leakage is associated with Alzheimer disease and the accumulation of amyloid beta-protein in perivascular astrocytes. J Neuropathol Exp Neurol. 2008;67:842–56. https://doi.org/10.1097/NEN.0b013e3181836a71.
130. Maat-Schieman MLC, Yamaguchi H, Hegeman-Kleinn IM, et al. Glial reactions and the clearance of amyloid beta protein in the brains of patients with hereditary cerebral hemorrhage with amyloidosis-Dutch type. Acta Neuropathol. 2004;107:389–98. https://doi.org/10.1007/s00401-004-0824-4.
131. Jablonski AM, Warren L, Usenovic M, et al. Astrocytic expression of the Alzheimer's disease risk allele, ApoEε4, potentiates neuronal tau pathology in multiple preclinical models. Sci Rep. 2021;11:3438. https://doi.org/10.1038/s41598-021-82901-1.
132. Liu C-C, Zhao N, Fu Y, et al. ApoE4 accelerates early seeding of amyloid pathology. Neuron. 2017;96:1024–1032.e3. https://doi.org/10.1016/j.neuron.2017.11.013.
133. Mahan TE, Wang C, Bao X, et al. Selective reduction of astrocyte apoE3 and apoE4 strongly reduces Aβ accumulation and plaque-related pathology in a mouse model of amyloidosis. Mol Neurodegener. 2022;17:13. https://doi.org/10.1186/s13024-022-00516-0.
134. Xiong M, Wang C, Gratuze M, et al. Astrocytic APOE4 removal confers cerebrovascular protection despite increased cerebral amyloid angiopathy. Mol Neurodegener. 2023;18:17. https://doi.org/10.1186/s13024-023-00610-x.
135. Zheng J-Y, Sun J, Ji C-M, et al. Selective deletion of apolipoprotein E in astrocytes ameliorates the spatial learning and memory deficits in Alzheimer's disease (APP/PS1) mice by inhibiting TGF-β/Smad2/STAT3 signaling. Neurobiol Aging. 2017;54:112–32. https://doi.org/10.1016/j.neurobiolaging.2017.03.002.
136. Chen F, Swartzlander DB, Ghosh A, et al. Clusterin secreted from astrocyte promotes excitatory synaptic transmission and ameliorates Alzheimer's disease neuropathology. Mol Neurodegener. 2021;16:5. https://doi.org/10.1186/s13024-021-00426-7.
137. Wojtas AM, Sens JP, Kang SS, et al. Astrocyte-derived clusterin suppresses amyloid formation in vivo. Mol Neurodegener. 2020;15:71. https://doi.org/10.1186/s13024-020-00416-1.
138. Zhou J, Singh N, Galske J, et al. BACE1 regulates expression of Clusterin in astrocytes for enhancing clearance of β-amyloid peptides. Mol Neurodegener. 2023;18:31. https://doi.org/10.1186/s13024-023-00611-w.
139. Liu C-C, Hu J, Zhao N, et al. Astrocytic LRP1 mediates brain Aβ clearance and impacts amyloid deposition. J Neurosci. 2017;37:4023–31. https://doi.org/10.1523/JNEUROSCI.3442-16.2017.
140. Basak JM, Verghese PB, Yoon H, et al. Low-density lipoprotein receptor represents an apolipoprotein E-independent pathway of Aβ uptake and degradation by astrocytes. J Biol Chem. 2012;287:13959–71. https://doi.org/10.1074/jbc.M111.288746.
141. Katsouri L, Georgopoulos S. Lack of LDL receptor enhances amyloid deposition and decreases glial response in an Alzheimer's disease mouse model. PLoS One. 2011;6:e21880. https://doi.org/10.1371/journal.pone.0021880.
142. Zhang H, Chen W, Tan Z, et al. A role of low-density lipoprotein receptor-related protein 4 (LRP4) in astrocytic Aβ clearance. J Neurosci. 2020;40:5347–61. https://doi.org/10.1523/JNEUROSCI.0250-20.2020.
143. Raha S, Ghosh A, Dutta D, et al. Activation of PPARα enhances astroglial uptake and degradation of β-amyloid. Sci Signal. 2021;14:eabg4747. https://doi.org/10.1126/scisignal.abg4747.
144. Xiao Q, Yan P, Ma X, et al. Enhancing astrocytic lysosome biogenesis facilitates Aβ clearance and attenuates amyloid plaque pathogenesis. J Neurosci. 2014;34:9607–20. https://doi.org/10.1523/JNEUROSCI.3788-13.2014.

145. Iram T, Trudler D, Kain D, et al. Astrocytes from old Alzheimer's disease mice are impaired in Aβ uptake and in neuroprotection. Neurobiol Dis. 2016;96:84–94. https://doi.org/10.1016/j.nbd.2016.08.001.
146. Lin Y-T, Seo J, Gao F, et al. APOE4 causes widespread molecular and cellular alterations associated with Alzheimer's disease phenotypes in human iPSC-derived brain cell types. Neuron. 2018;98:1141–1154.e7. https://doi.org/10.1016/j.neuron.2018.05.008.
147. Yin K-J, Cirrito JR, Yan P, et al. Matrix metalloproteinases expressed by astrocytes mediate extracellular amyloid-beta peptide catabolism. J Neurosci. 2006;26:10939–48. https://doi.org/10.1523/JNEUROSCI.2085-06.2006.
148. Jiang Q, Lee CYD, Mandrekar S, et al. ApoE promotes the proteolytic degradation of Abeta. Neuron. 2008;58:681–93. https://doi.org/10.1016/j.neuron.2008.04.010.
149. Kim E, Kim H, Jedrychowski MP, et al. Irisin reduces amyloid-β by inducing the release of neprilysin from astrocytes following downregulation of ERK-STAT3 signaling. Neuron. 2023;111:3619–3633.e8. https://doi.org/10.1016/j.neuron.2023.08.012.
150. McAlpine CS, Park J, Griciuc A, et al. Astrocytic interleukin-3 programs microglia and limits Alzheimer's disease. Nature. 2021;595:701–6. https://doi.org/10.1038/s41586-021-03734-6.
151. Yamamoto K, Tanei Z-I, Hashimoto T, et al. Chronic optogenetic activation augments aβ pathology in a mouse model of Alzheimer disease. Cell Rep. 2015;11:859–65. https://doi.org/10.1016/j.celrep.2015.04.017.
152. Ikeda K, Haga C, Akiyama H, et al. Coexistence of paired helical filaments and glial filaments in astrocytic processes within ghost tangles. Neurosci Lett. 1992;148:126–8. https://doi.org/10.1016/0304-3940(92)90820-w.
153. Irwin DJ, Cohen TJ, Grossman M, et al. Acetylated tau, a novel pathological signature in Alzheimer's disease and other tauopathies. Brain. 2012;135:807–18. https://doi.org/10.1093/brain/aws013.
154. Probst A, Ulrich J, Heitz PU. Senile dementia of Alzheimer type: astroglial reaction to extracellular neurofibrillary tangles in the hippocampus. An immunocytochemical and electron-microscopic study. Acta Neuropathol. 1982;57:75–9. https://doi.org/10.1007/BF00688880.
155. Kovacs GG, Ferrer I, Grinberg LT, et al. Aging-related tau astrogliopathy (ARTAG): harmonized evaluation strategy. Acta Neuropathol. 2016;131:87–102. https://doi.org/10.1007/s00401-015-1509-x.
156. Forrest SL, Lee S, Nassir N, et al. Cell-specific MAPT gene expression is preserved in neuronal and glial tau cytopathologies in progressive supranuclear palsy. Acta Neuropathol. 2023;146:395–414. https://doi.org/10.1007/s00401-023-02604-x.
157. Holmes BB, DeVos SL, Kfoury N, et al. Heparan sulfate proteoglycans mediate internalization and propagation of specific proteopathic seeds. Proc Natl Acad Sci USA. 2013;110:E3138–47. https://doi.org/10.1073/pnas.1301440110.
158. Rauch JN, Luna G, Guzman E, et al. LRP1 is a master regulator of tau uptake and spread. Nature. 2020;580:381–5. https://doi.org/10.1038/s41586-020-2156-5.
159. Saroja SR, Gorbachev K, Julia T, et al. Astrocyte-secreted glypican-4 drives APOE4-dependent tau hyperphosphorylation. Proc Natl Acad Sci USA. 2022;119:e2108870119. https://doi.org/10.1073/pnas.2108870119.
160. Wang P, Ye Y. Filamentous recombinant human Tau activates primary astrocytes via an integrin receptor complex. Nat Commun. 2021;12:95. https://doi.org/10.1038/s41467-020-20322-w.
161. Martini-Stoica H, Cole AL, Swartzlander DB, et al. TFEB enhances astroglial uptake of extracellular tau species and reduces tau spreading. J Exp Med. 2018;215:2355–77. https://doi.org/10.1084/jem.20172158.
162. Noori A, Jayakumar R, Moturi V, et al. Alzheimer DataLENS: an open data analytics portal for Alzheimer's Disease research. J Alzheimers Dis. 2024; https://doi.org/10.3233/JAD-230884.
163. Sheehan PW, Nadarajah CJ, Kanan MF, et al. An astrocyte BMAL1-BAG3 axis protects against alpha-synuclein and tau pathology. Neuron. 2023;111:2383–2398.e7. https://doi.org/10.1016/j.neuron.2023.05.006.

164. Bandyopadhyay U, Sridhar S, Kaushik S, et al. Identification of regulators of chaperone-mediated autophagy. Mol Cell. 2010;39:535–47. https://doi.org/10.1016/j.molcel.2010.08.004.
165. Endicott SJ, Ziemba ZJ, Beckmann LJ, et al. Inhibition of class I PI3K enhances chaperone-mediated autophagy. J Cell Biol. 2020;219:e202001031. https://doi.org/10.1083/jcb.202001031.
166. Maté de Gérando A, d'Orange M, Augustin E, et al. Neuronal tau species transfer to astrocytes and induce their loss according to tau aggregation state. Brain. 2021;144:1167–82. https://doi.org/10.1093/brain/awab011.
167. Richetin K, Steullet P, Pachoud M, et al. Tau accumulation in astrocytes of the dentate gyrus induces neuronal dysfunction and memory deficits in Alzheimer's disease. Nat Neurosci. 2020;23:1567–79. https://doi.org/10.1038/s41593-020-00728-x.
168. Eltom K, Mothes T, Libard S, et al. Astrocytic accumulation of tau fibrils isolated from Alzheimer's disease brains induces inflammation, cell-to-cell propagation and neuronal impairment. Acta Neuropathol Commun. 2024;12:34. https://doi.org/10.1186/s40478-024-01745-8.
169. DeVos SL, Goncharoff DK, Chen G, et al. Antisense reduction of tau in adult mice protects against seizures. J Neurosci. 2013;33:12887–97. https://doi.org/10.1523/JNEUROSCI.2107-13.2013.
170. Crimins JL, Rocher AB, Luebke JI. Electrophysiological changes precede morphological changes to frontal cortical pyramidal neurons in the rTg4510 mouse model of progressive tauopathy. Acta Neuropathol. 2012;124:777–95. https://doi.org/10.1007/s00401-012-1038-9.
171. García-Cabrero AM, Guerrero-López R, Giráldez BG, et al. Hyperexcitability and epileptic seizures in a model of frontotemporal dementia. Neurobiol Dis. 2013;58:200–8. https://doi.org/10.1016/j.nbd.2013.06.005.
172. Gomez-Murcia V, Sandau U, Ferry B, et al. Hyperexcitability and seizures in the THY-Tau22 mouse model of tauopathy. Neurobiol Aging. 2020;94:265–70. https://doi.org/10.1016/j.neurobiolaging.2020.06.004.
173. Przybyla M, van Eersel J, van Hummel A, et al. Onset of hippocampal network aberration and memory deficits in P301S tau mice are associated with an early gene signature. Brain. 2020;143:1889–904. https://doi.org/10.1093/brain/awaa133.
174. Wu JW, Hussaini SA, Bastille IM, et al. Neuronal activity enhances tau propagation and tau pathology in vivo. Nat Neurosci. 2016;19:1085–92. https://doi.org/10.1038/nn.4328.
175. Yamada K, Holth JK, Liao F, et al. Neuronal activity regulates extracellular tau in vivo. J Exp Med. 2014;211:387–93. https://doi.org/10.1084/jem.20131685.
176. Calafate S, Buist A, Miskiewicz K, et al. Synaptic contacts enhance cell-to-cell tau pathology propagation. Cell Rep. 2015;11:1176–83. https://doi.org/10.1016/j.celrep.2015.04.043.
177. Chun H, Im H, Kang YJ, et al. Severe reactive astrocytes precipitate pathological hallmarks of Alzheimer's disease via H_2O_2- production. Nat Neurosci. 2020;23:1555–66. https://doi.org/10.1038/s41593-020-00735-y.
178. Mann CN, Devi SS, Kersting CT, et al. Astrocytic α2-Na+/K+ ATPase inhibition suppresses astrocyte reactivity and reduces neurodegeneration in a tauopathy mouse model. Sci Transl Med. 2022;14:eabm4107. https://doi.org/10.1126/scitranslmed.abm4107.
179. Guttenplan KA, Weigel MK, Prakash P, et al. Neurotoxic reactive astrocytes induce cell death via saturated lipids. Nature. 2021;599:102–7. https://doi.org/10.1038/s41586-021-03960-y.
180. Zhang L, Jia Z, Wu Q, et al. Alleviating symptoms of neurodegenerative disorders by astrocyte-specific overexpression of TMEM164 in mice. Nat Metab. 2023;5:1787–802. https://doi.org/10.1038/s42255-023-00887-8.
181. Galea E, Weinstock LD, Larramona-Arcas R, et al. Multi-transcriptomic analysis points to early organelle dysfunction in human astrocytes in Alzheimer's disease. Neurobiol Dis. 2022;166:105655. https://doi.org/10.1016/j.nbd.2022.105655.

182. Ventura R, Harris KM. Three-dimensional relationships between hippocampal synapses and astrocytes. J Neurosci. 1999;19:6897–906. https://doi.org/10.1523/JNEUROSCI.19-16-06897.1999.
183. Christopherson KS, Ullian EM, Stokes CCA, et al. Thrombospondins are astrocyte-secreted proteins that promote CNS synaptogenesis. Cell. 2005;120:421–33. https://doi.org/10.1016/j.cell.2004.12.020.
184. Chung W-S, Clarke LE, Wang GX, et al. Astrocytes mediate synapse elimination through MEGF10 and MERTK pathways. Nature. 2013;504:394–400. https://doi.org/10.1038/nature12776.
185. Taddei RN, Perbet R, Mate de Gerando A, et al. Tau oligomer-containing synapse elimination by microglia and astrocytes in Alzheimer disease. JAMA Neurol. 2023;80:1209–21. https://doi.org/10.1001/jamaneurol.2023.3530.
186. Tzioras M, Daniels MJD, Davies C, et al. Human astrocytes and microglia show augmented ingestion of synapses in Alzheimer's disease via MFG-E8. Cell Rep Med. 2023;4:101175. https://doi.org/10.1016/j.xcrm.2023.101175.
187. Bilousova T, Melnik M, Miyoshi E, et al. Apolipoprotein E/Amyloid-β complex accumulates in Alzheimer disease cortical synapses via apolipoprotein E receptors and is enhanced by APOE4. Am J Pathol. 2019;189:1621–36. https://doi.org/10.1016/j.ajpath.2019.04.010.
188. Chung W-S, Verghese PB, Chakraborty C, et al. Novel allele-dependent role for APOE in controlling the rate of synapse pruning by astrocytes. Proc Natl Acad Sci USA. 2016;113:10186–91. https://doi.org/10.1073/pnas.1609896113.
189. Sanchez-Mico MV, Jimenez S, Gomez-Arboledas A, et al. Amyloid-β impairs the phagocytosis of dystrophic synapses by astrocytes in Alzheimer's disease. Glia. 2021;69:997–1011. https://doi.org/10.1002/glia.23943.
190. Bezzi P, Gundersen V, Galbete JL, et al. Astrocytes contain a vesicular compartment that is competent for regulated exocytosis of glutamate. Nat Neurosci. 2004;7:613–20. https://doi.org/10.1038/nn1246.
191. de Ceglia R, Ledonne A, Litvin DG, et al. Specialized astrocytes mediate glutamatergic gliotransmission in the CNS. Nature. 2023;622:120–9. https://doi.org/10.1038/s41586-023-06502-w.
192. Jourdain P, Bergersen LH, Bhaukaurally K, et al. Glutamate exocytosis from astrocytes controls synaptic strength. Nat Neurosci. 2007;10:331–9. https://doi.org/10.1038/nn1849.
193. Rossi D, Brambilla L, Valori CF, et al. Defective tumor necrosis factor-alpha-dependent control of astrocyte glutamate release in a transgenic mouse model of Alzheimer disease. J Biol Chem. 2005;280:42088–96. https://doi.org/10.1074/jbc.M504124200.
194. Balu DT, Pantazopoulos H, Huang CCY, et al. Neurotoxic astrocytes express the d-serine synthesizing enzyme, serine racemase, in Alzheimer's disease. Neurobiol Dis. 2019;130:104511. https://doi.org/10.1016/j.nbd.2019.104511.
195. Folorunso OO, Harvey TL, Brown SE, et al. The D-serine biosynthetic enzyme serine racemase is expressed by reactive astrocytes in the amygdala of human and a mouse model of Alzheimer's disease. Neurosci Lett. 2023;792:136958. https://doi.org/10.1016/j.neulet.2022.136958.
196. Wu Z, Guo Z, Gearing M, Chen G. Tonic inhibition in dentate gyrus impairs long-term potentiation and memory in an Alzheimer's [corrected] disease model. Nat Commun. 2014;5:4159. https://doi.org/10.1038/ncomms5159.
197. Jo S, Yarishkin O, Hwang YJ, et al. GABA from reactive astrocytes impairs memory in mouse models of Alzheimer's disease. Nat Med. 2014;20:886–96. https://doi.org/10.1038/nm.3639.
198. Yoon B-E, Woo J, Chun Y-E, et al. Glial GABA, synthesized by monoamine oxidase B, mediates tonic inhibition. J Physiol. 2014;592:4951–68. https://doi.org/10.1113/jphysiol.2014.278754.

199. Gomez-Arboledas A, Davila JC, Sanchez-Mejias E, et al. Phagocytic clearance of presynaptic dystrophies by reactive astrocytes in Alzheimer's disease. Glia. 2018;66:637–53. https://doi.org/10.1002/glia.23270.
200. Hirbec H, Déglon N, Foo LC, et al. Emerging technologies to study glial cells. Glia. 2020;68:1692–728. https://doi.org/10.1002/glia.23780.
201. Lee H-G, Wheeler MA, Quintana FJ. Function and therapeutic value of astrocytes in neurological diseases. Nat Rev Drug Discov. 2022;21:339–58. https://doi.org/10.1038/s41573-022-00390-x.

Index

E. Blanco-Suarez, I. Farhy-Tselnicker (eds.), *Astrocyte-Neuron Interactions in Health and Disease*, Advances in Neurobiology 39,
https://doi.org/10.1007/978-3-031-64839-7